Cranial Visions

CRANIAL VISIONS

Exploring The Skull Through Artistic Interpretation

Editor: Jinxi Caddel

Design & Layout: Jamie Parker & Mike DeVries

Artist Research & Collections: Jeff Johnson

Cover Art: Shawn Barber

Cover Text: Ryan Smith

ISBN: 978-0-615-39053-6

Printed and bound in China

MEMENTO
PUBLISHING

www.MDTattoos.com
www.JJTattoos.com
www.JinxiBoo.com
www.MementoPublishing.com

Dear Readers,

As artists, we constantly marvel at the wealth of creativity flowing around us. Both in the tattoo industry and the art world in general, the continual sources of inspiration are immense and keep us working hard to move to the next level. In the midst of such ingeniousness, we began noticing how often certain subjects were studied and explored through artwork; most notably, the skull.

It was this revelation, along with our own appreciation, that lead us to begin collecting cranium-themed creations of all kinds. Our goal was to display and honor this seemingly infinitely interpreted motif and the artists who have chosen to add their own perspective of the skull in all its glory.

With submissions from all around the globe, we hope that the collection we have gathered will be both enjoyable and inspirational to readers. The imagination and creativity discovered was certainly impressive to us, and we are thrilled to have the opportunity to share it with you.

There are so many people who have made this project possible and we want to thank each of the artists who contributed their time and their artwork. In addition, a special thanks goes out to:

Jeff Gogue, Shawn Barber, Volko Merschky, Jinxi Caddel, Jamie Parker, Sean Herman, Carlos Rojas, Juan Salgado, Thomas Kynst, Bob Tyrrell, Nick Baxter, Josh Duffy, Nathan Kostechko, Tom Strom, Craig Driscoll, Alex De Pase, Christian Perez, Bugs, Kari Barba, Jeremiah Barba, Durb Morrison, Nikko Hurtado, Carlos Torres, Guy Aitchison, Aaron Cain, and Alex Grey.

We hope that you enjoy the Cranial Visions journey as much as we have appreciated working with the talented artists whose work you will find within the forthcoming pages. Here's to the skull. Though it is diversely interpreted and holds symbolically different representations to many, it's something we all have in common. Through these pages we celebrate the cranium and the wealth of creativity it holds inside.

With Gratitude,

Mike DeVries
Jeff Johnson

· CONTENTS ·

·INTRODUCTION·

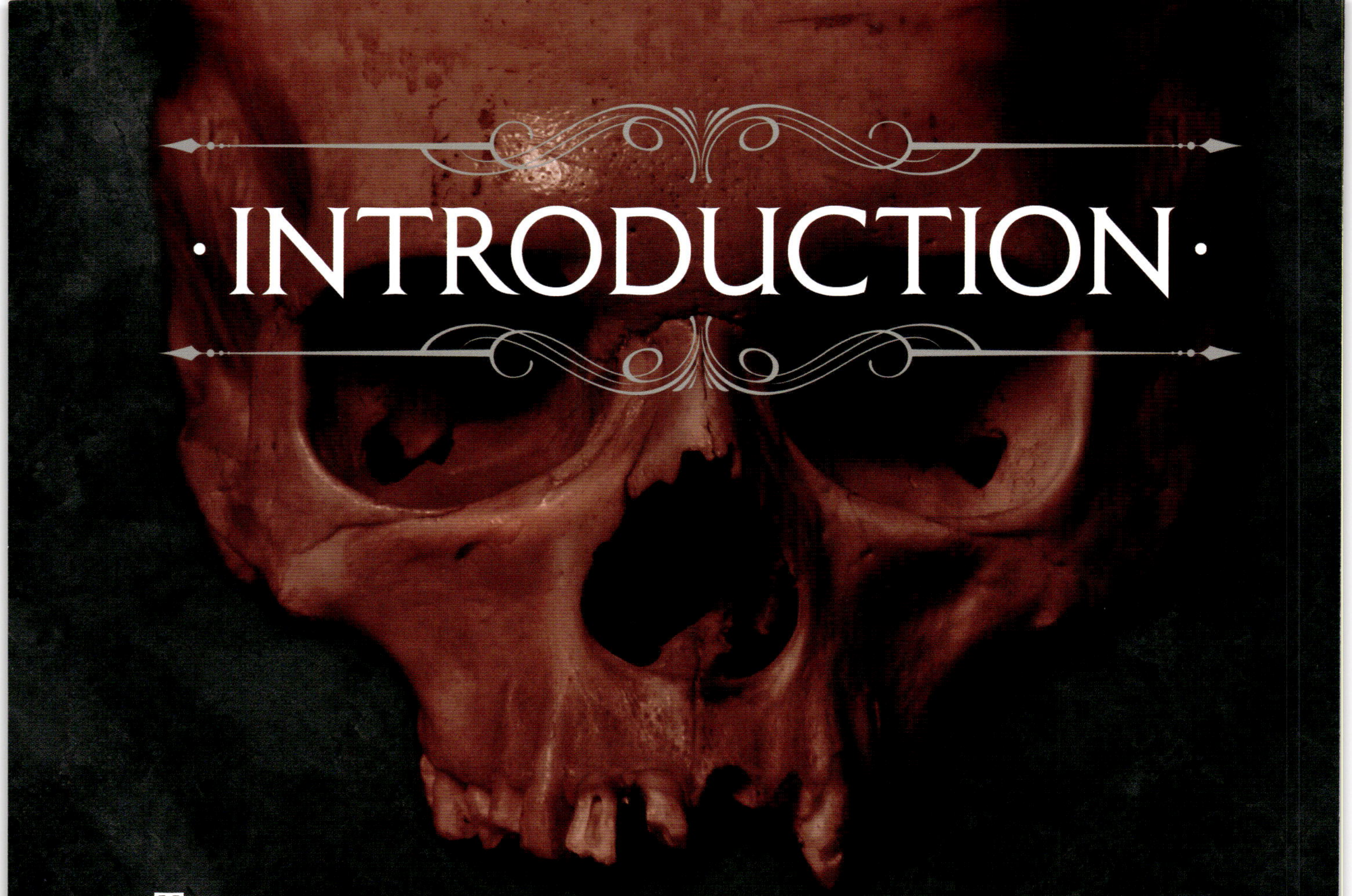

The skull has no tongue, nor communication skills; no flesh, nor life - yet it speaks volumes. Skulls are obviously vital to us all. They are an essential element in the human skeleton and while they certainly do keep our brains protected, they also incite the mind inside each brain with creativity of all sorts.

Skulls have long held symbolic meanings, with representations of mortality, religion, power, protection, eternal life, fear, rebellion and more. Used as icons in secret societies, on pirate ships, as toxic warning symbols, and ancient iconography, the usage of the skull has a rich history.

Skulls have been utilized in human rituals since ancient times. Paleolithic burial sites have revealed skulls of both animals and humans; and while they are commonly seen as a symbol of death, they are often known to represent the rebirth of life and a pathway to "higher knowledge" as well.

Shakespeare employed the use of skulls in his famous Hamlet soliloquy, as the Prince of Denmark holds the skull of the beloved court jester, Yorick, and expresses sorrow, meditating on the fragility of life.

Cultures all over the world have used skulls to symbolize both life and death. Stone sculptures and carved skulls from lava or jade were made by the Aztecs as representations and worship of death, to bring about the promise of resurrection. These designs were commonly seen on scrolls, clothing, pots, jewelry, hieroglyphics, and other decorations.

Mexican art history is also rich with creations of the skull-related theme, as they are well known for the Day of the Dead (Dia de los Muertos) celebration and beautiful sugar skulls. In the 17th century, sugar art was brought to the New World by Italian missionaries. During Easter, sugar angels and lambs were made to decorate the alters inside the Catholic Church. Because Mexico, was rich in sugar production, yet too poor to buy fancy imported European church decorations, they quickly discovered how to make sugar art for their religious festivals from the friars of the church. Sugar skulls represented a departed soul, had the name written on the forehead and were placed on the gravestone to honor the return of a spirit. Today, the sugar skull tradition continues and is widely portrayed in folk art creations of all sorts.

In Mexico, Central America and South America, thirteen crystal skulls of ancient origin have been found, spurring on a great deal of intrigue with modern day archaeologists. These skulls, found near the ancient ruins of Mayan and Aztec civilizations are mysterious and controversial, with some of those found believed to be between 5,000 and 36,000 years old. There are some who believe that the crystal skulls hold healing, magical powers; while others are certain that they are counterfeits. Falling in line with the usual intrigue of the skull, the controversy seems bound to continue, and the true origin might always remain a mystery.

Renaissance art commonly employed the use of skulls to convey symbolic meanings. These skull pieces were known as "Vanitas" paintings and portrayed the commonly held humanist beliefs of the time that man was the center of all things. Leonardo Da Vinci illustrated numerous cranium-themed sketches as he examined the structure of the skull and its components. He was well known for geometric proportion and anatomical accuracy in his work, and numerous drawings by Da Vinci have been found, featuring the skull from all angles, degrees, and dimensions.

The well-known American female artist, Georgia O'Keeffe, routinely used animal skulls and bones in her paintings. O'Keeffe was fascinated by skulls that she routinely came upon during her research and exploration of the Southwest landscape. These pieces were thought to represent the death and destruction of the American landscape by some; and as a tribute to animals and the Western land by others. She would also commonly use flowers alongside the animal skull paintings, once remarking that she did so after sorting through a bunch of artificial flowers: "Someone came to the kitchen door. As I went to answer the door, I stuck a pink rose in the eye socket of a horse's skull. And when I came back the rose looked pretty fine, so I thought I would just go with that."

Surrealist legend, Salvador Dali, also utilized skulls in his work. In his piece, "The Face of War," a large skull contains within the mouth and eyes, additional skulls; and inside of those smaller skulls are more skulls. In a 1951 collaboration with photographer, Philippe Halsman, Dali is seen beside several nude women whose bodies form a skull, entitled "In Voluptas Mors."

In more contemporary work, American artist Alex Grey's visionary art has inspired people of all walks of life with his paintings that feature the human body with stunning precision. These, of course, include pieces involving the skull as well as the dimension of the human energy field.

Another visionary who includes cranium homage in his work is Swiss surrealism painter and sculptor, H.R. Giger. Influenced early on by Dali and Ernst Fuchs, his work uses unique biomechanical configurations that blend human bodies with machines. Skulls and skeletons run rampant in his artwork and have inspired countless painters, artisans, and tattooists.

The usage of skulls in artwork today is commonplace, though not all pieces produced can be described as "common." In 2007, Damien Hirst, British artist and provocateur completed his piece entitled, "For the Love of God," a diamond-encrusted (8,601 flawless diamonds to be exact) platinum human skull, which sold for $100 million. Hirst noted that the idea for the sculpture came from an Aztec turquoise skull he had seen at the British Museum.

It's no secret that skull motifs are a favorite theme in the tattoo world, and these ink designs come in all shapes, sizes, colors, and levels of intricacy. Not only are they common body art designs, but the cranial theme also graces many clothing lines; as well as collectibles, keepsakes, and jewelry lines.

The long tradition of celebrating the skull through unique interpretations was the basis for Cranial Visions. The collection of pictorial imagery gathered together for this project celebrates the cranium theme in all of its glory. An accumulation of artwork with diverse themes, bold statements, and comprehensive creativity, all with one specific centrality...the skull.

TATTOOS

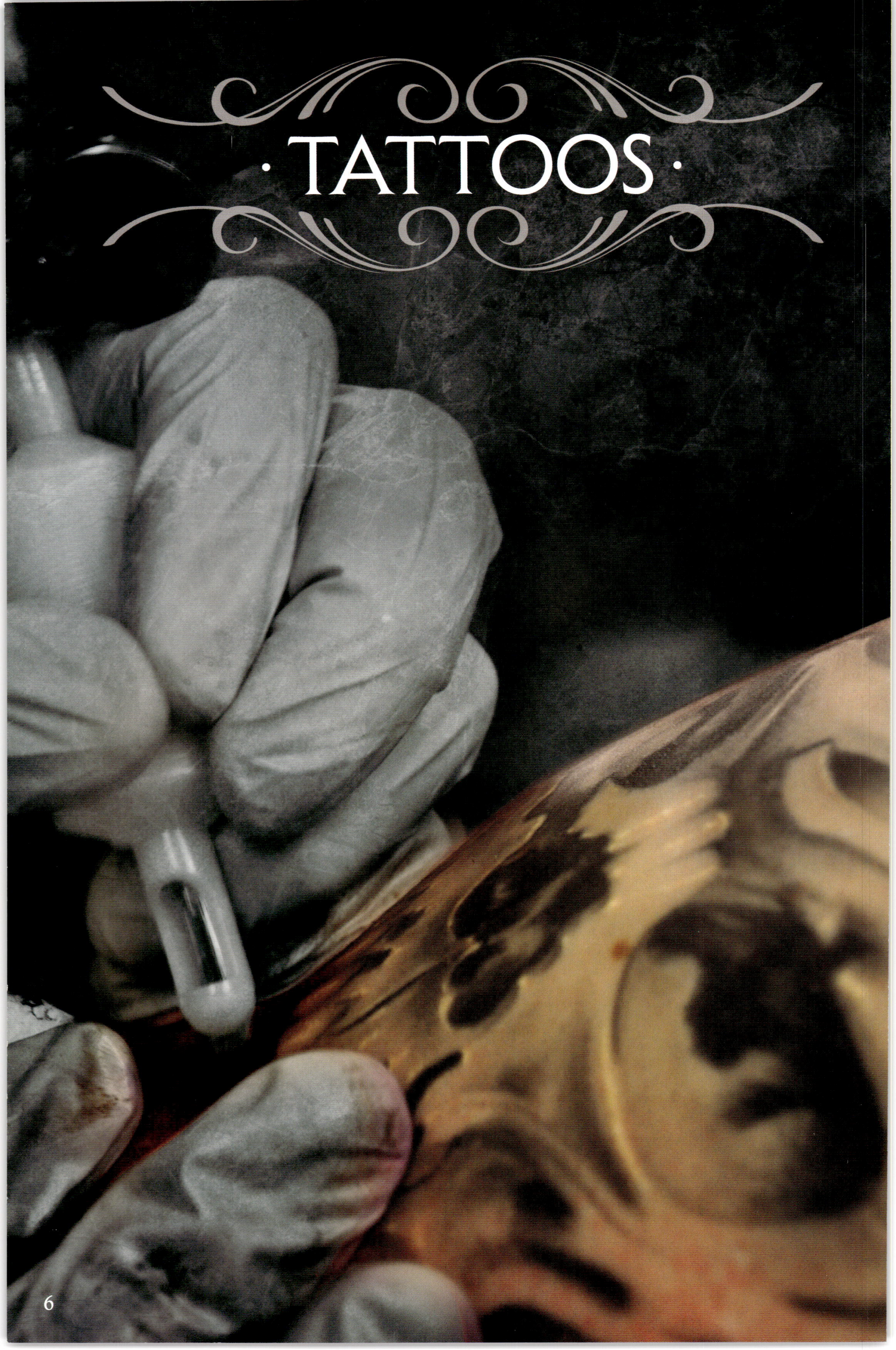

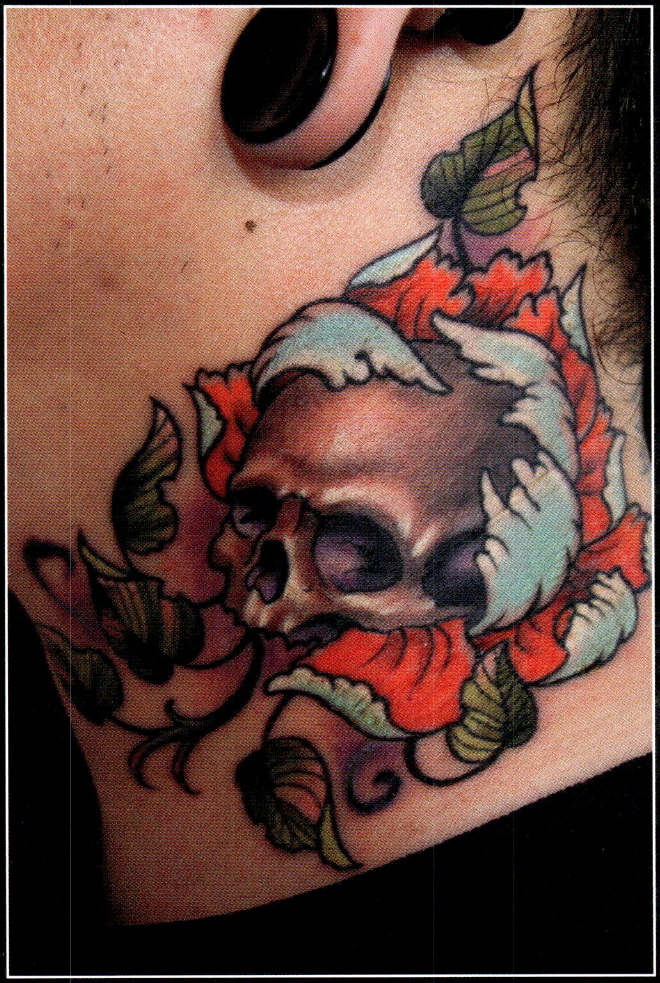

Cory Norris

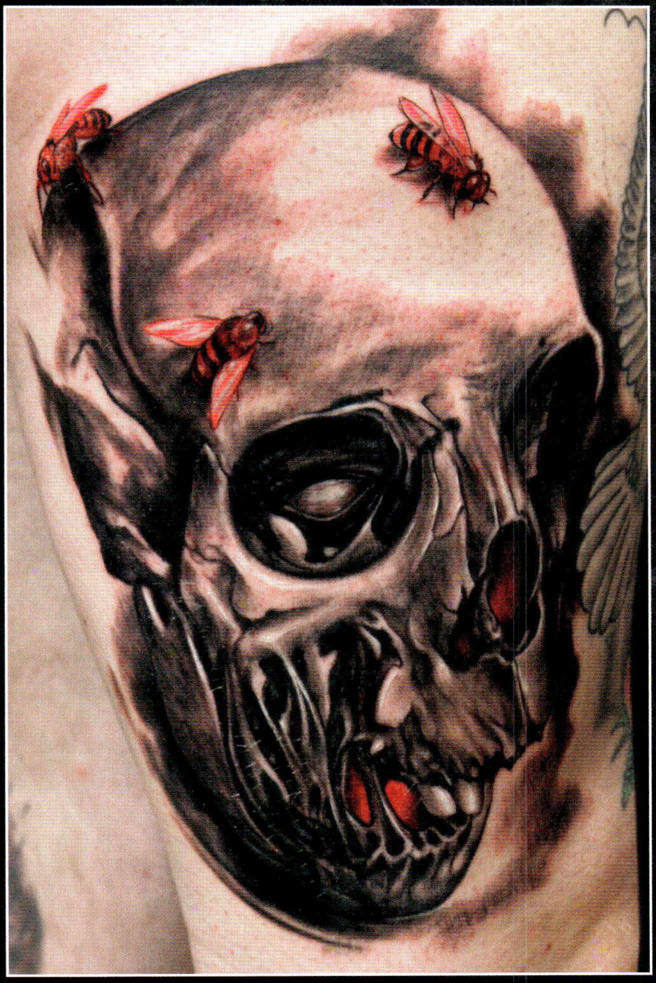

Thomas Kynst

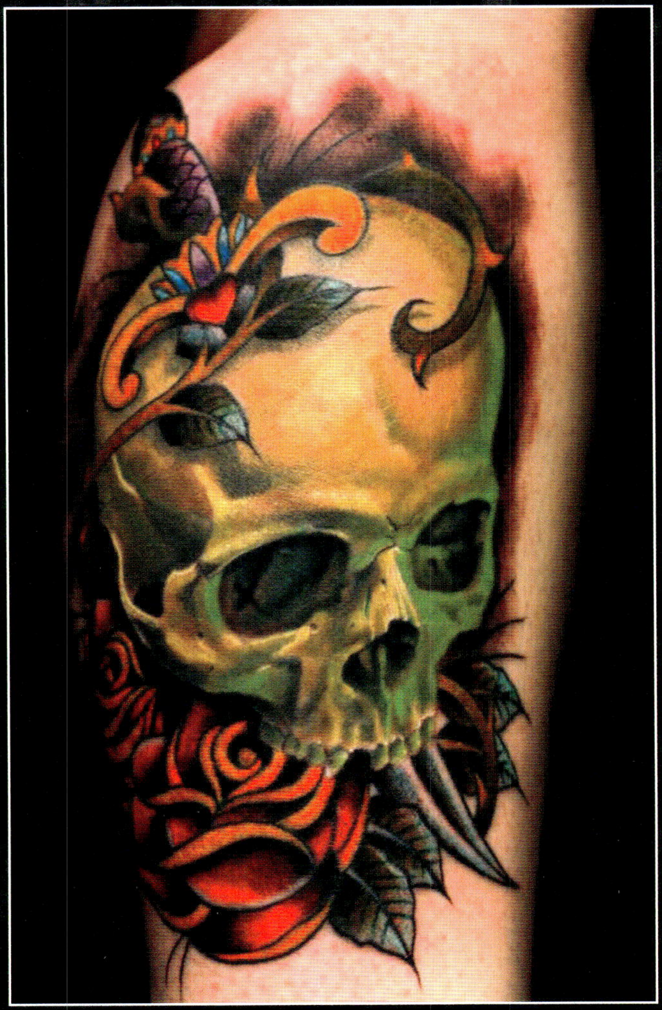

Jason Morrow

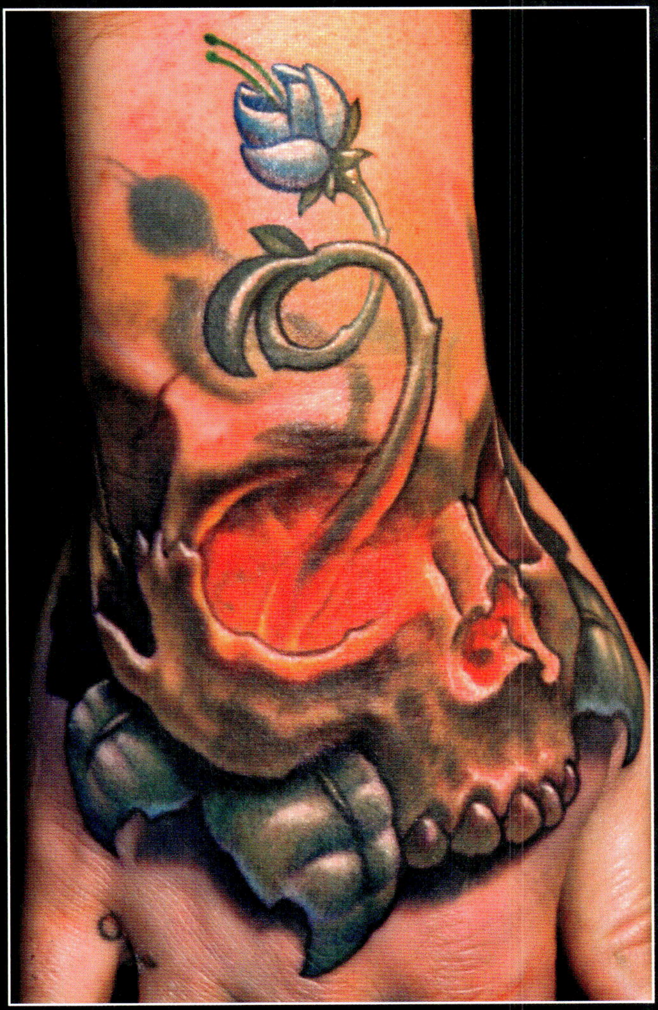

Nick Baxter

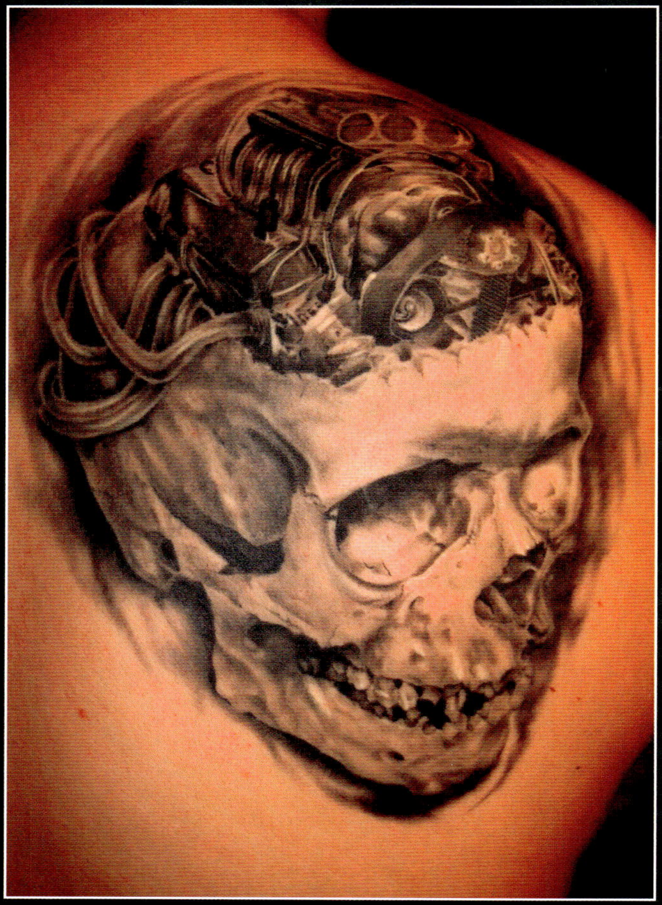

Jack Matisse

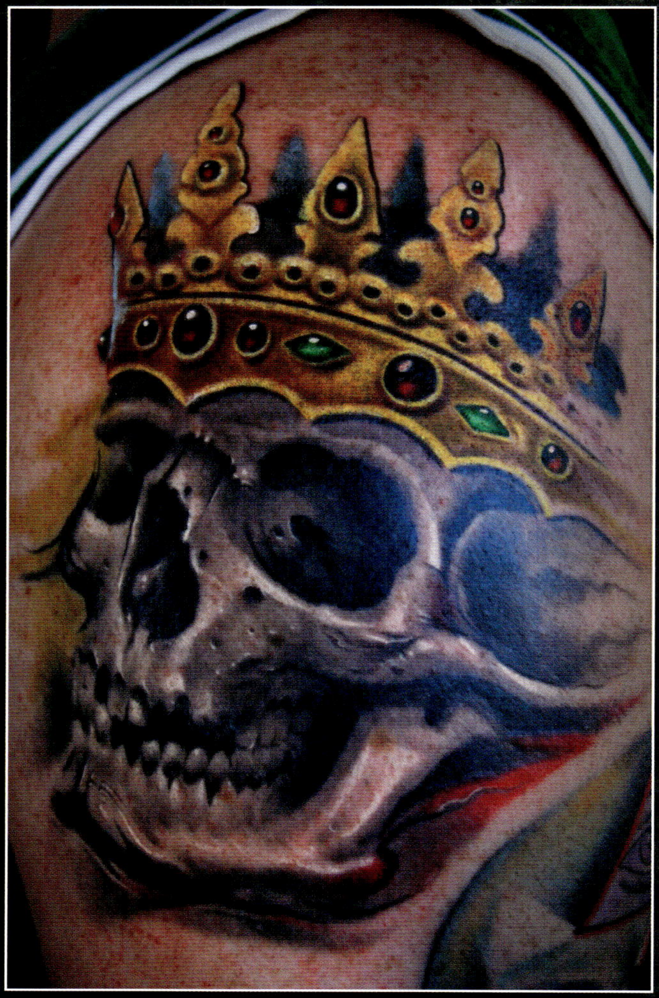

Cory Norris

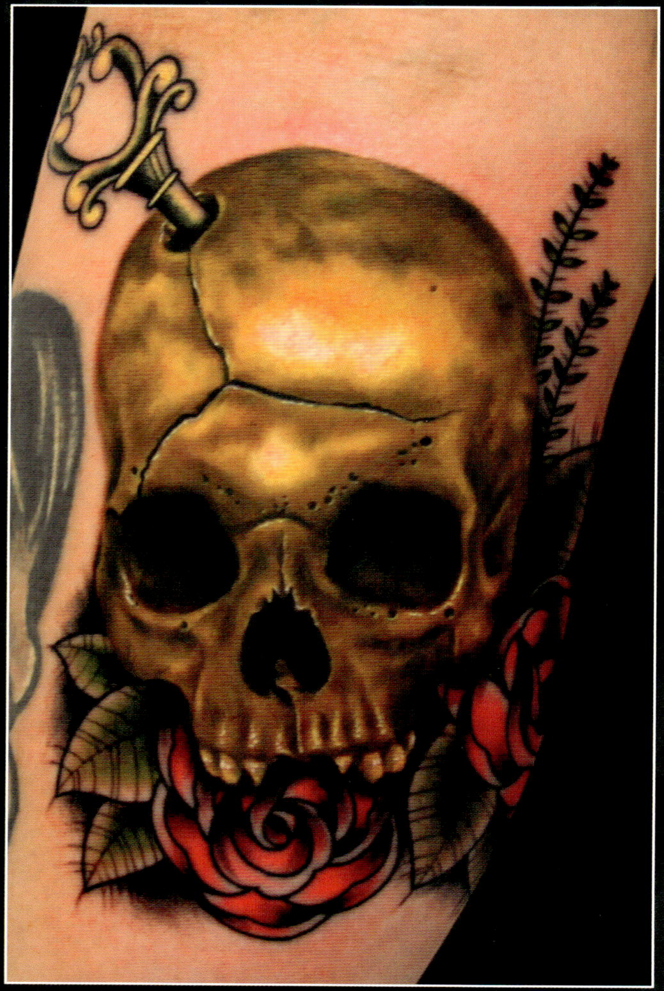

Mick Squires

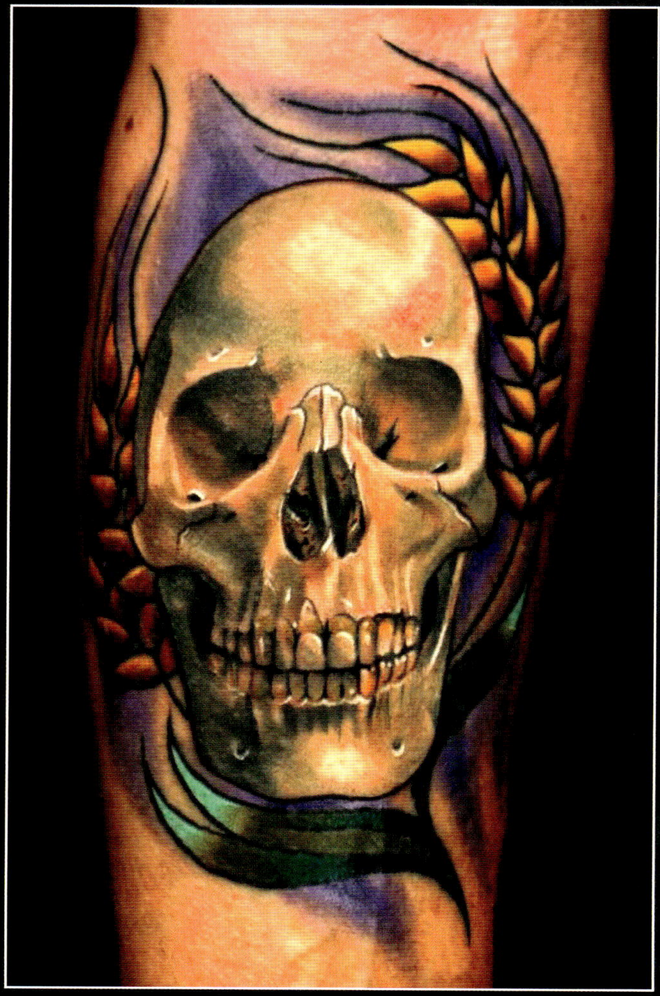

Jason Morrow

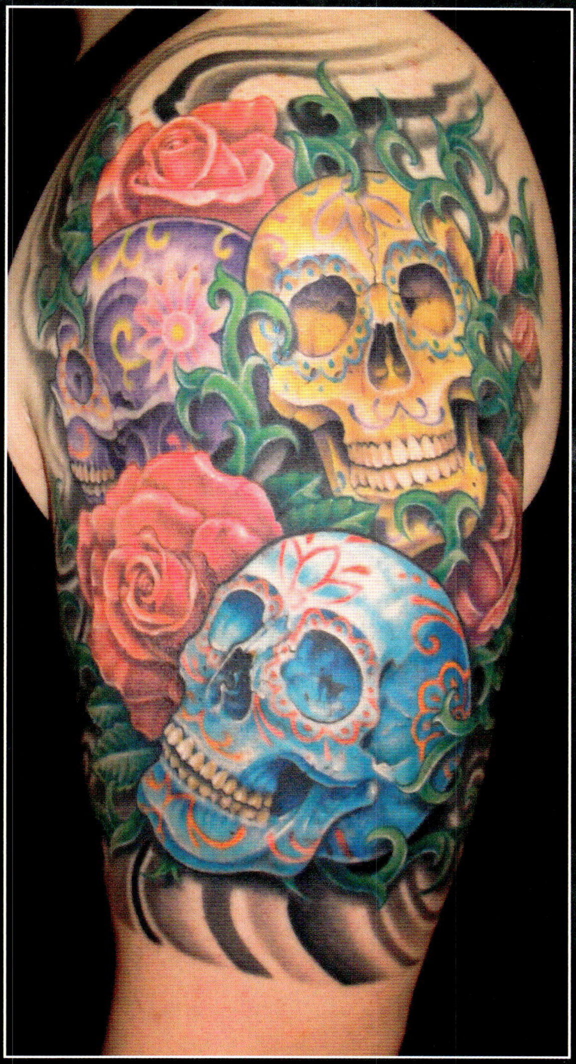

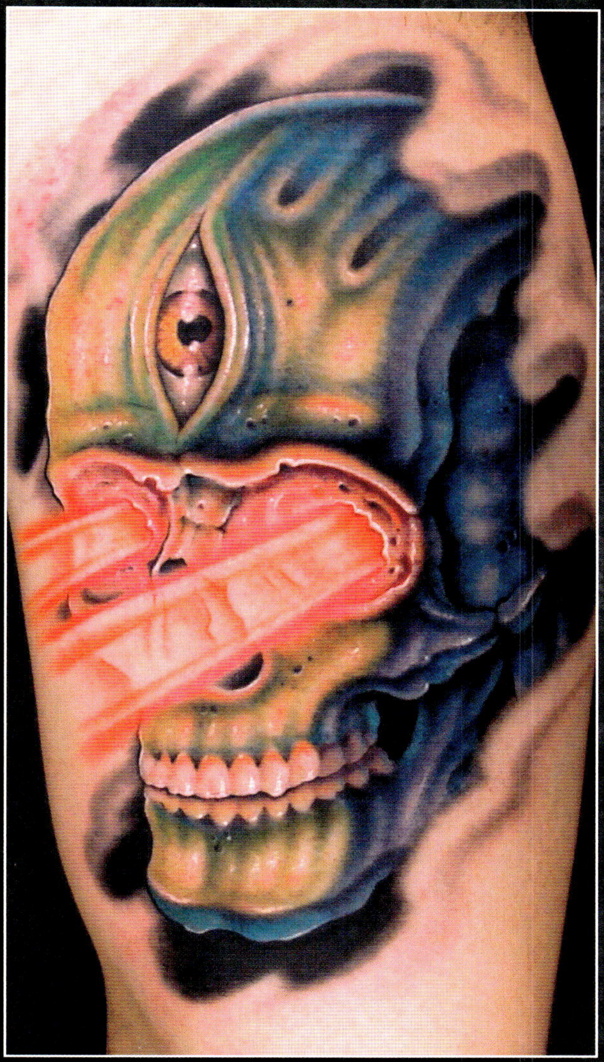

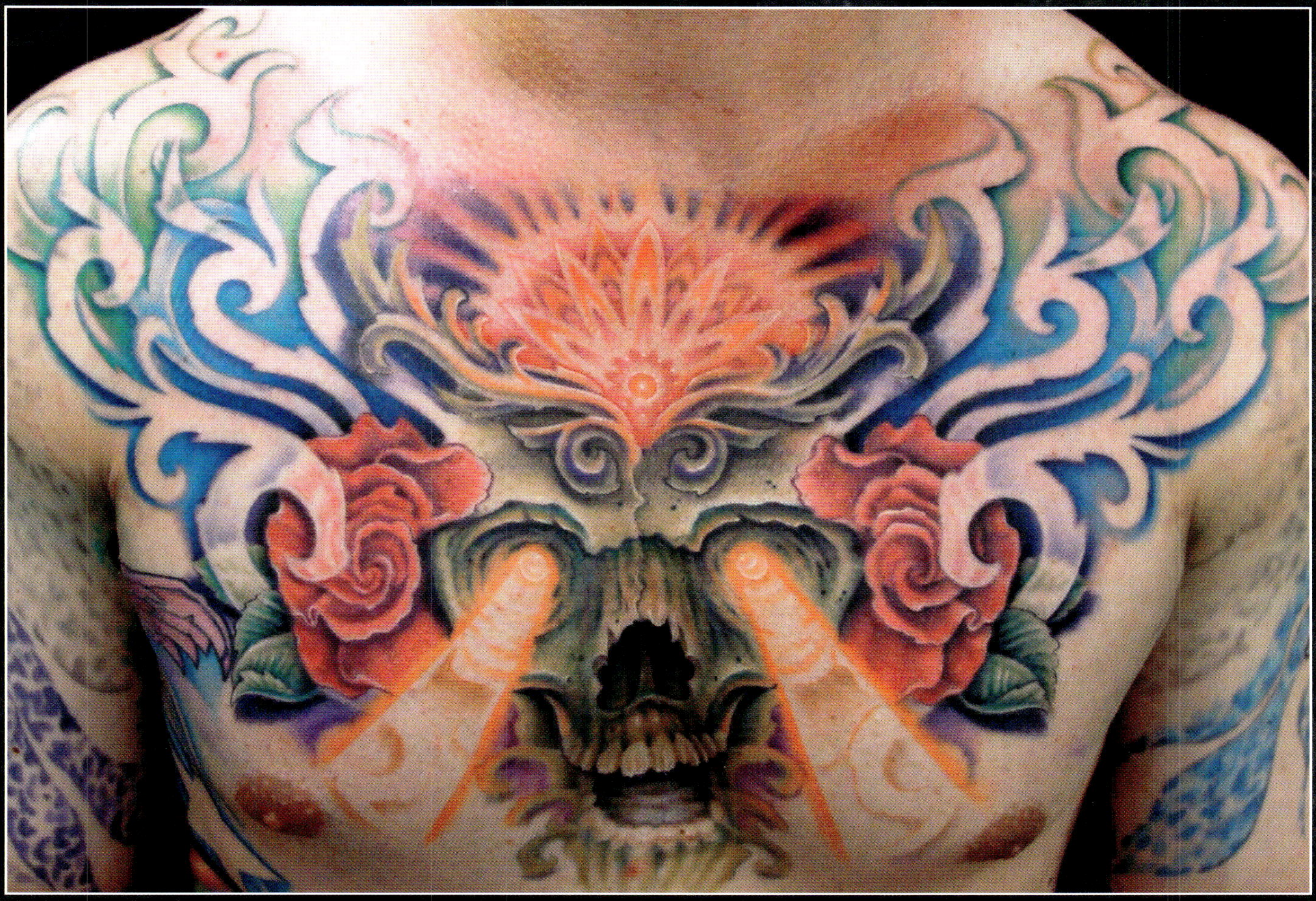

Juan Salgado

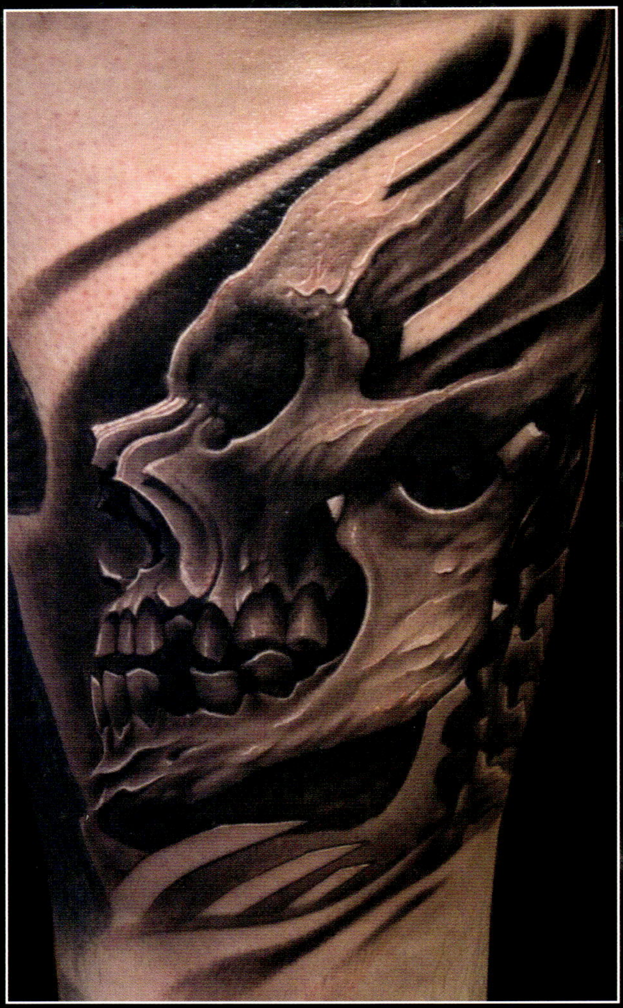

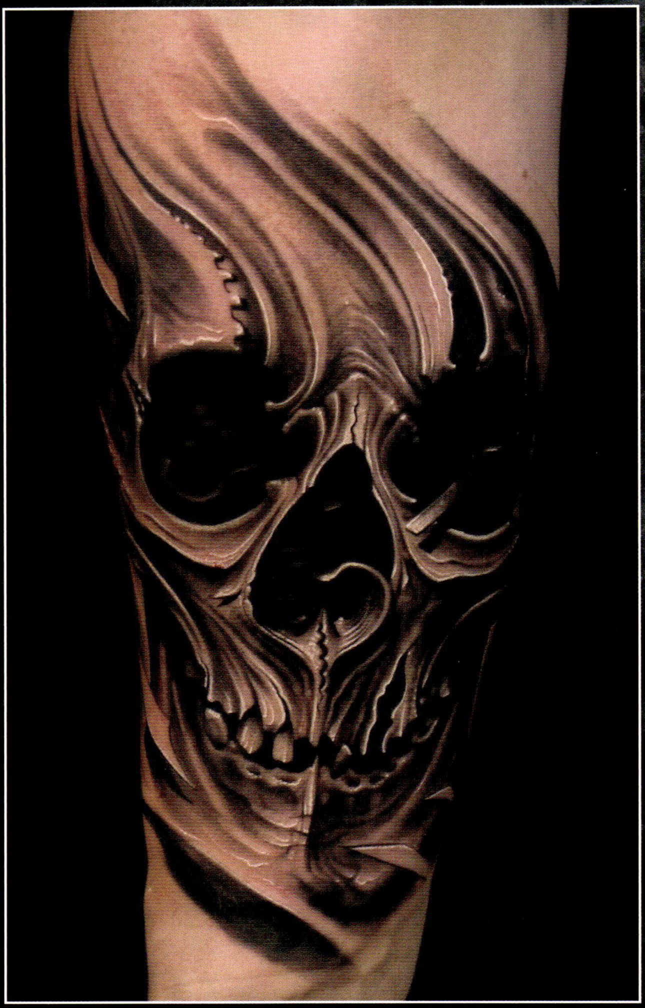

Victor Portugal

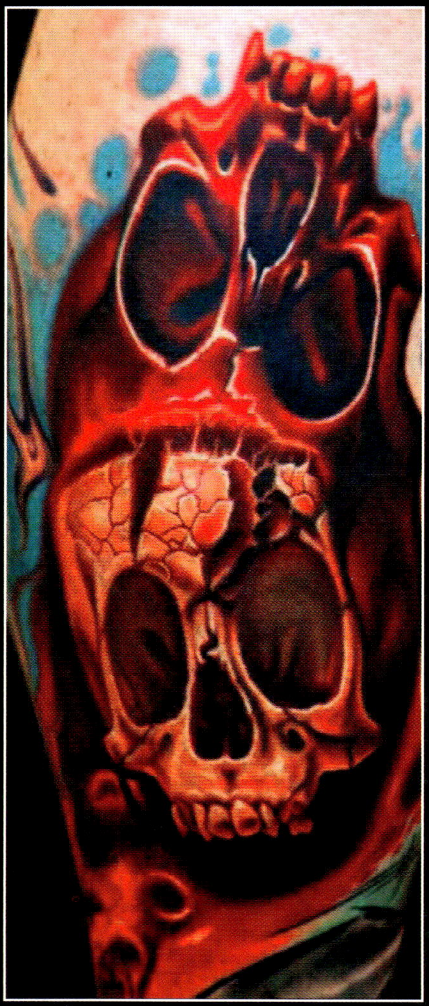

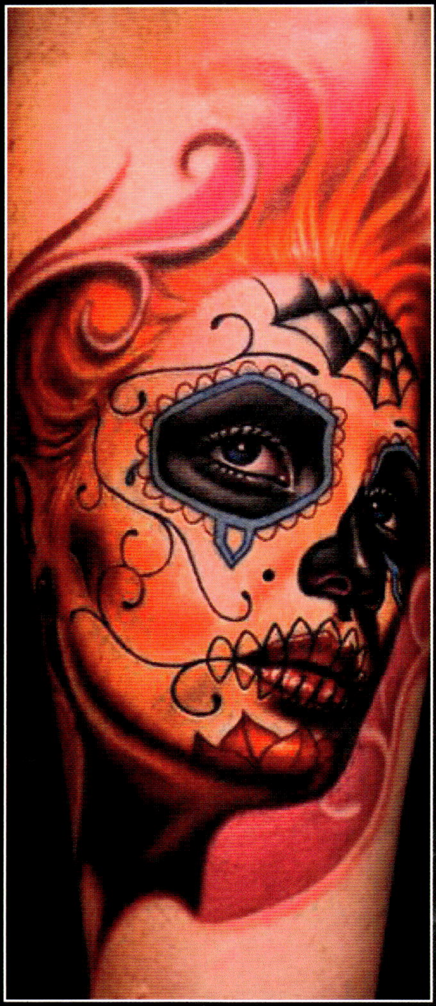

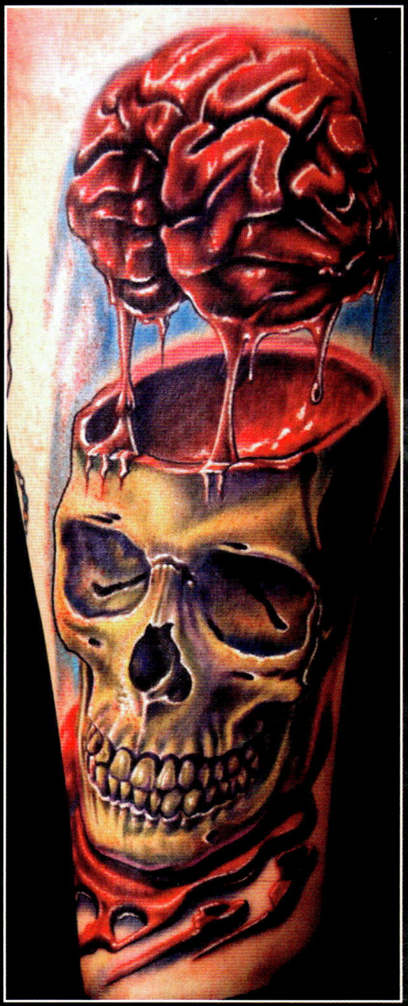

Nikko Hurtado

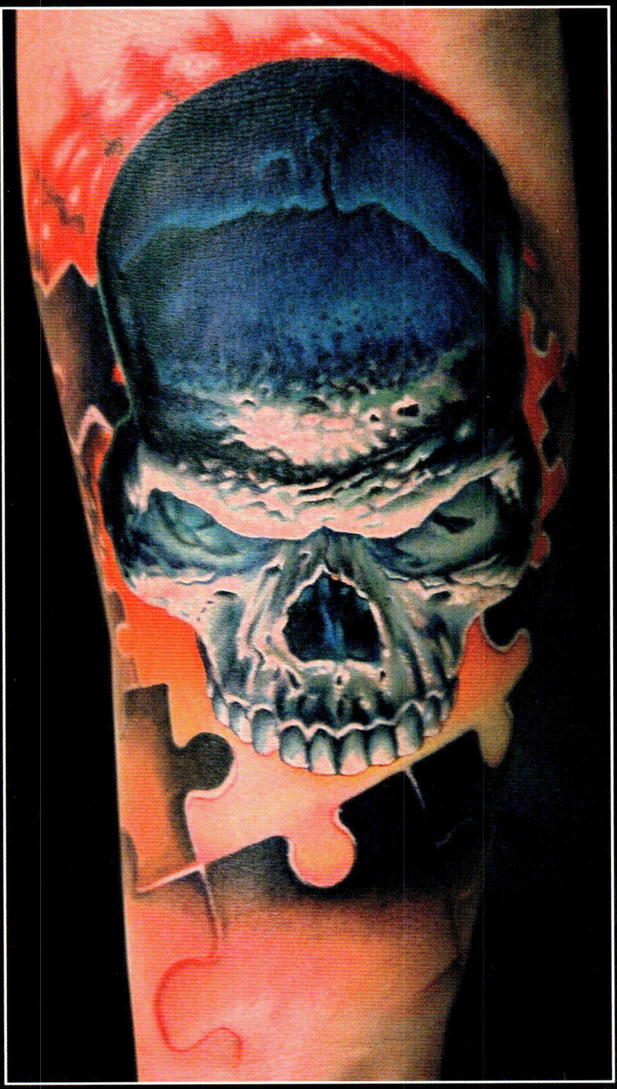

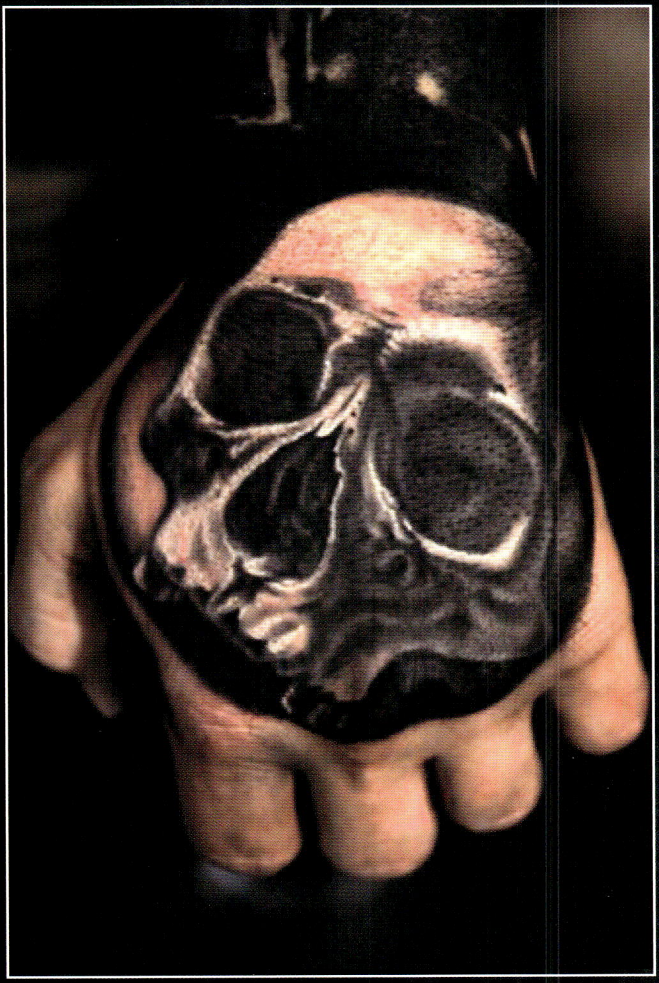

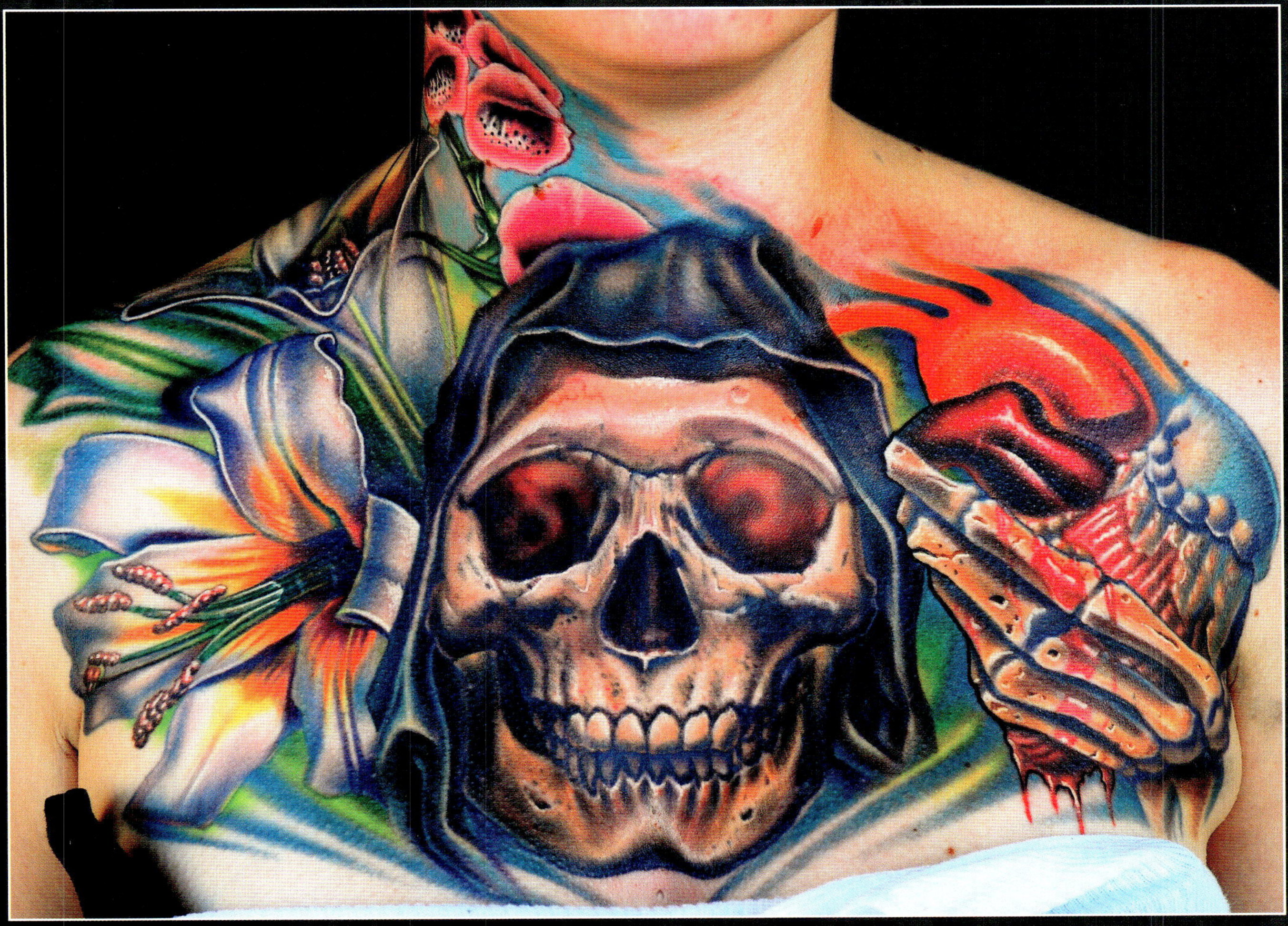

Nikko Hurtado

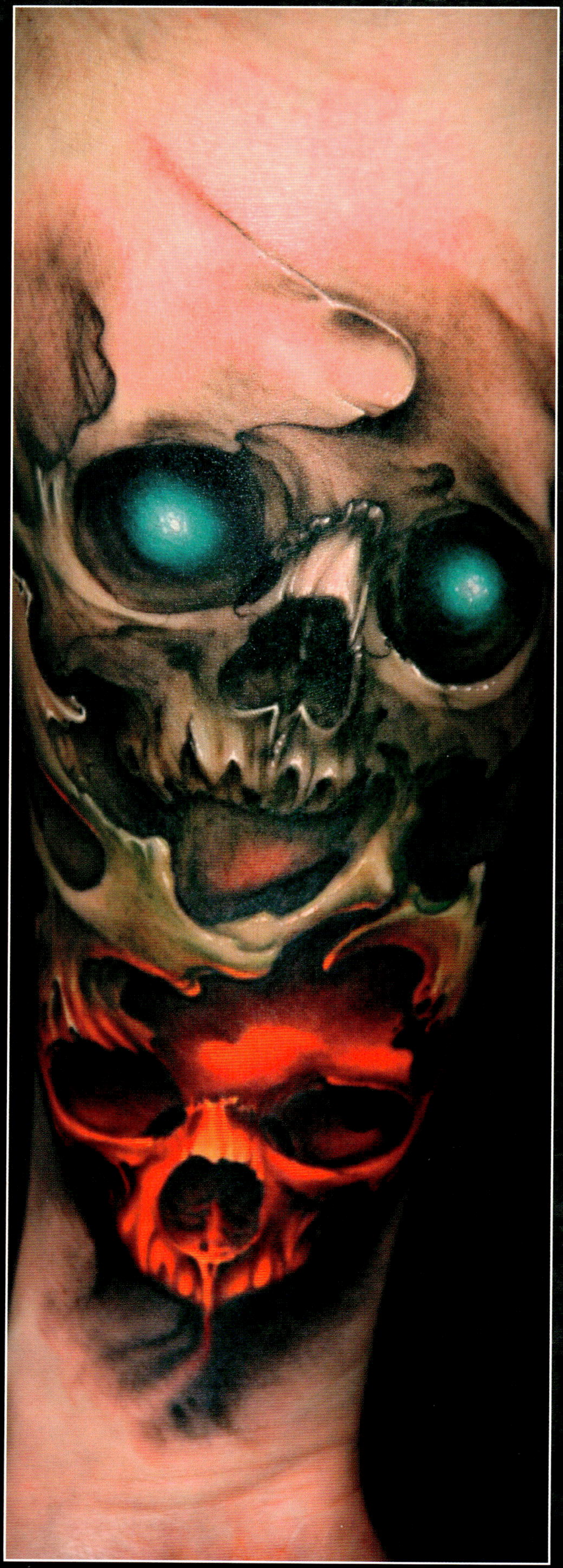

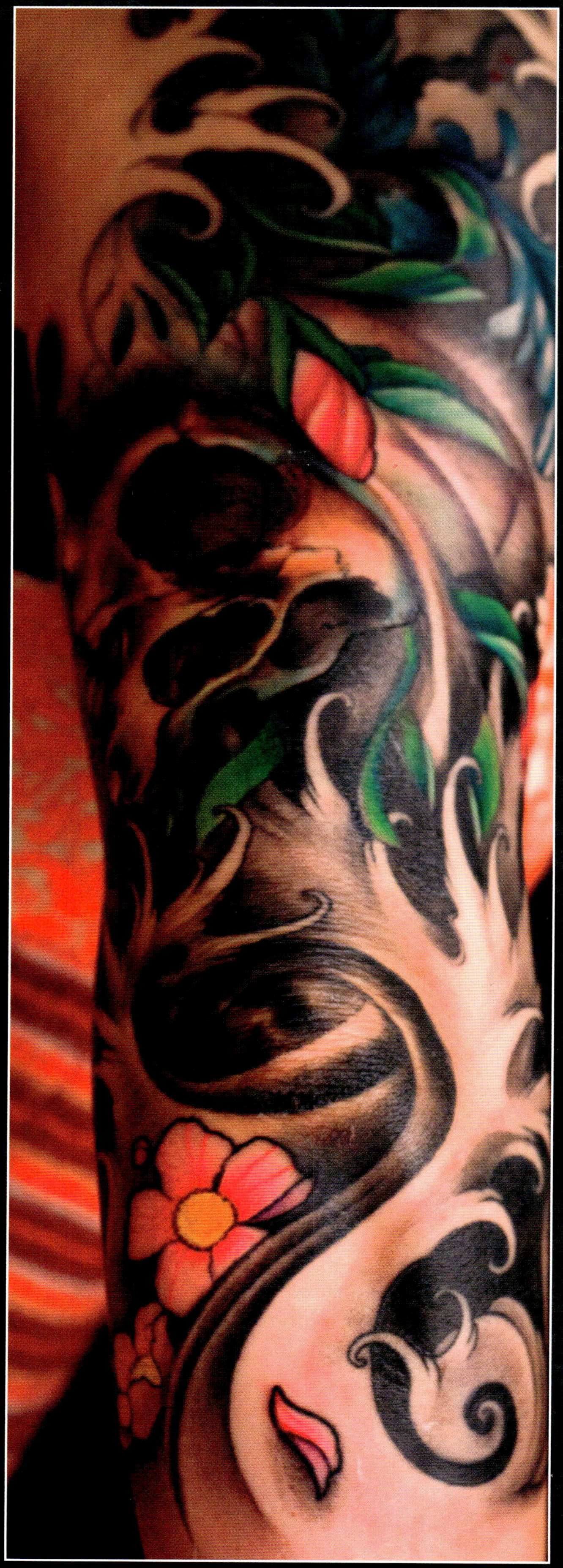

Jeff Gogué

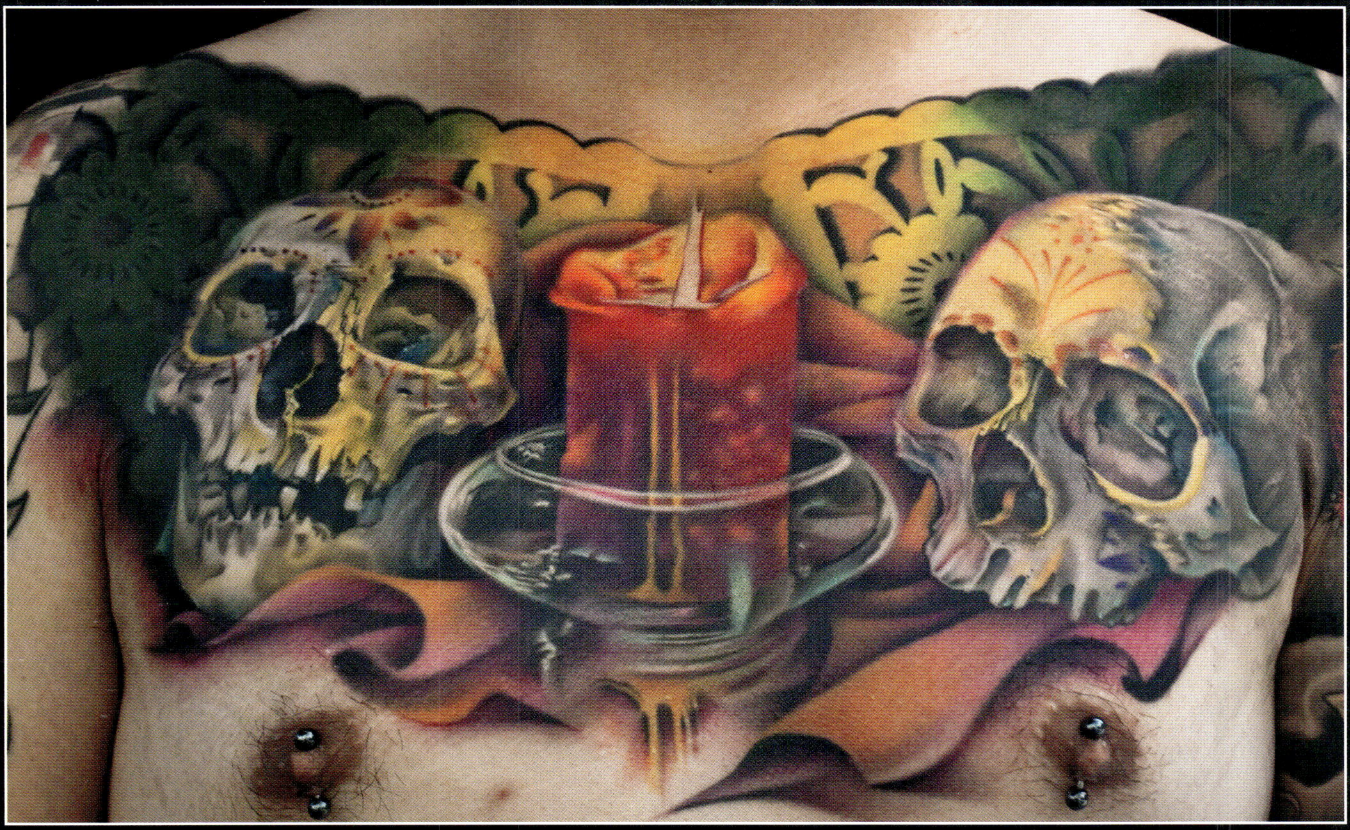

Nick Chaboya

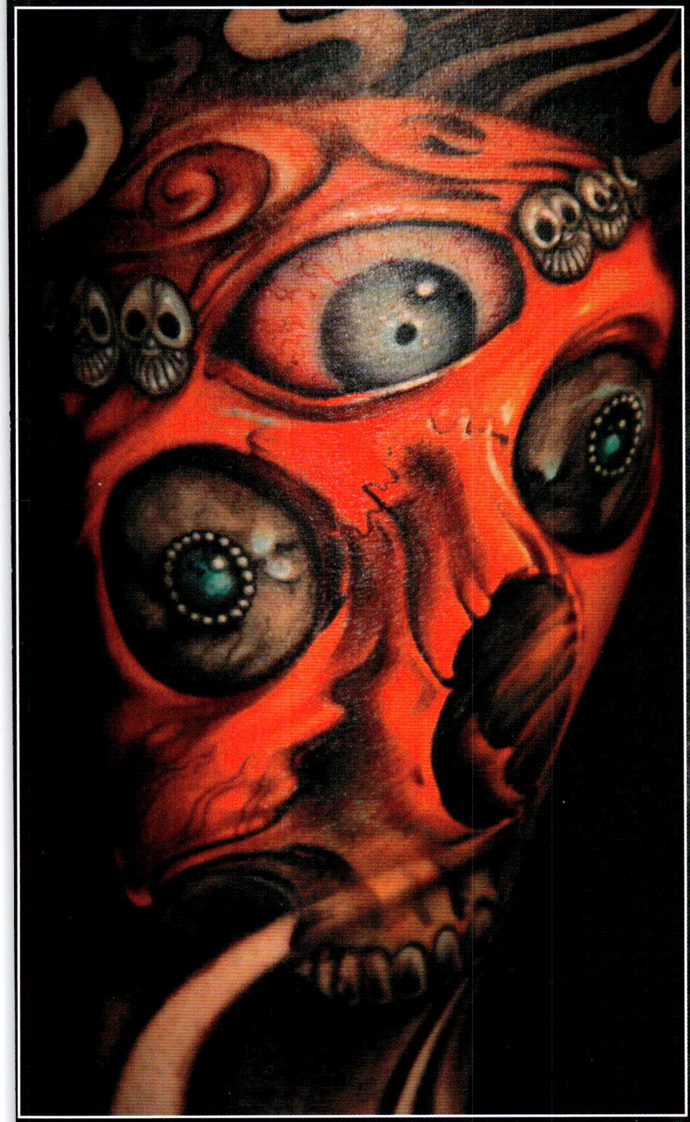

Jeff Gogué

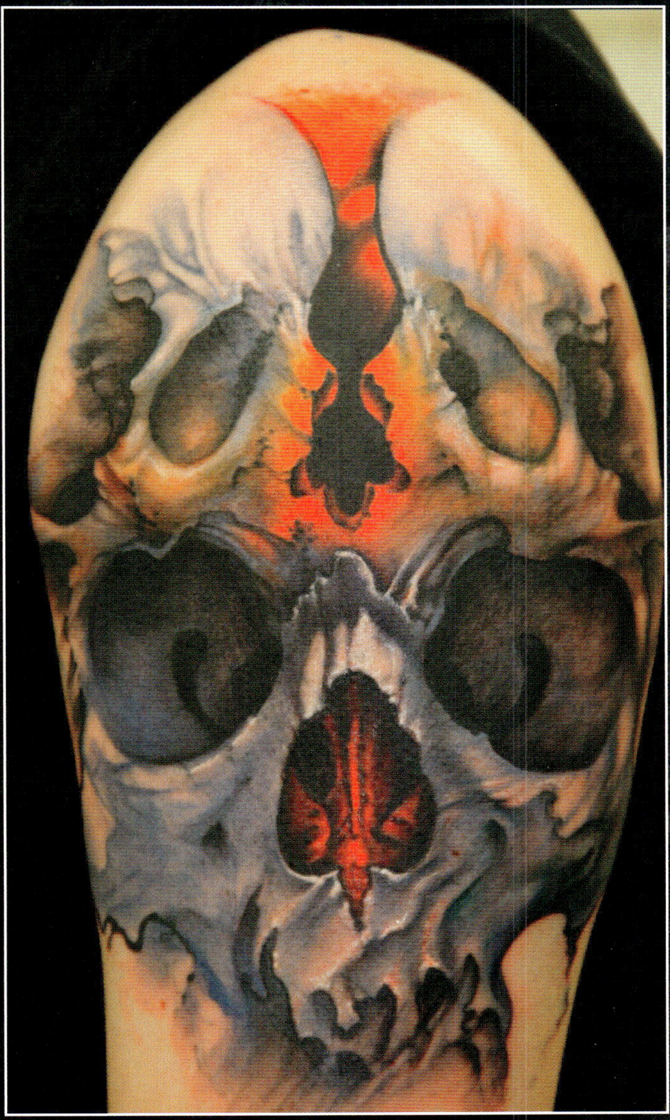

Jeff Gogué

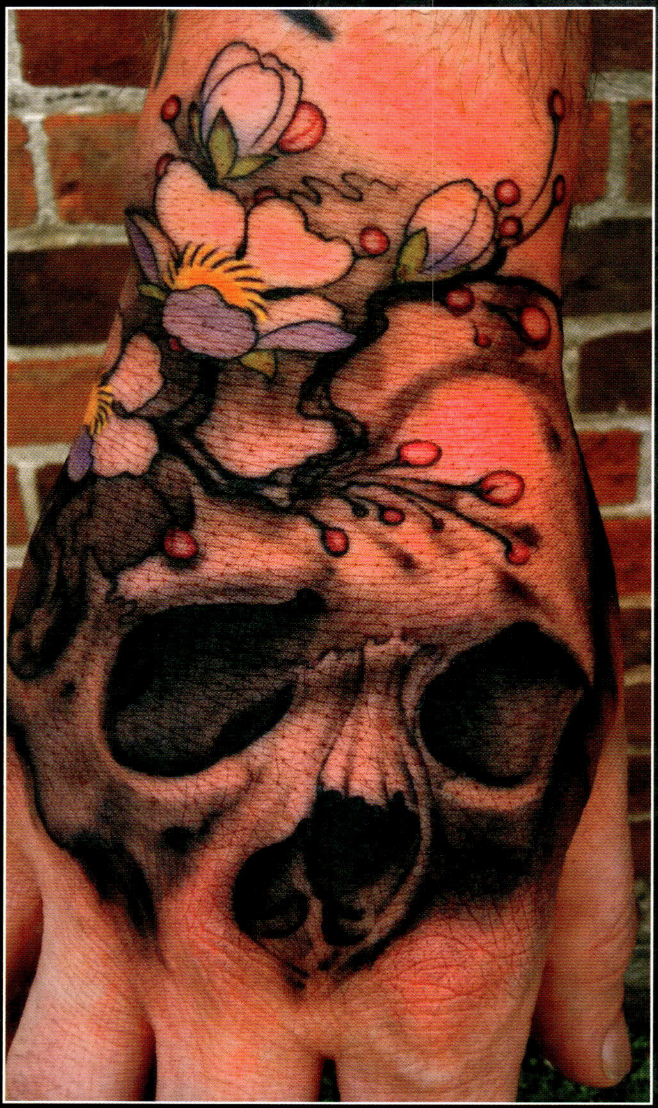

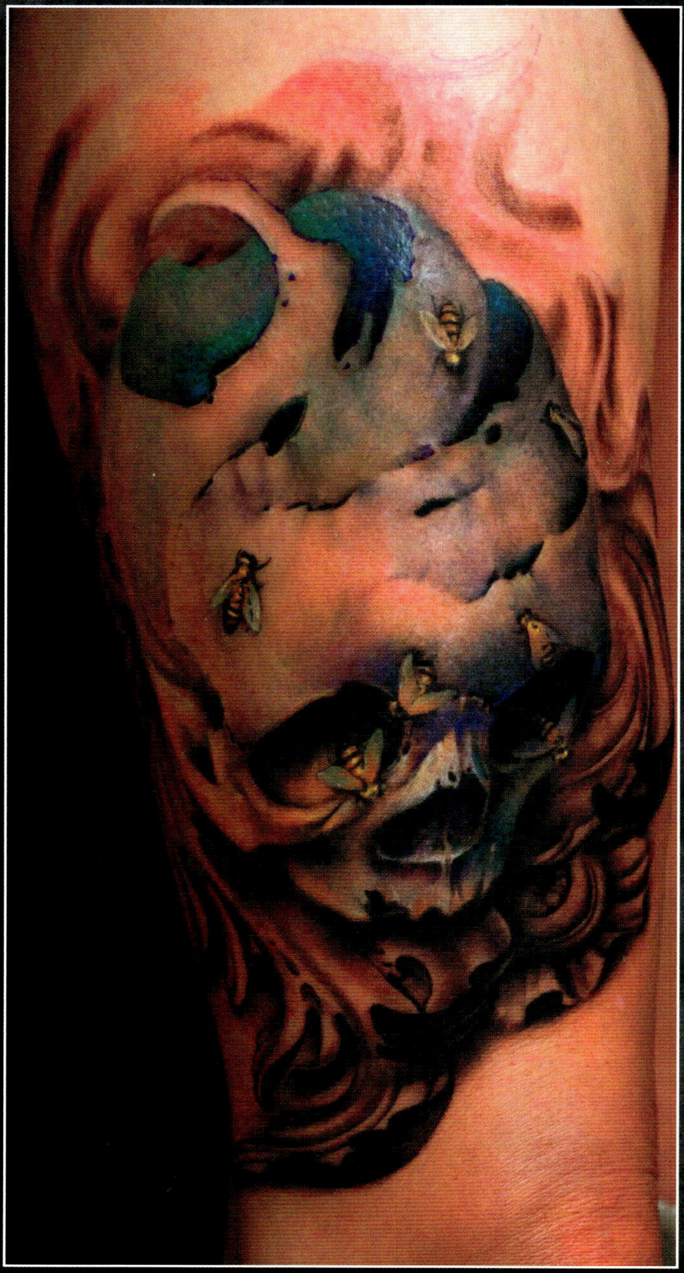

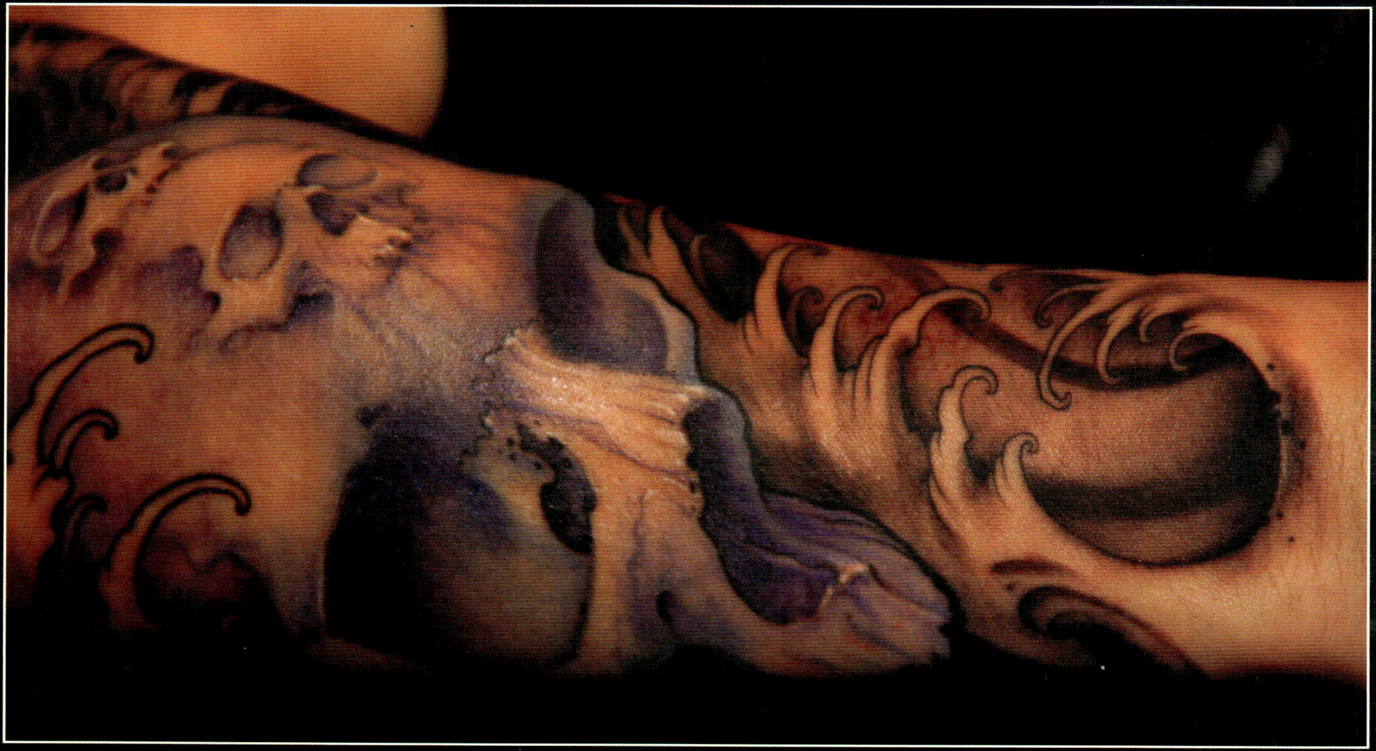

Jeff Gogué

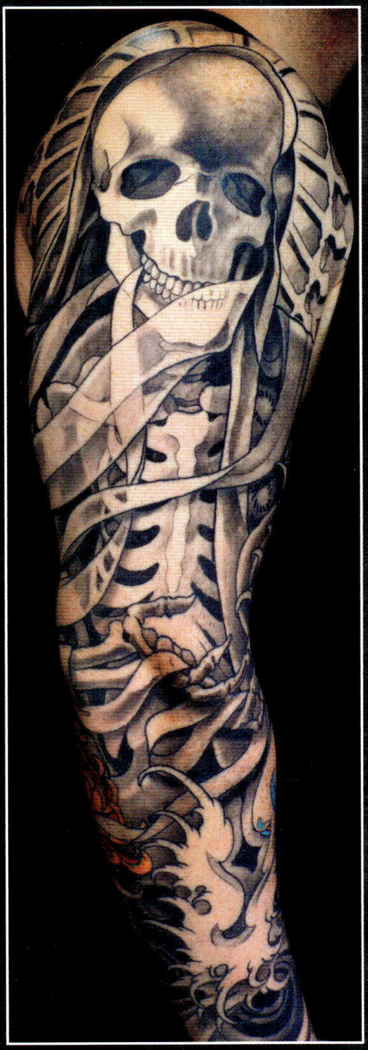
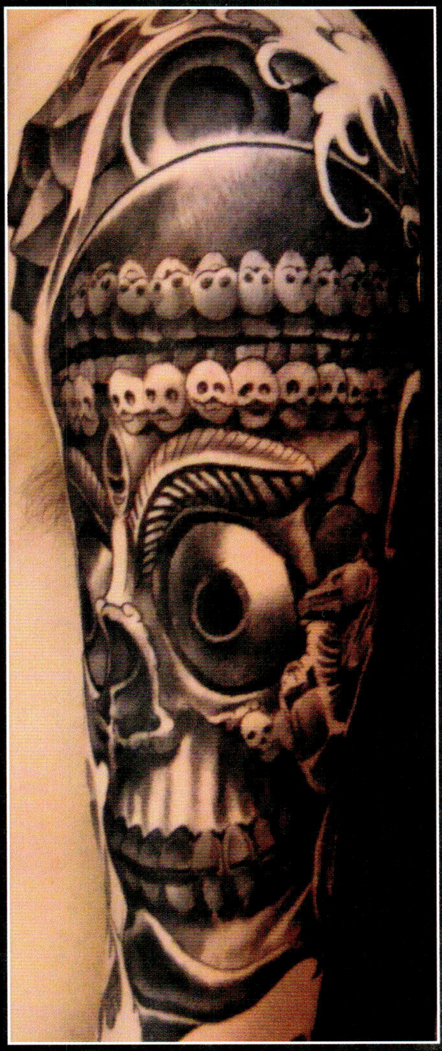
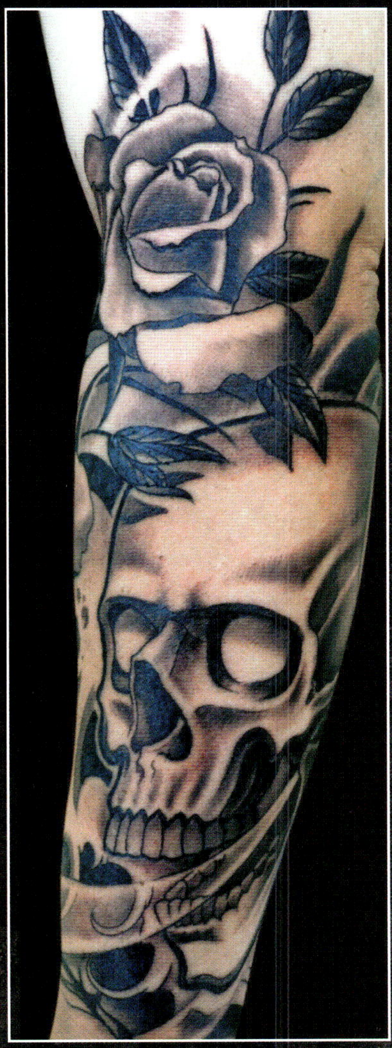
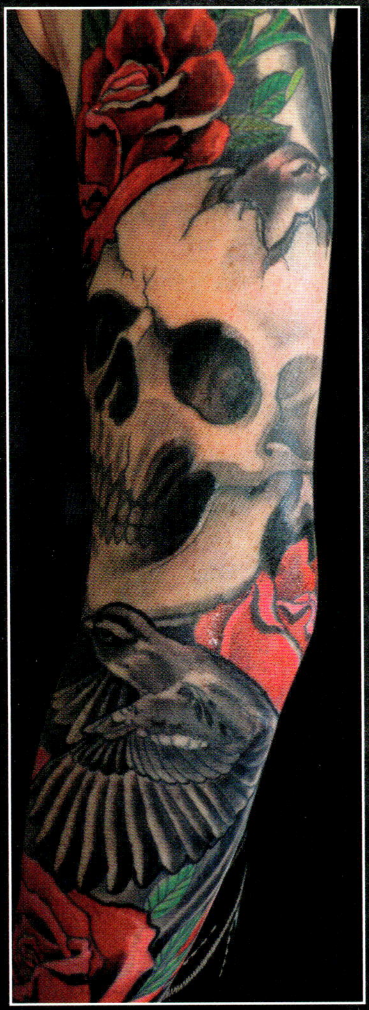
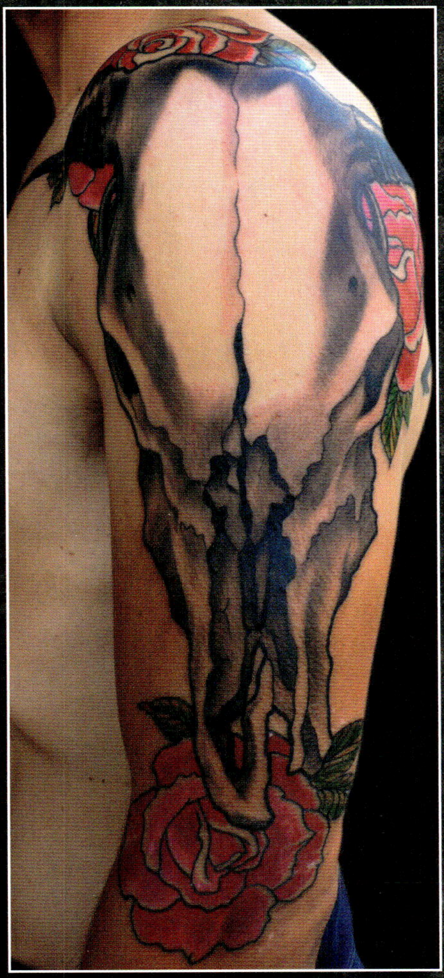
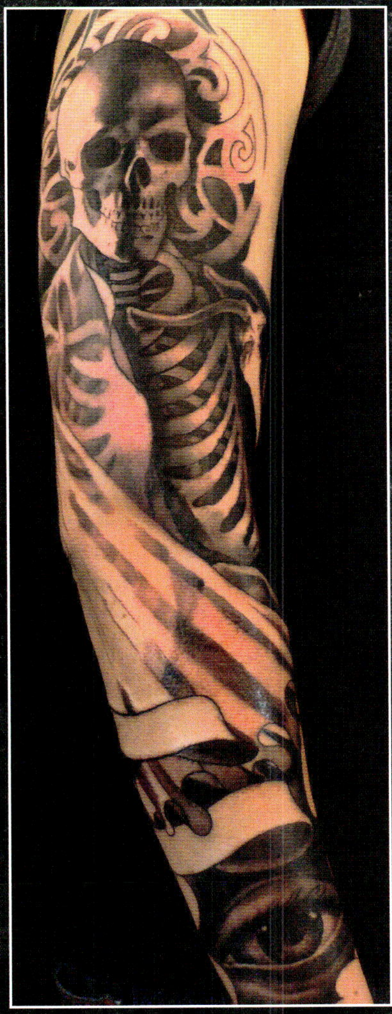

Kore Flatmo

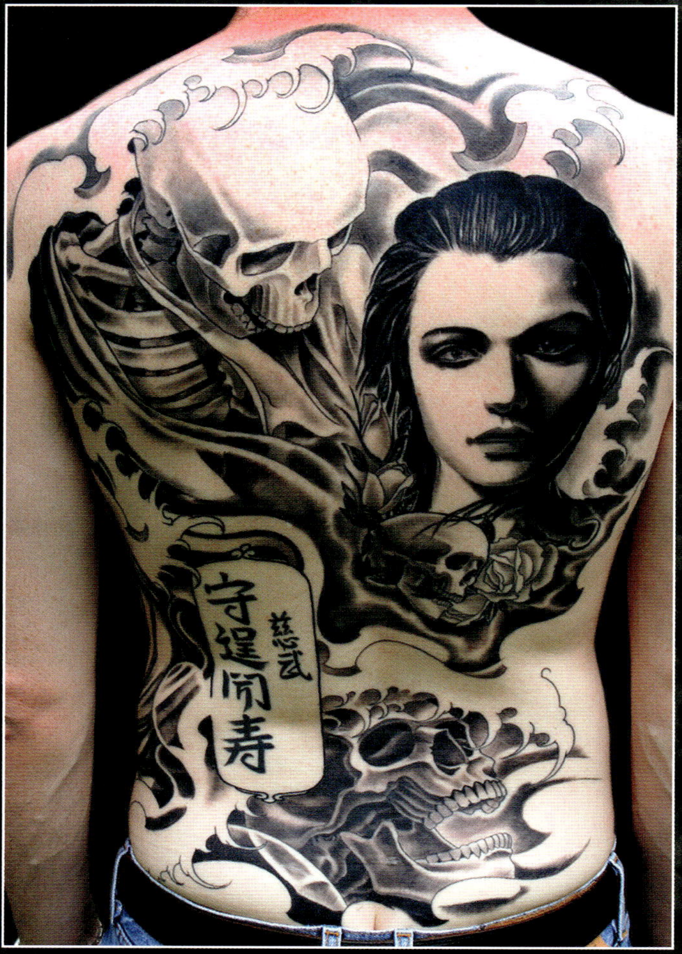

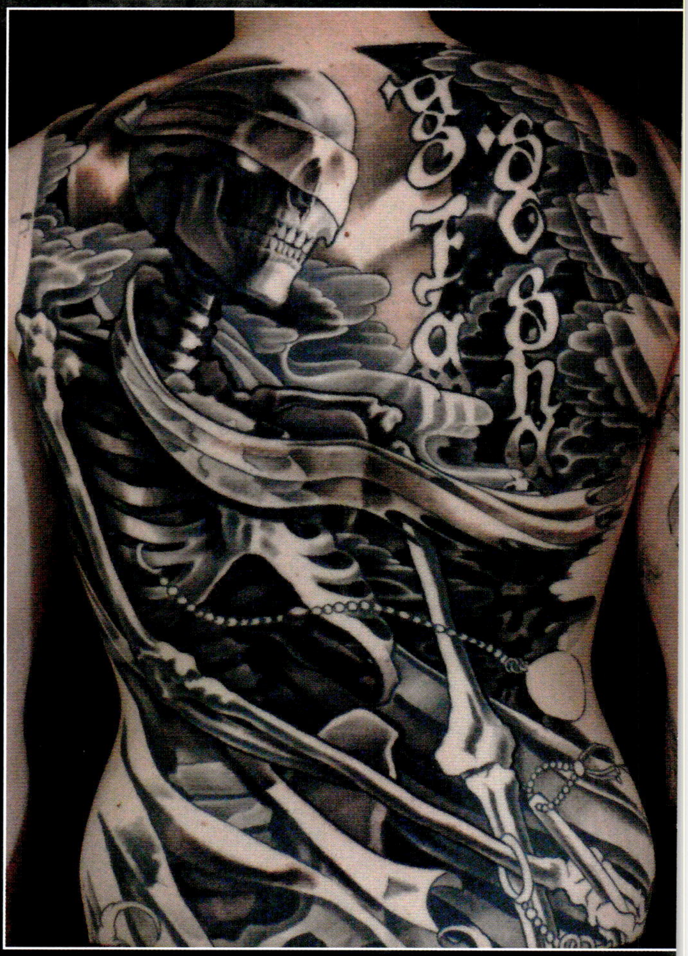

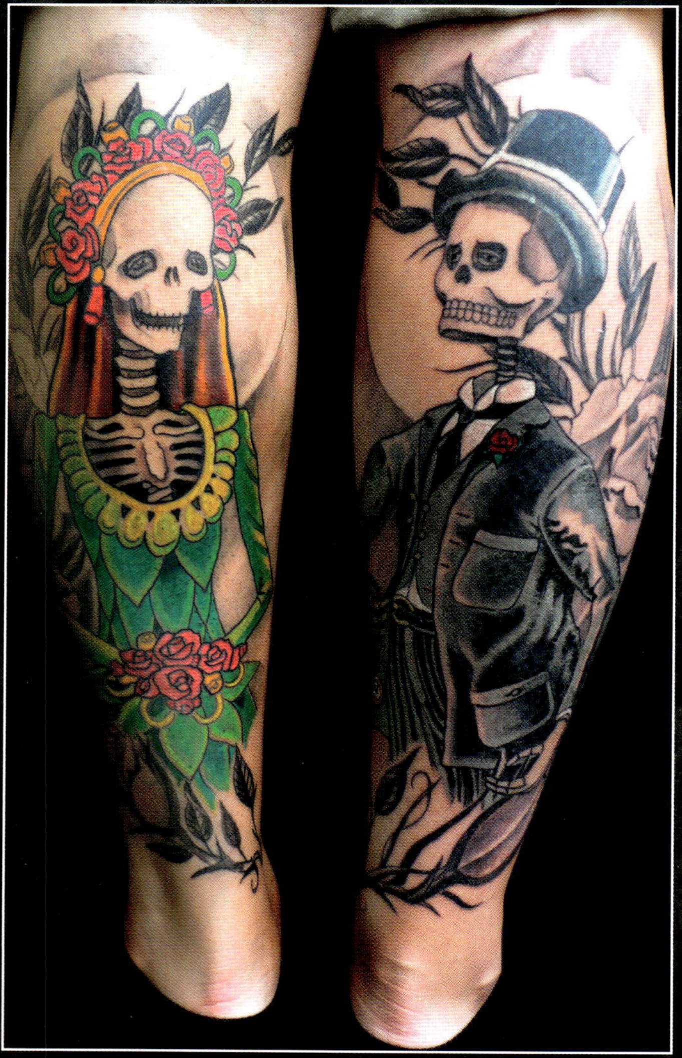

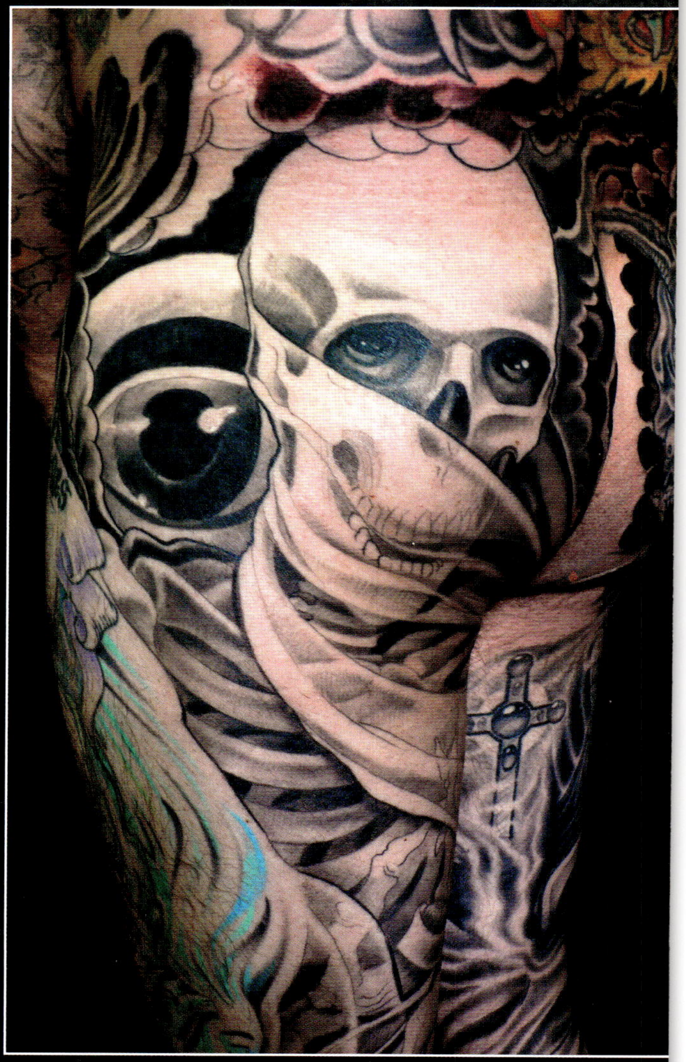

Kore Flatmo

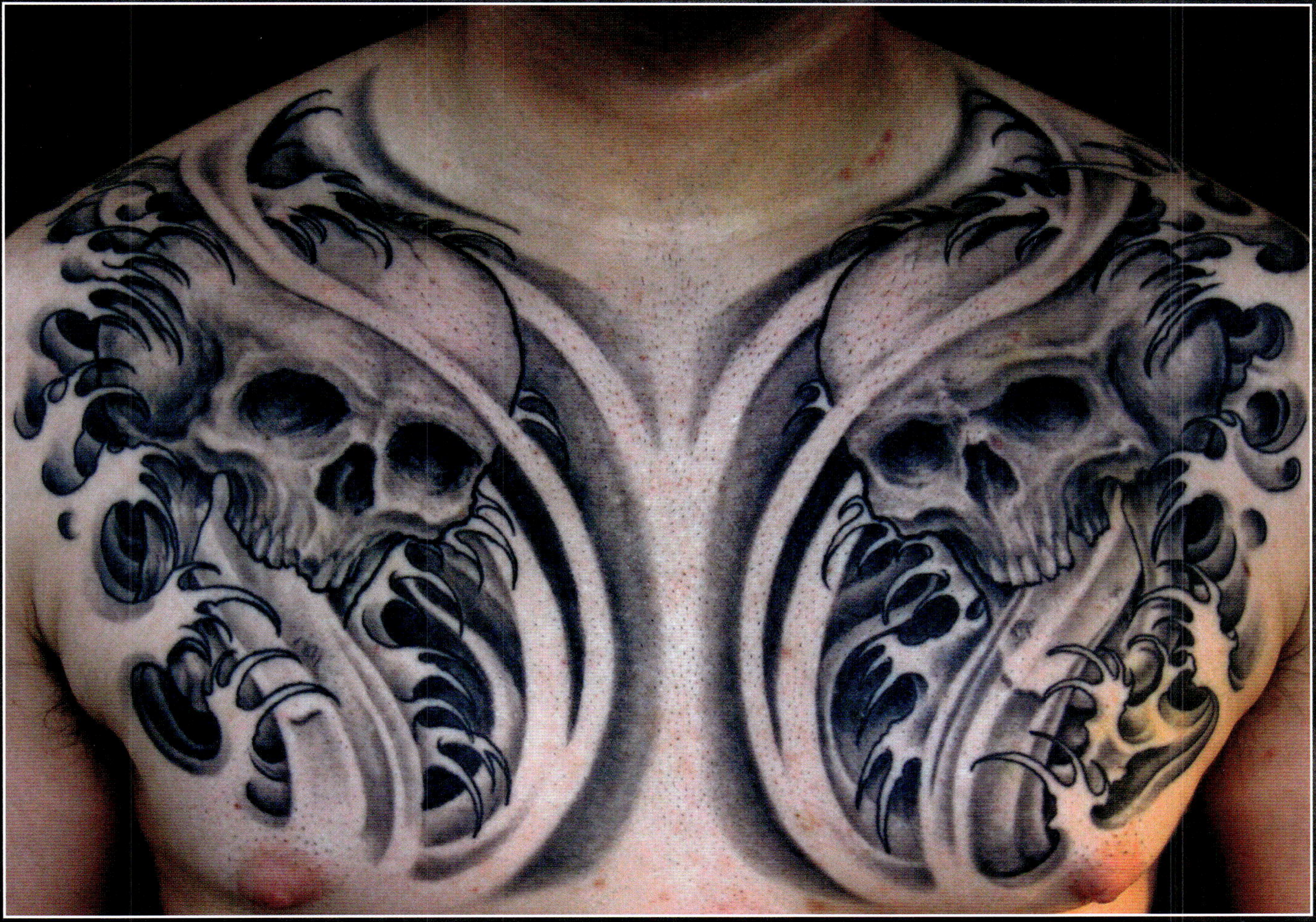

Jeremiah Barba

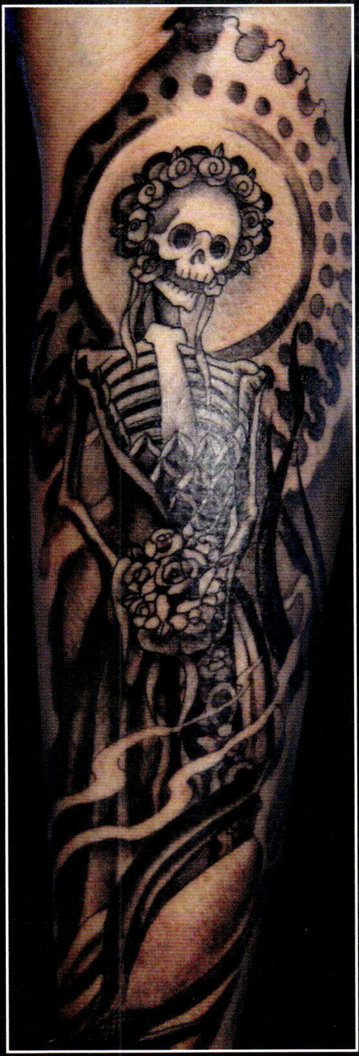

Kore Flatmo

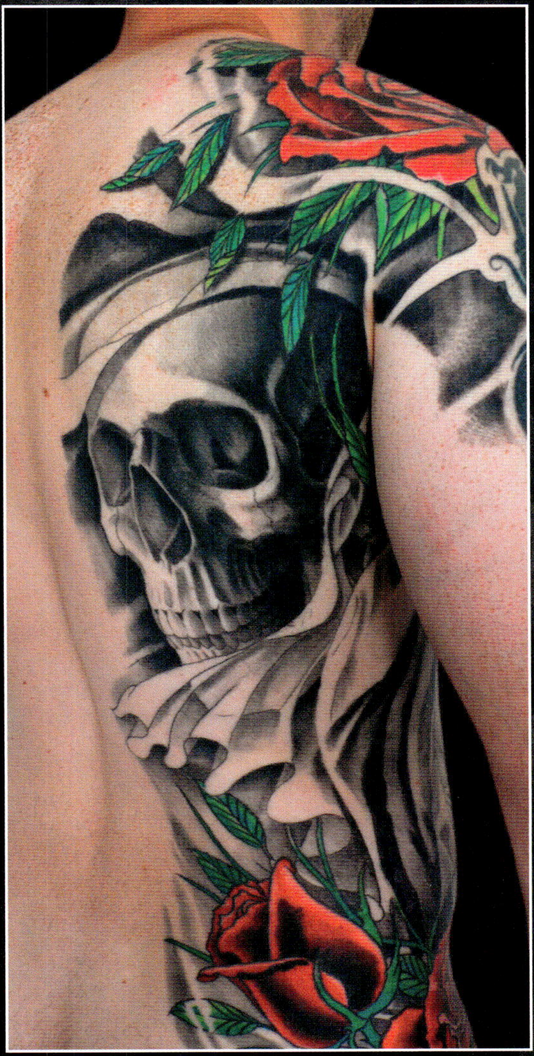

Kore Flatmo

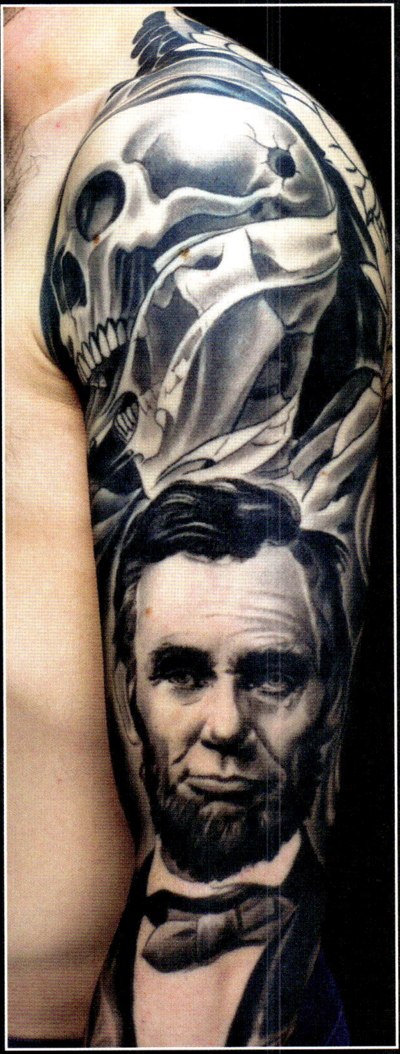

Kore Flatmo

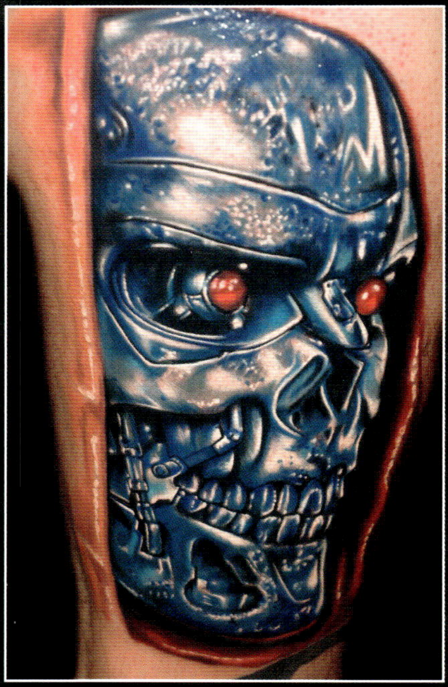

Roman Abrego

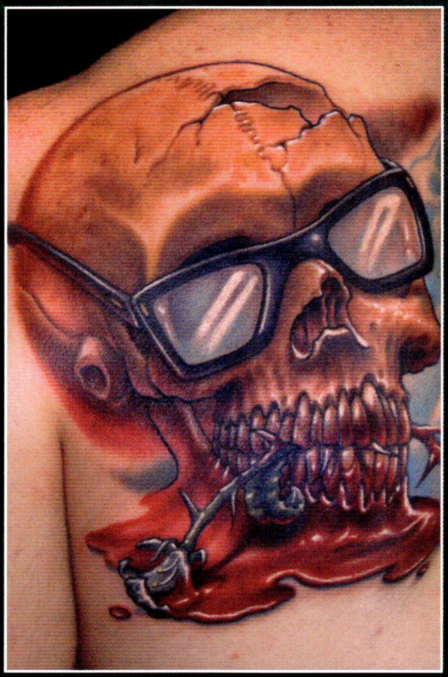

Nick Baxter

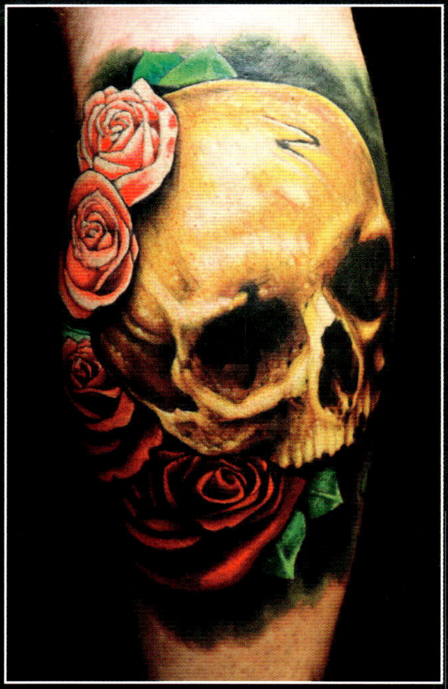

Chris Blinston

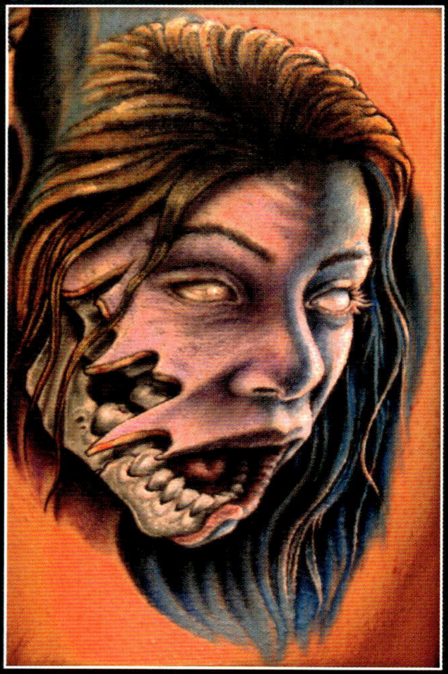

Stefano Alcantara

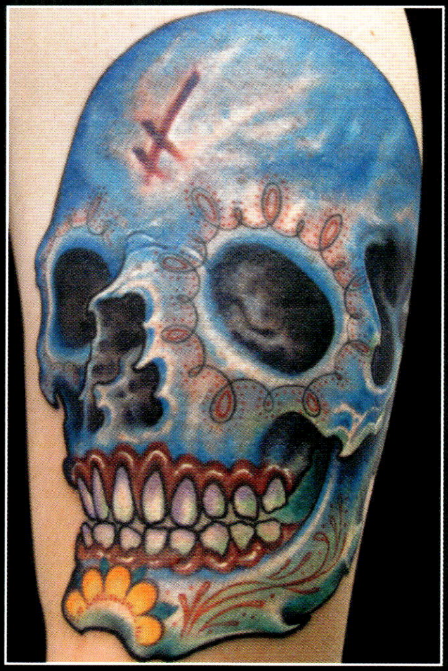

Nathan Kostechko

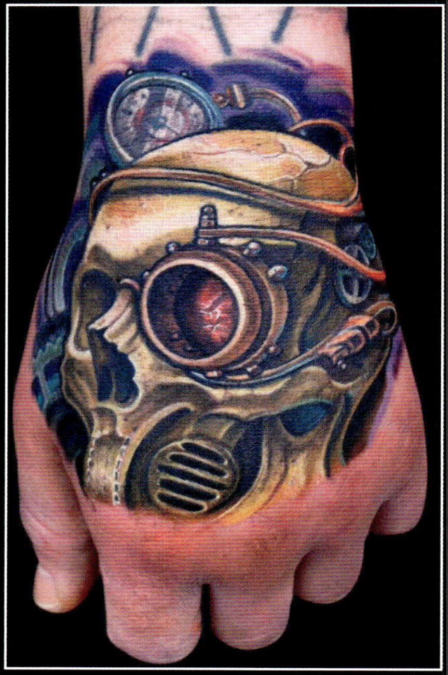

Nate Beavers

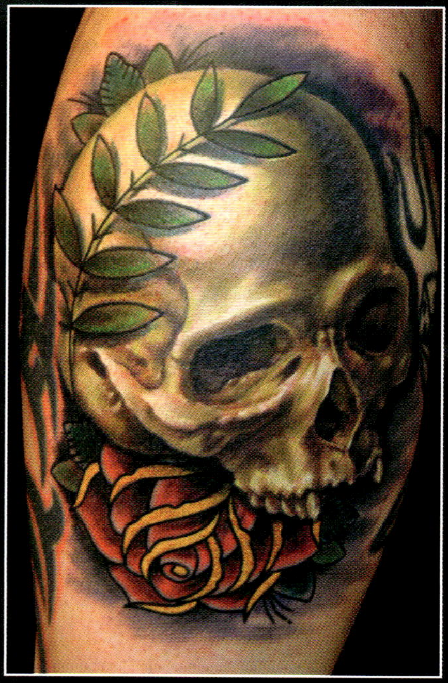

Nate Beavers

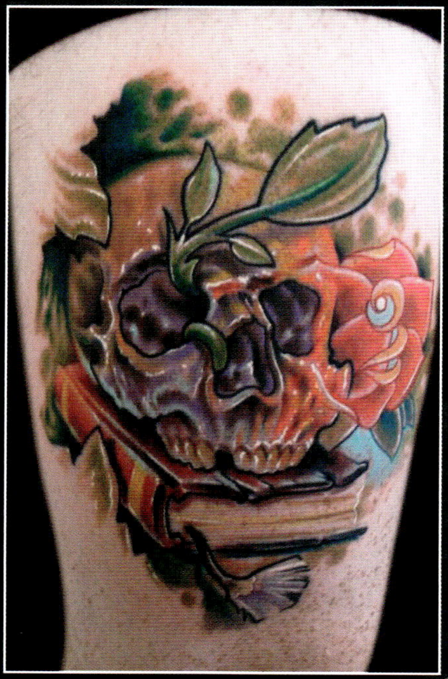

Tim Muller

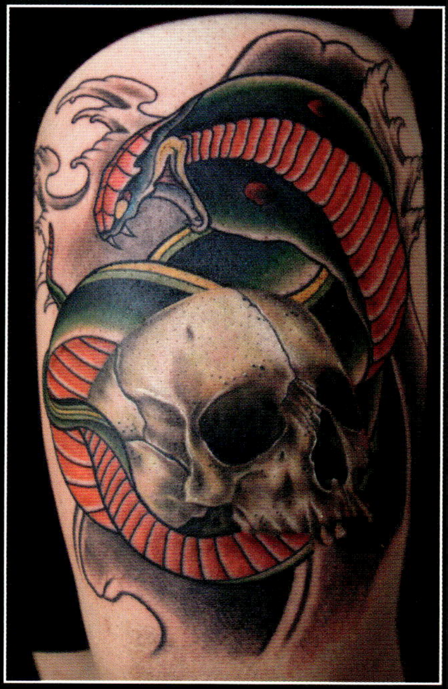

Mick Squires

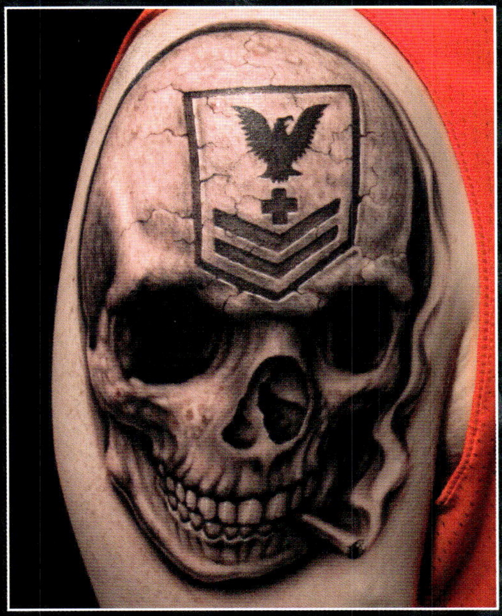
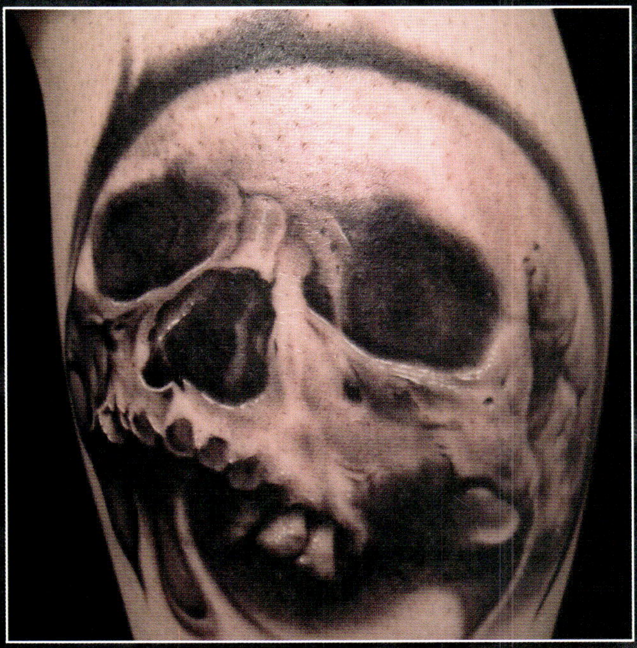
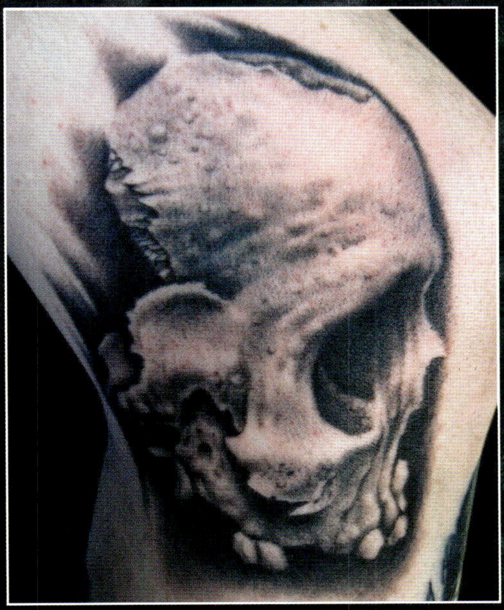
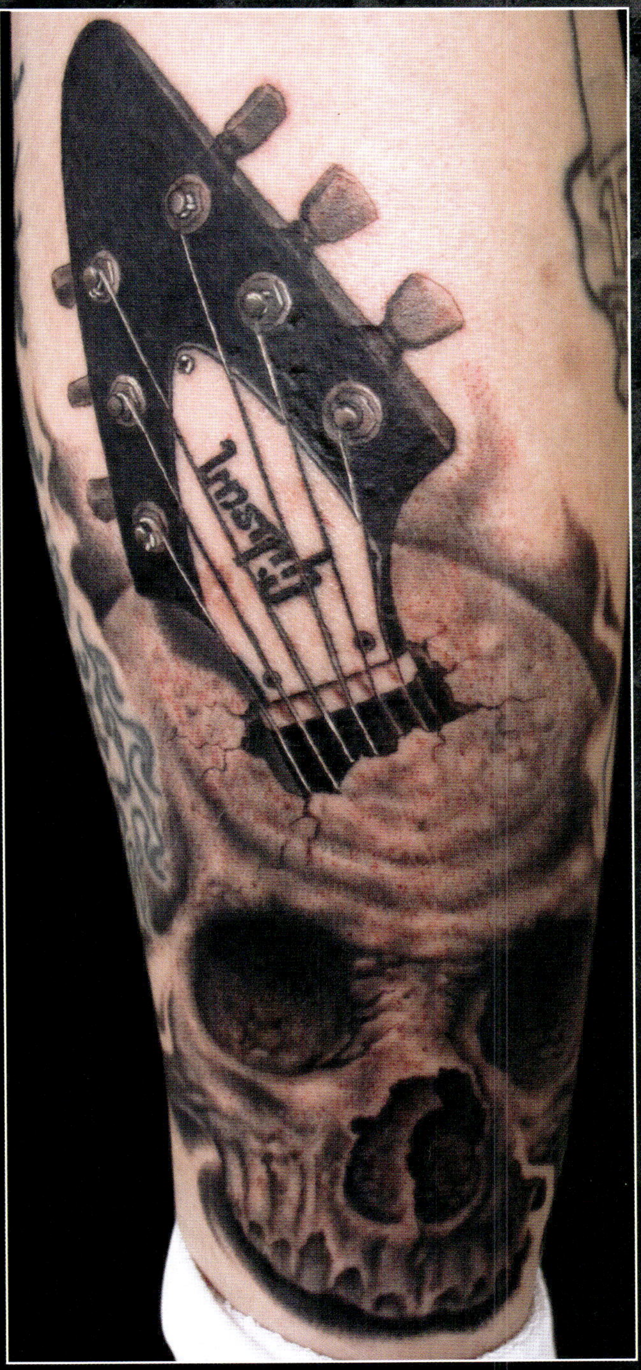
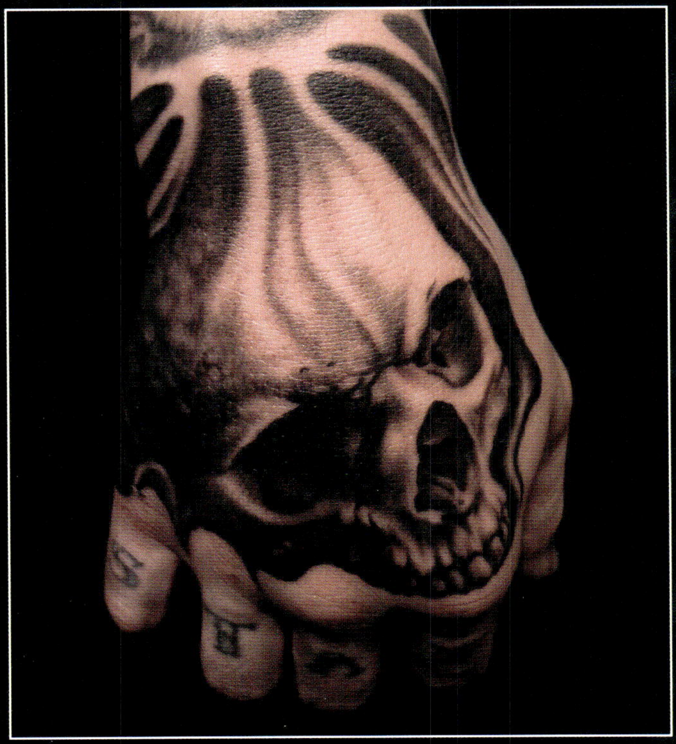

Bob Tyrrell

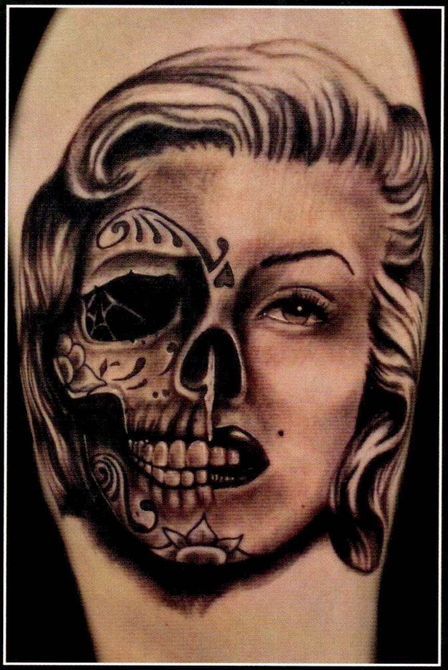
Anthony Zamora

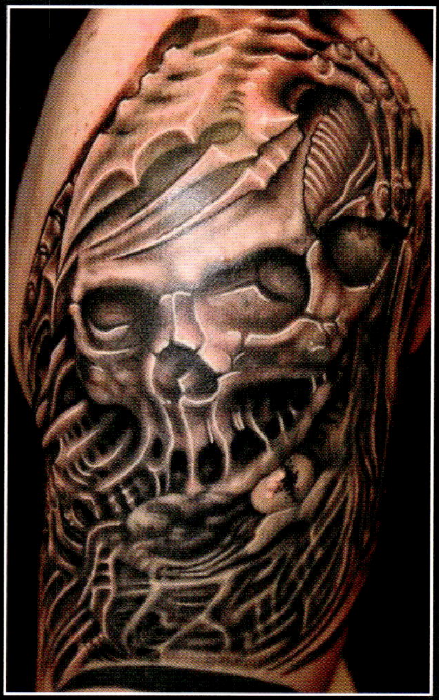
Jason Morrow

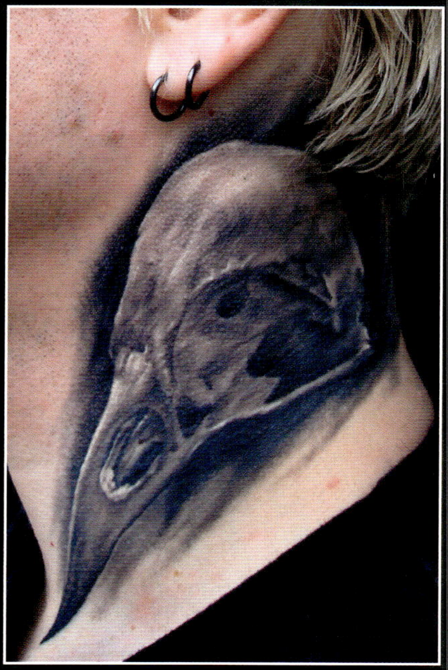
Benjamin Moss

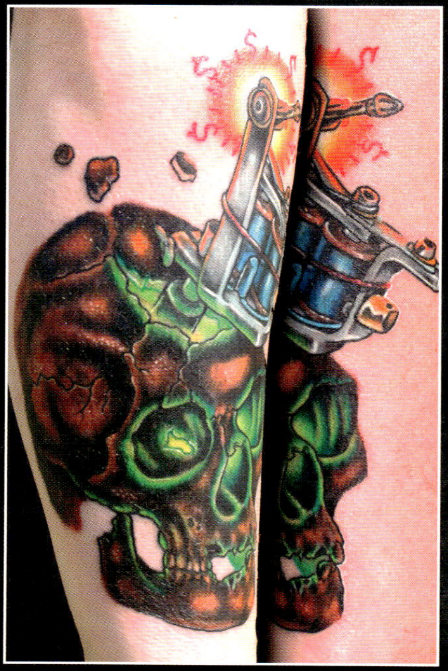
Mattie Macabre

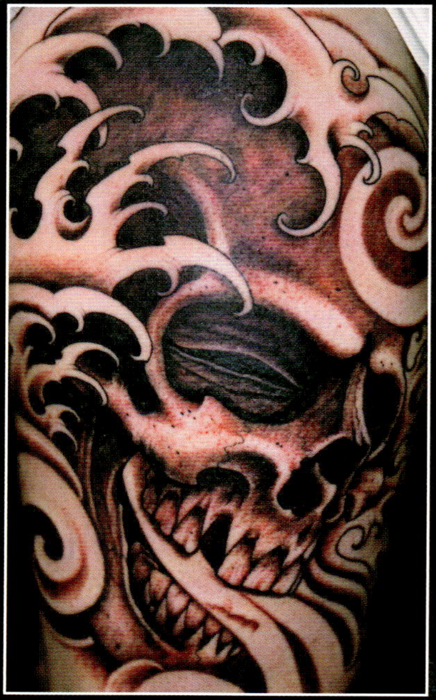
Ben Corn

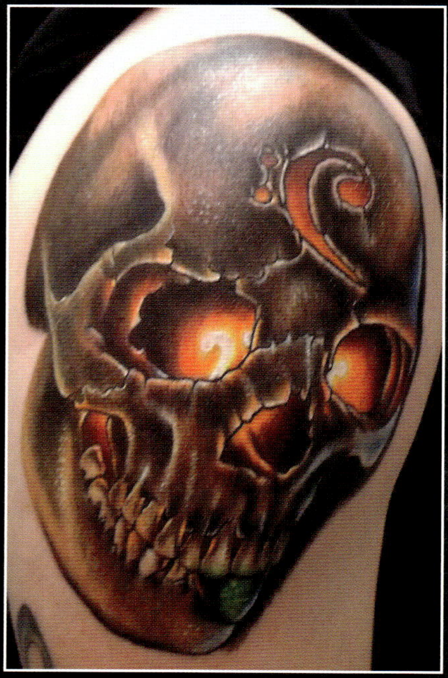
Mattie Macabre

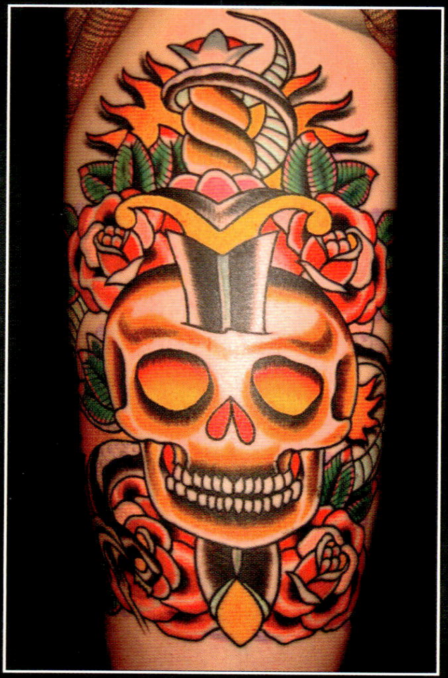
Myke Chambers

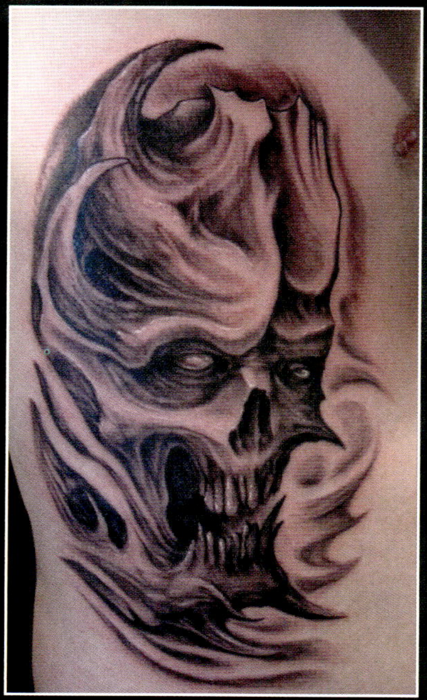
Jeremiah Barba

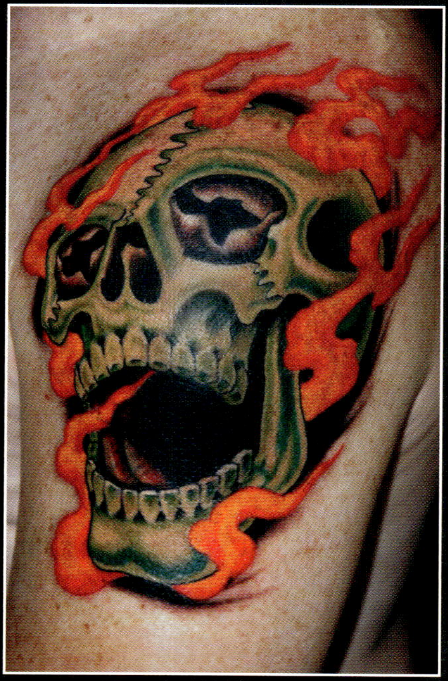
Brady Willmott

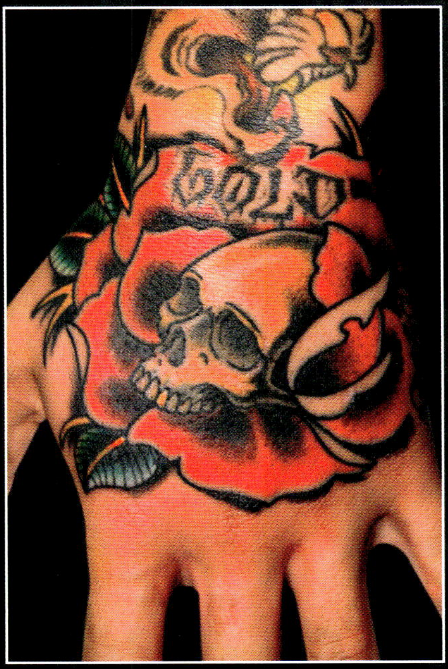

Jim Sylvia

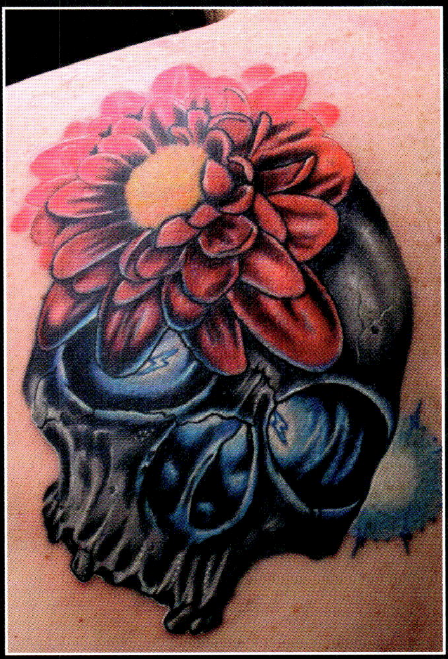

Mattie Macabre

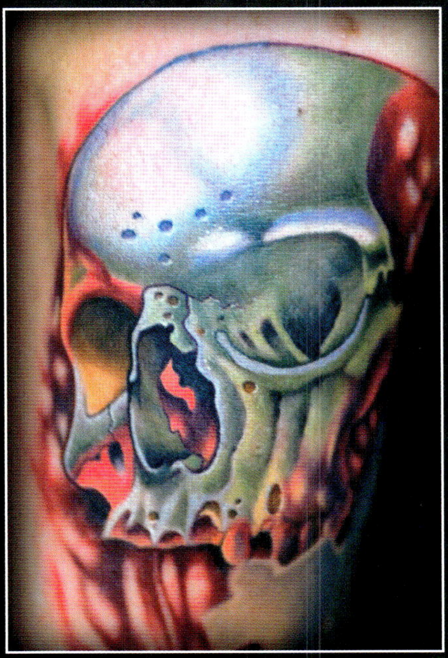

Keith Ciaramello

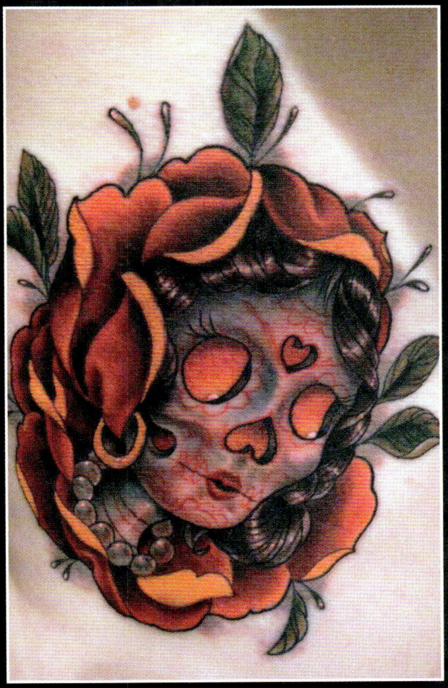

Pete Vaca

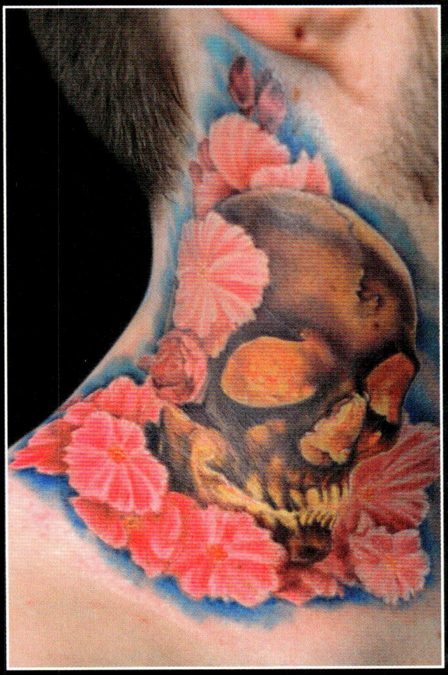

Steve Morris

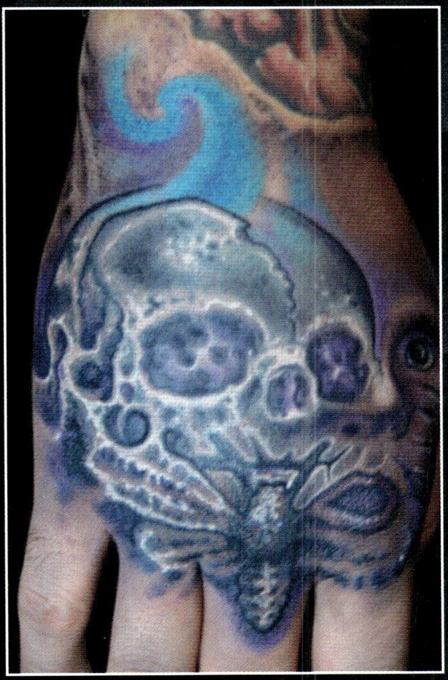

Nathan Kostechko

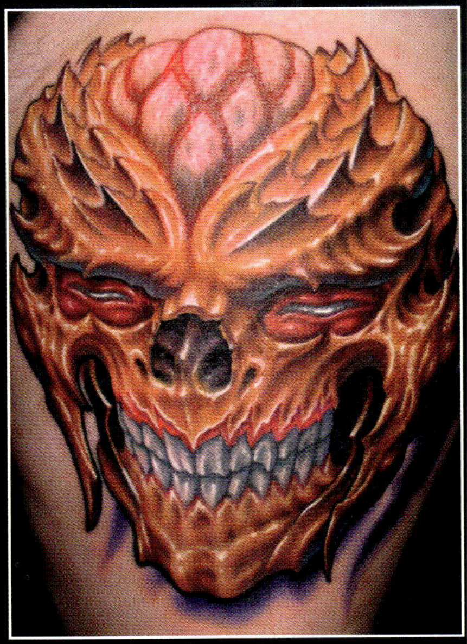

Sean Zee

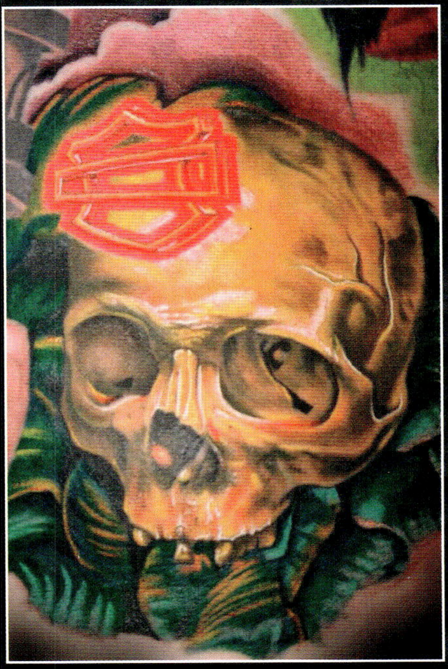

Jason Morrow

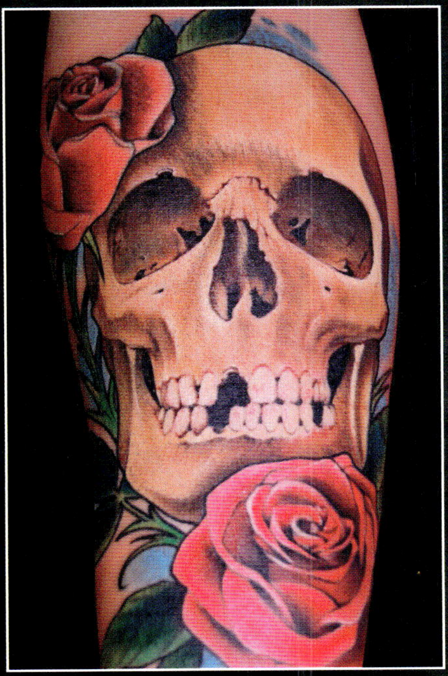

Trond Kjørkleiv

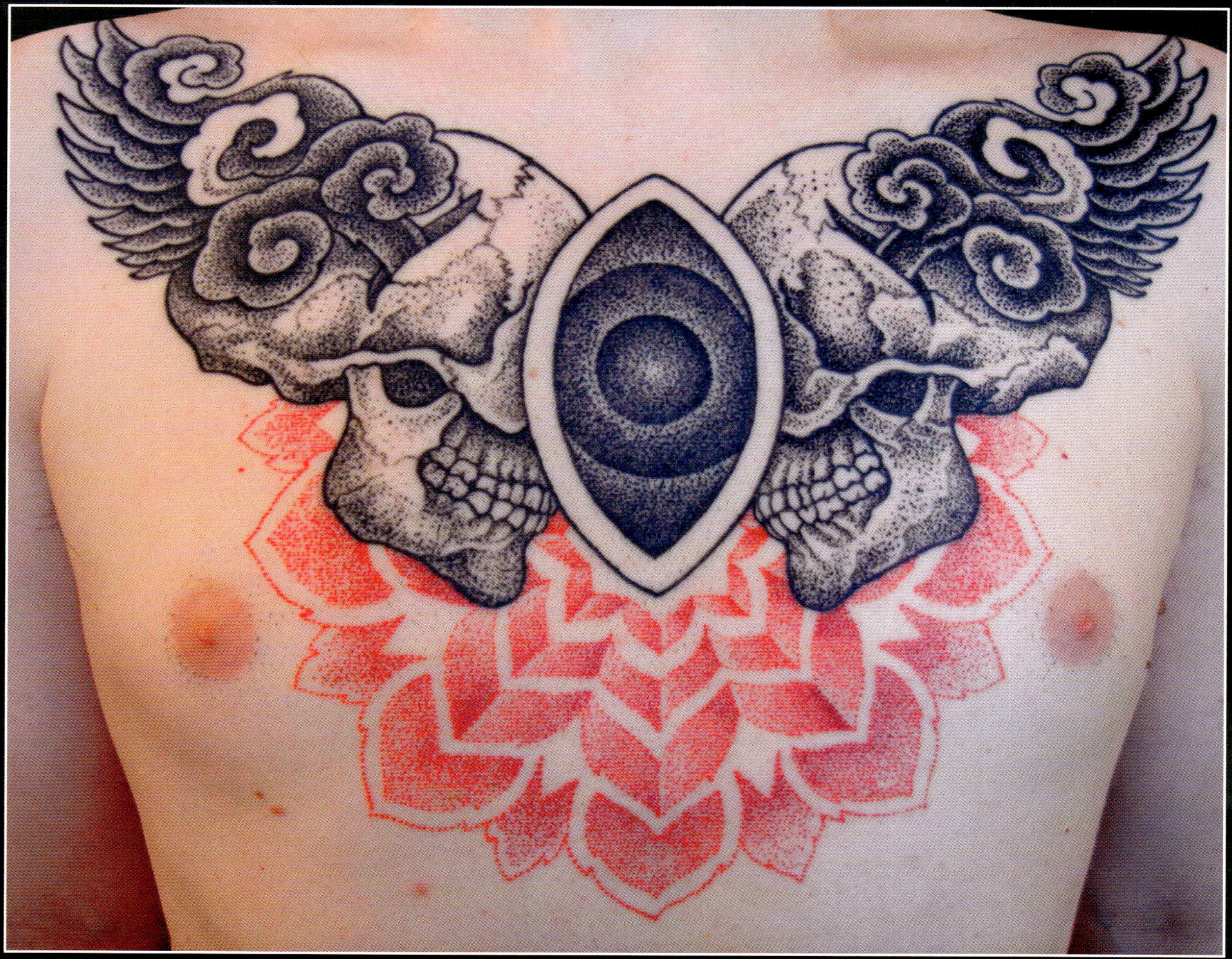

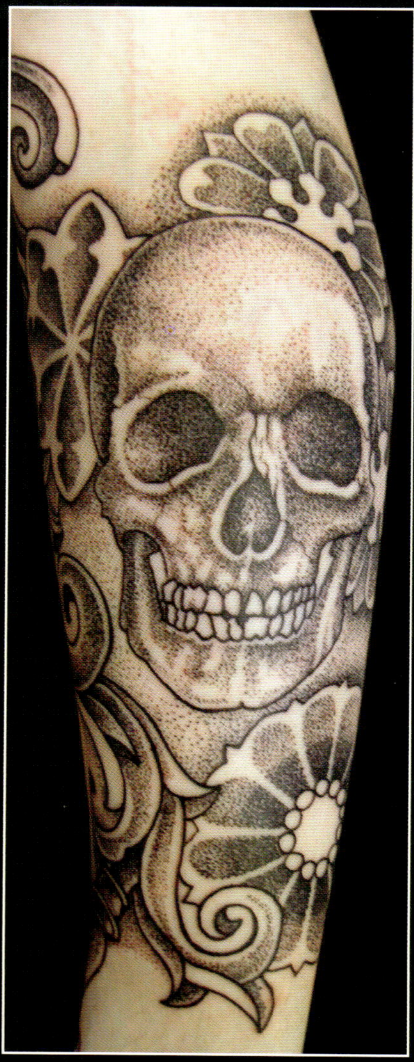

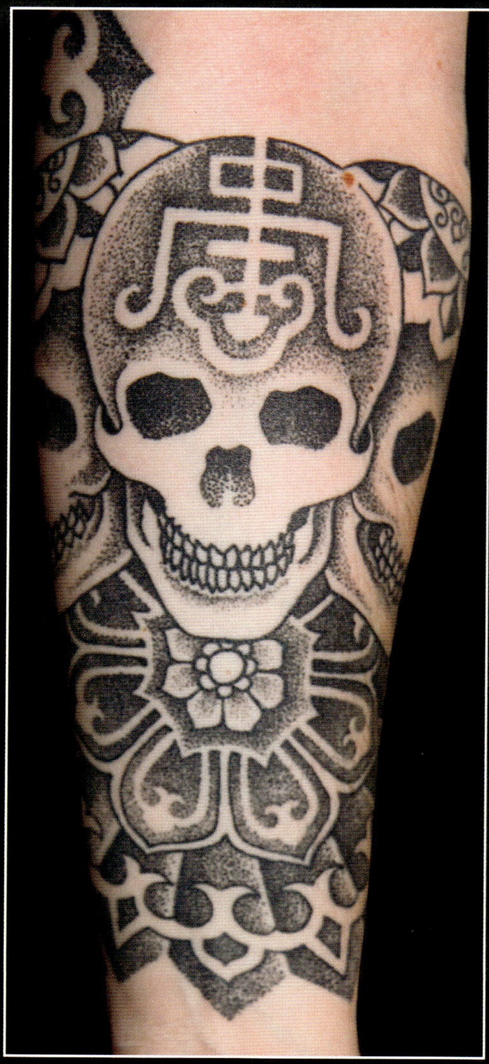

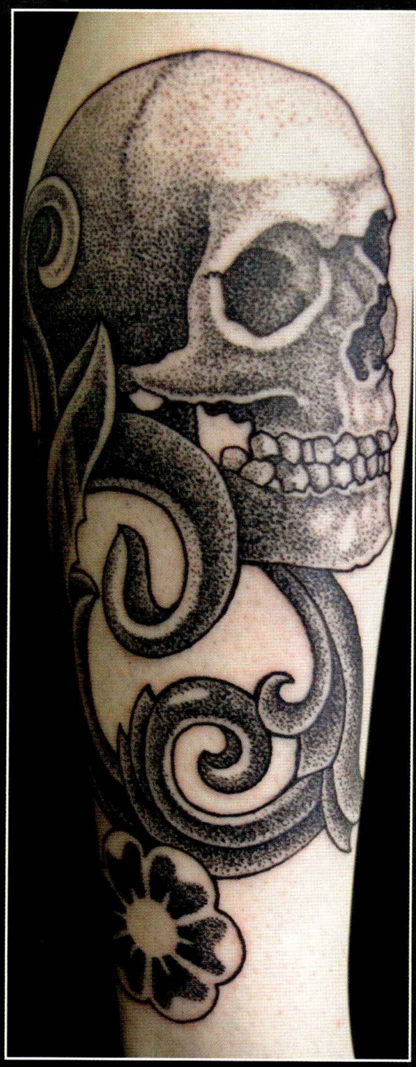

Daniel DiMattia

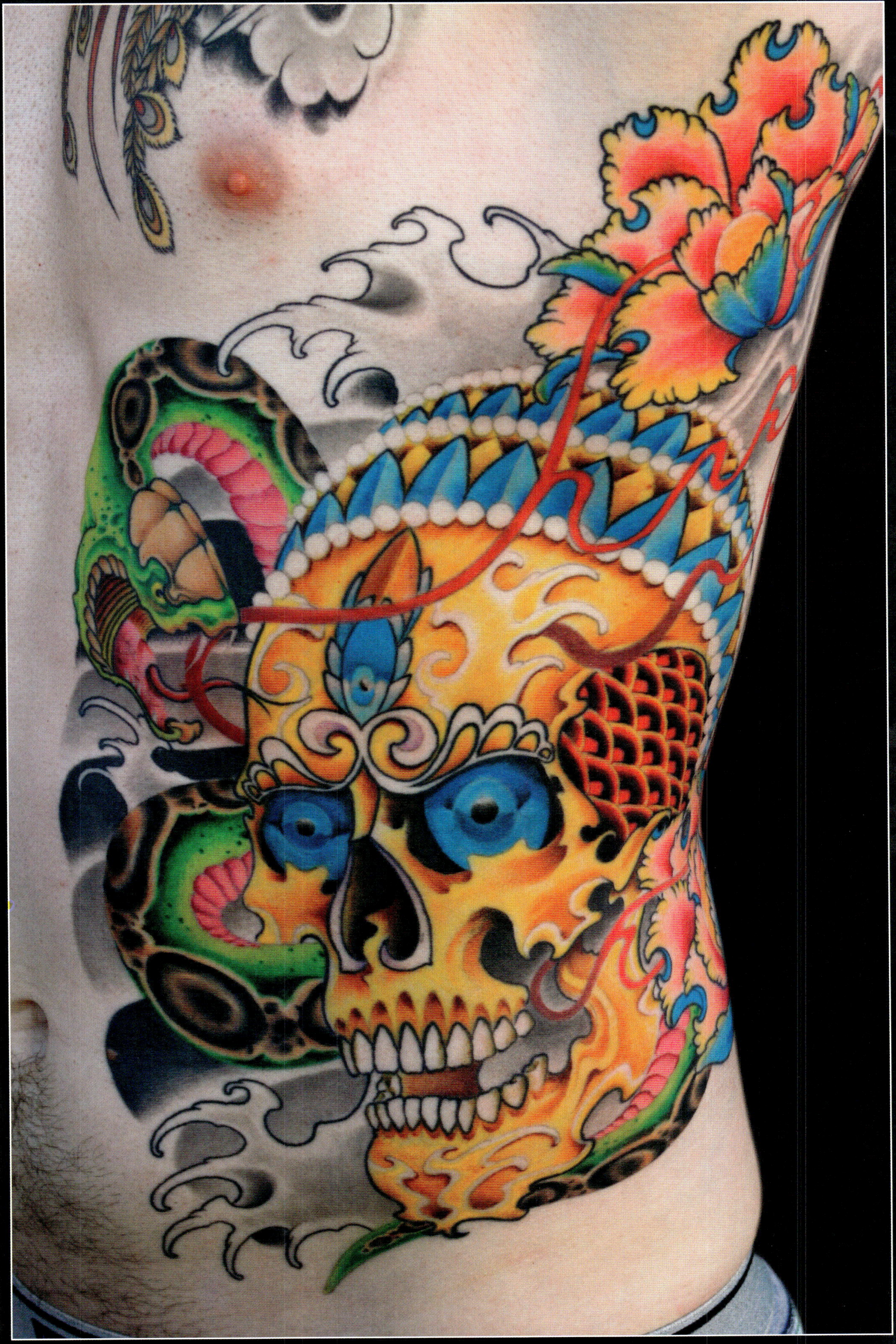

Tim McEvoy

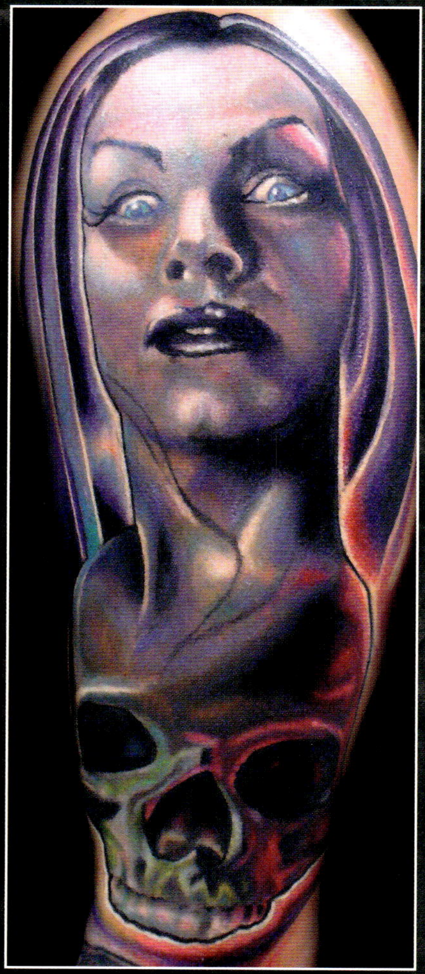

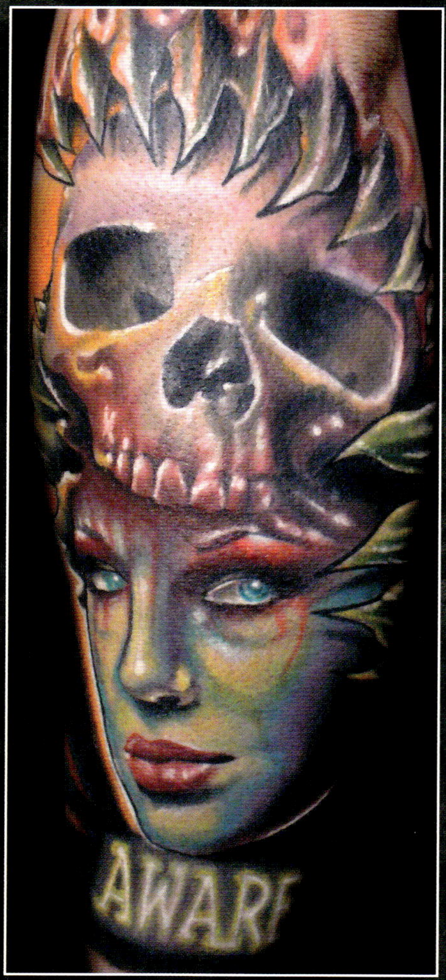

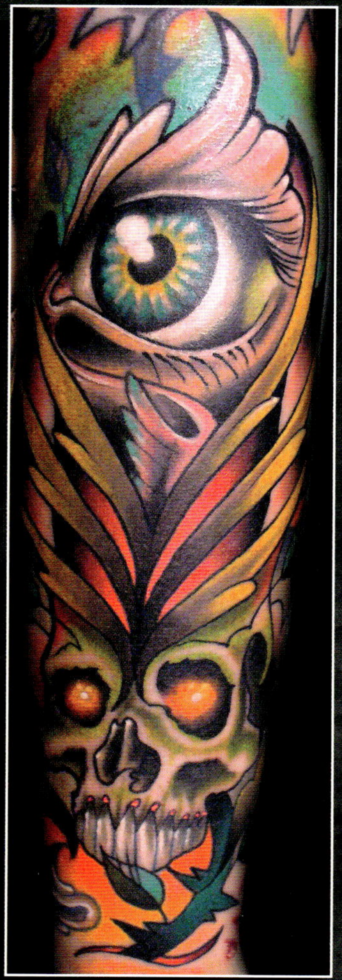

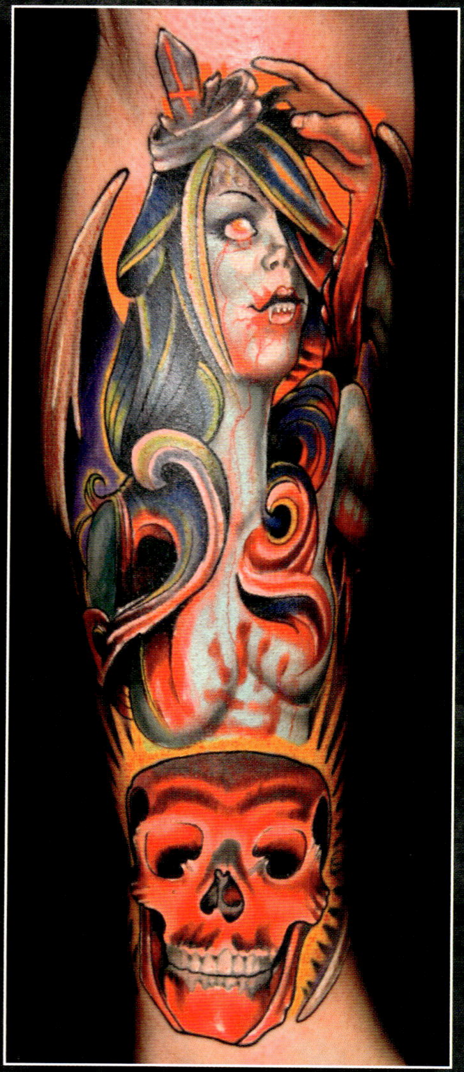

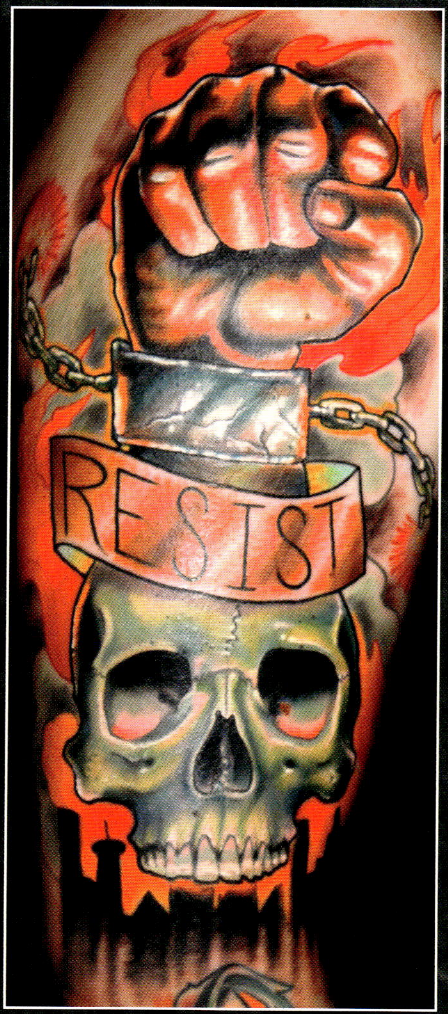

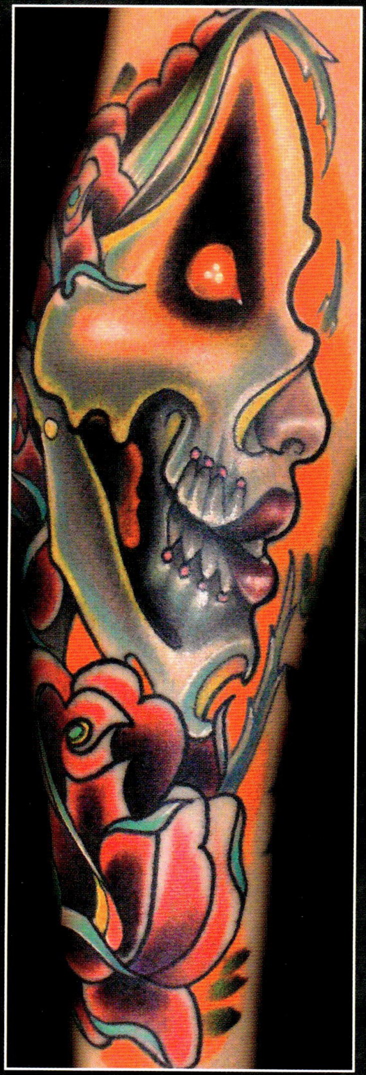

Sean Herman

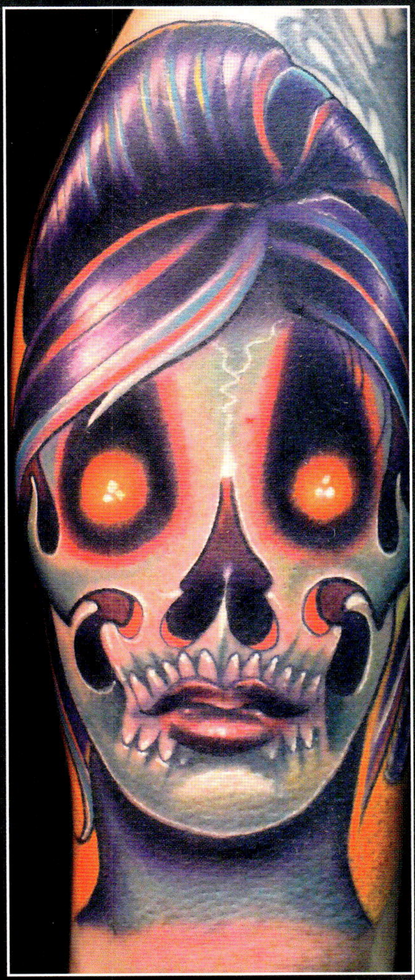
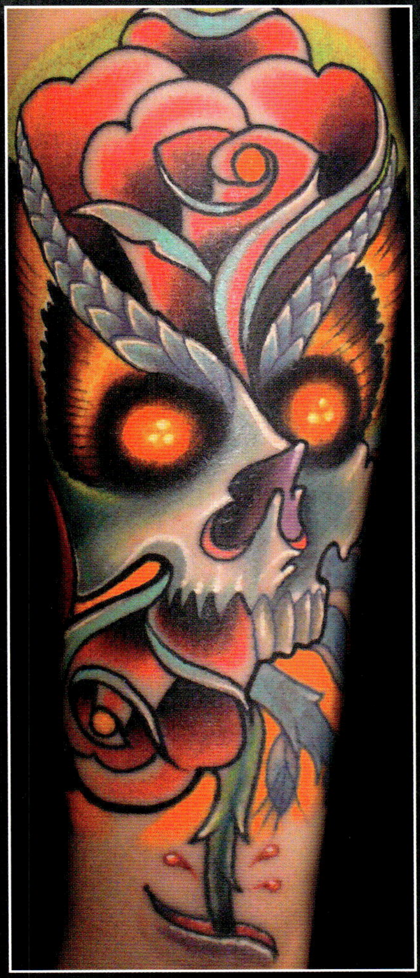
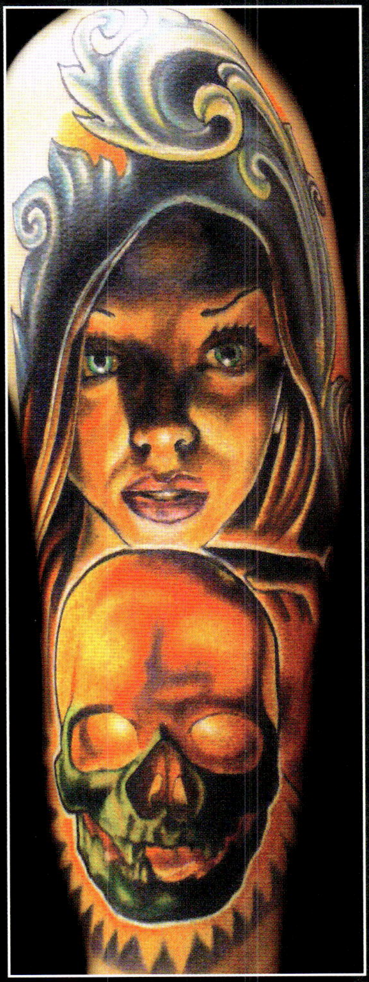
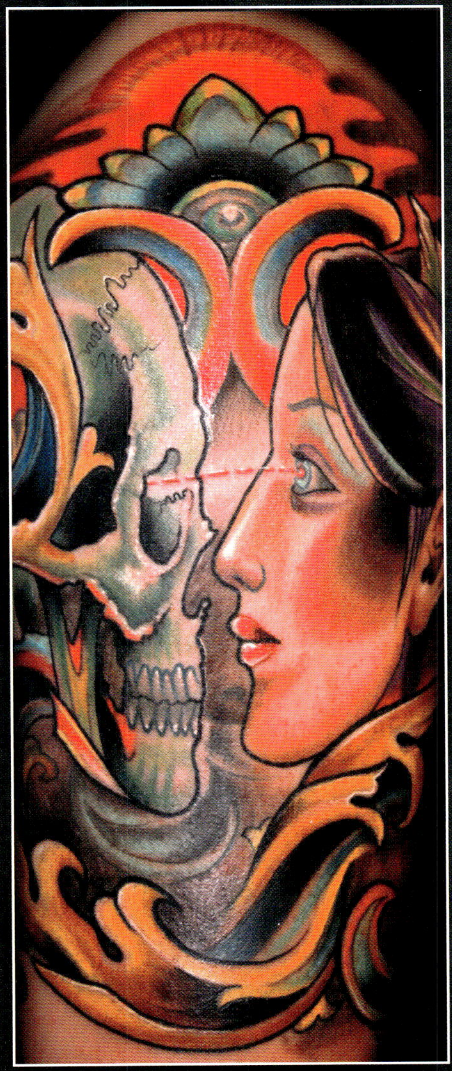
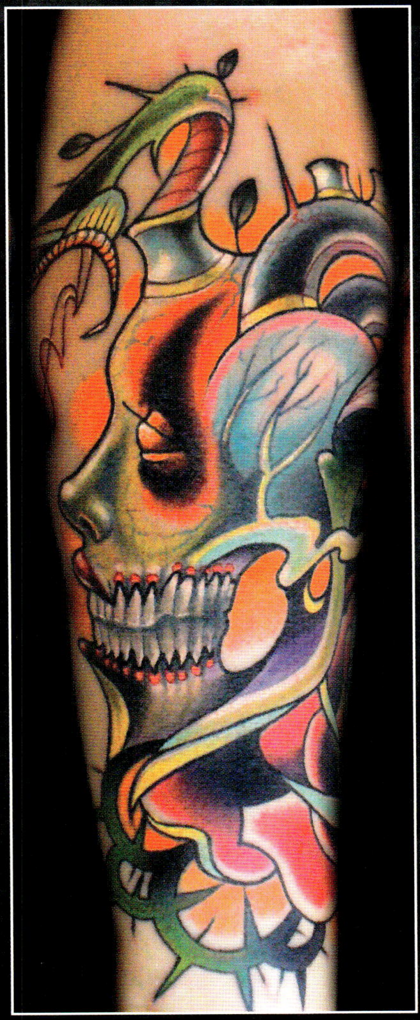
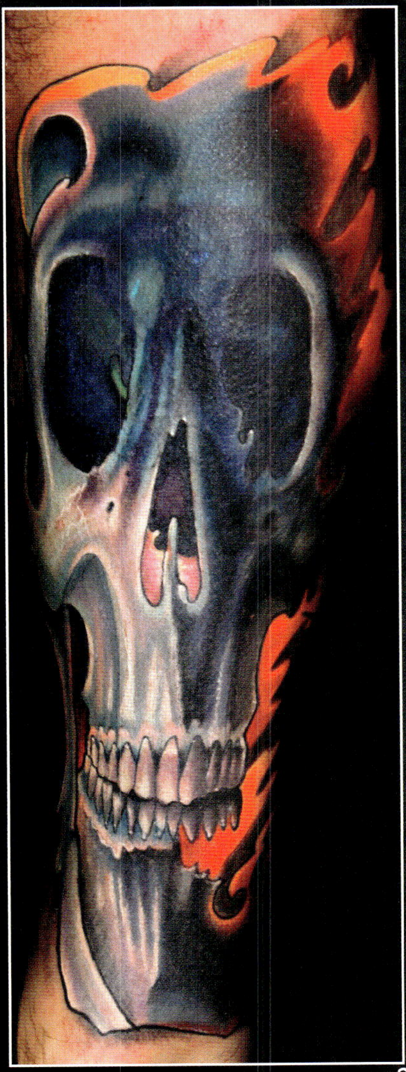

Sean Herman

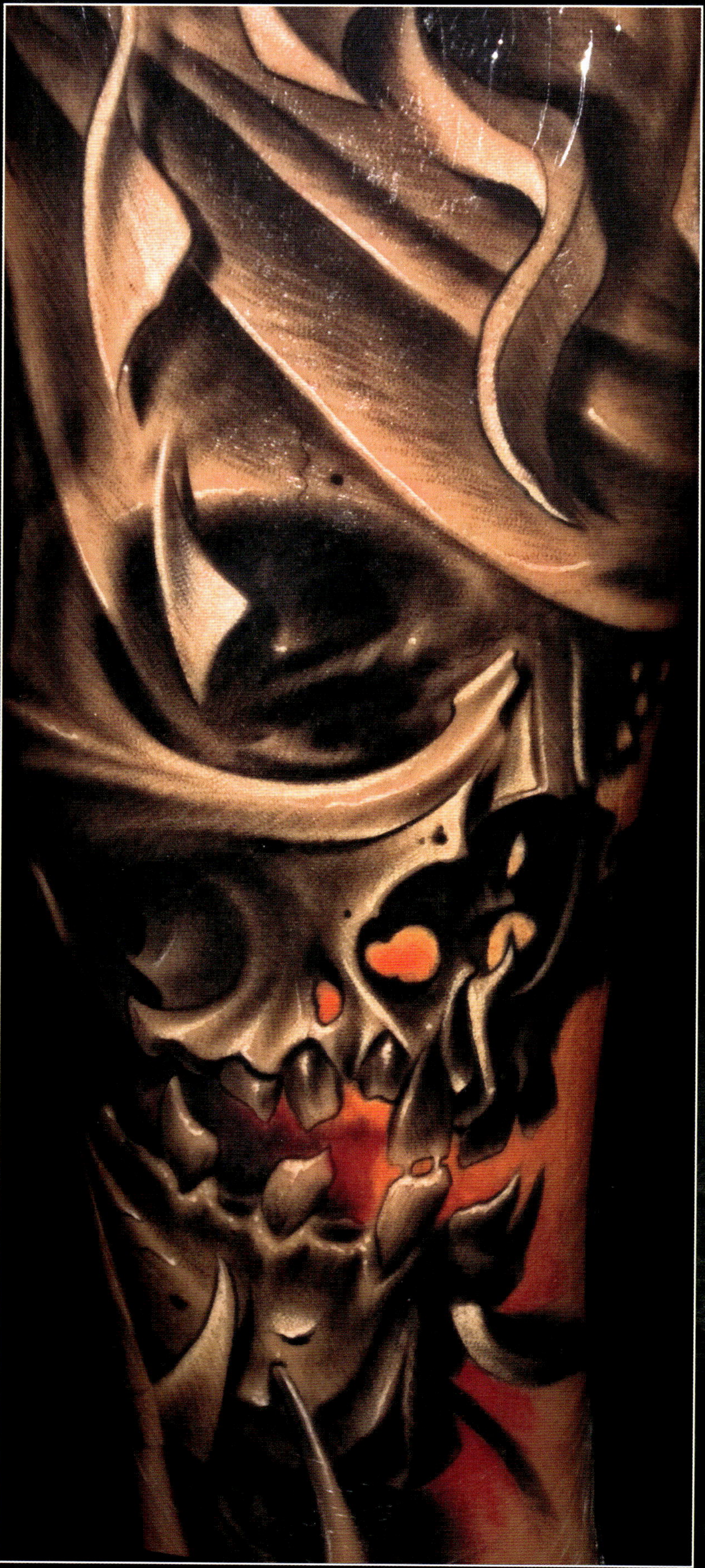

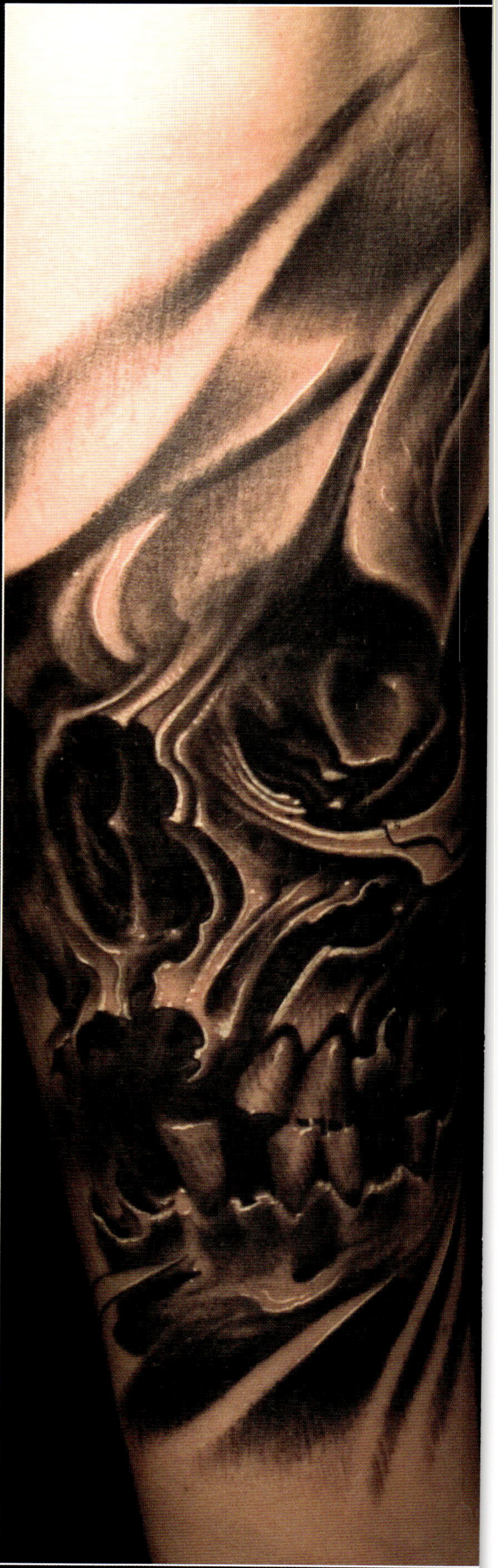

Victor Portugal

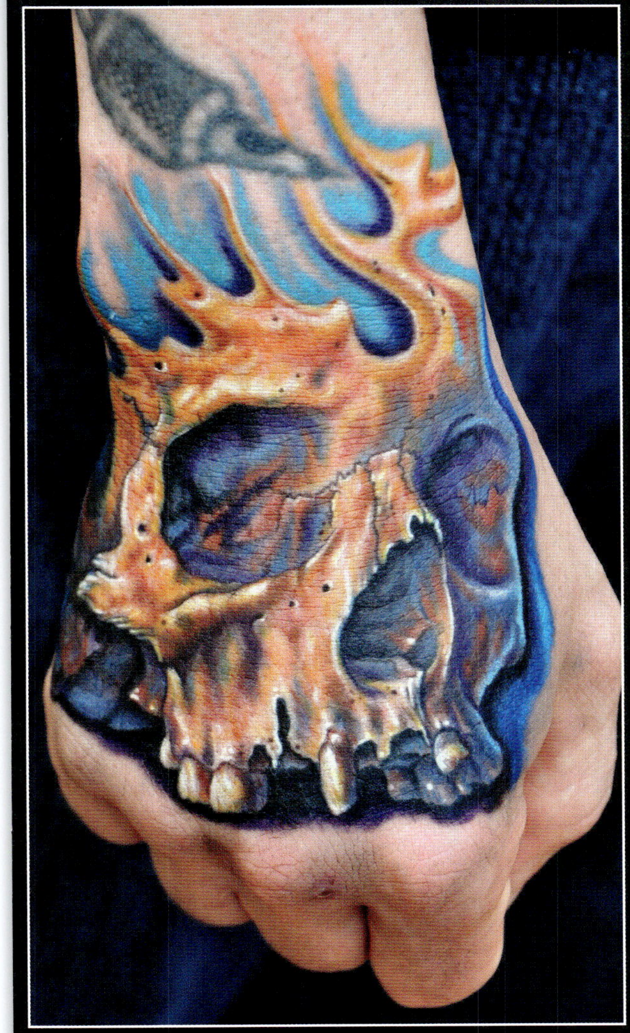

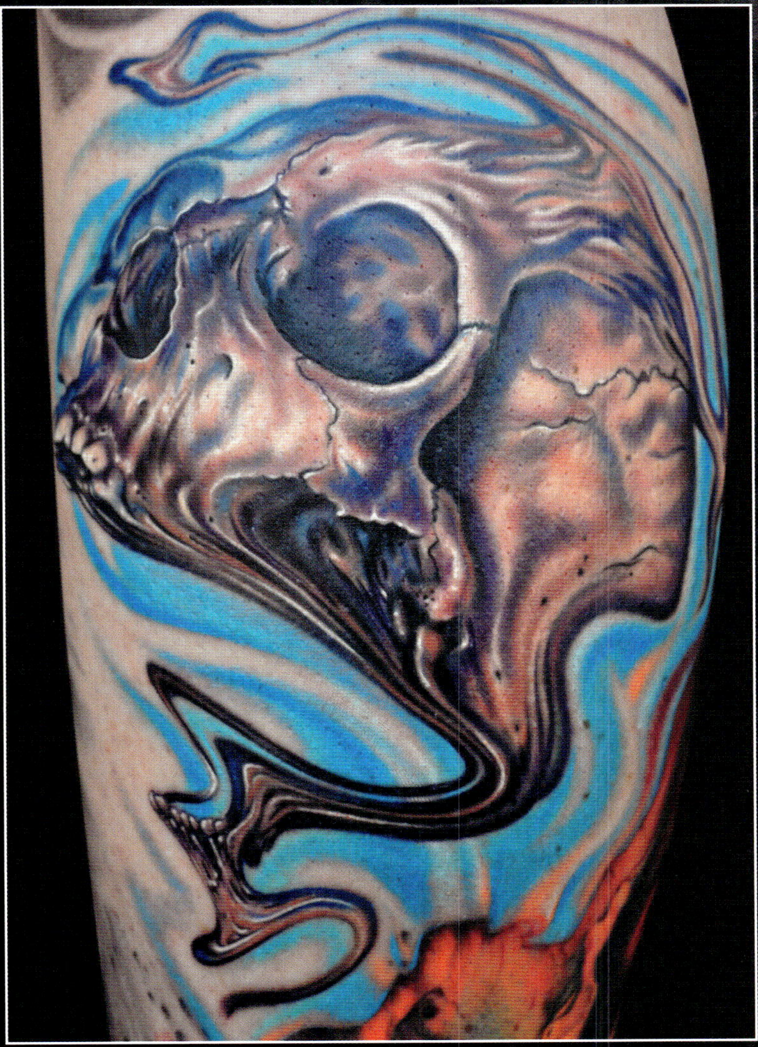

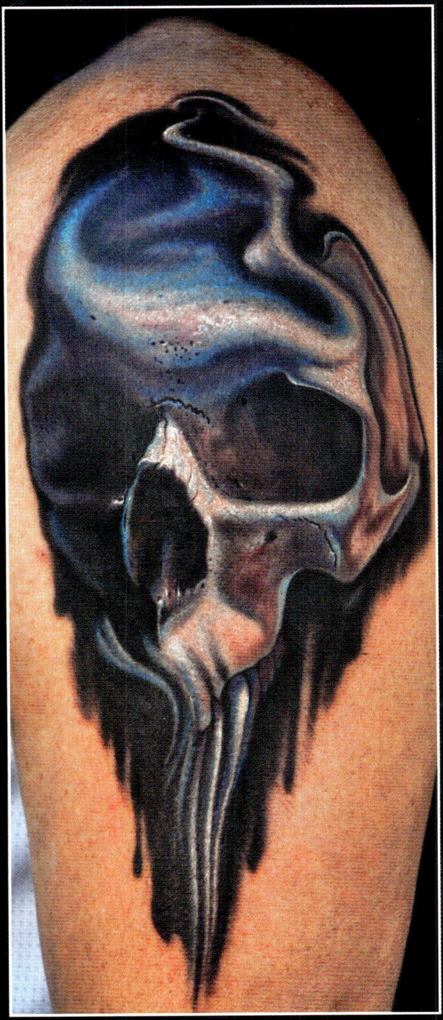

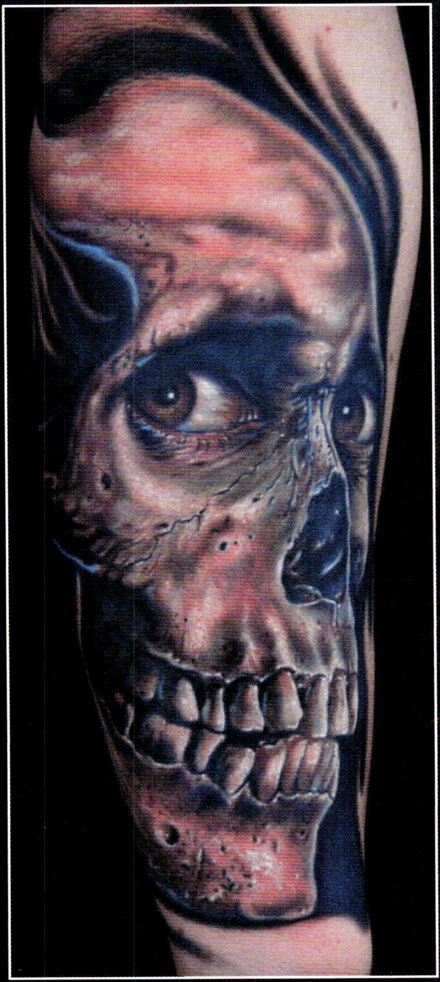

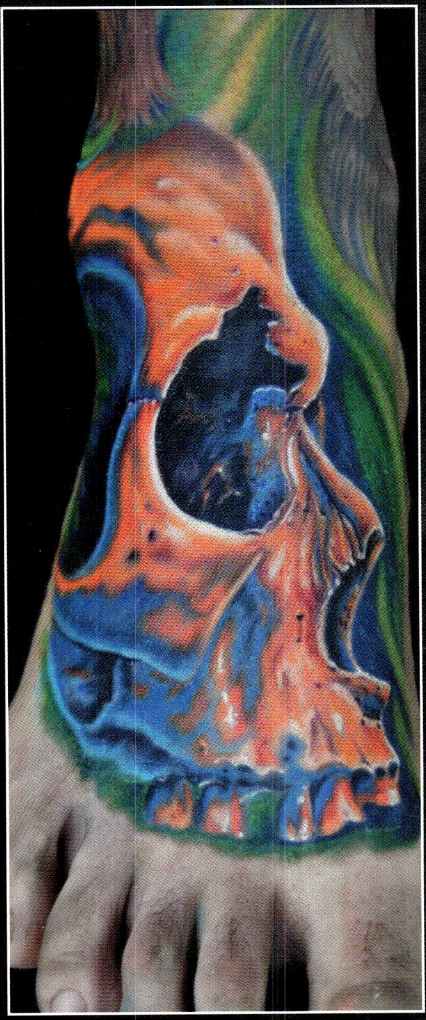

Mike DeVries

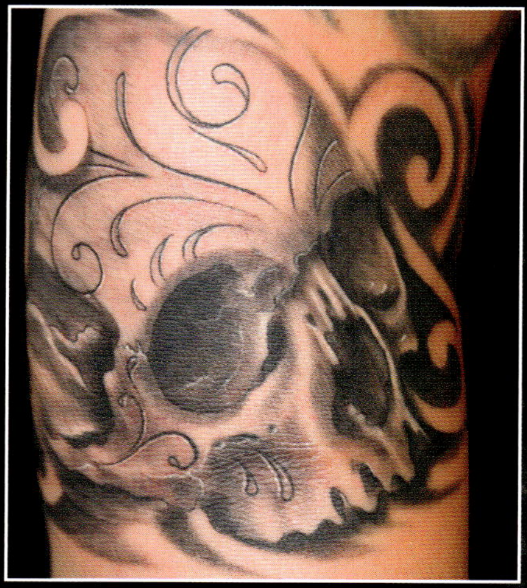
Mike Godfrey

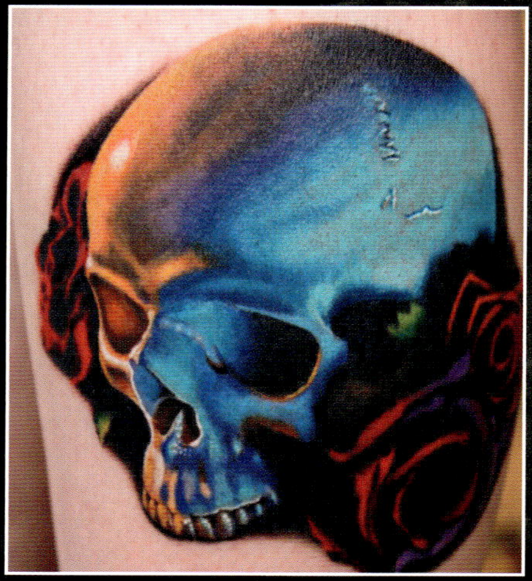
Josh Phillips

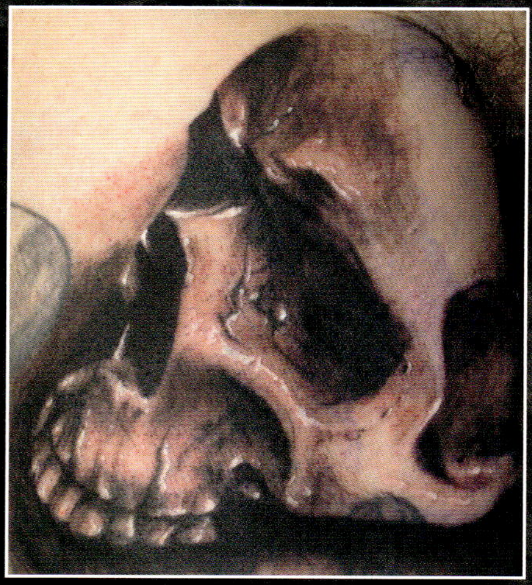
John Sweeney

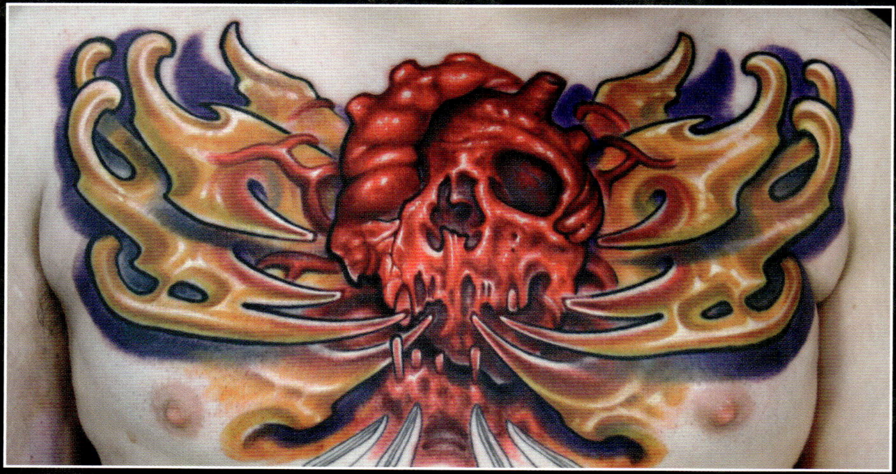
Nathan Kostechko

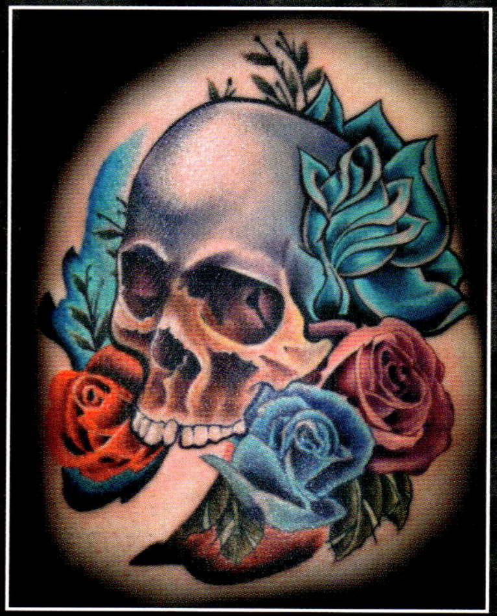
Fabian De Gaillande

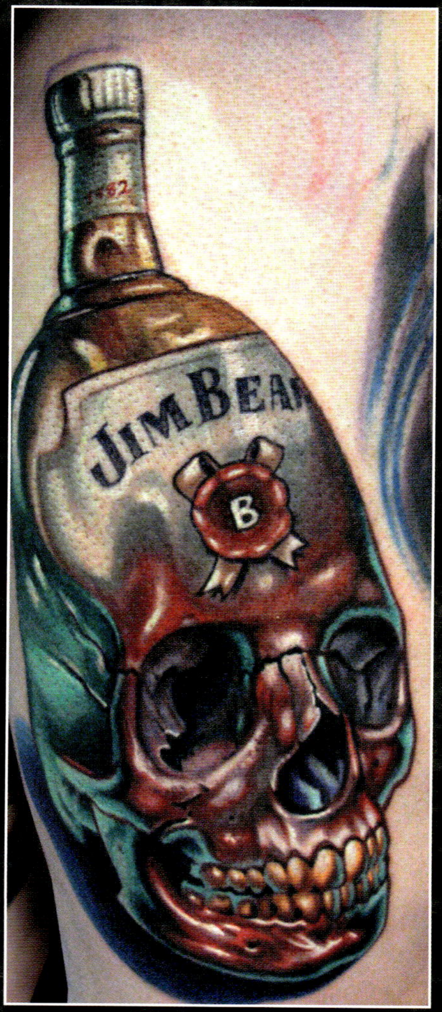

Christian Perez

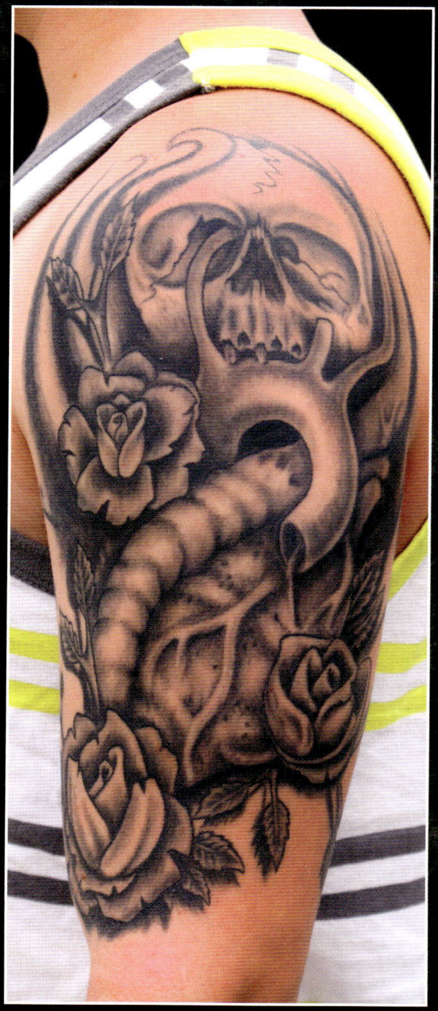
Jeff Johnson

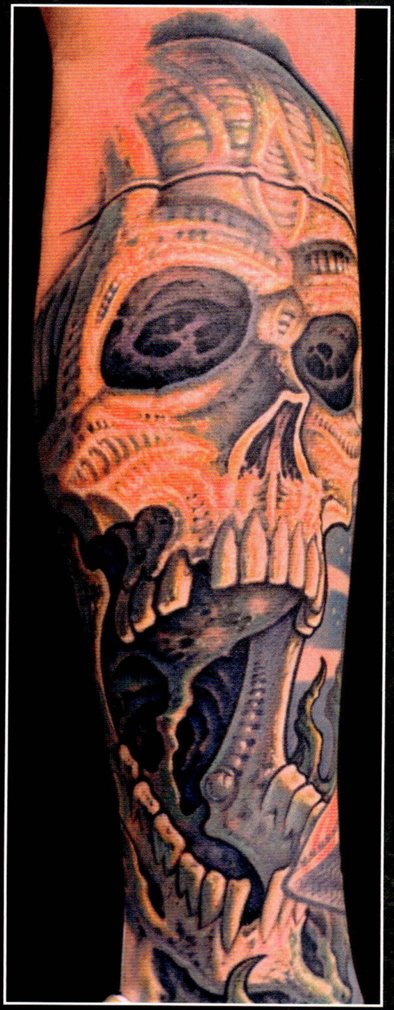
Guy Aitchison

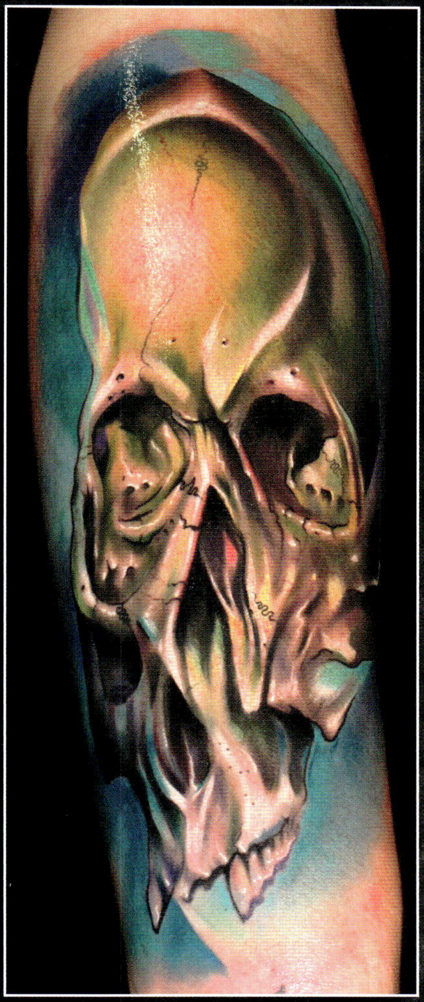

Byron Drechsler

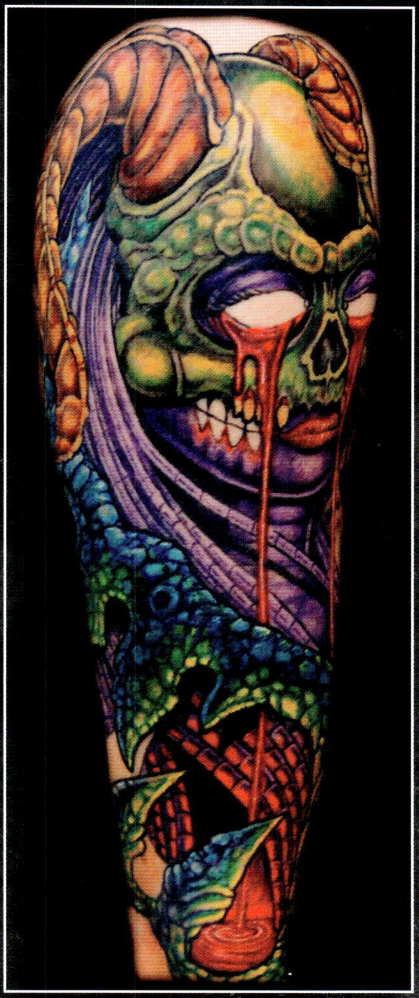

Brett Herman

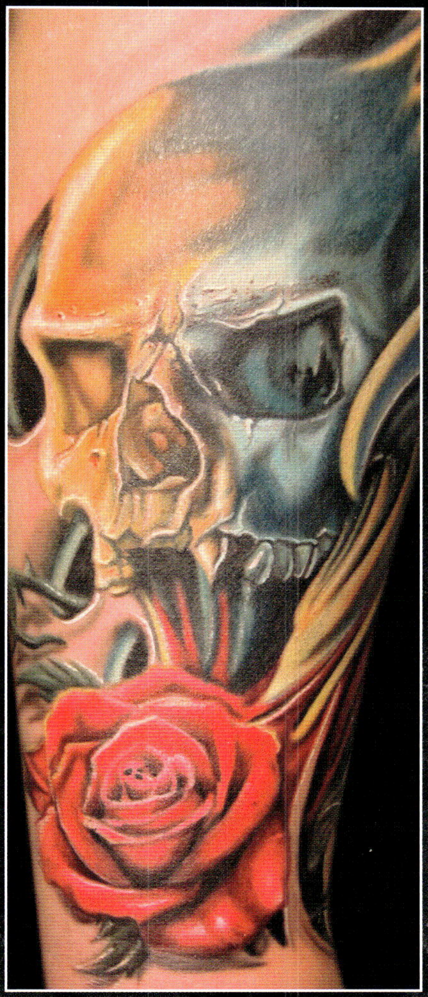

Frank la Natra

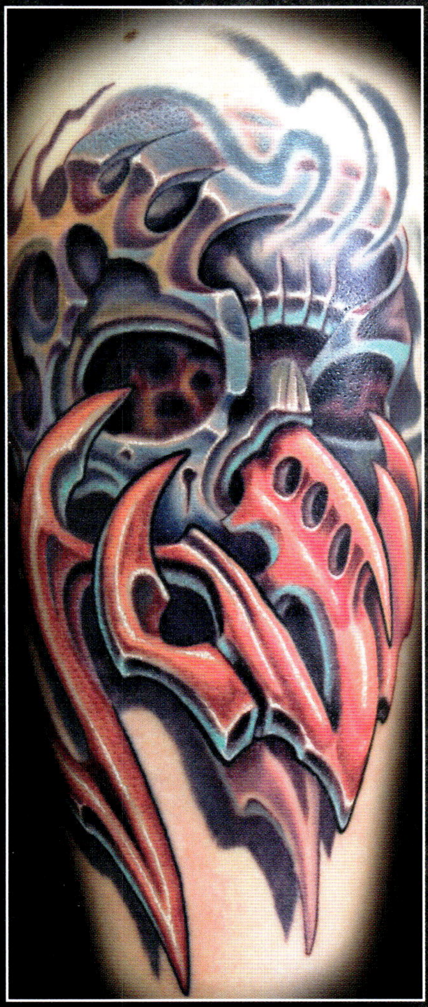

Marvin Silva

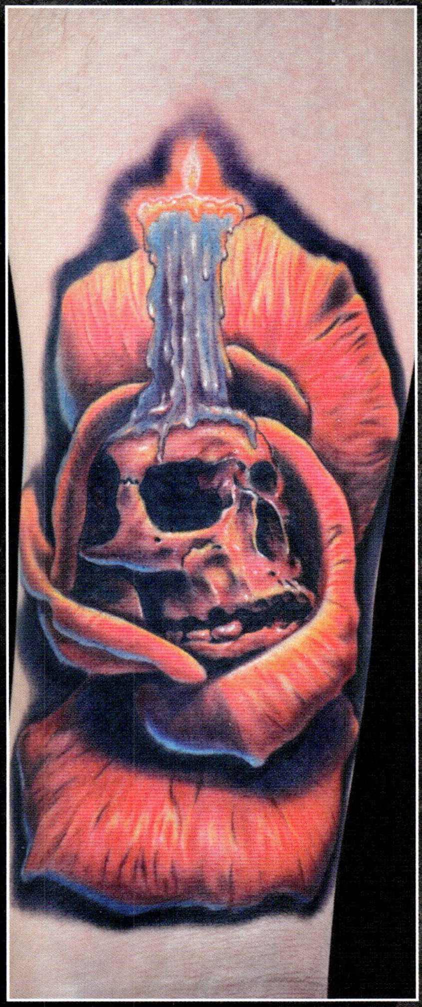

Mike DeVries

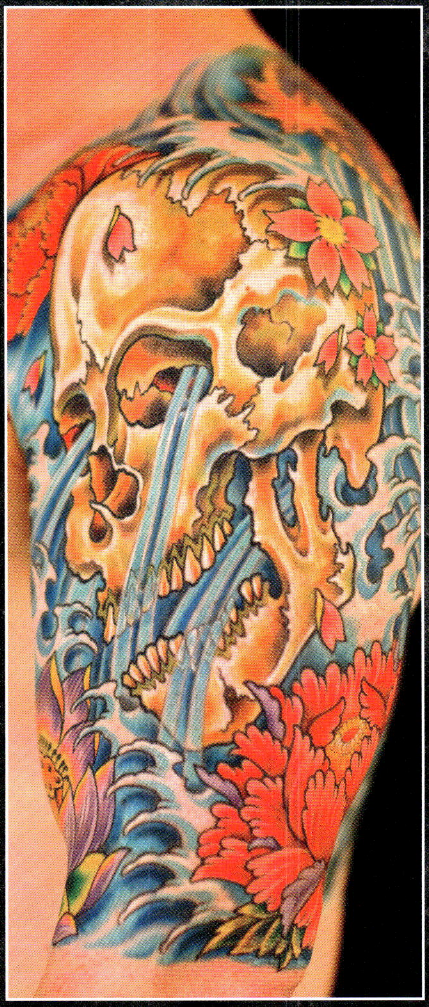

Durb Morrison

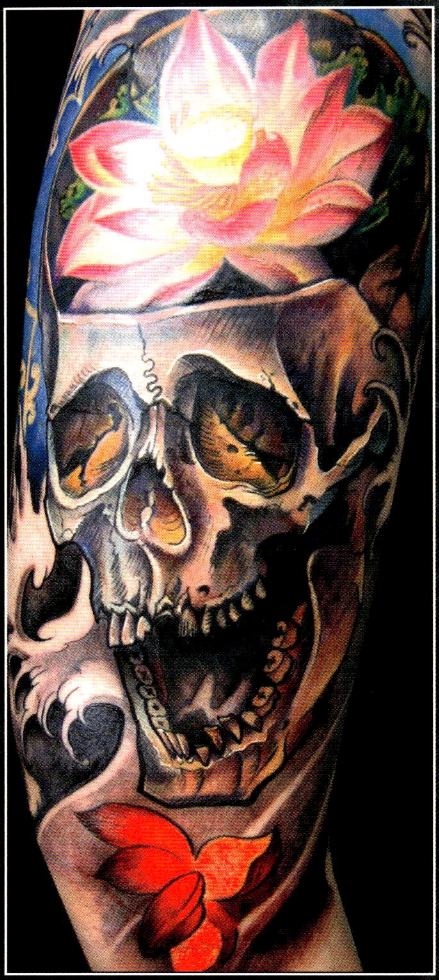
Paolo Acuna

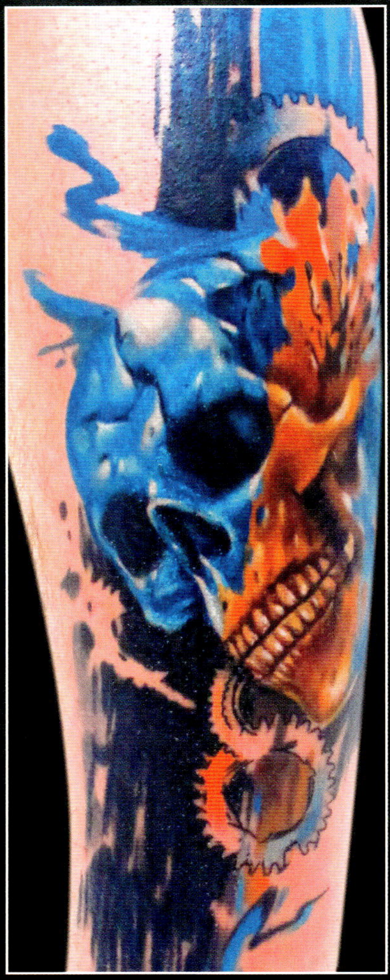
Orge Kalodimas

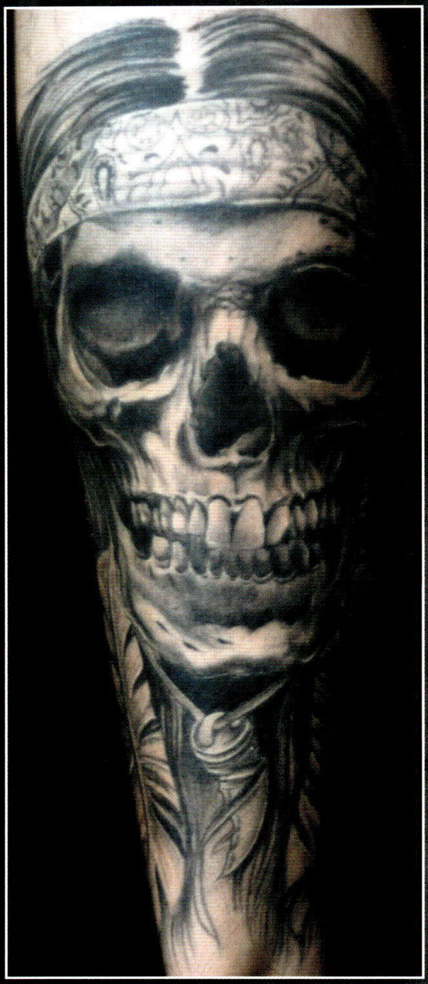
John Montgomery

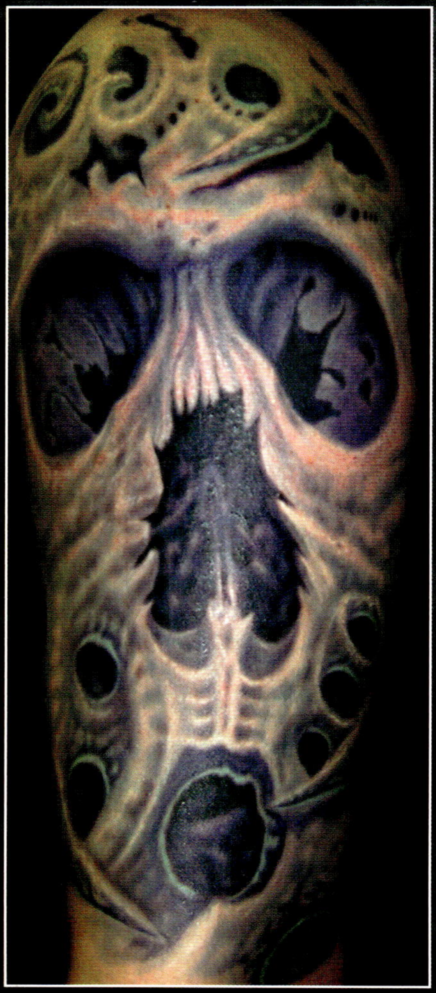
Josh Ruff

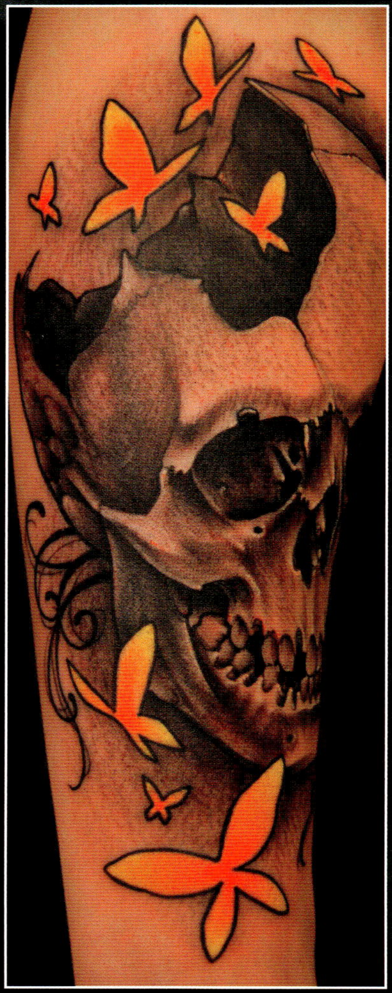
Trond Kjørkleiv

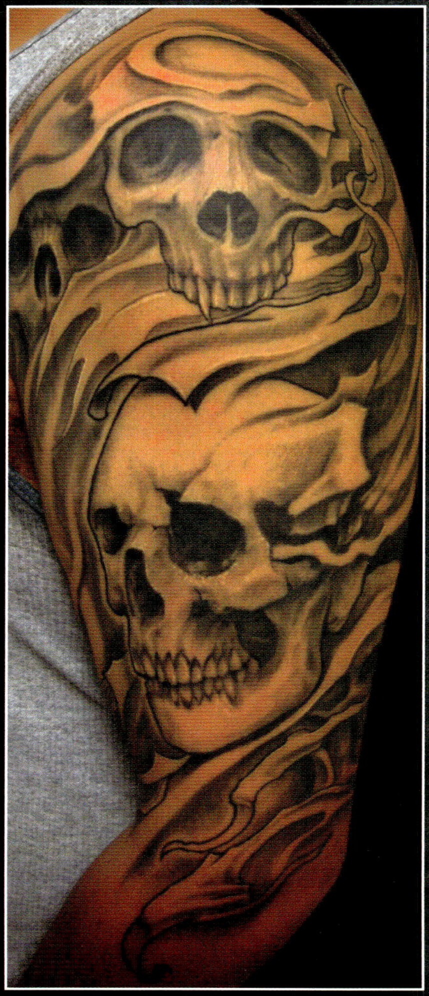
Jeremiah Barba

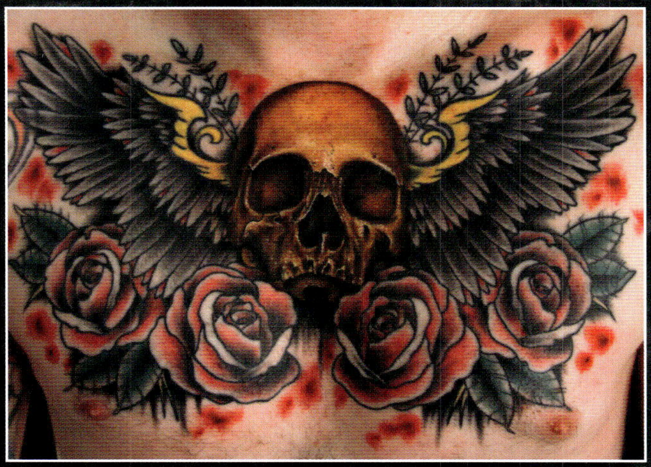
Mick Squires

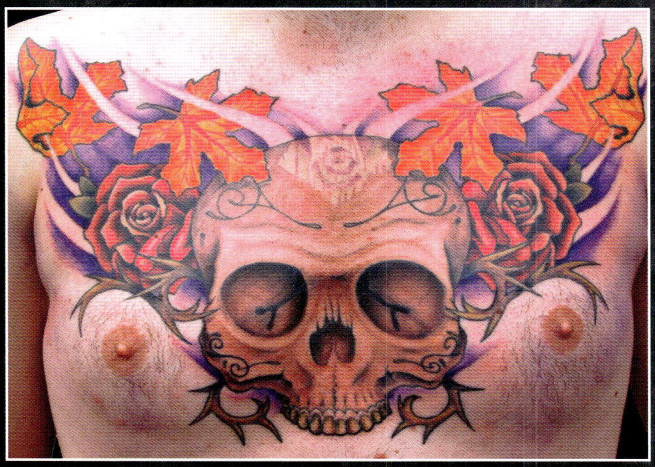
Zaq Weaver

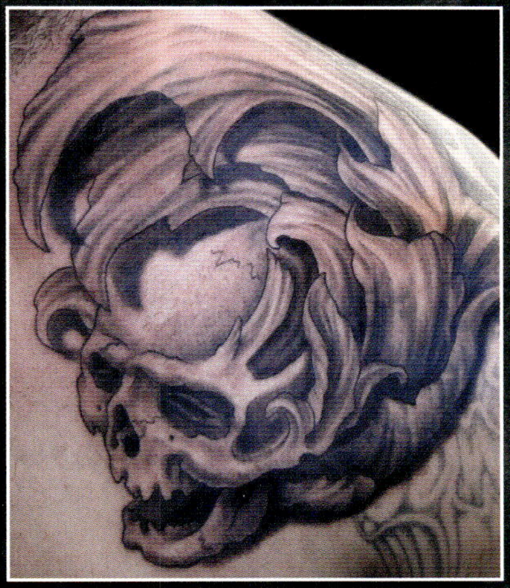
Bill Carlson

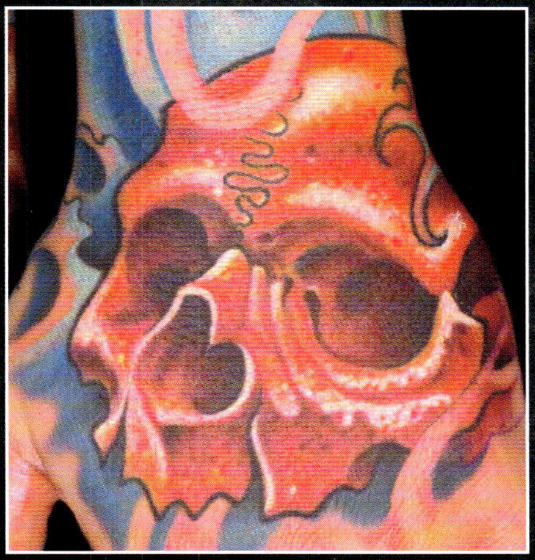
Jeff Ensminger

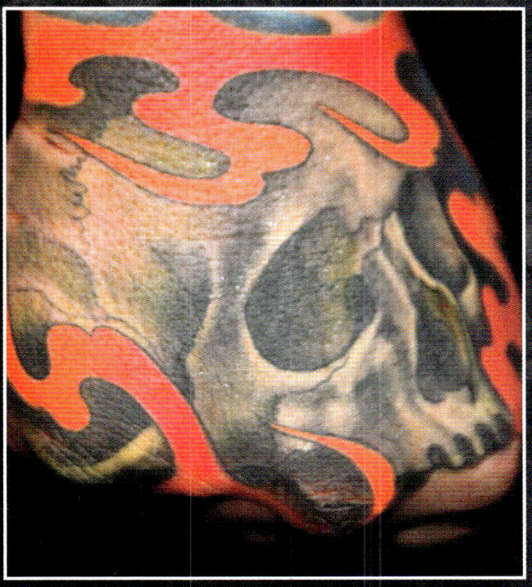
Mike Godfrey

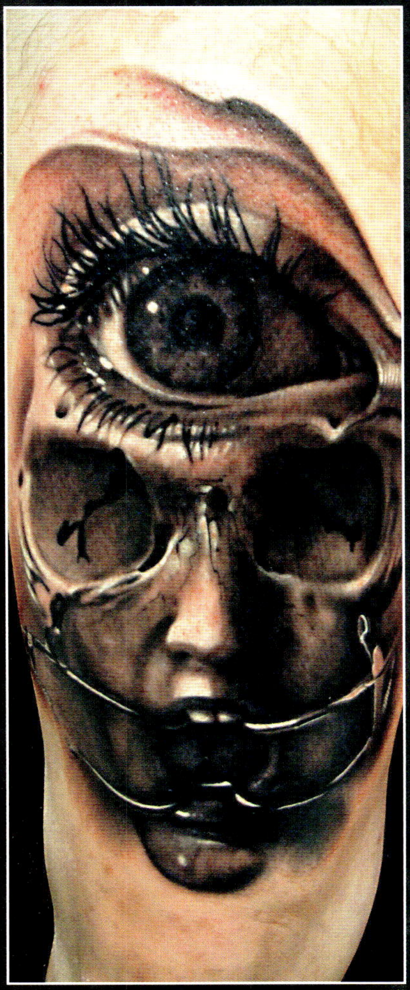
Nick Chaboya

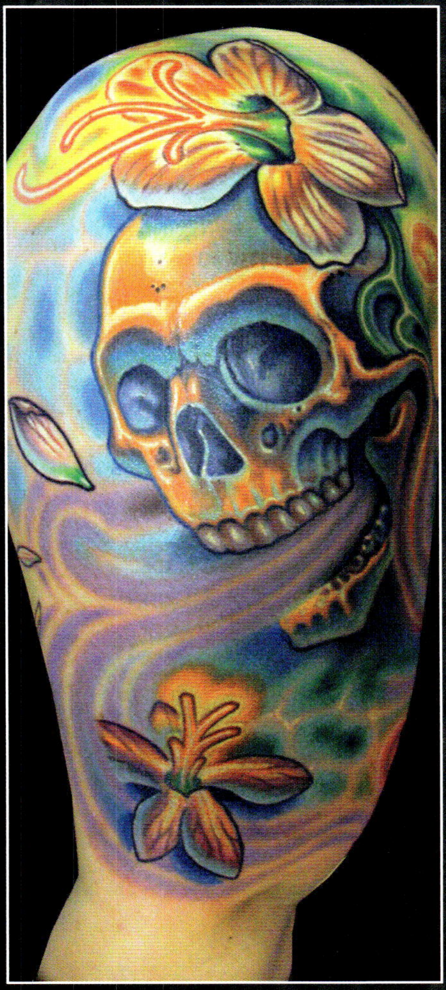
Christian Perez

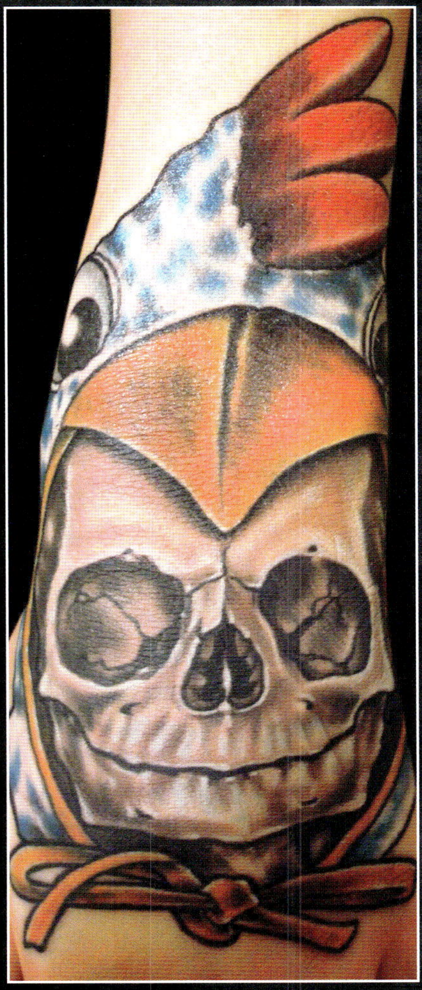
James White

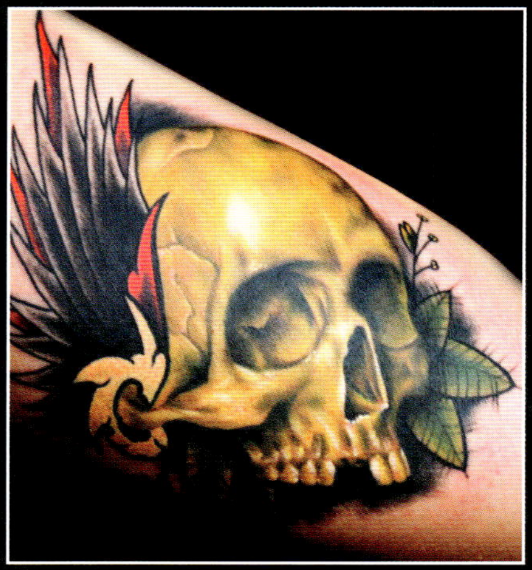
Mick Squires

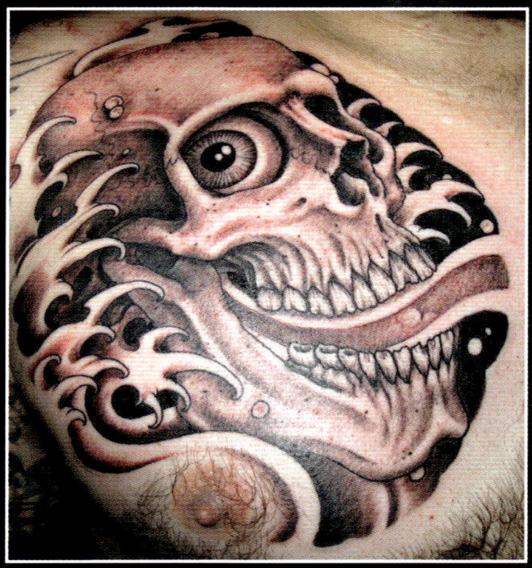
Ben Corn

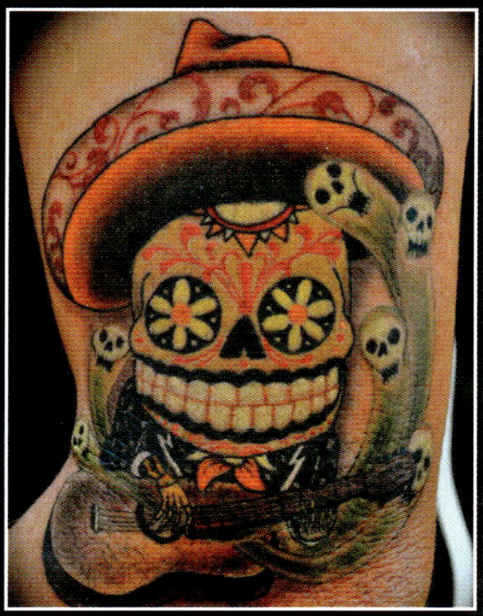
Derek Harrison

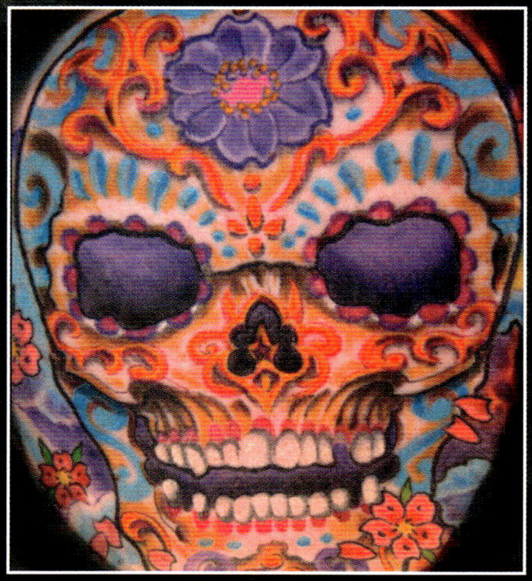
Danny Black

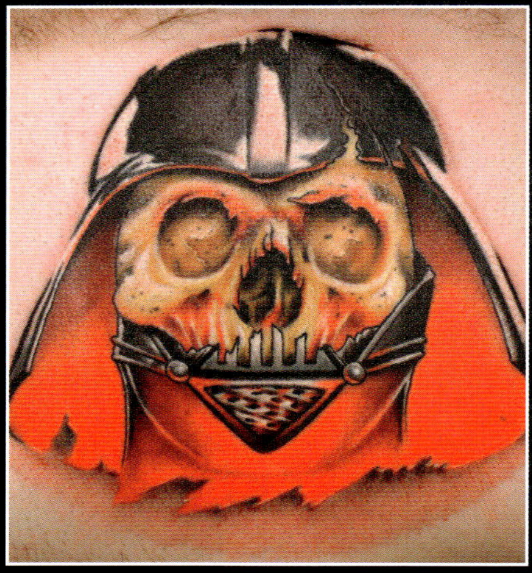
Sterling Nilsen

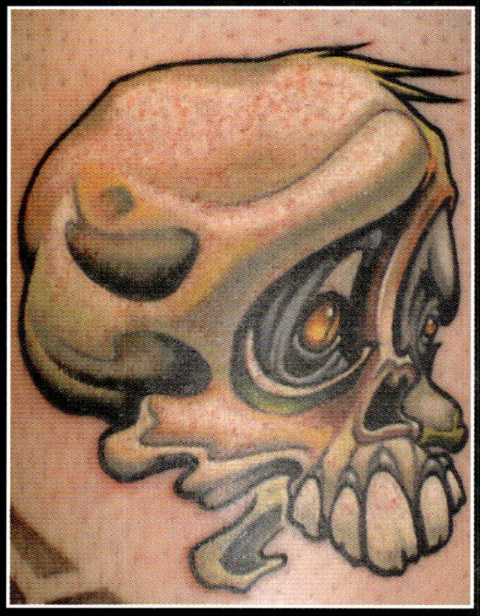
Jesse Smith

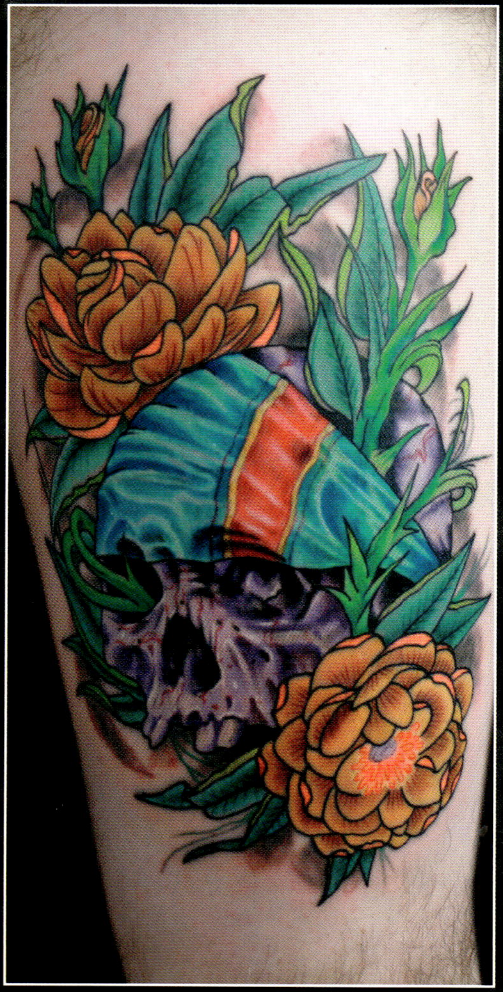
Dj & Jen Minor

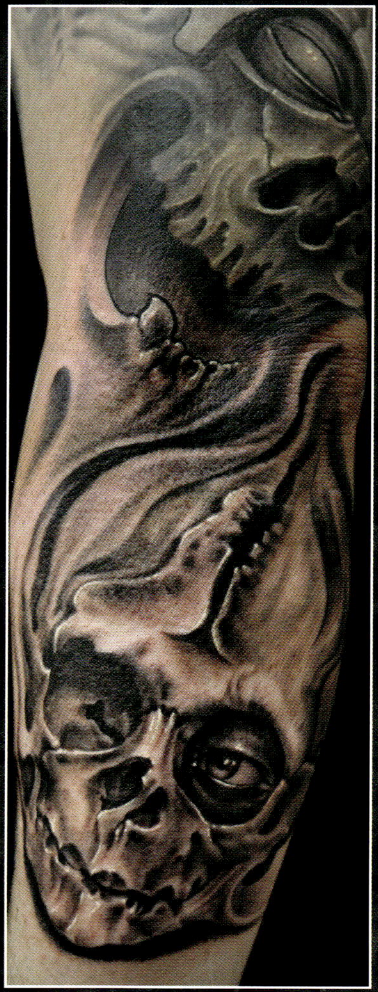
Josh Duffy

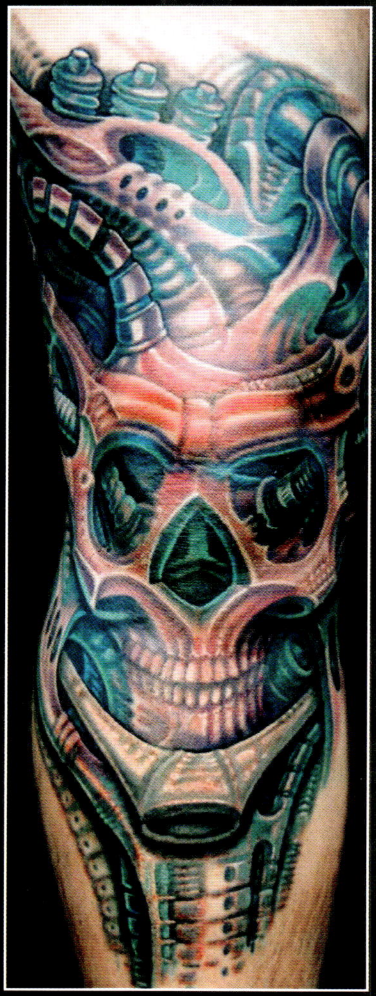
Don Mcdonald

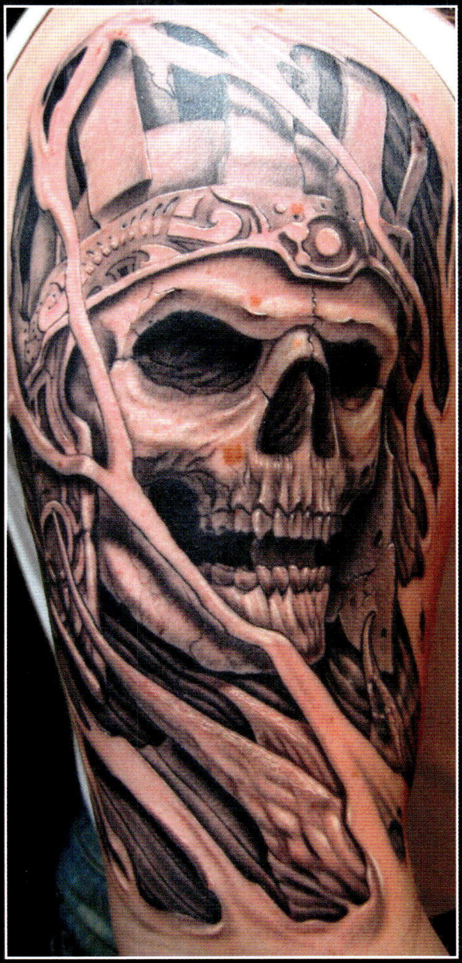

Sven Gnida

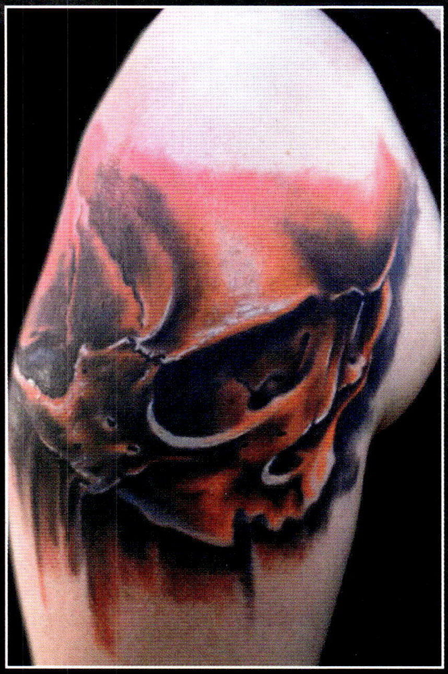

Ty McEwen

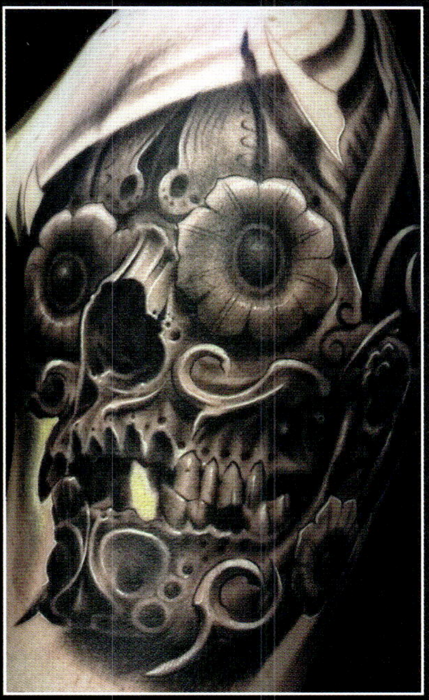

Victor Portugal

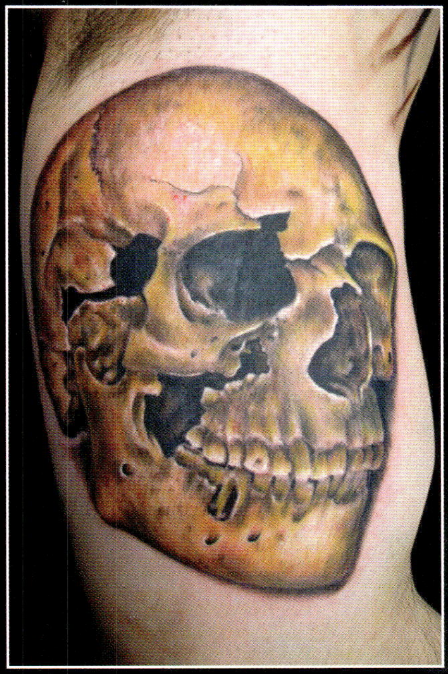

Nick Fredrich

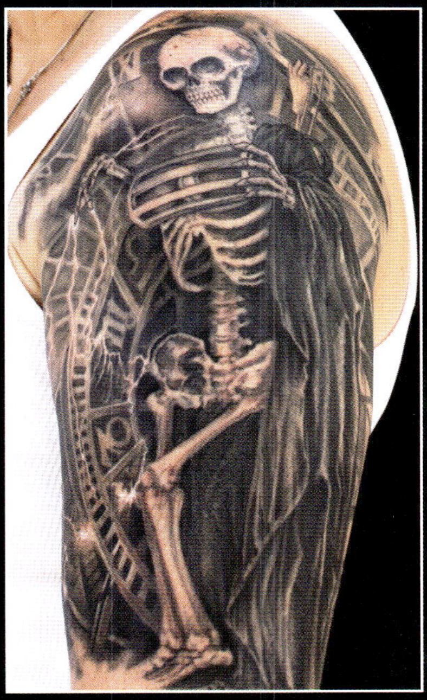

Kari Barba

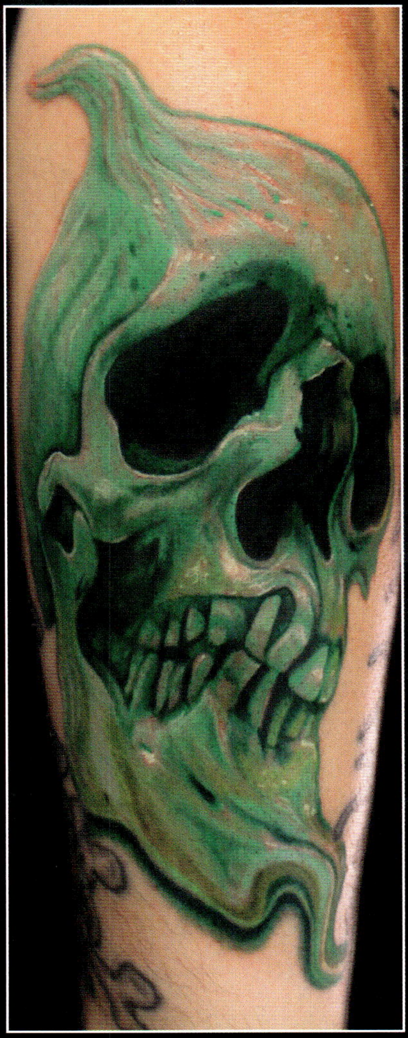

Nate Heim

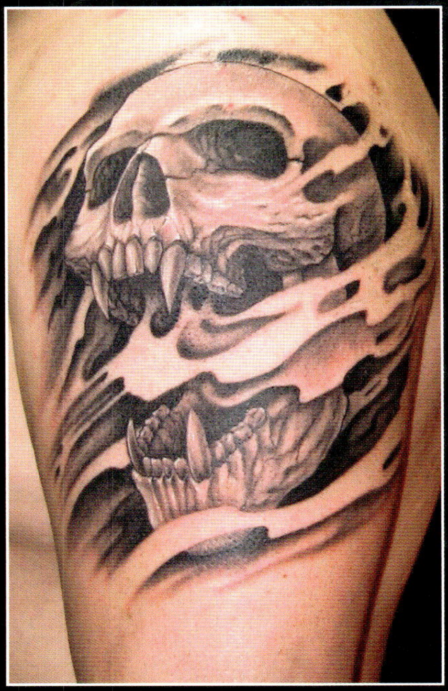

Sven Gnida

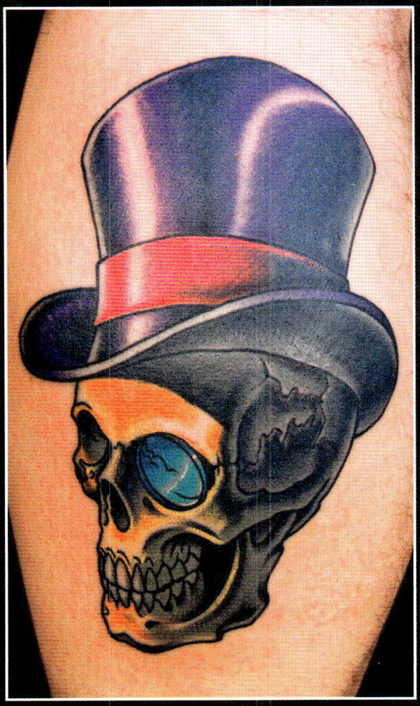

Jim Sylvia

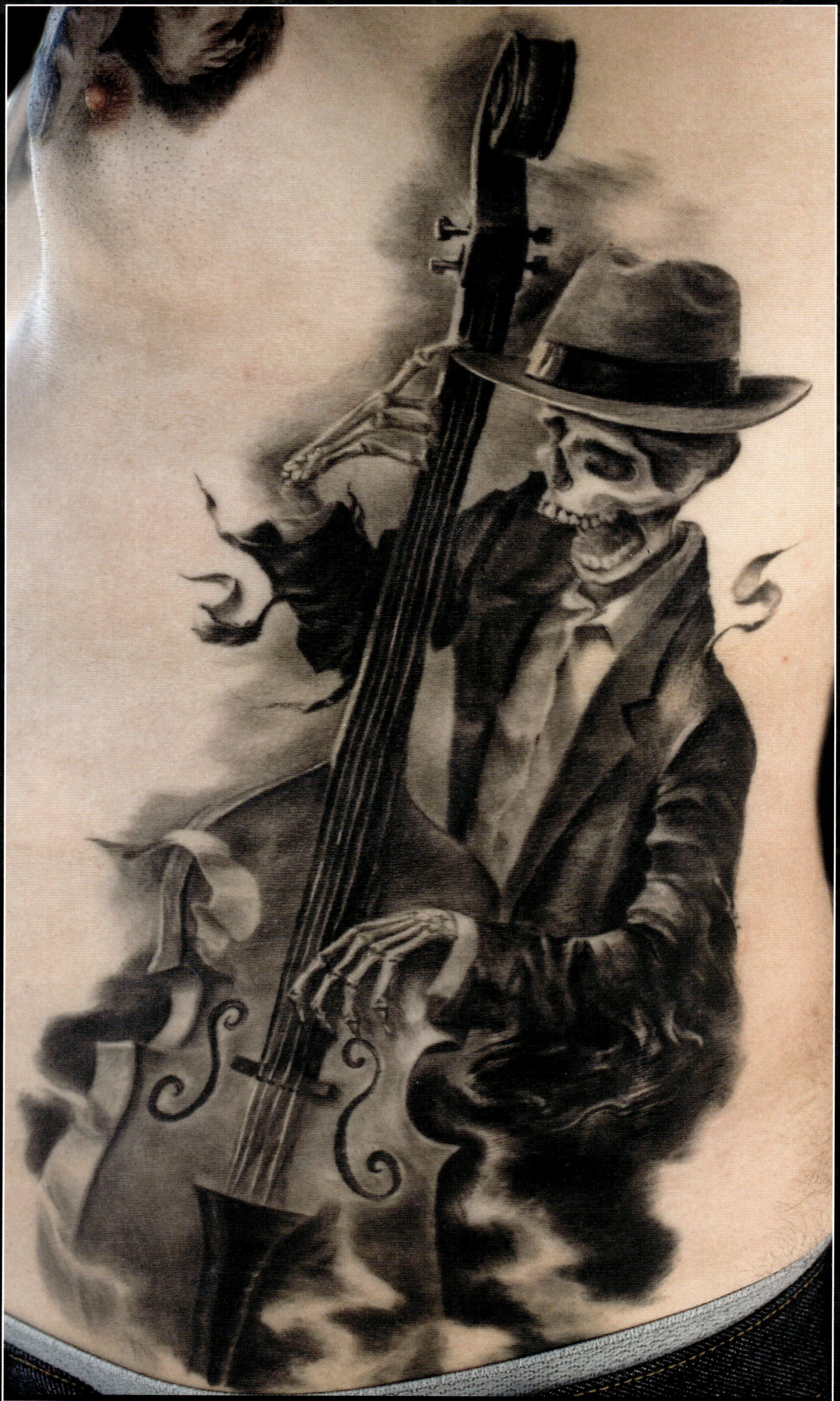

Carlos Torres

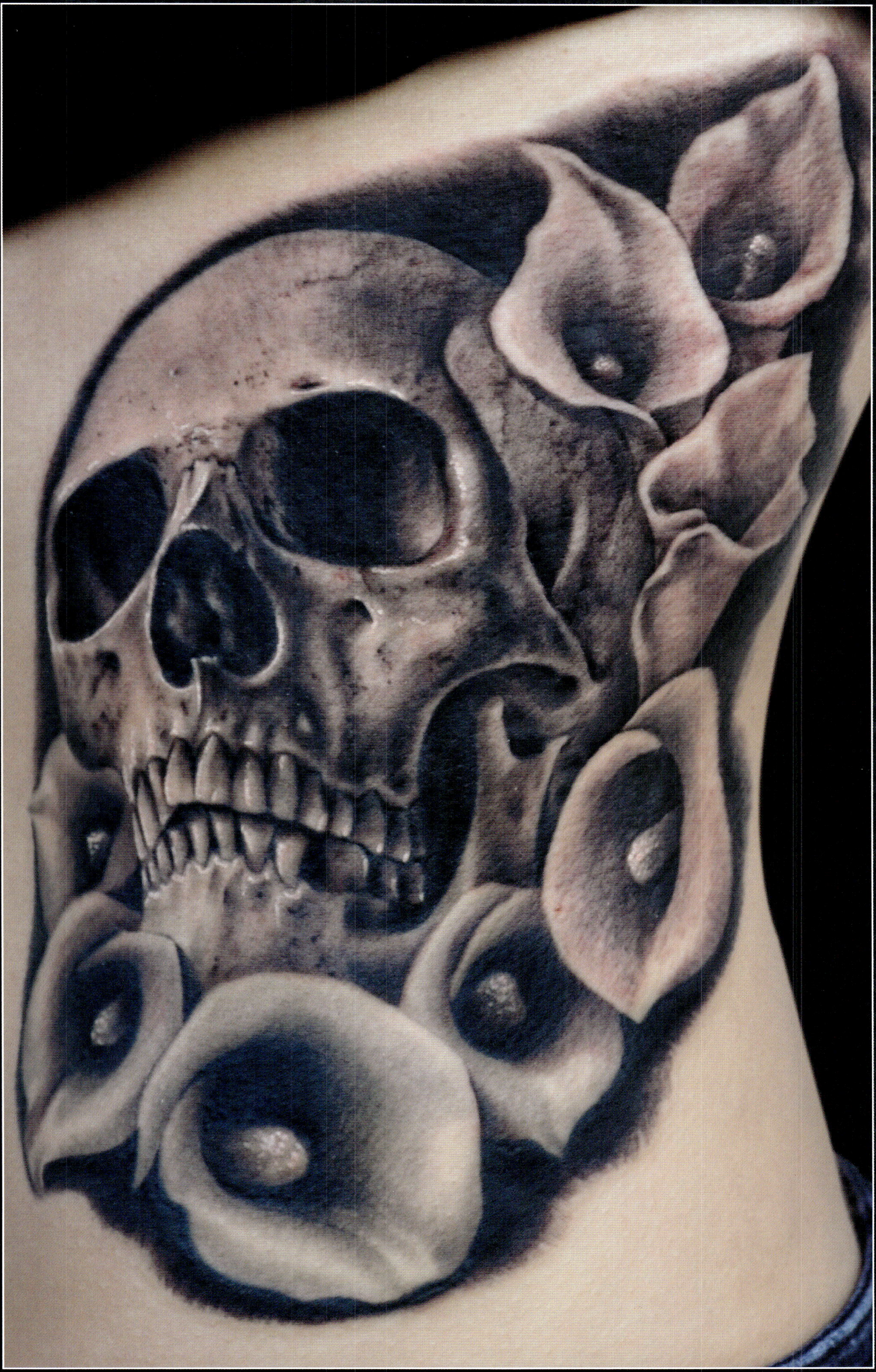

Josh Duffy

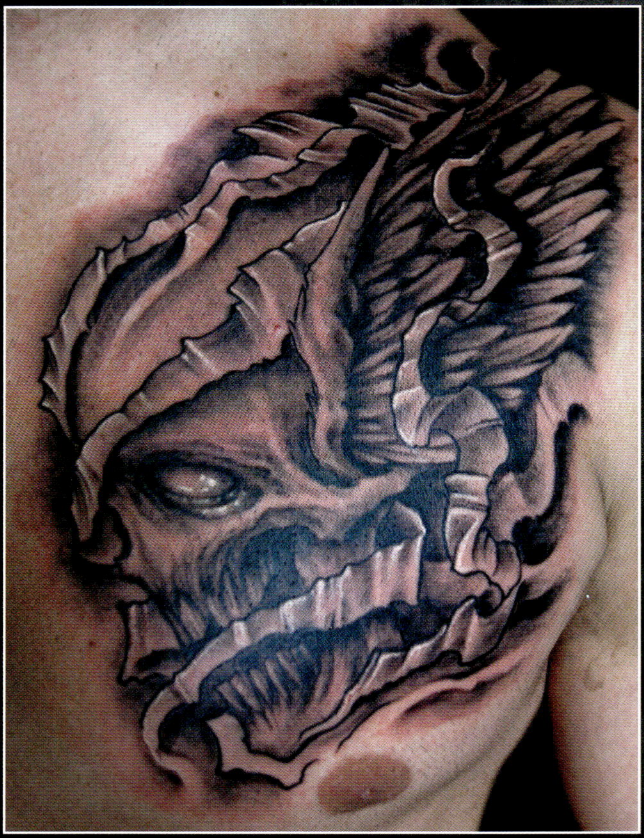

Jeremiah Barba

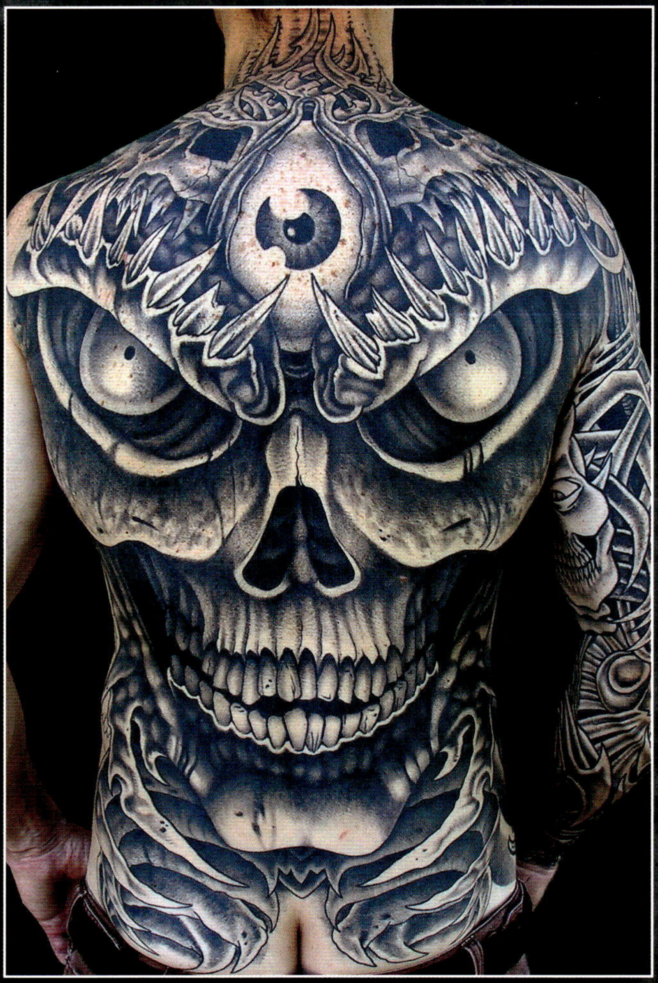

Edwin Shaffer

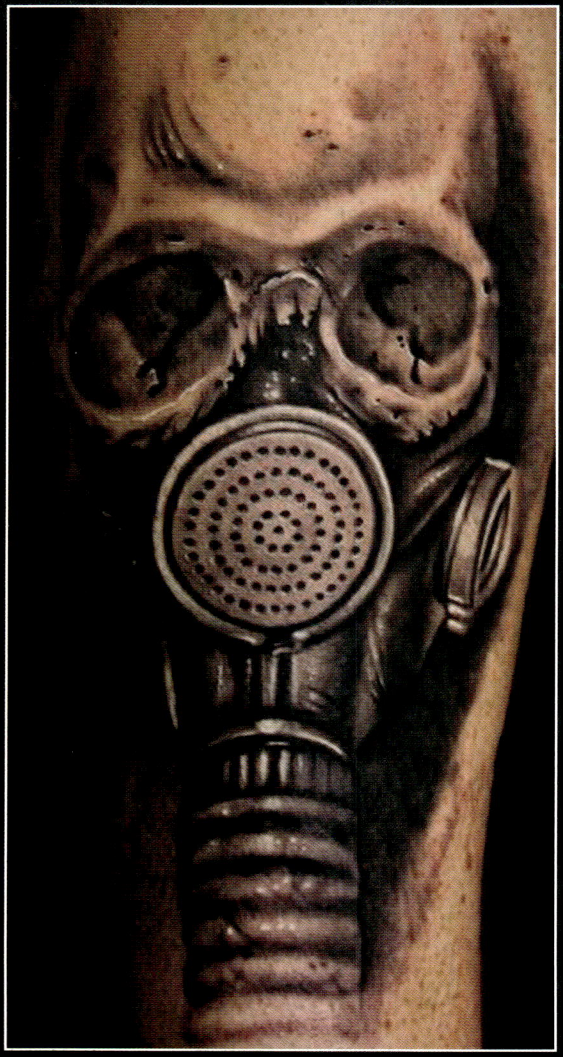

Cory Norris

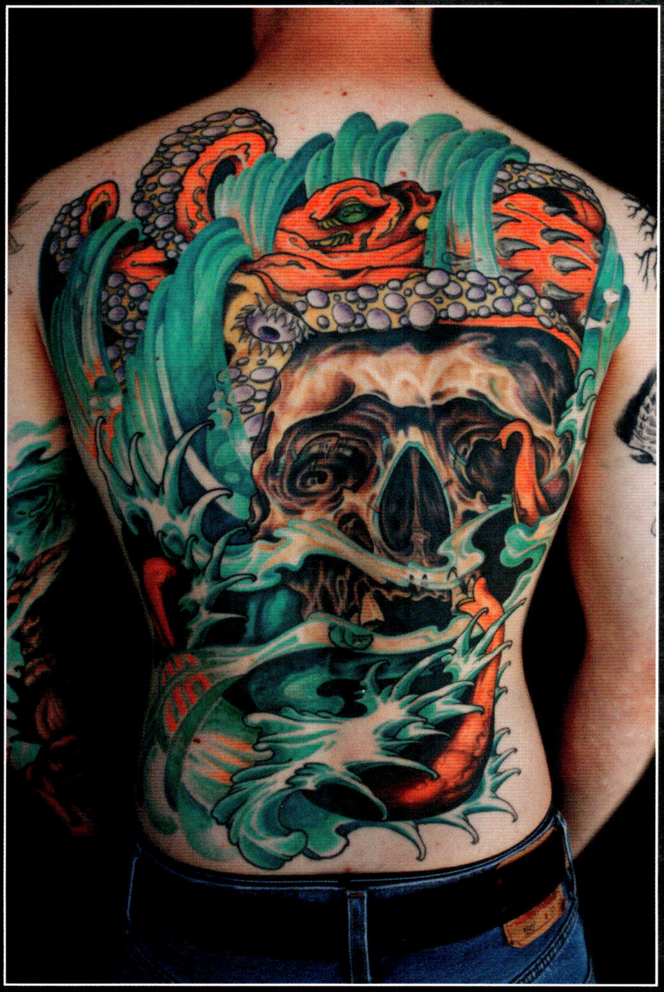

Dj Minor

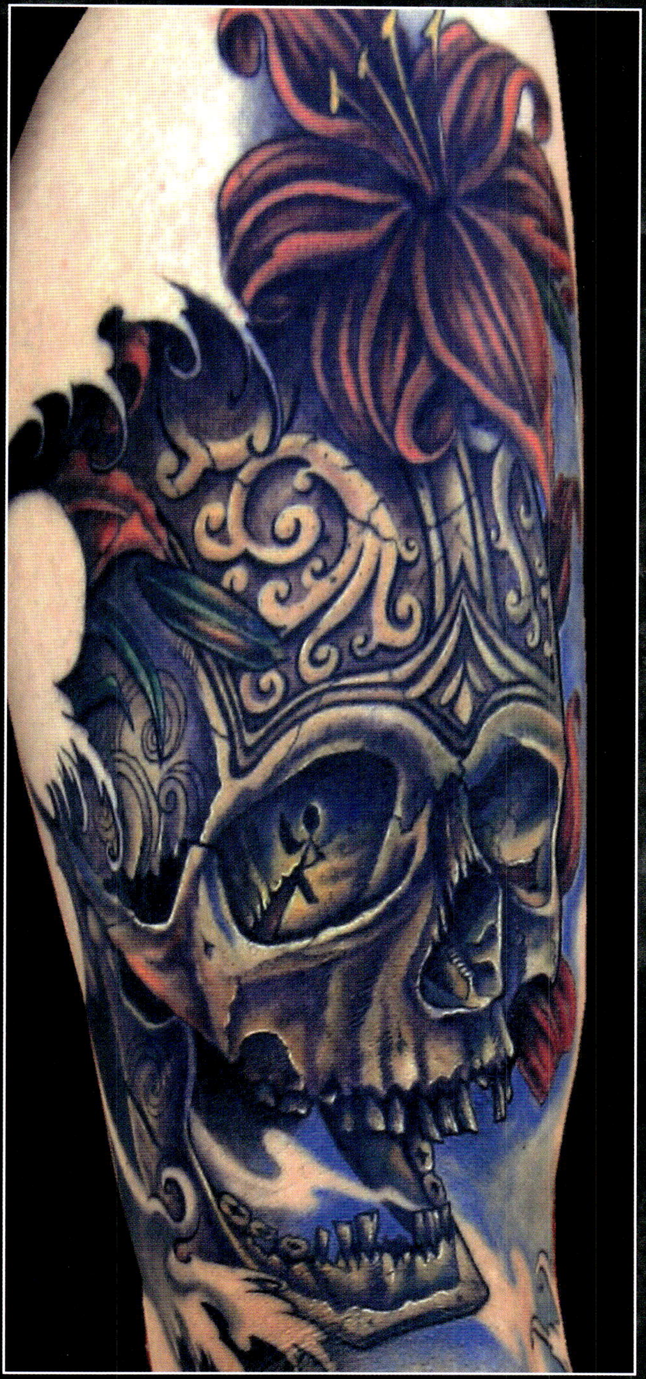

Paolo Acuna

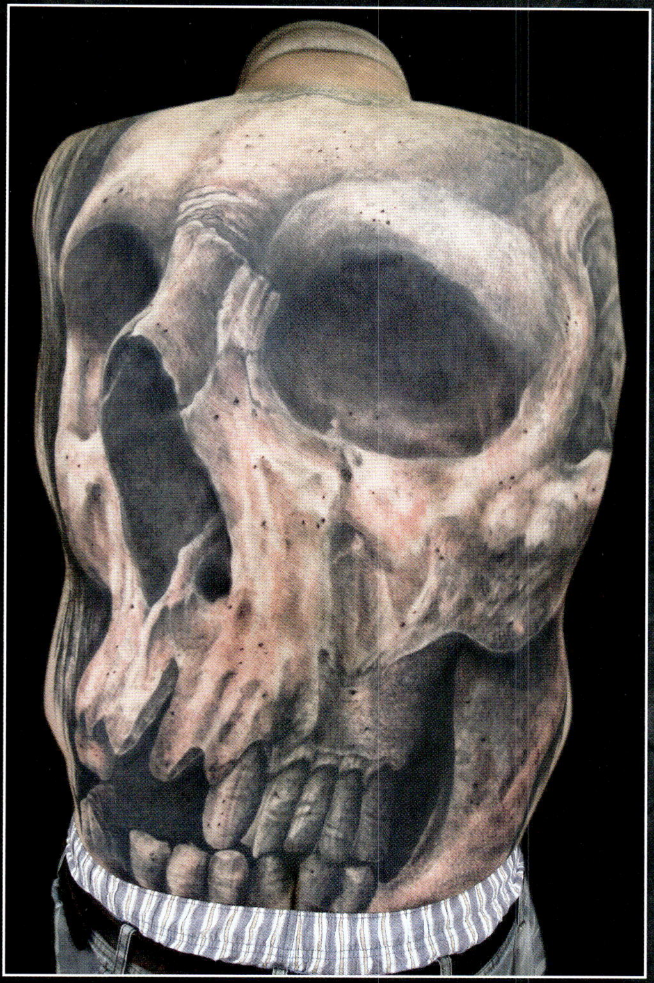

Bill Carlson

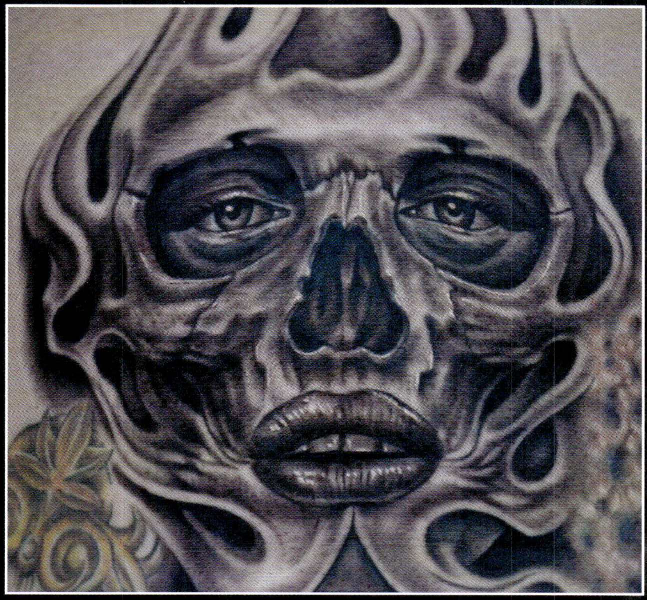

Josh Duffy

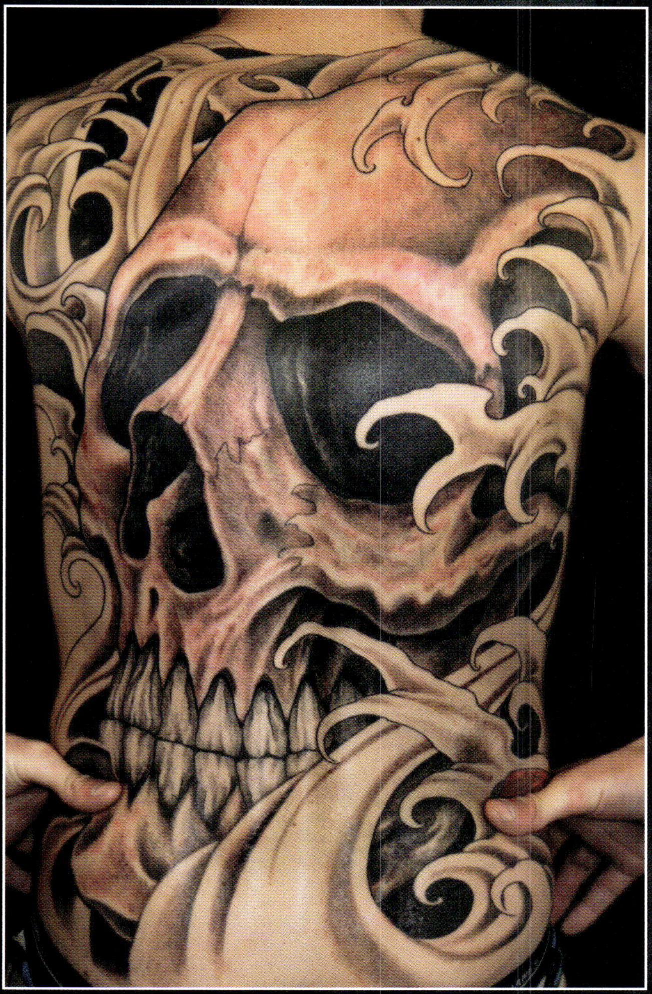

Bill Carlson

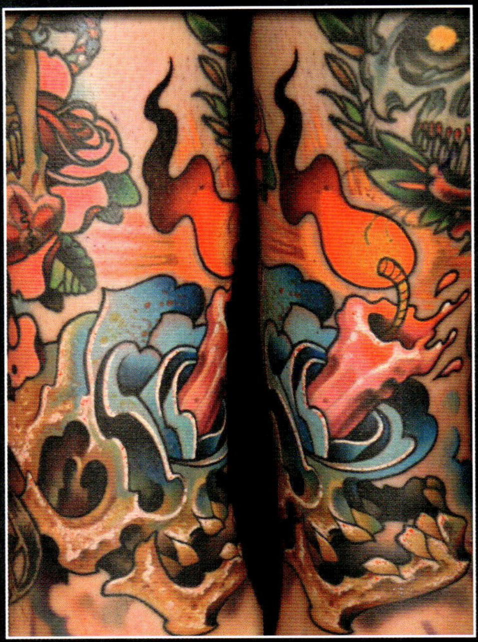
Jeff Ensminger

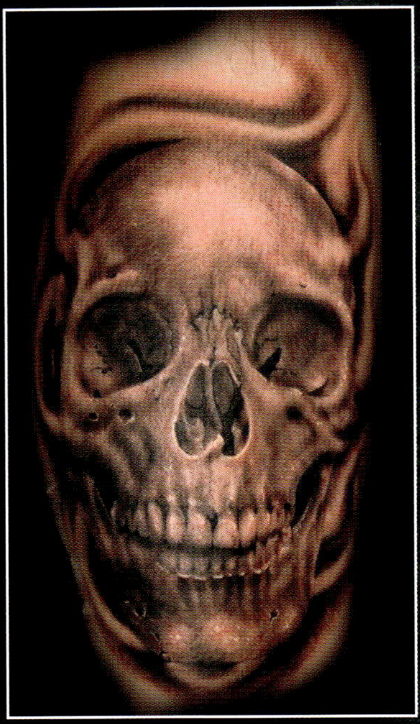
James Haun

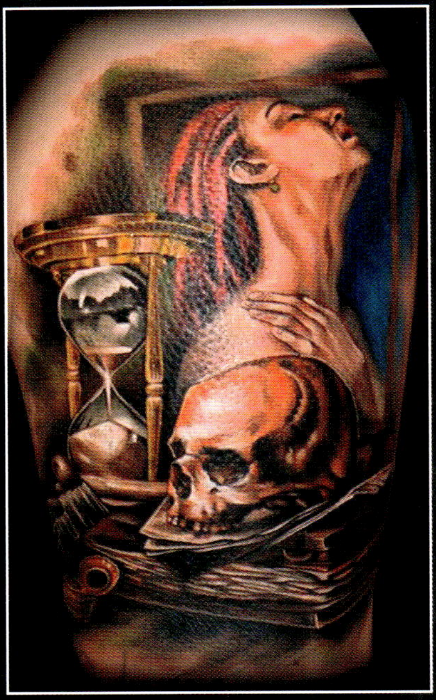
Chris Blinston

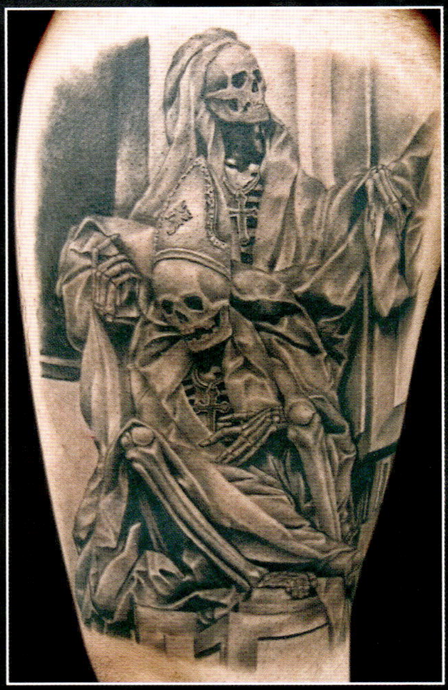
Carlos Rojas

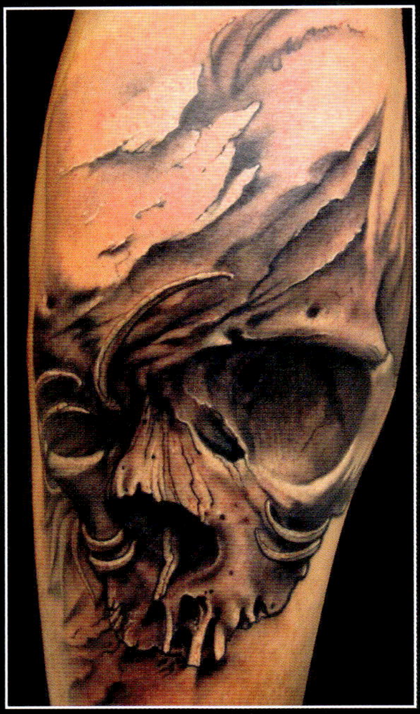
Nick Chaboya

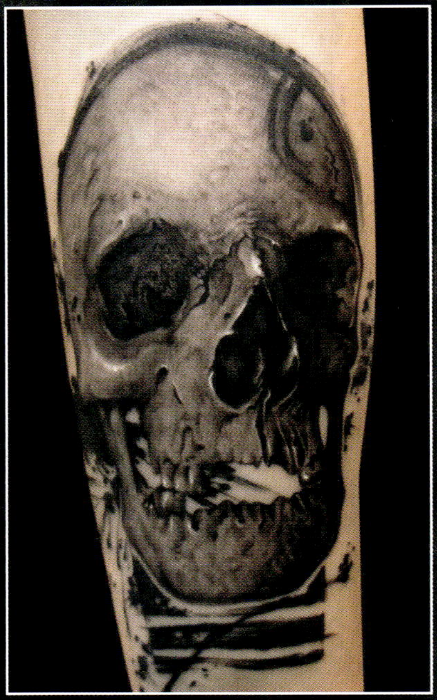
Elvin Yong

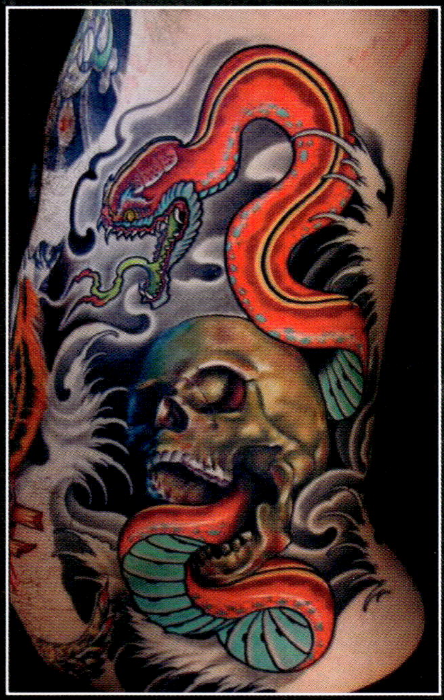
Cory Norris

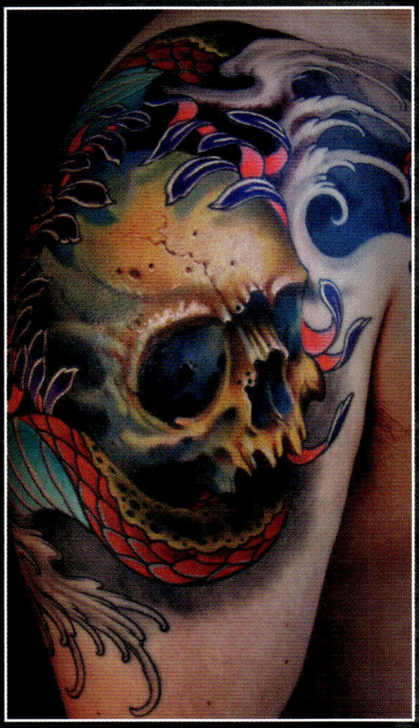
Cory Norris

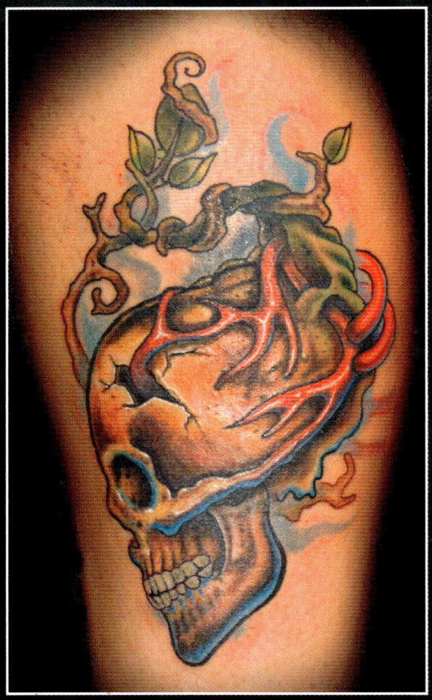
Dave Barton

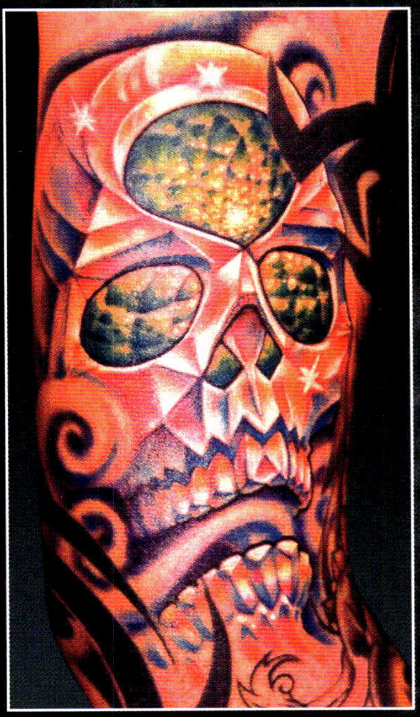

Guy Aitchison

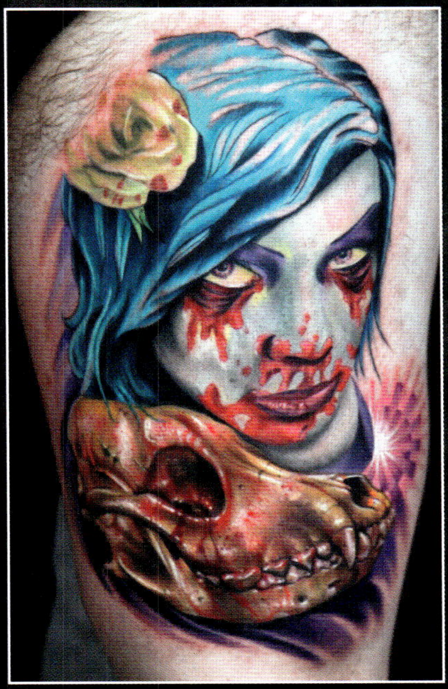

Dj Minor & Nate Beavers

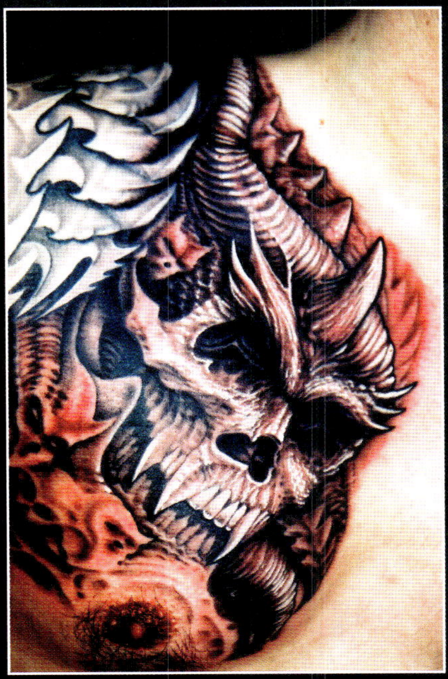

Don Mcdonald

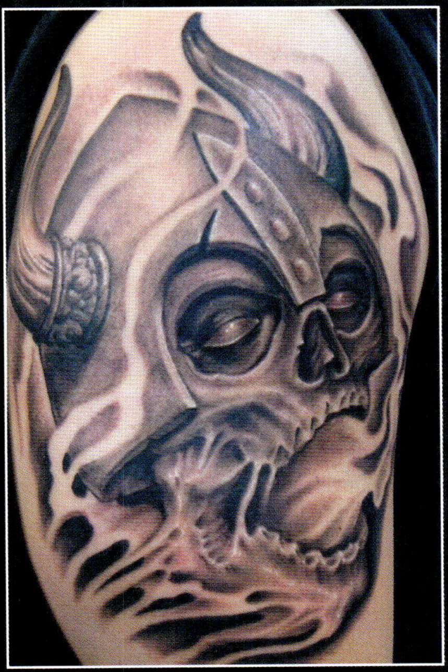

Josh Duffy

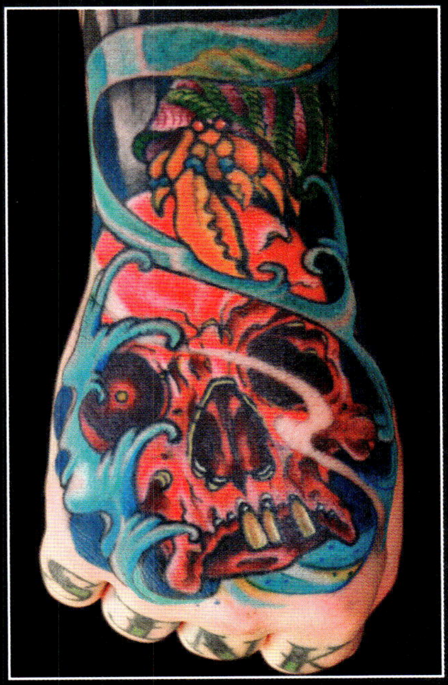

Dj Minor

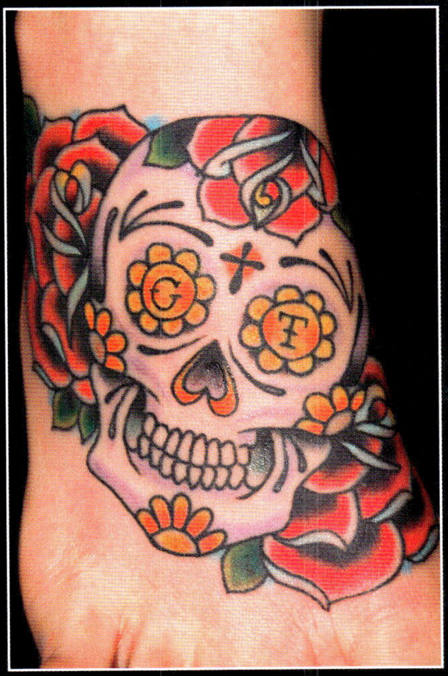

Jim Sylvia

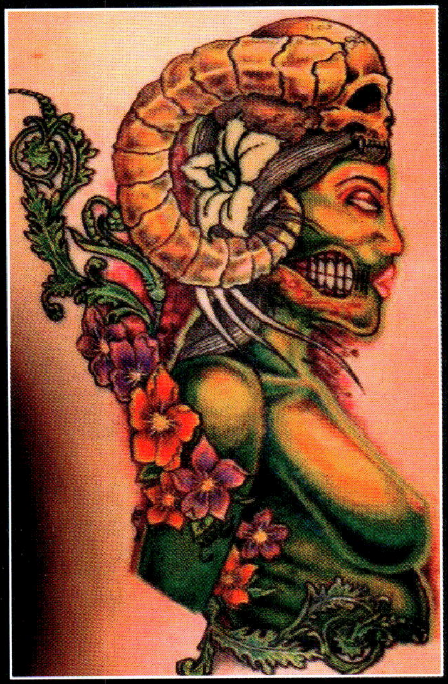

Brett Herman

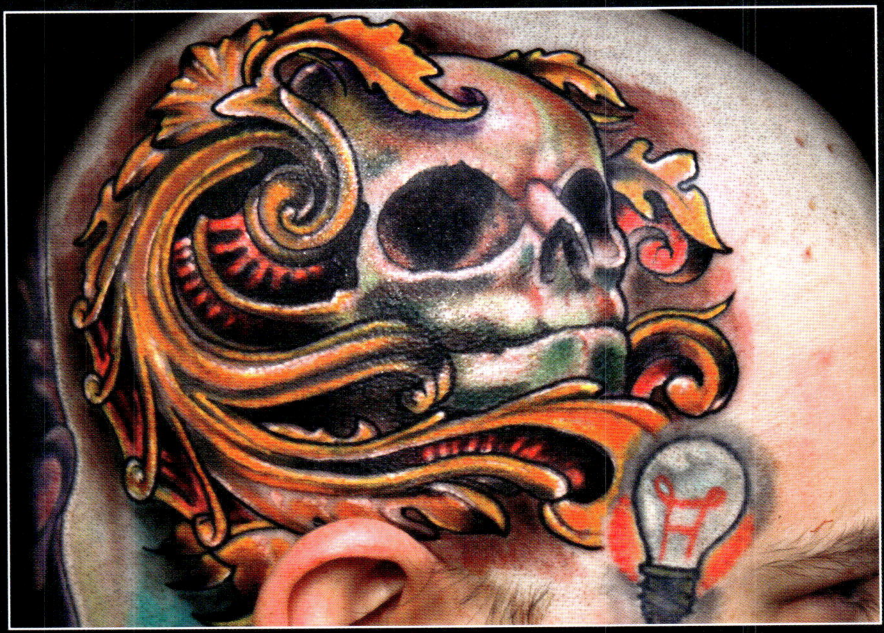

Sean Herman

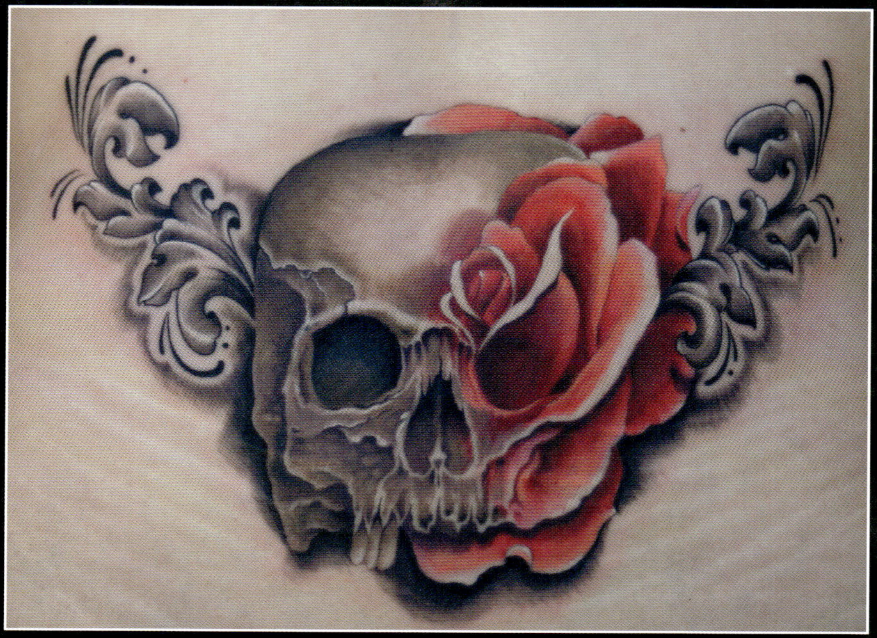
Tim McEvoy

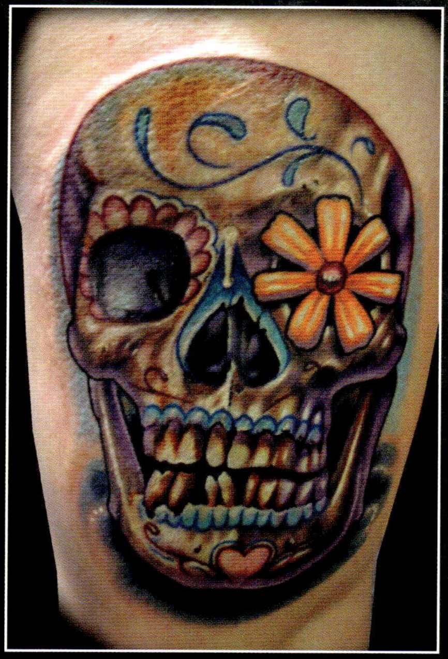
Christian Perez

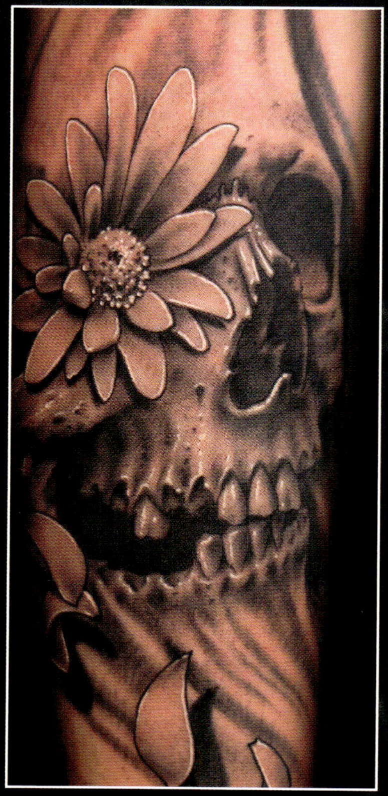
Josh Duffy

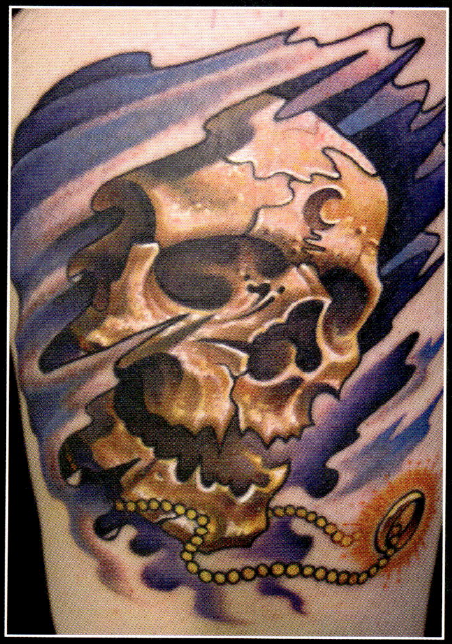
Jeff Ensminger

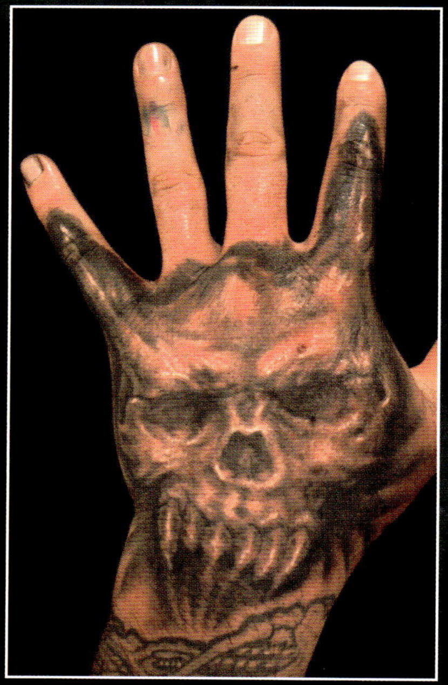
Benjamin Moss

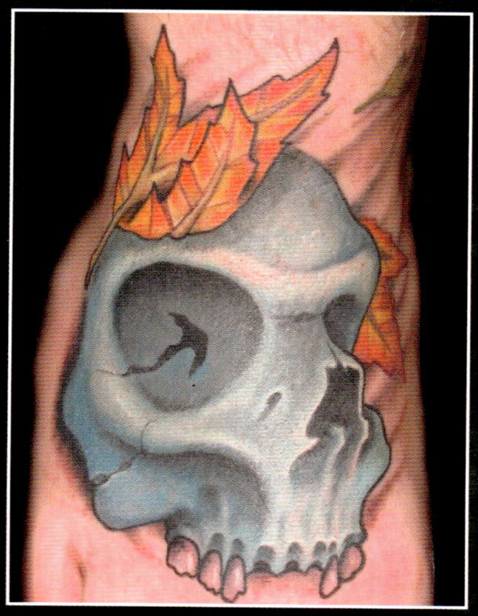
Zaq Weaver

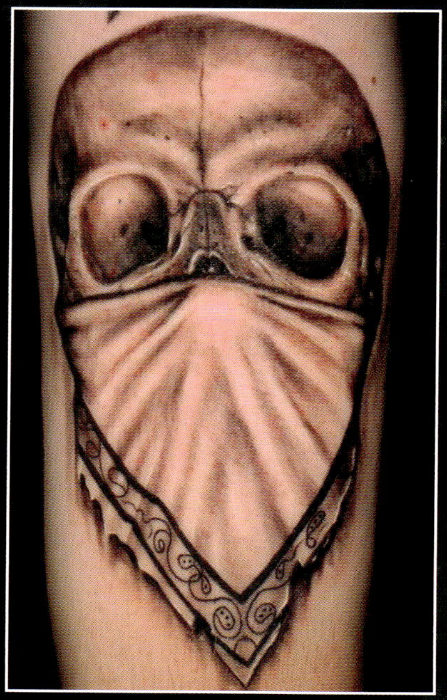
Josh Ruff

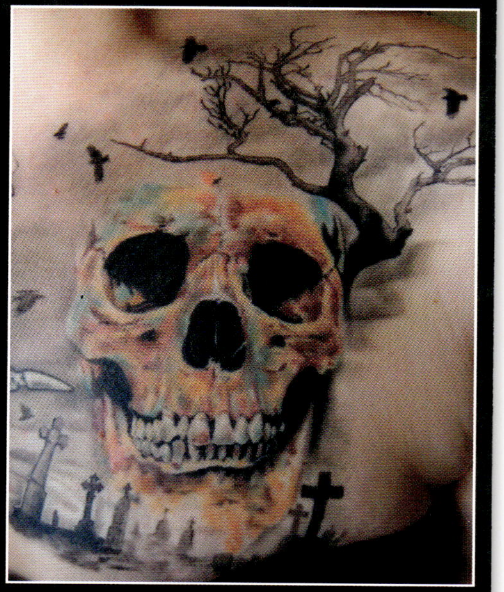
Cory Cudney

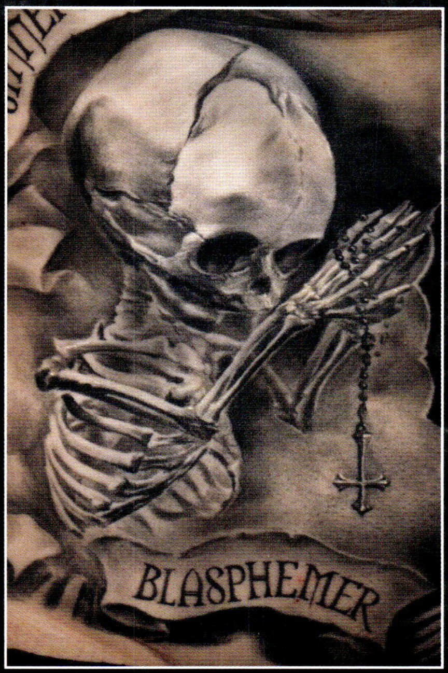

Carlos Torres

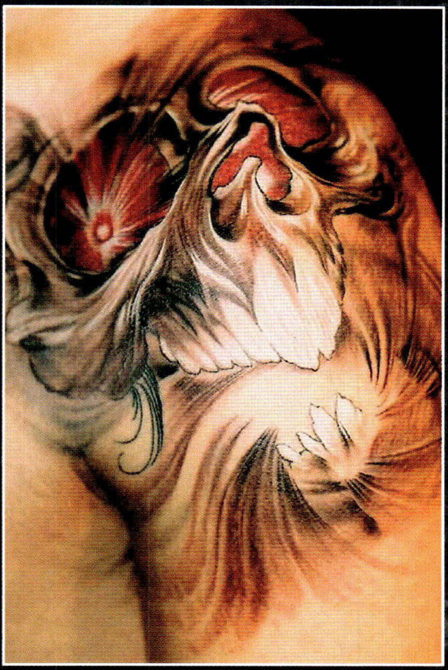
Brian Brenner

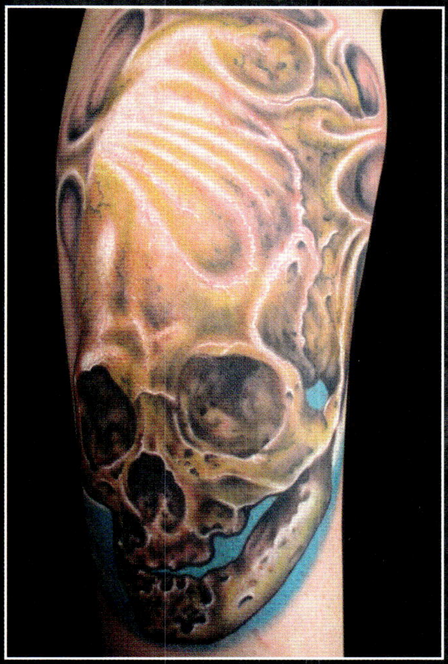
Nick Fredrich

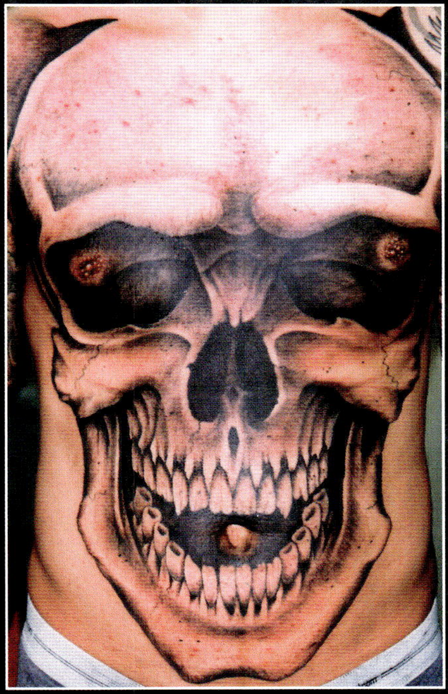
Ben Corn

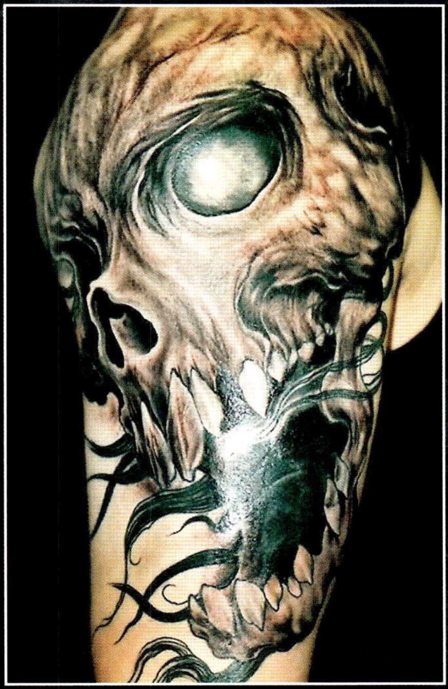
Brian Brenner

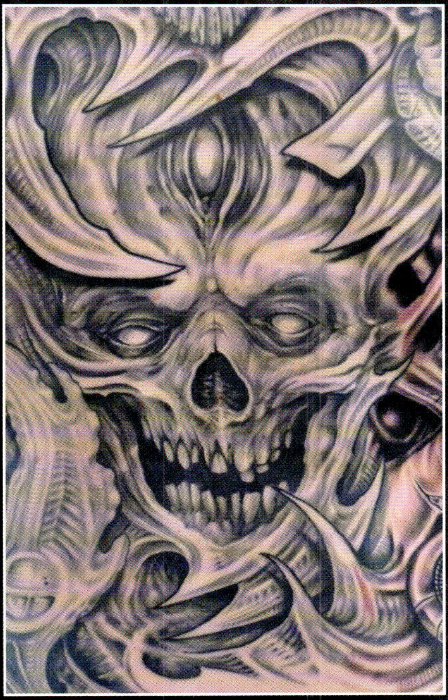
Jeremiah Barba

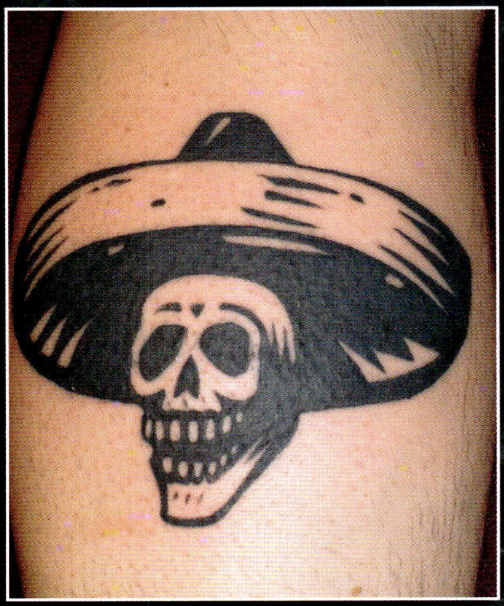
Gerardo Garcia

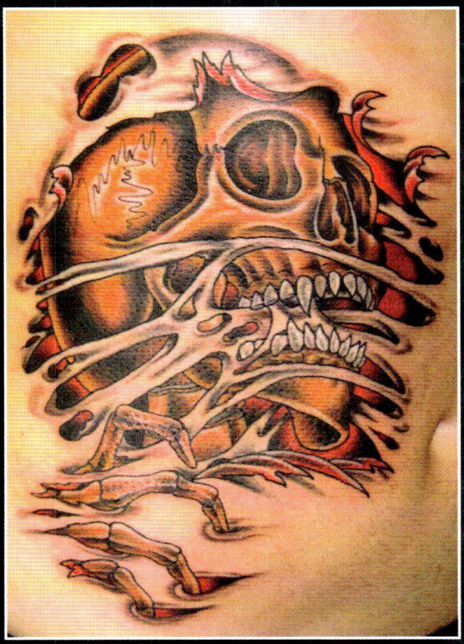
Nick Phillips

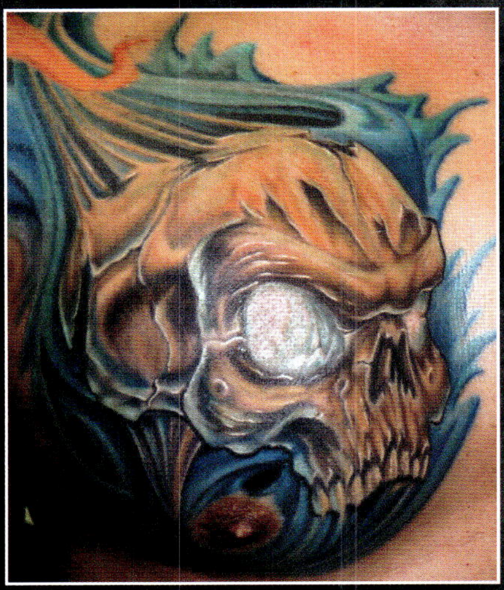
Frank La Natra

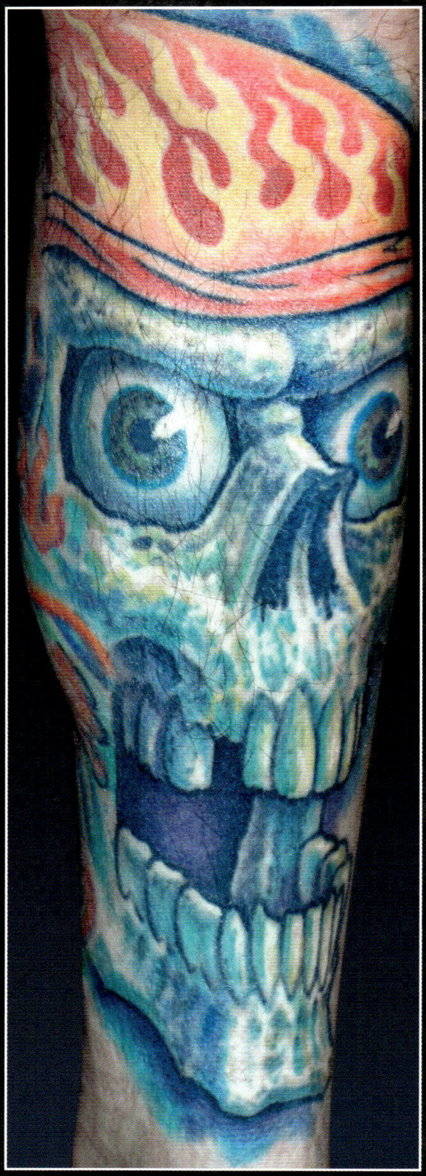

Guy Aitchison

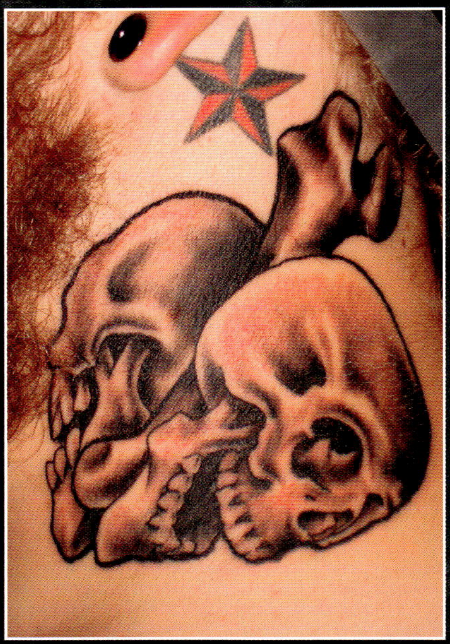

Brady Willmott

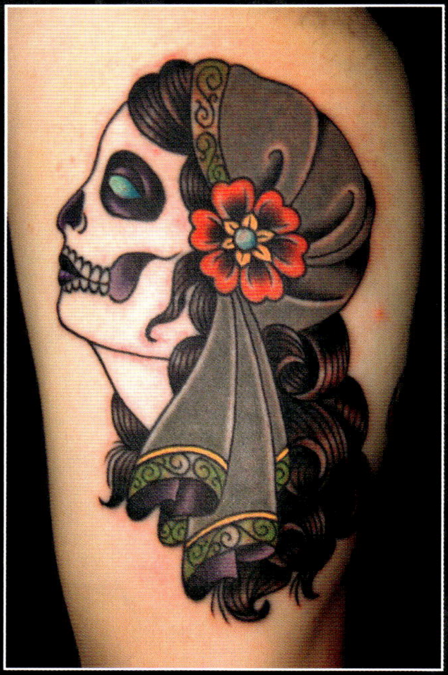

Jim Sylvia

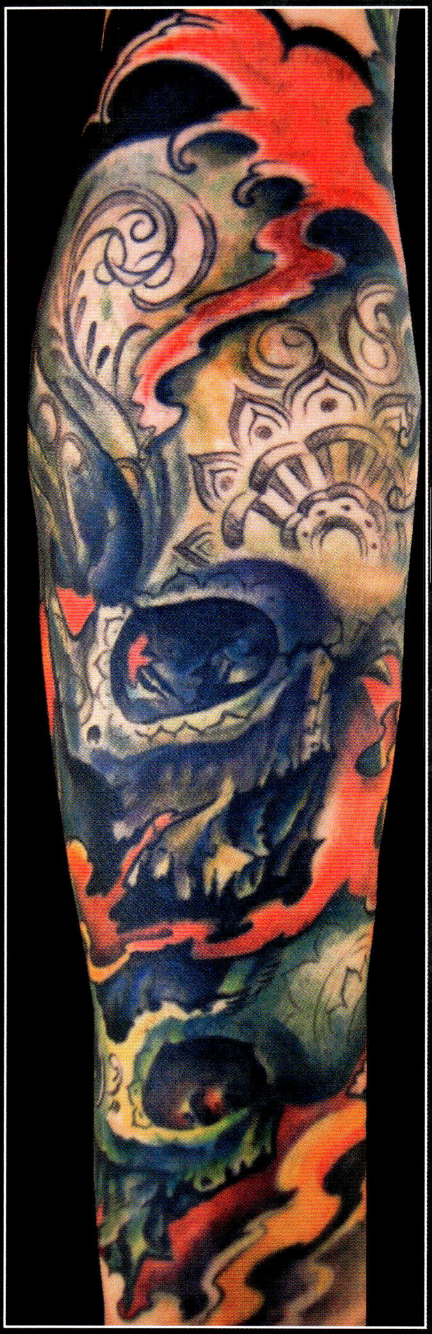

Paolo Acuna

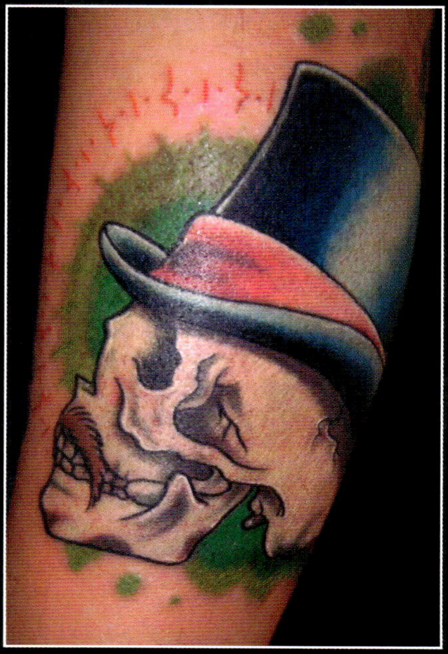

Cody Fortson

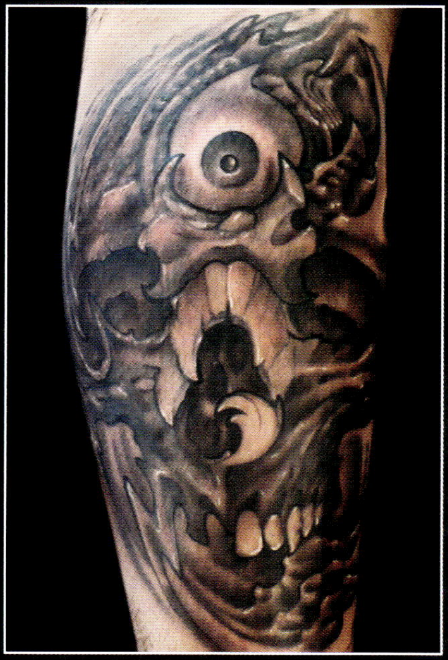

Alex Ruiz

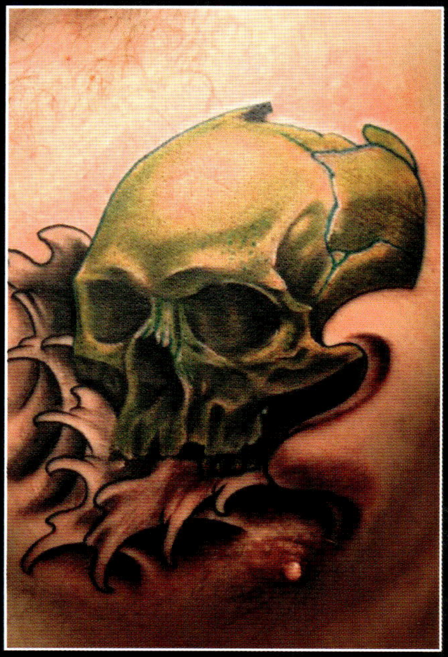

Jason Morrow

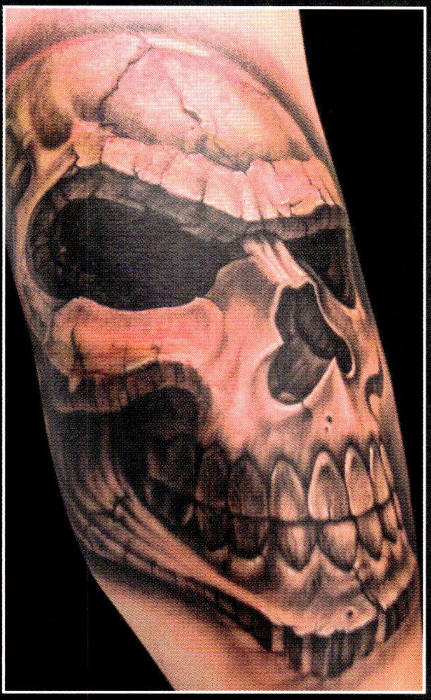

Stefano Alcantara

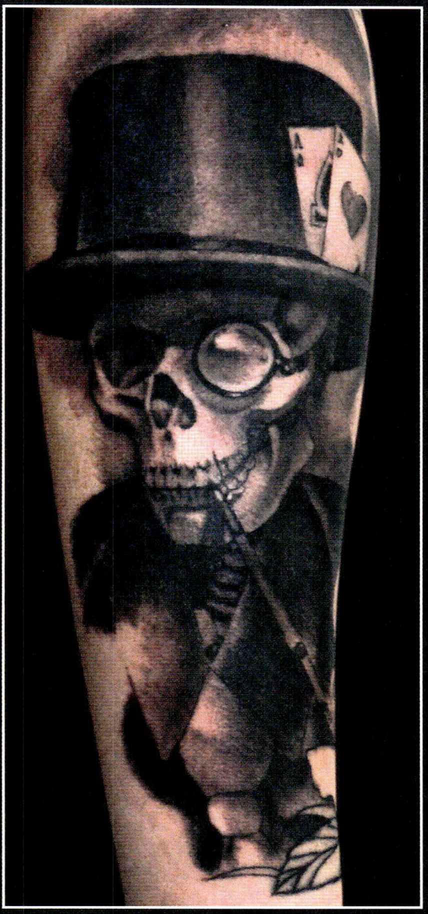

Kore Flatmo

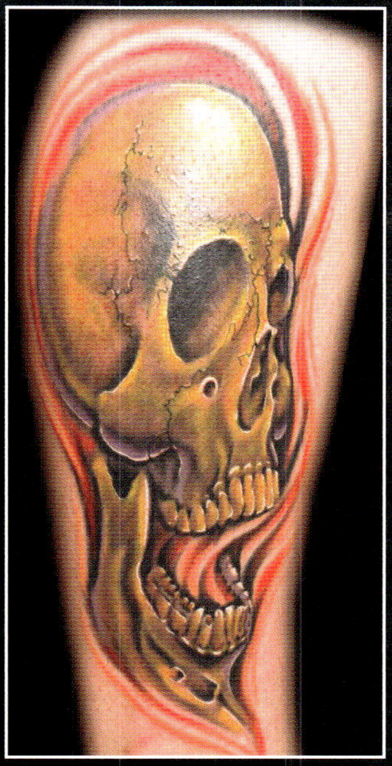

James Haun

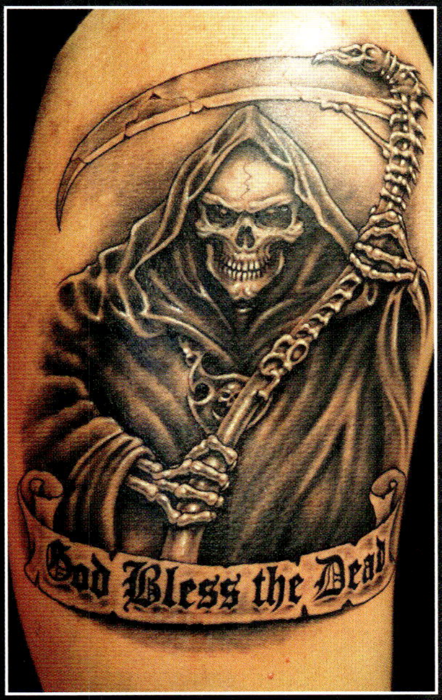

Jay Wheeler

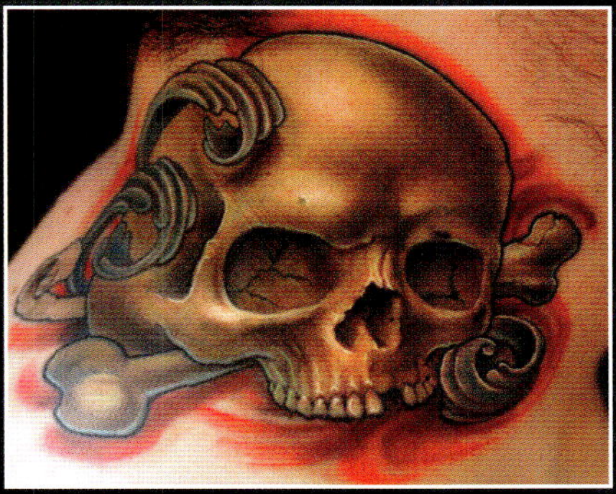

Nate Beavers

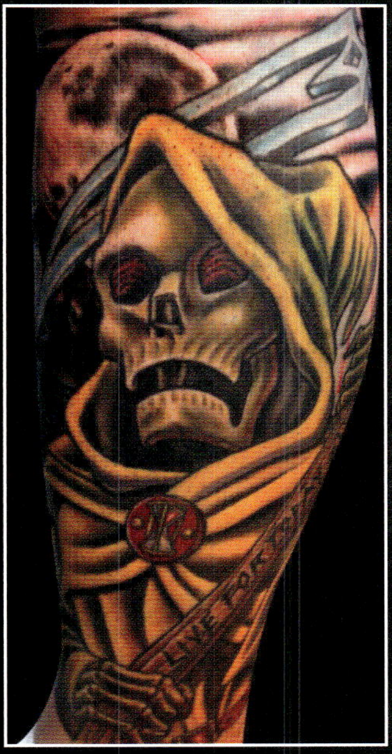

Larry Brogan

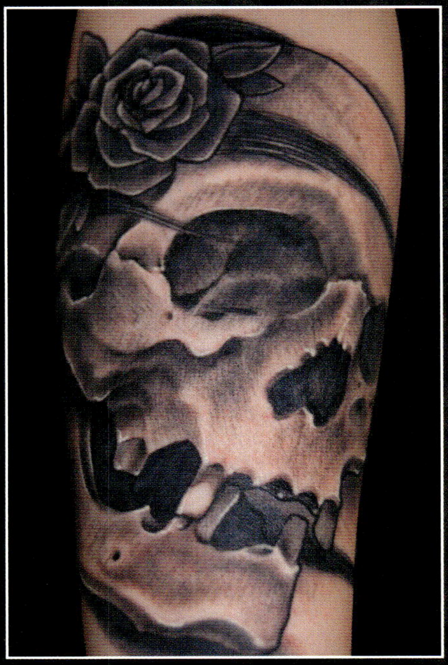

Trond Kjørkleiv

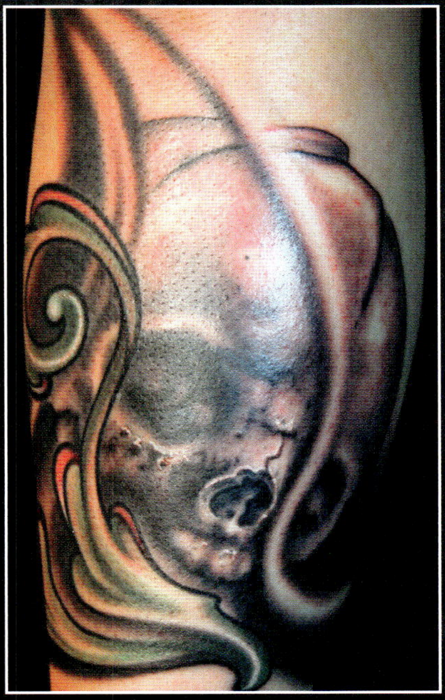

Nathan Kostechko

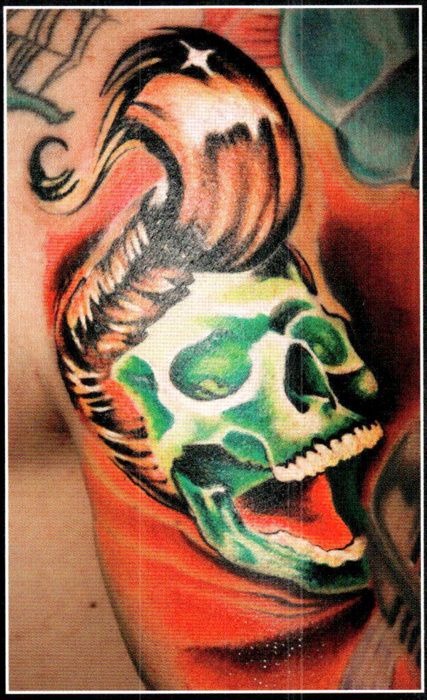

Fabian De Gaillande

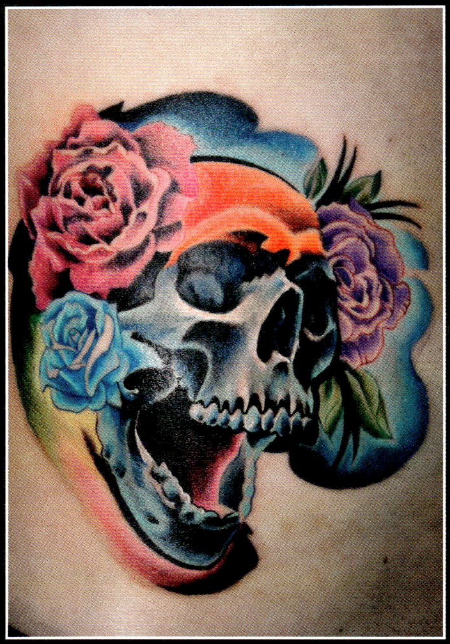

Fabian De Gaillande

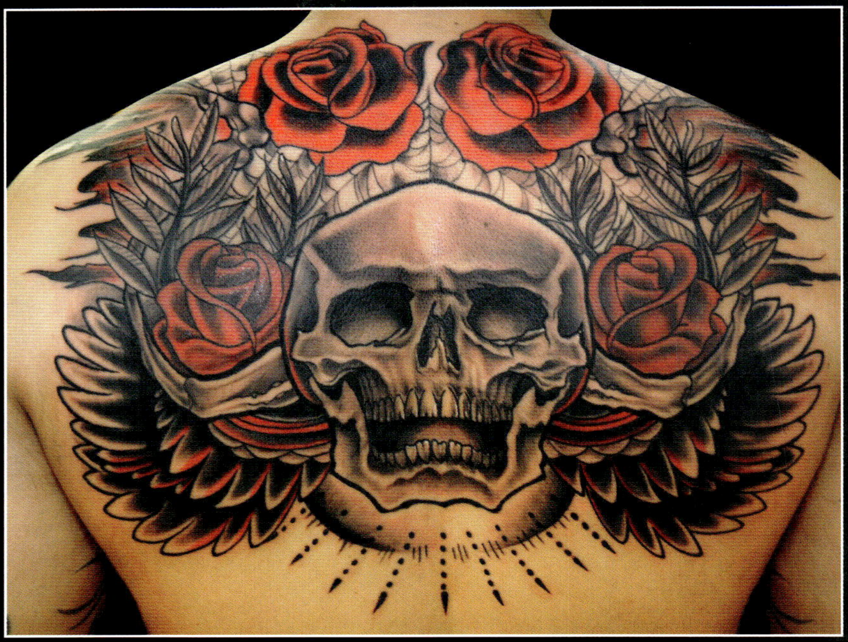

Mario Rosenau & Aric Taylor

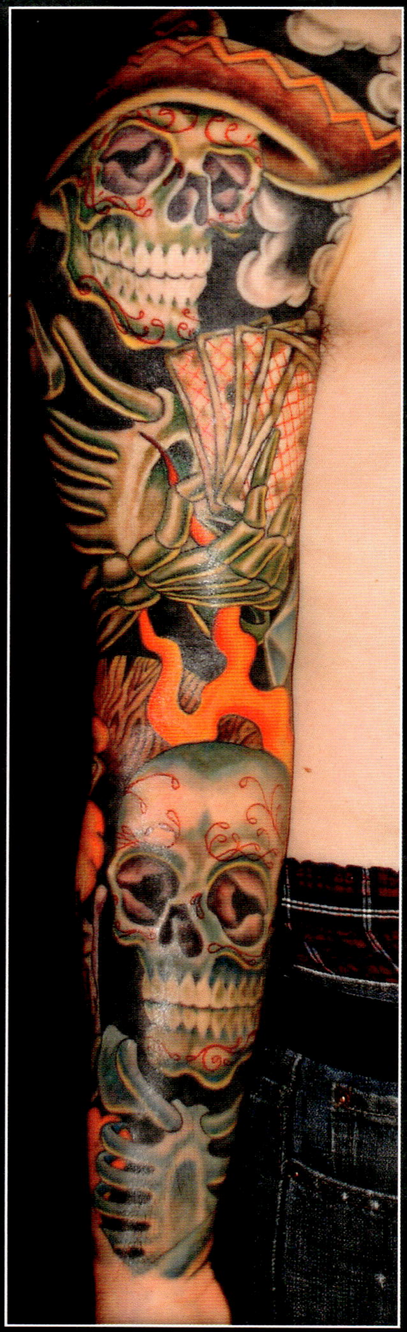

Brady Willmott

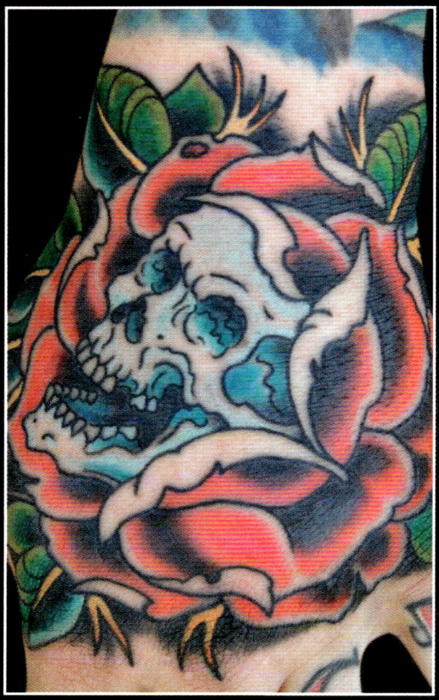

Jim Sylvia

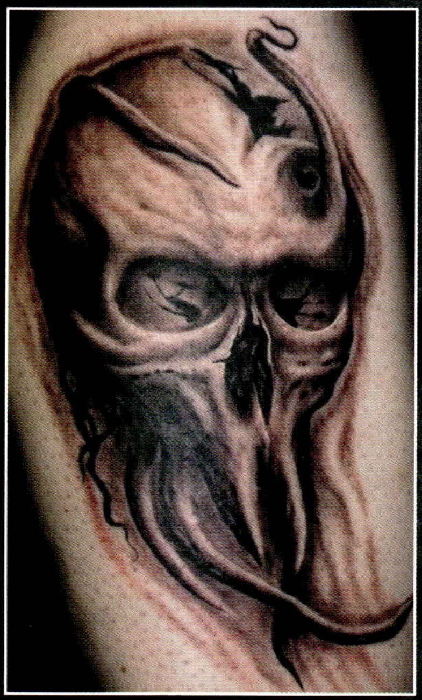

Josh Ruff

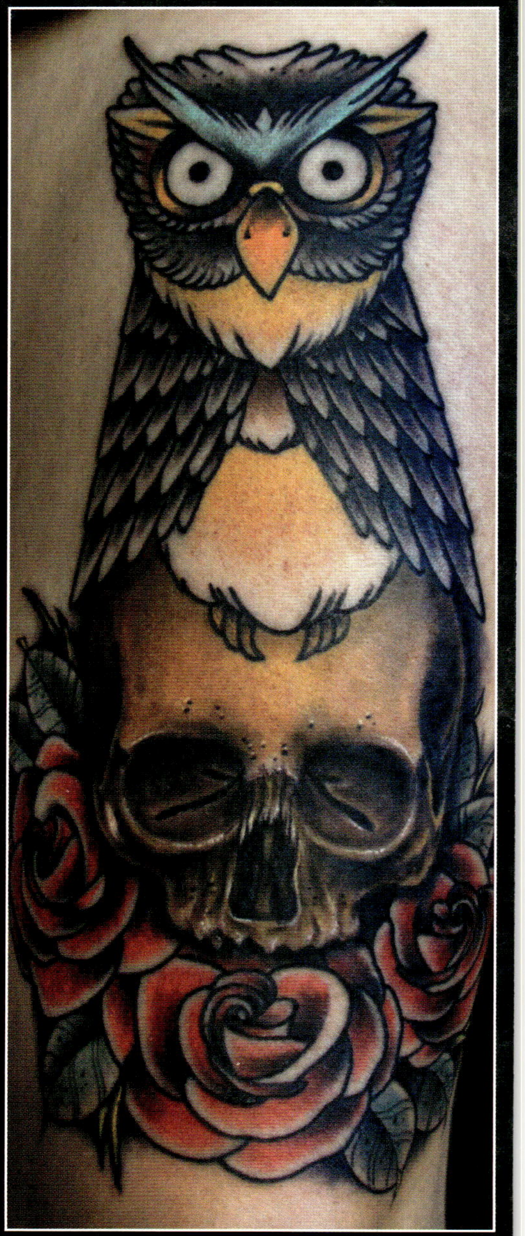

Mick Squires

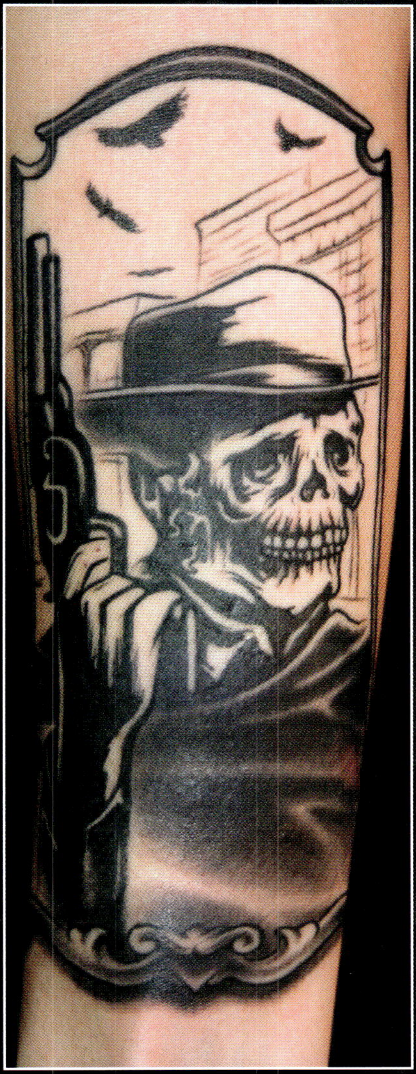

Nathan Kostechko

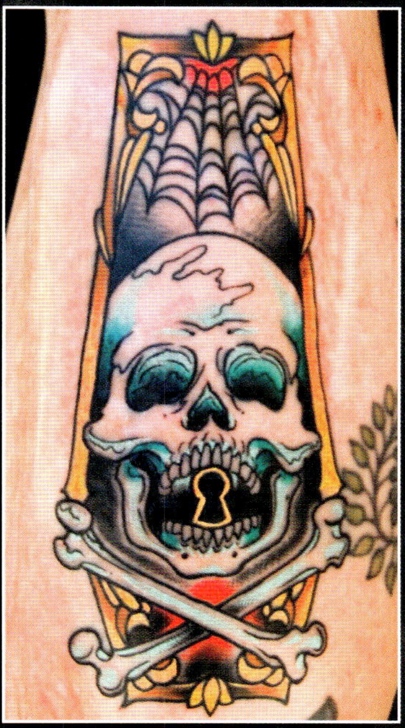

Jim Sylvia

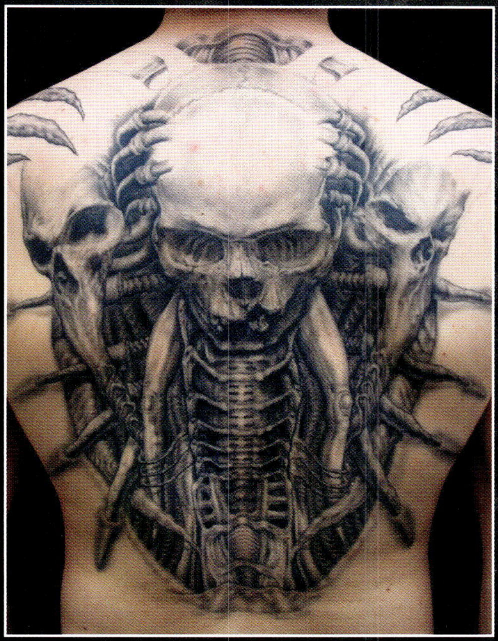

Benjamin Moss

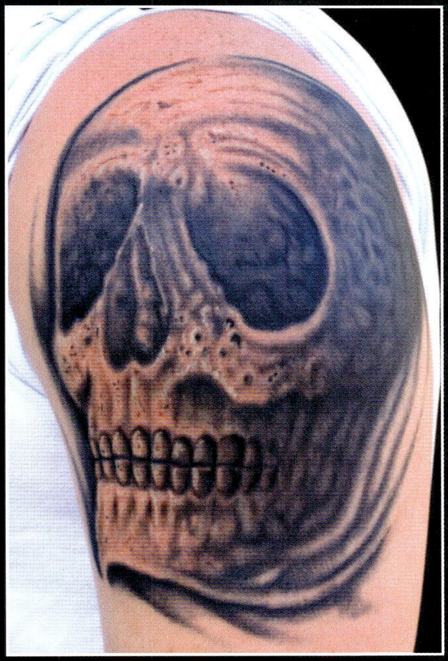

Jeff Johnson

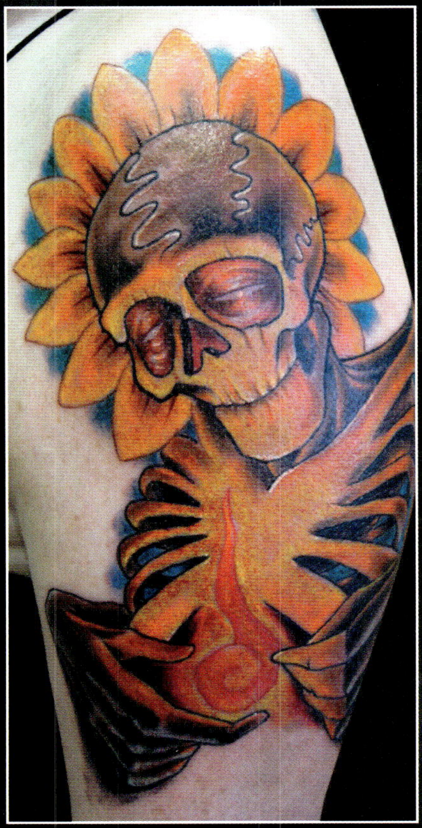

Pete Vaca

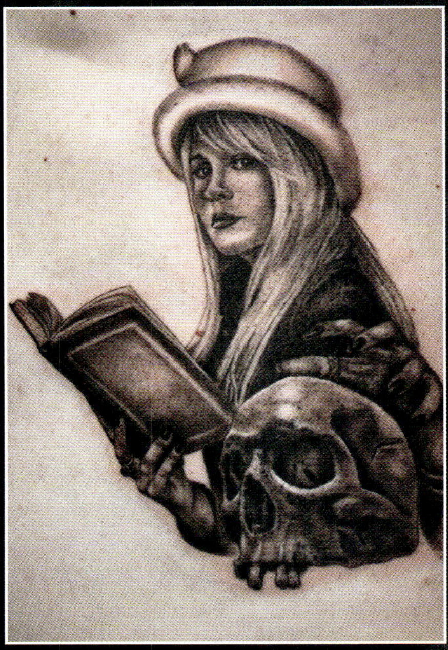

Larry Brogan

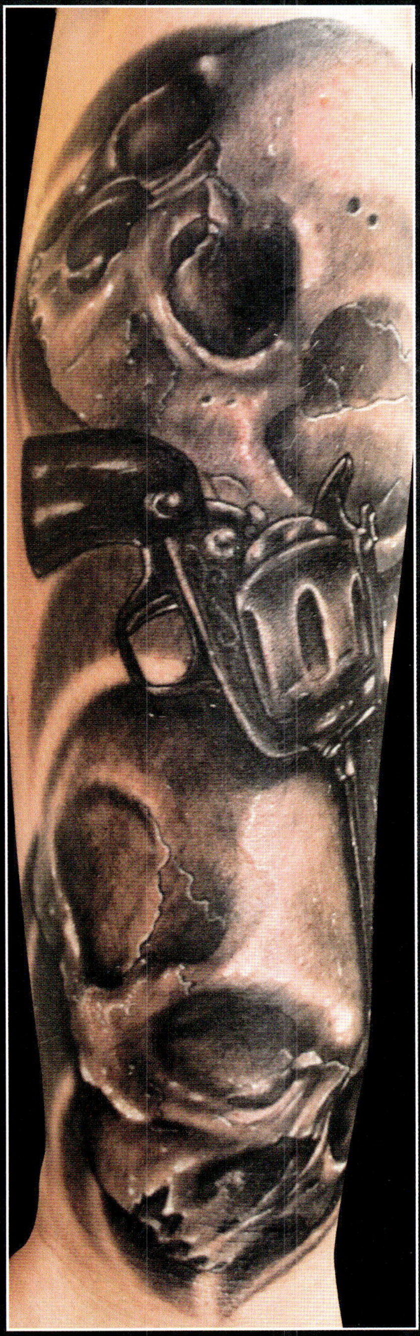

Mike Godfrey

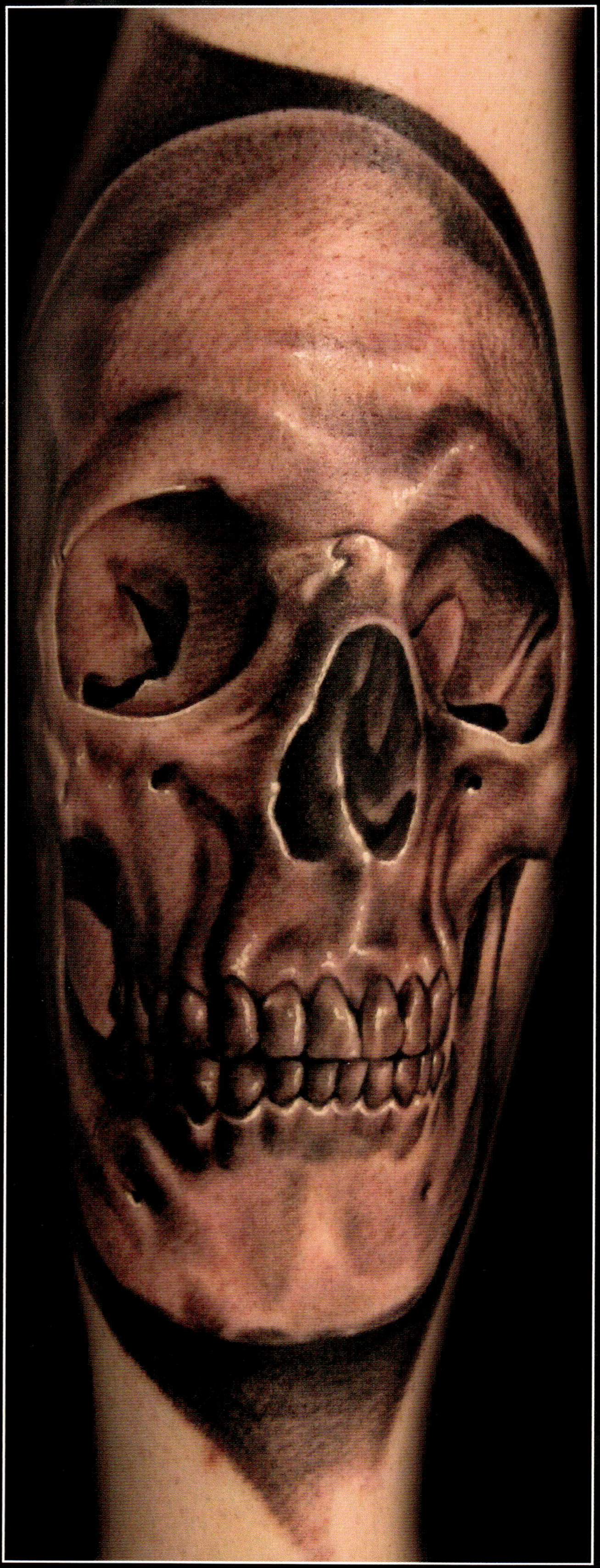
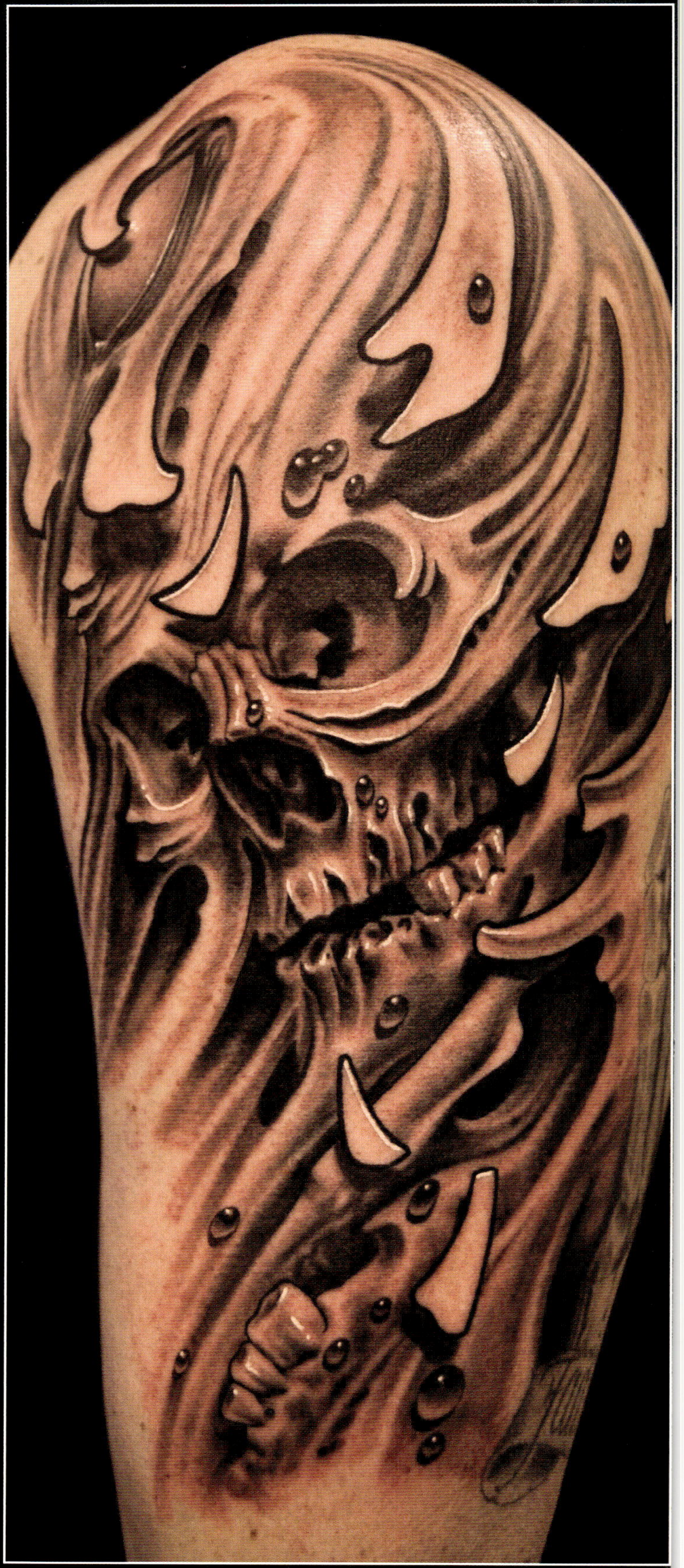

Josh Duffy

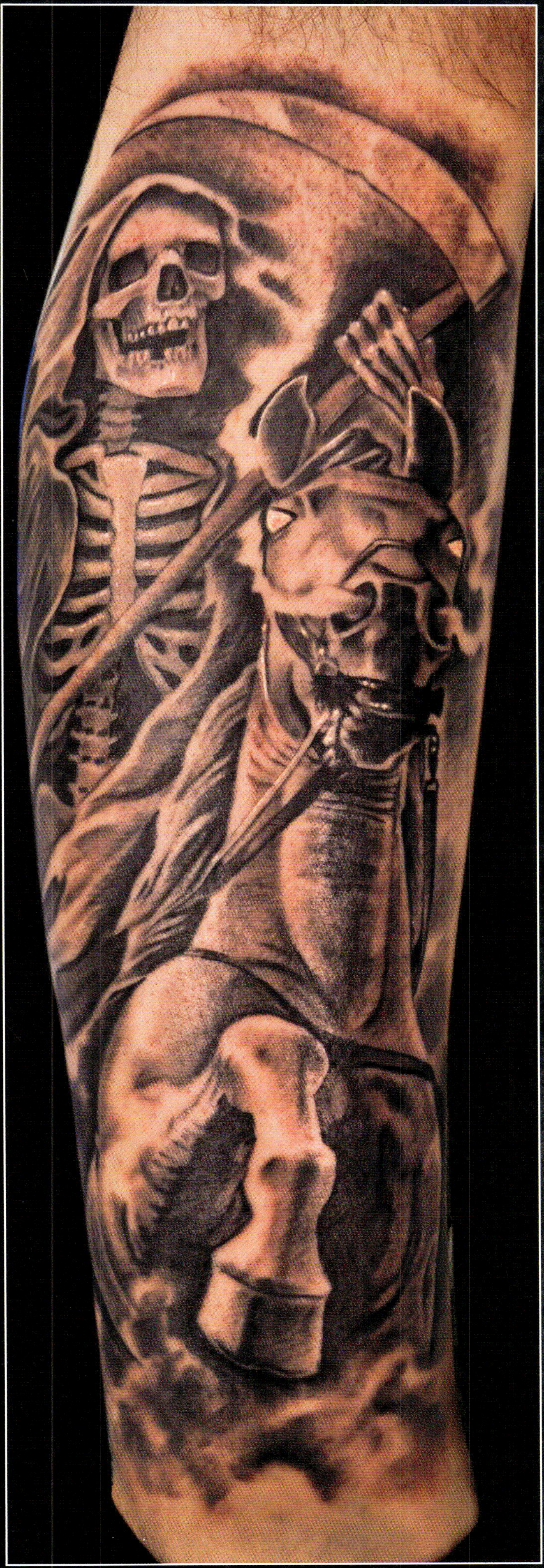

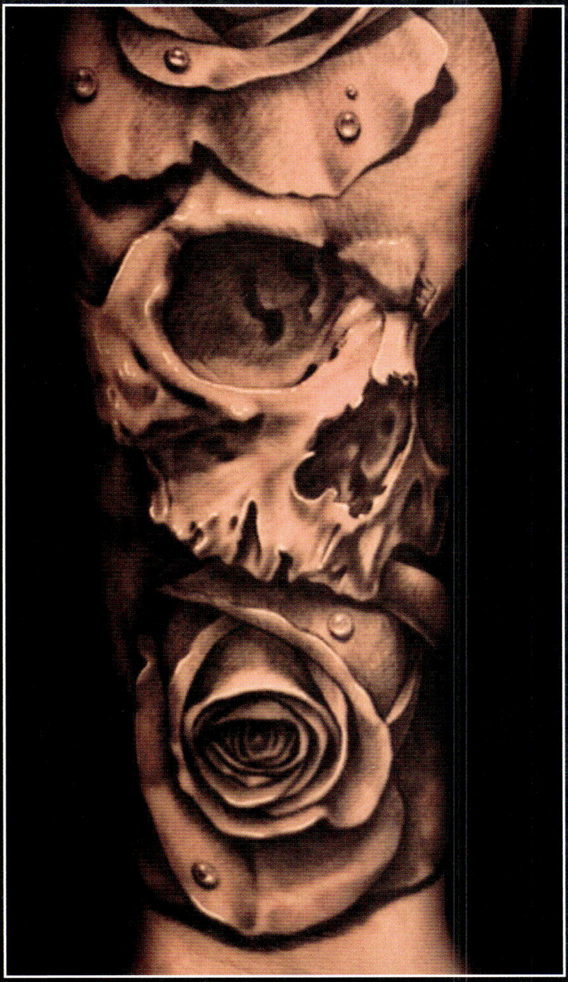

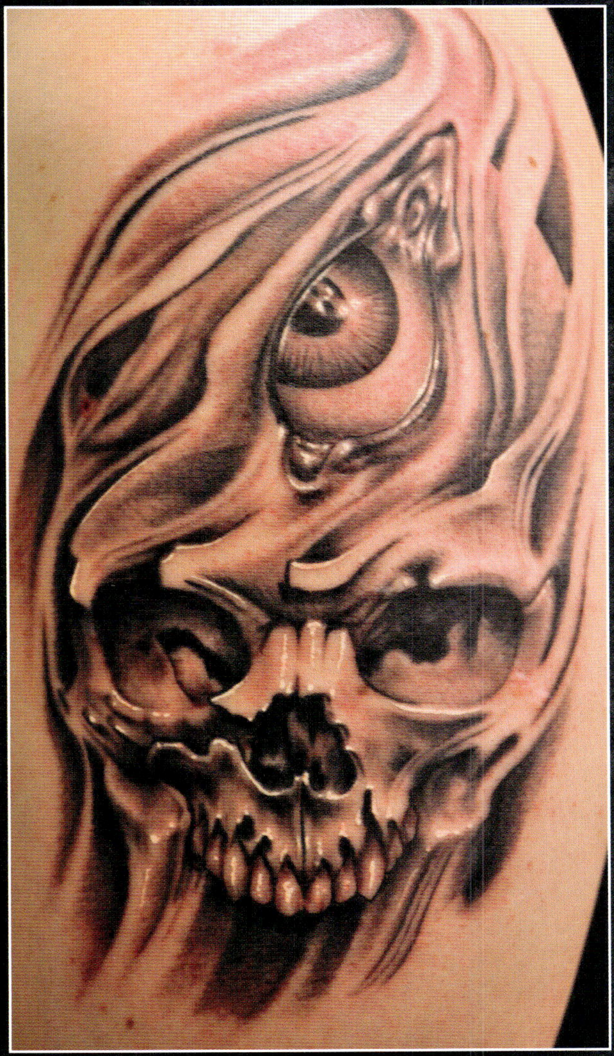

Josh Duffy

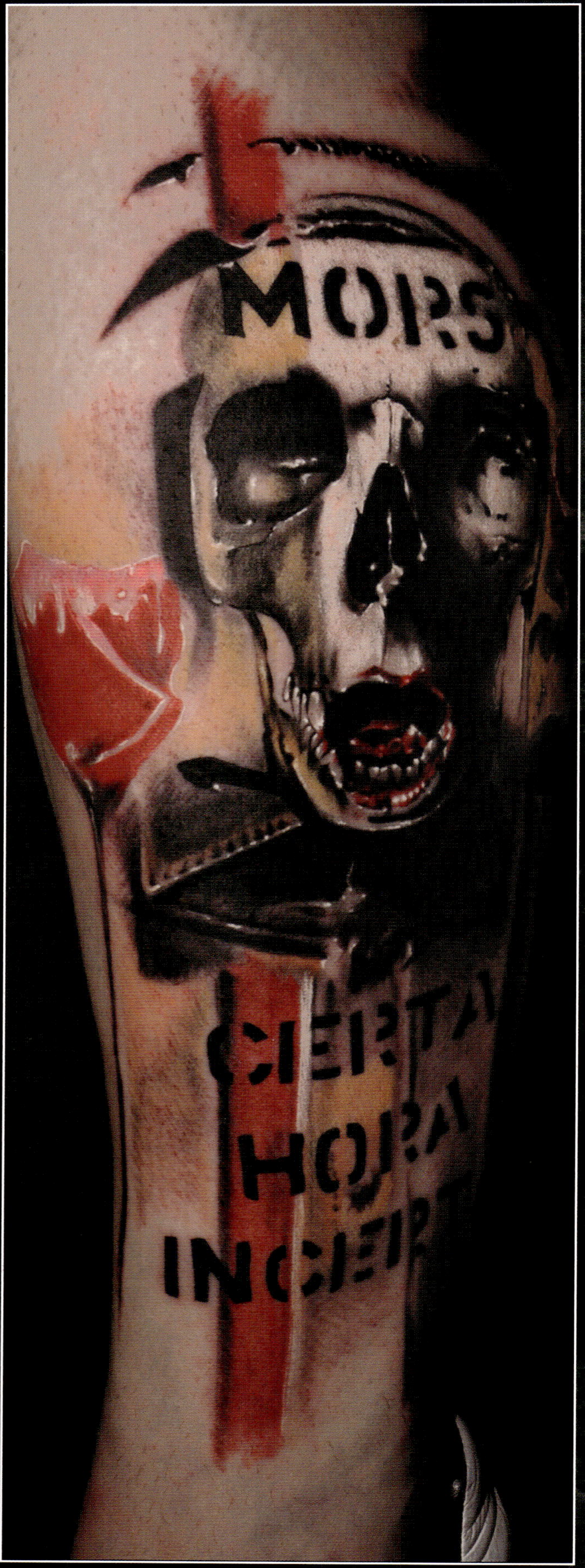

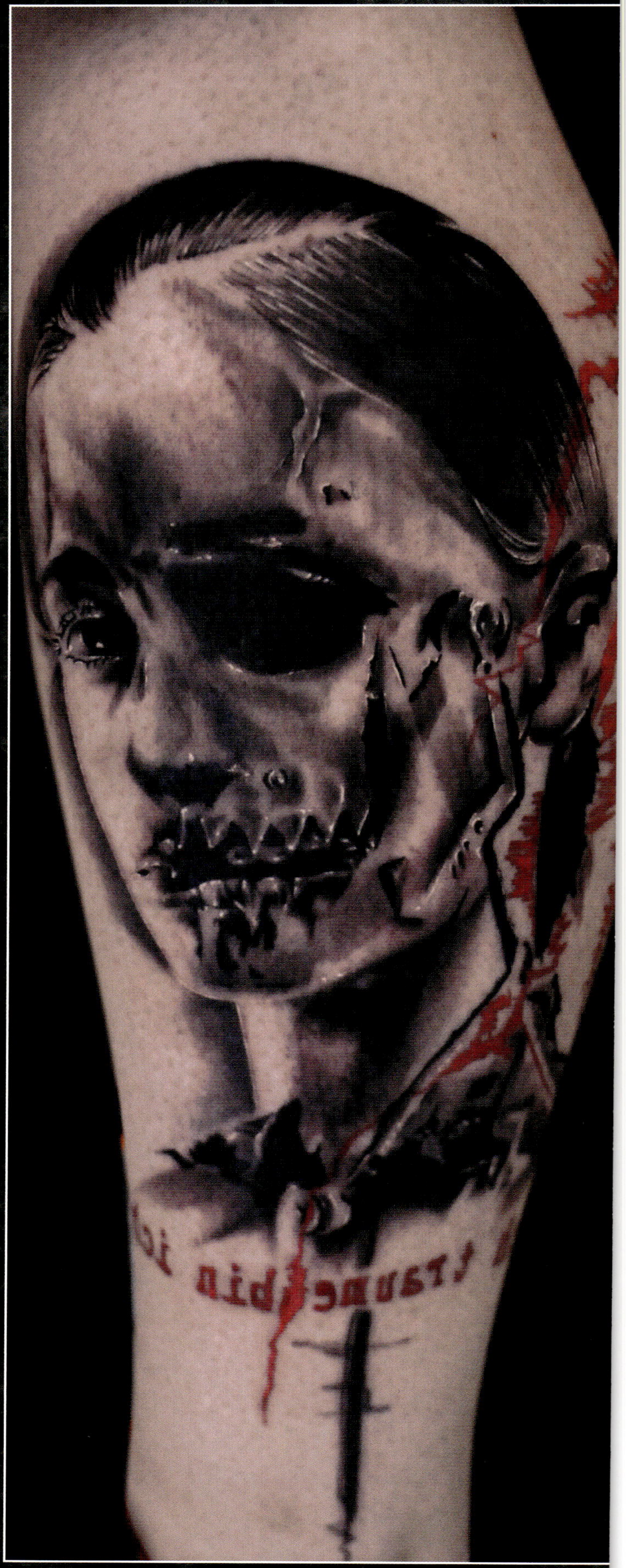

Volko Merschky & Simone Pfaff

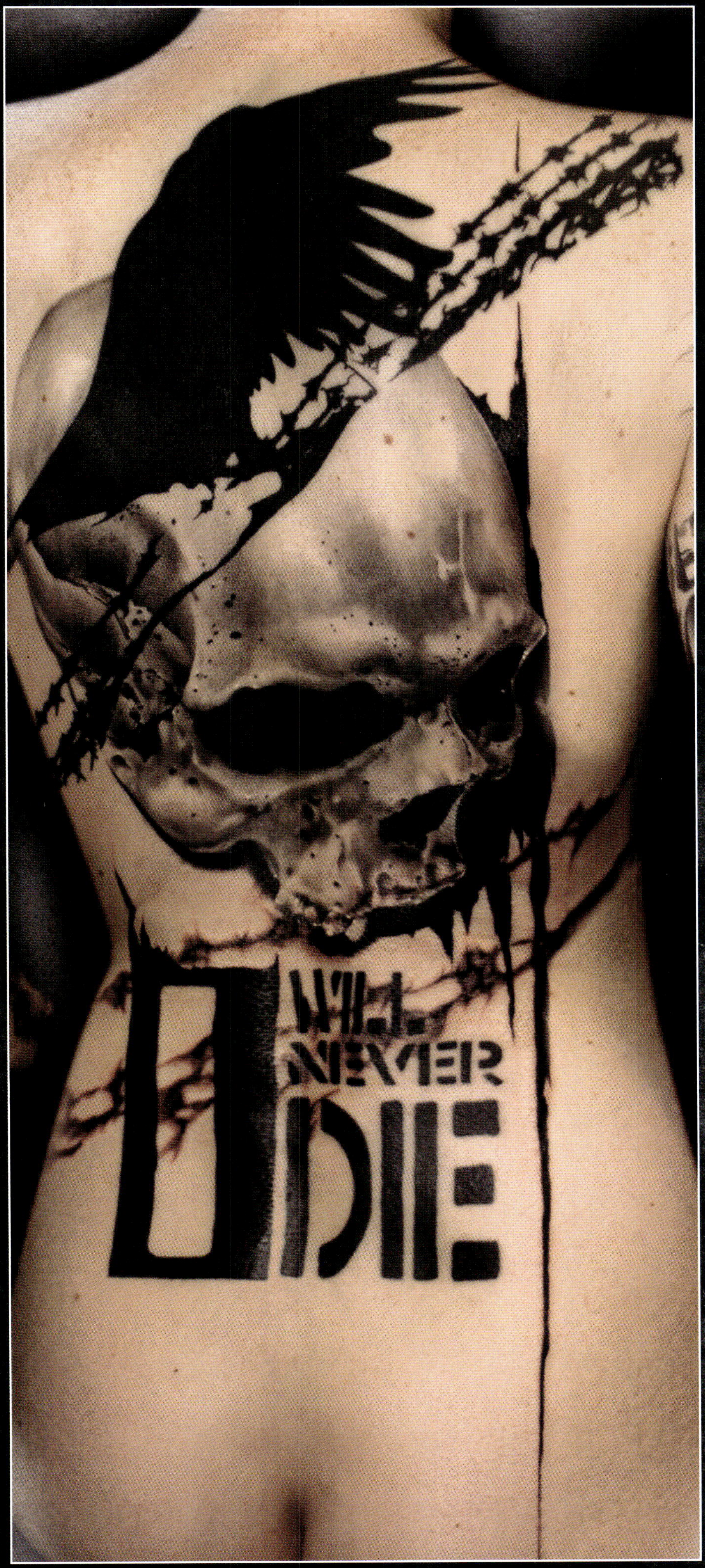

Volko Merschky & Simone Pfaff

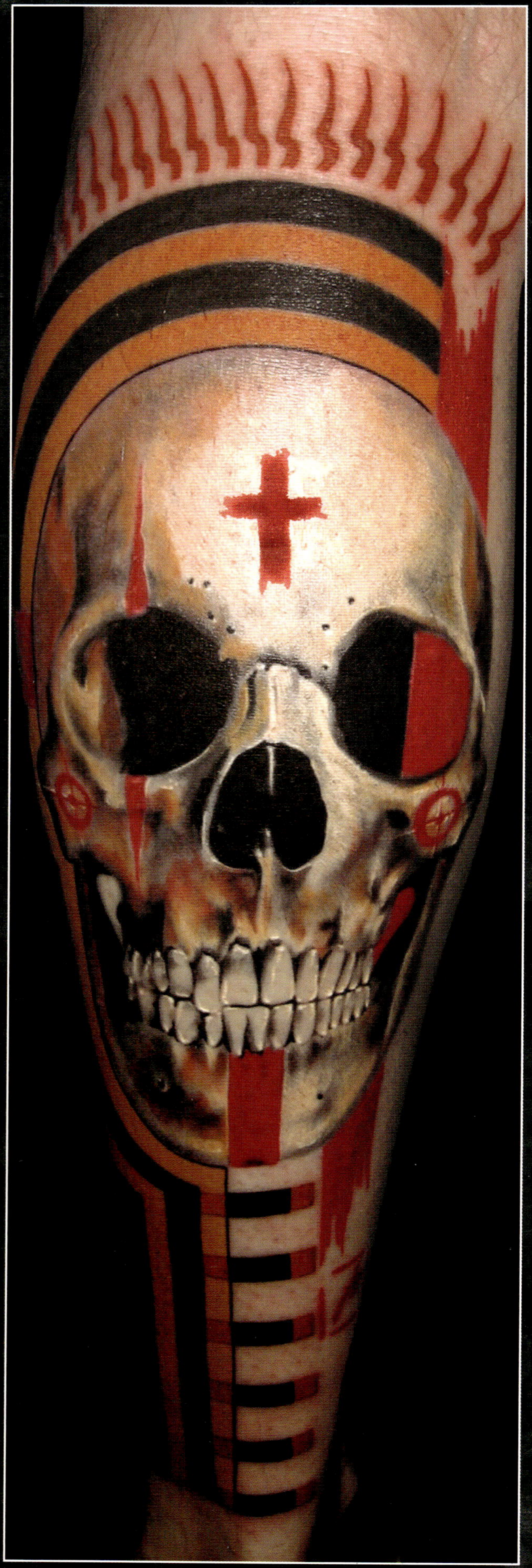

Volko Merschky

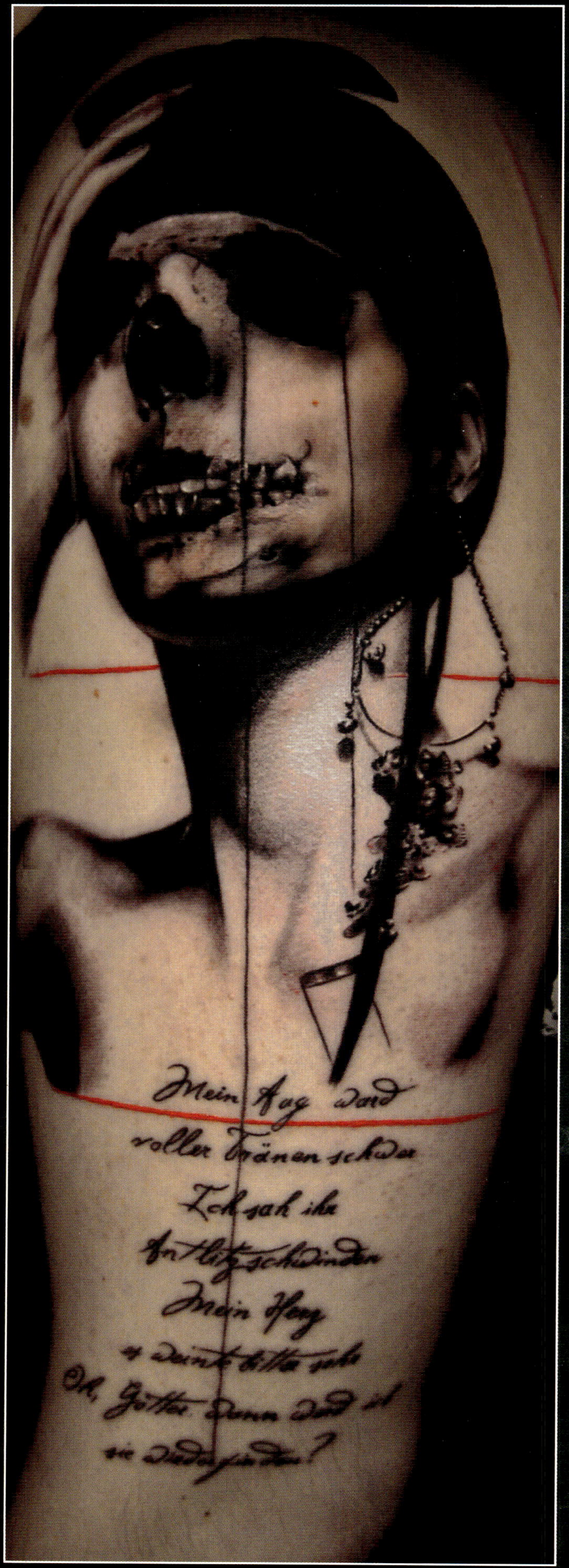

Volko Merschky & Simone Pfaff

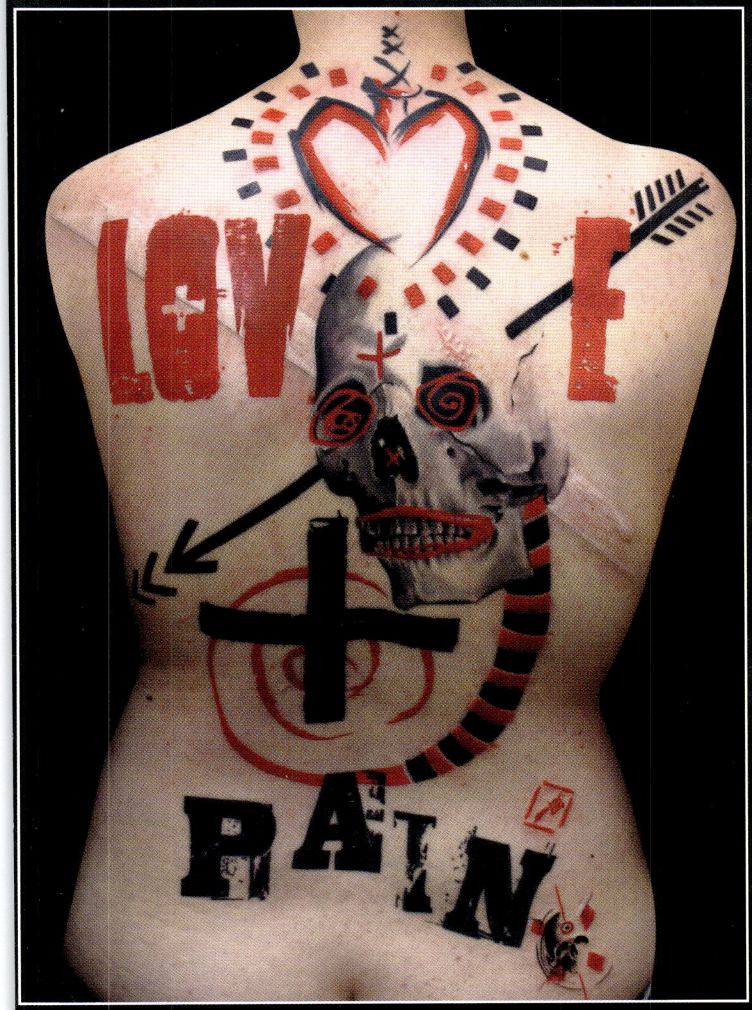
LOV E
PAIN

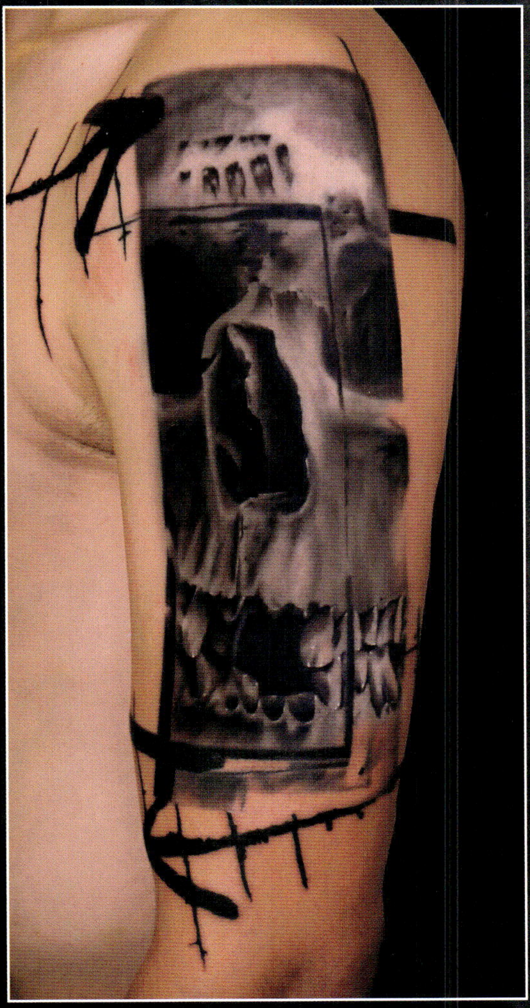

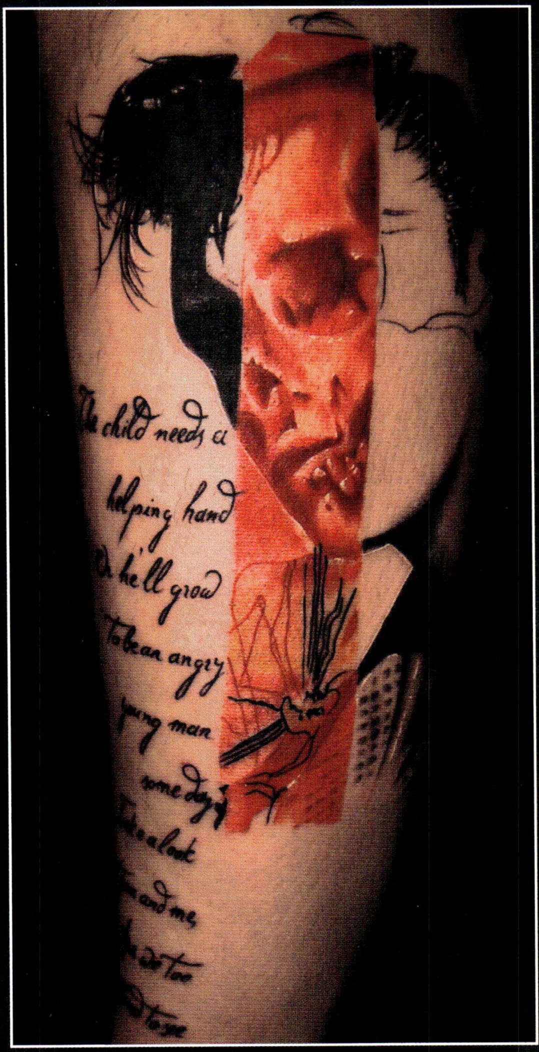
The child needs a
helping hand
Or he'll grow
to be an angry
young man
some day

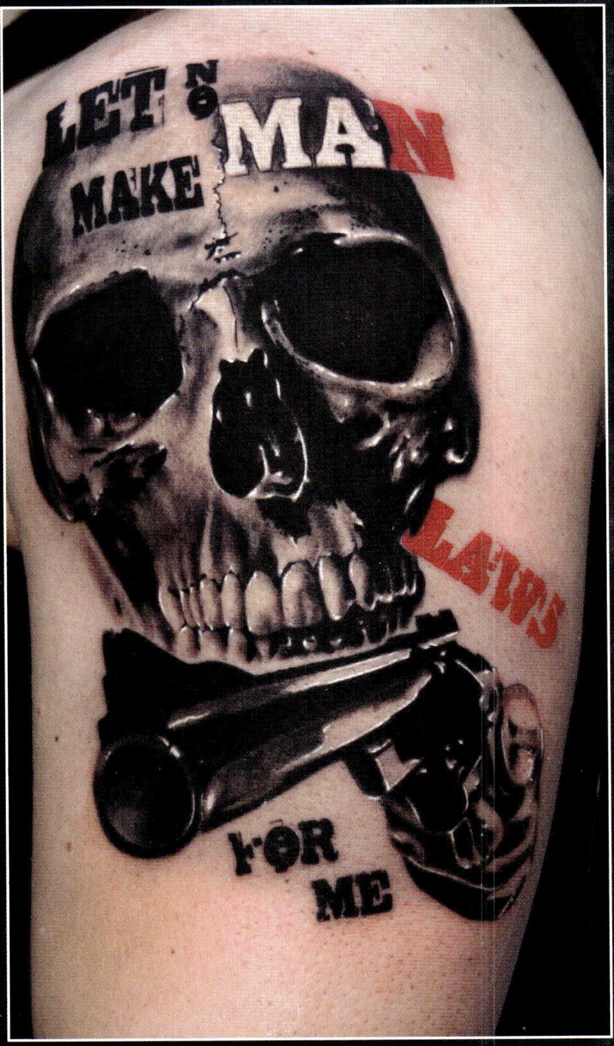
LET NO MAN
MAKE
LAWS
FOR
ME

· PAINTINGS ·

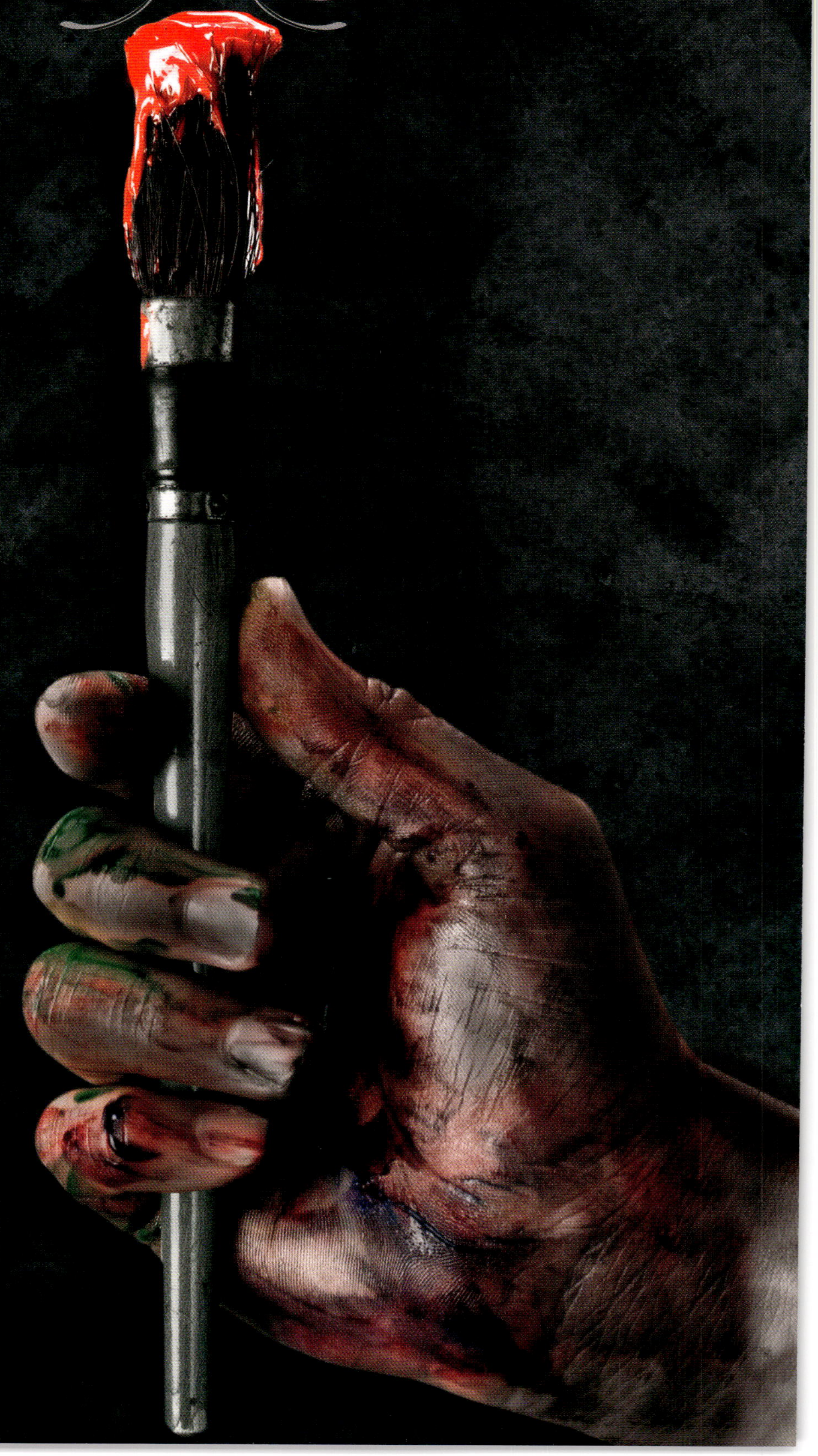

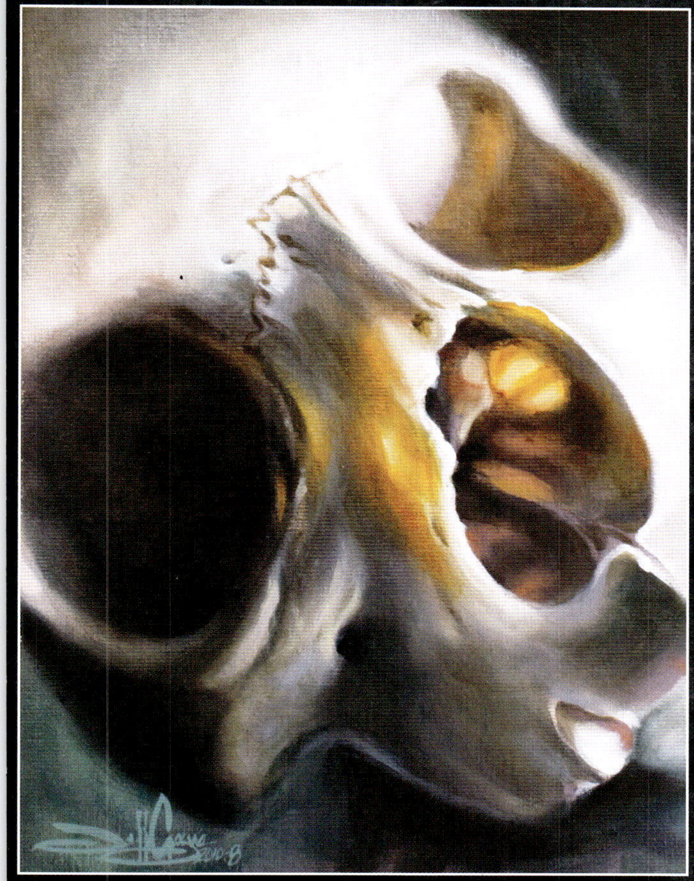

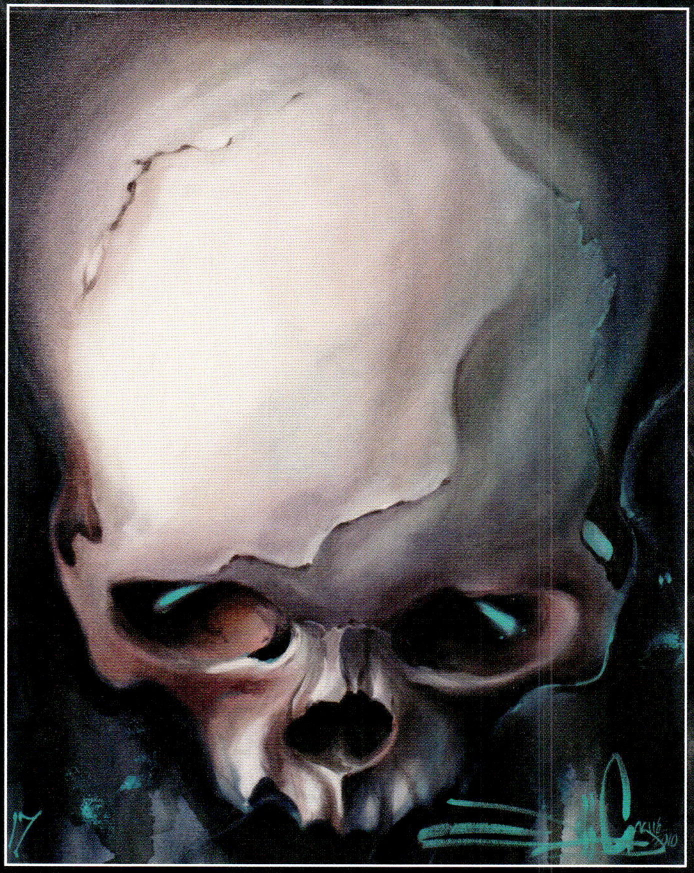

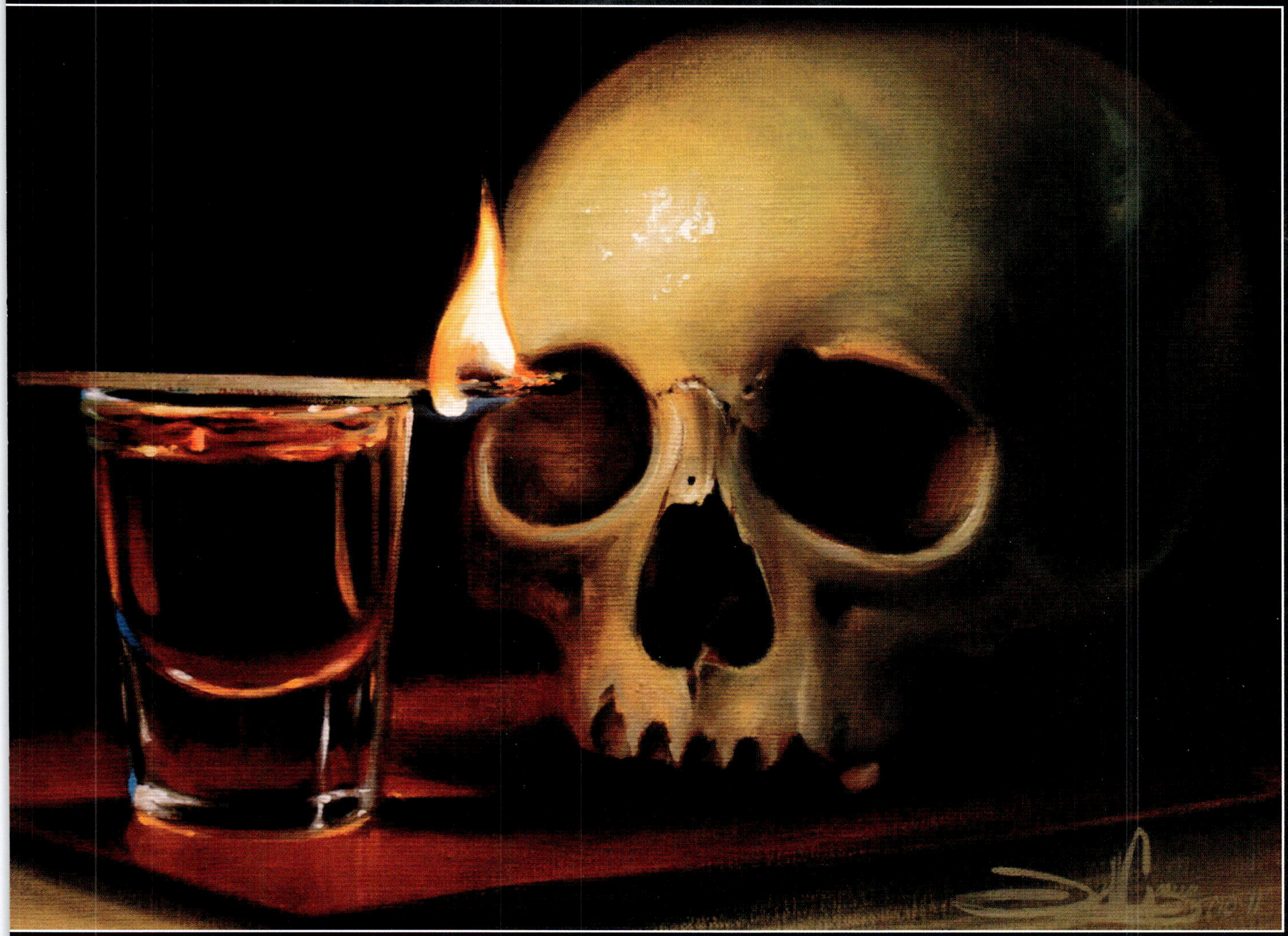

Jeff Gogué

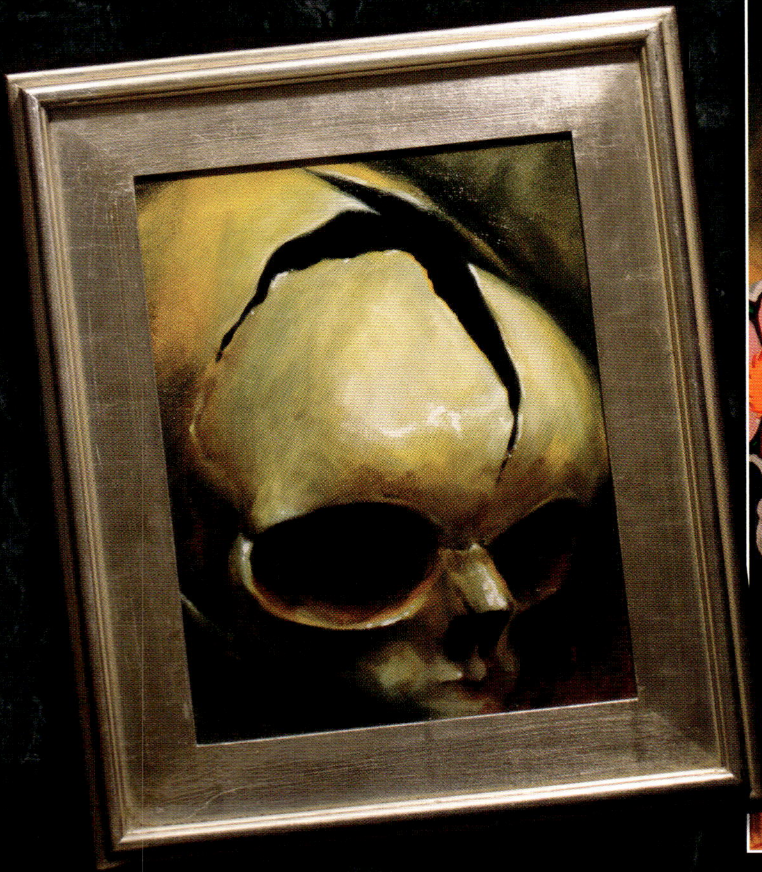

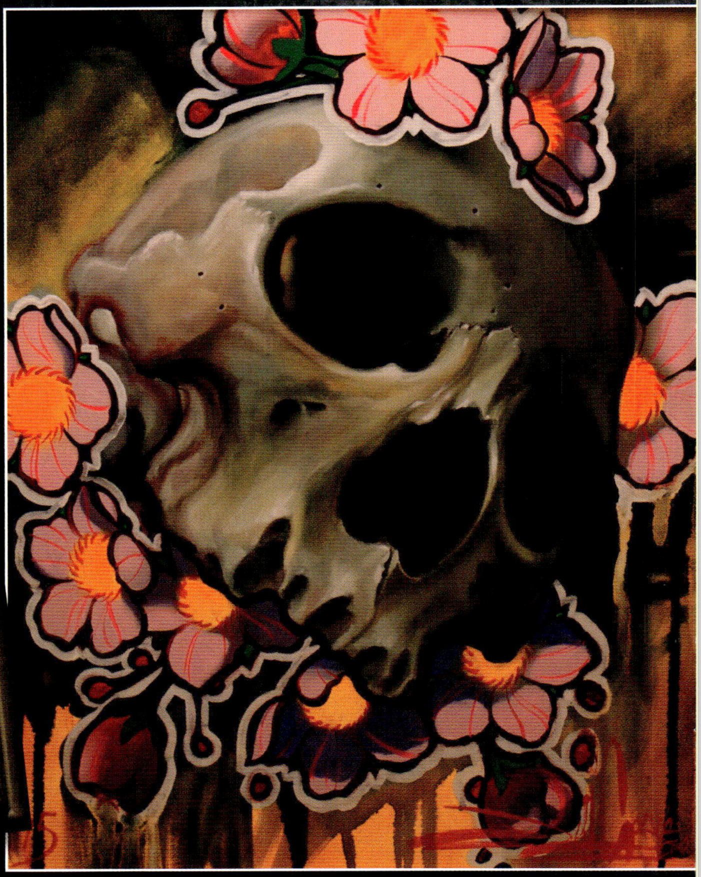

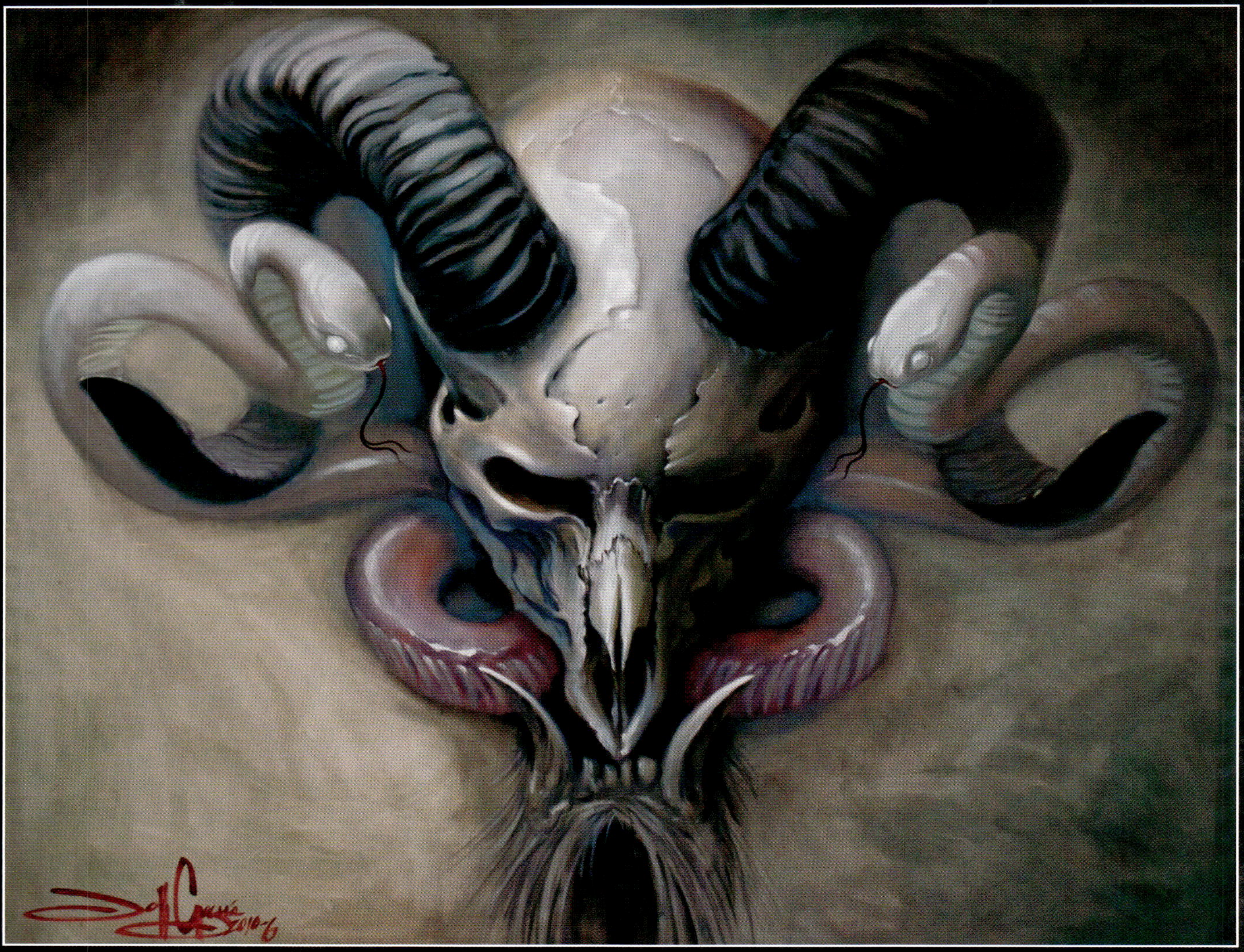

Jeff Gogué

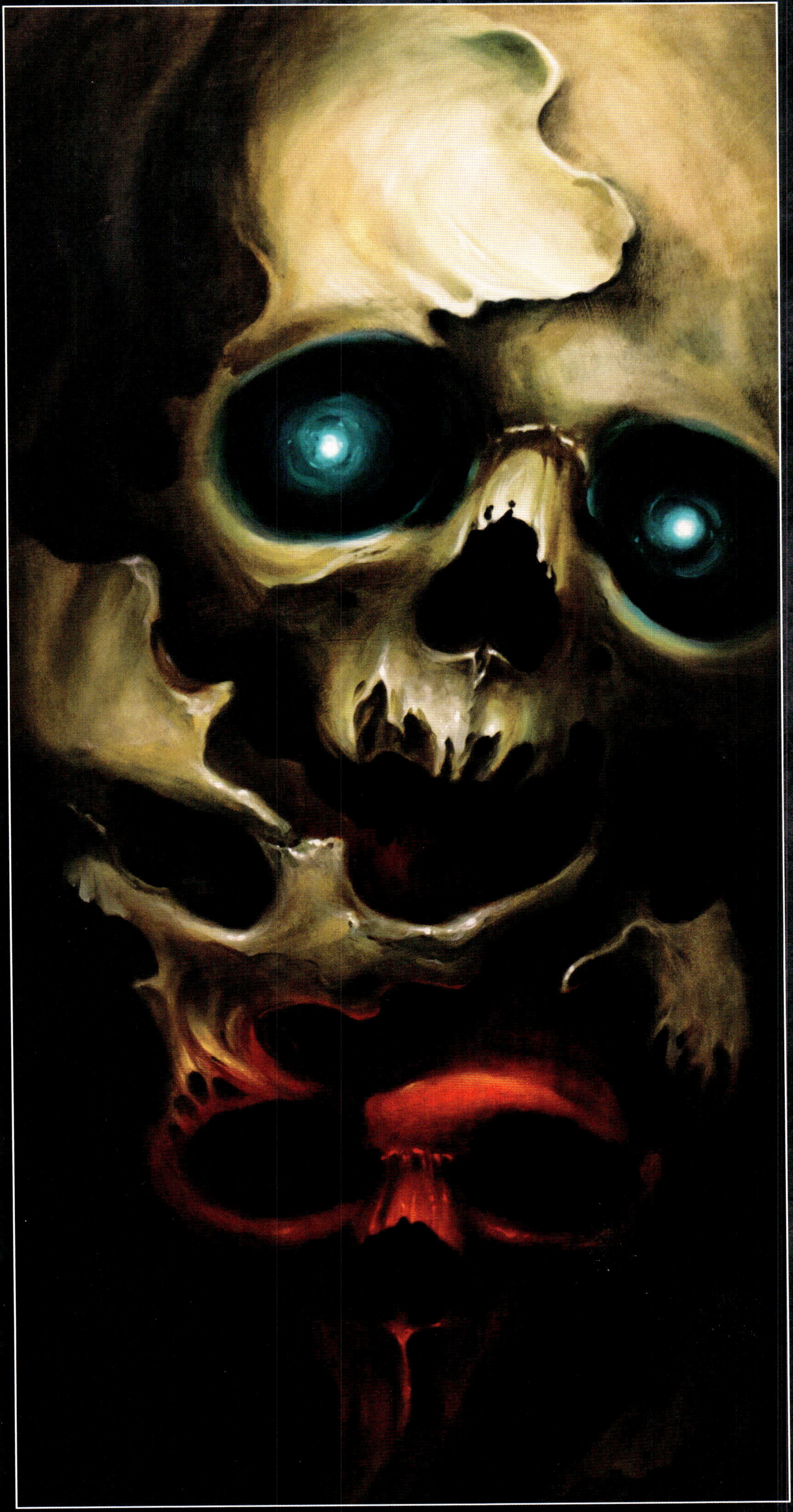

Jeff Gogué

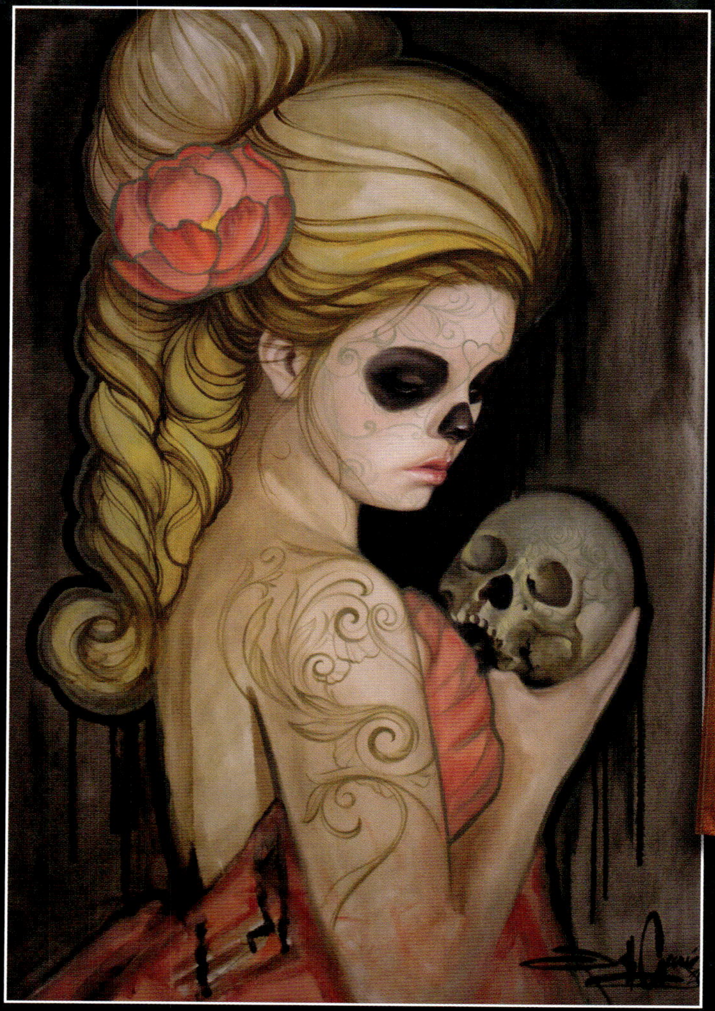

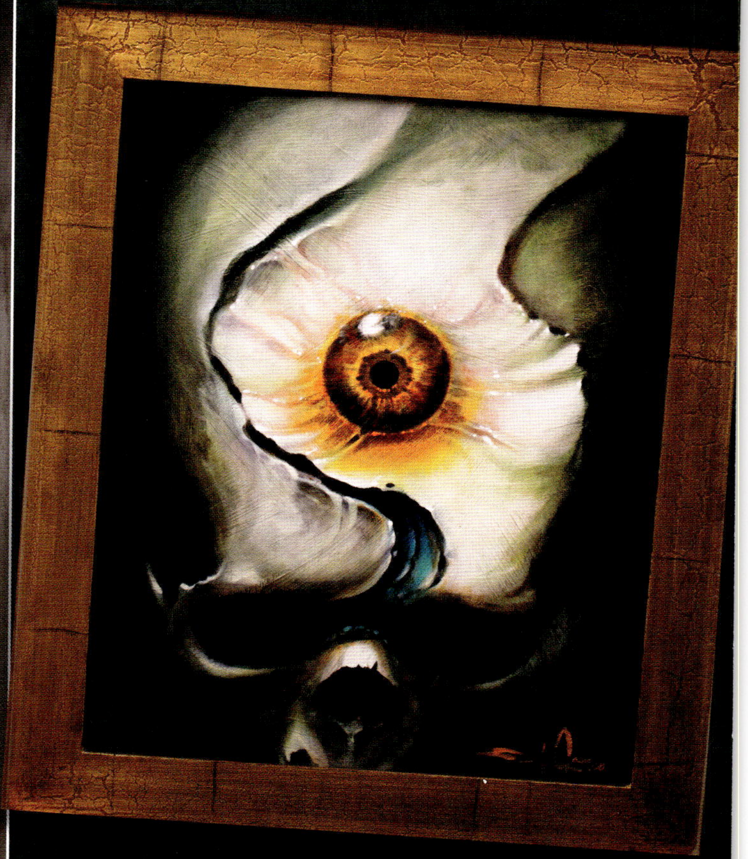

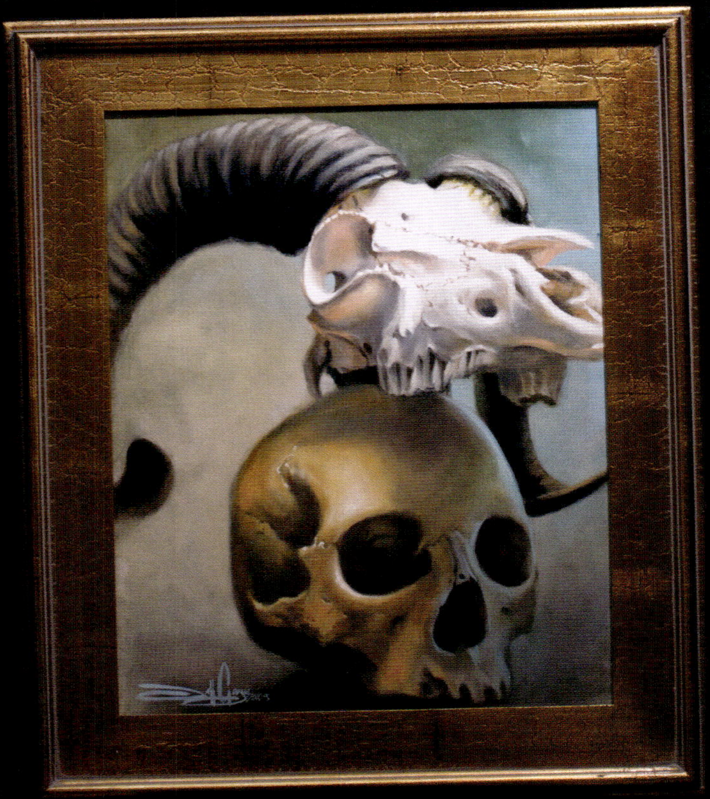

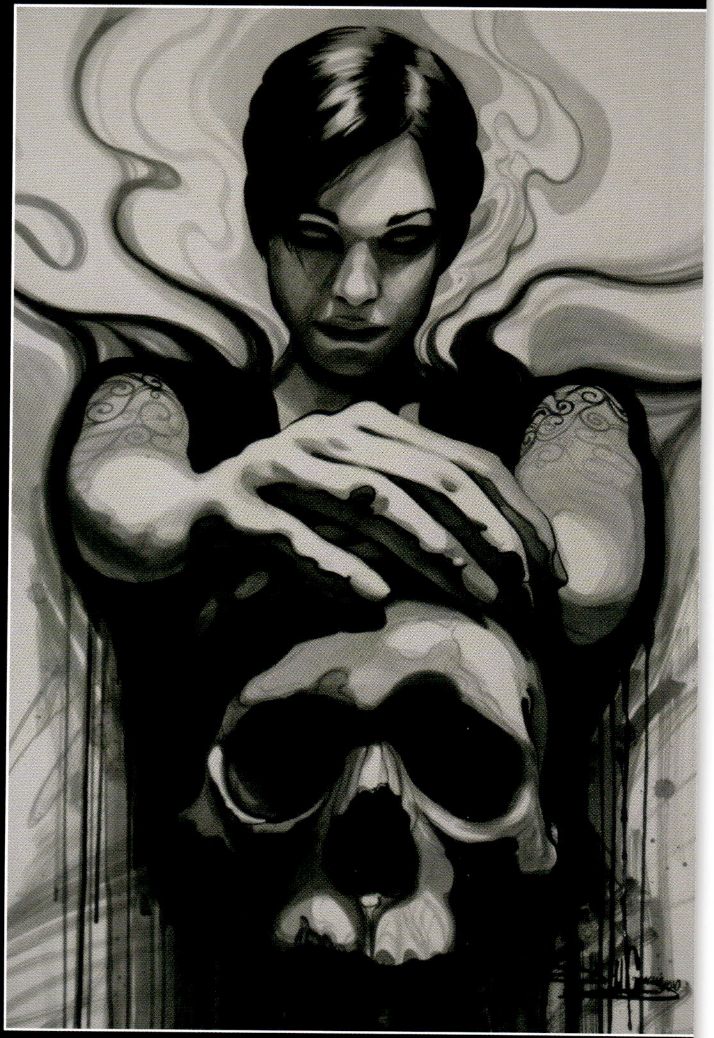

Jeff Gogué

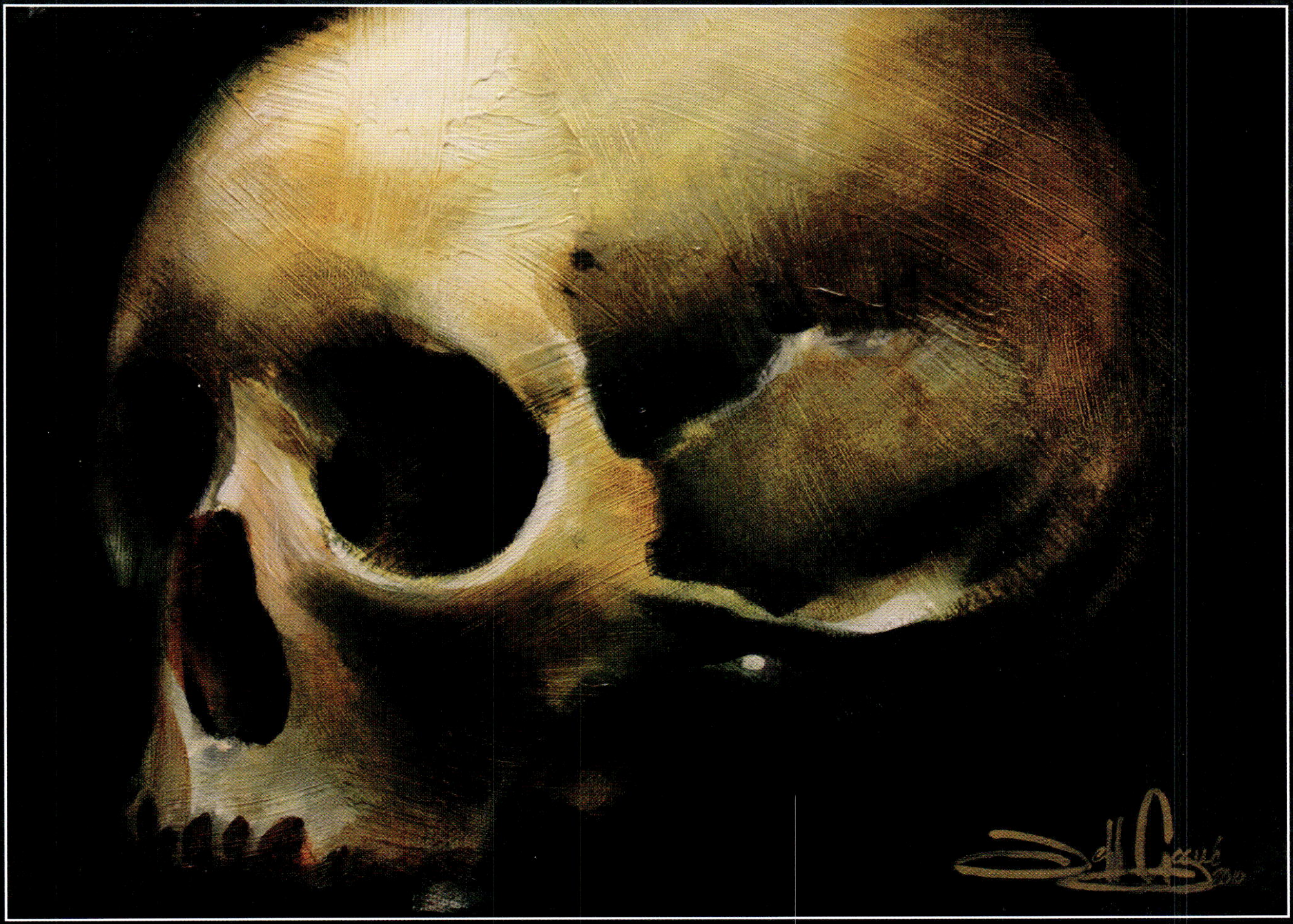

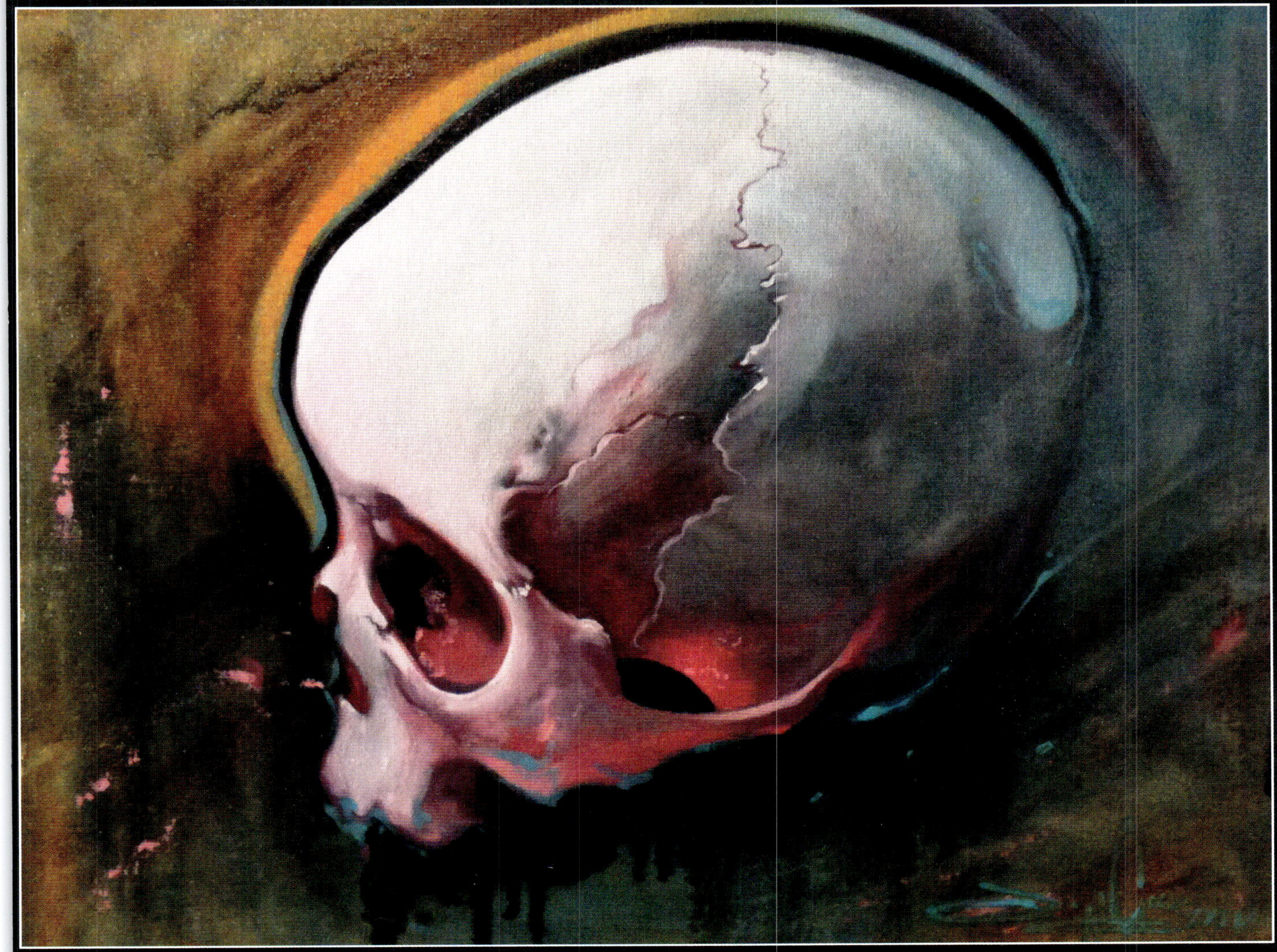

Jeff Gogué

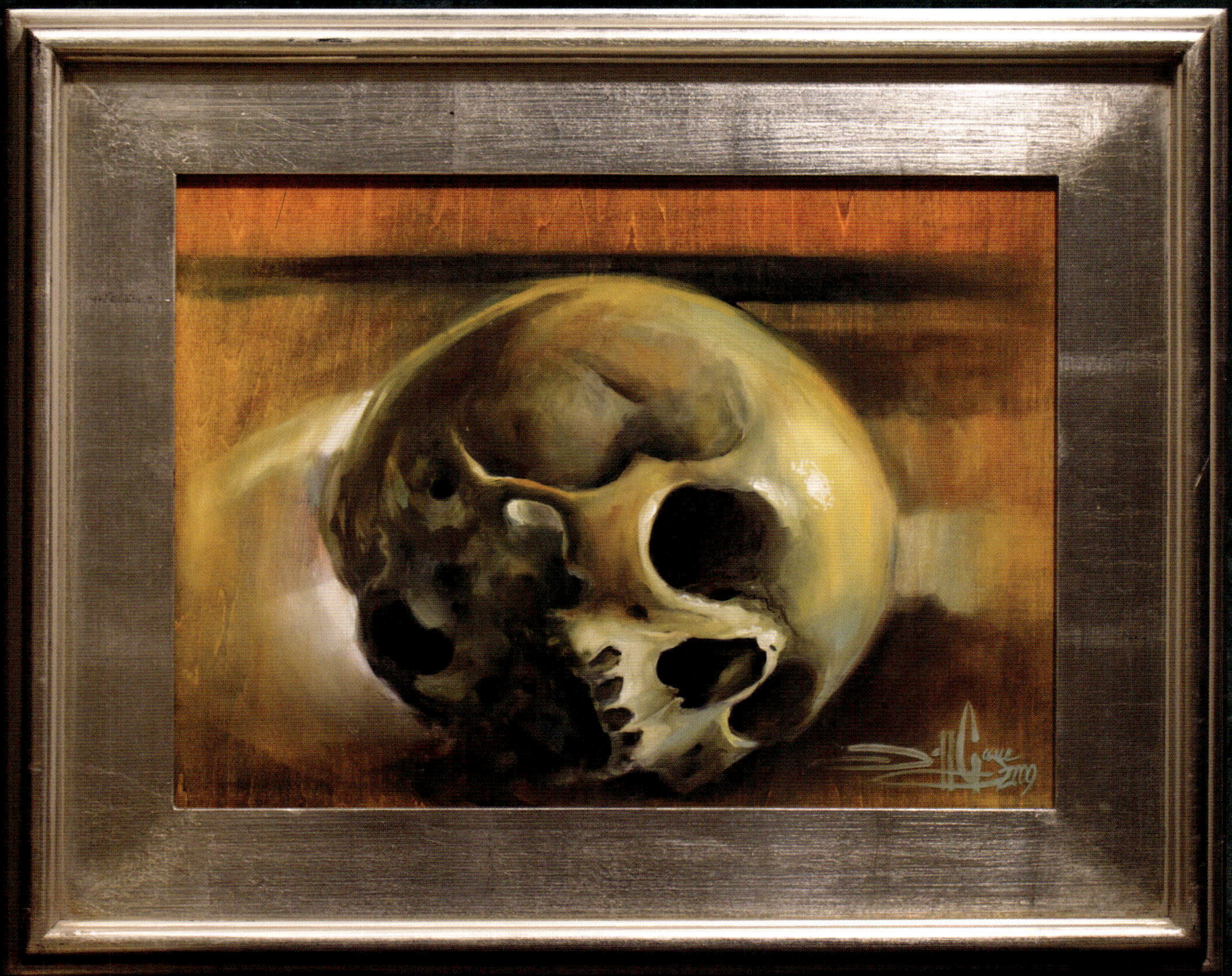

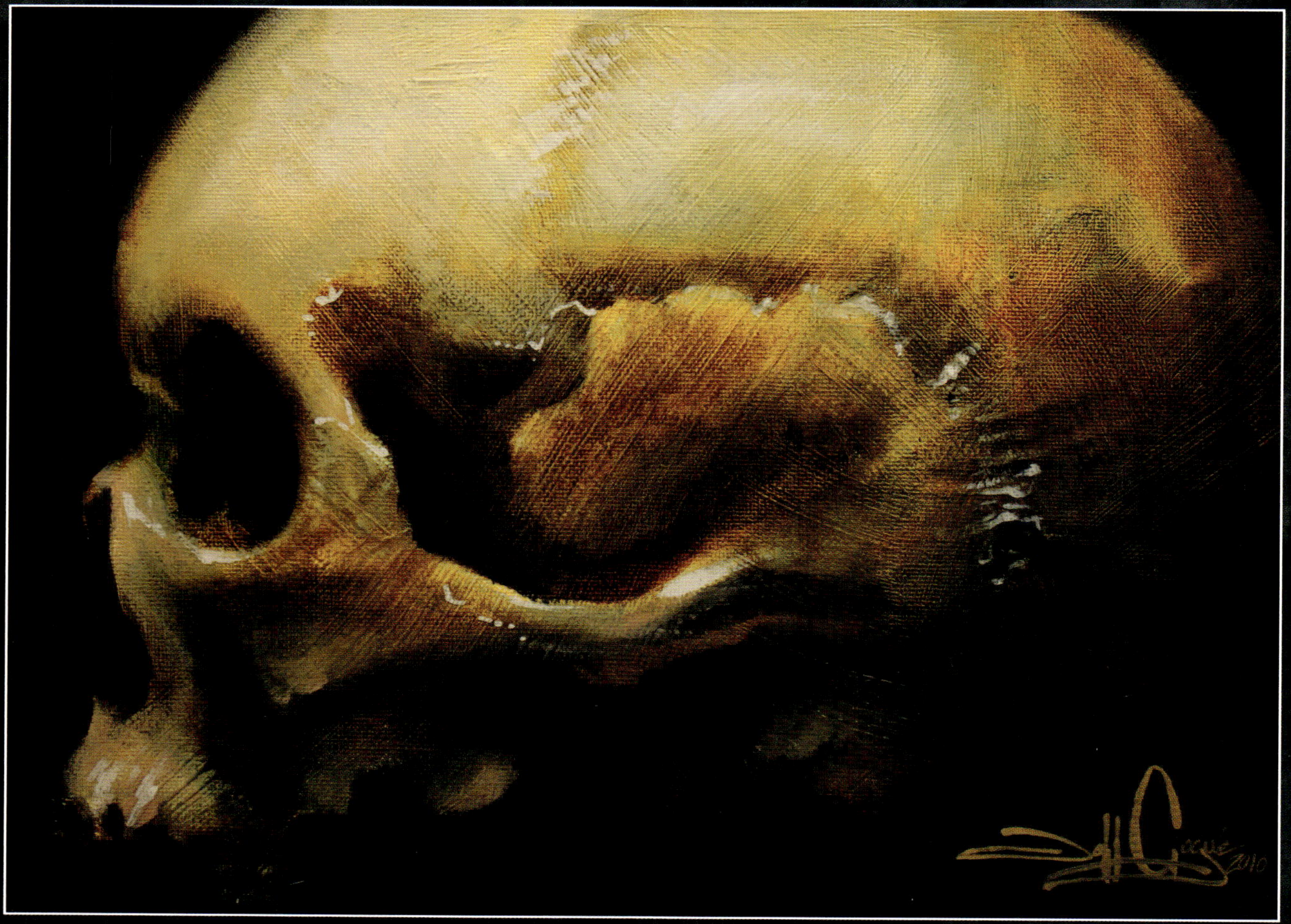

Jeff Gogué

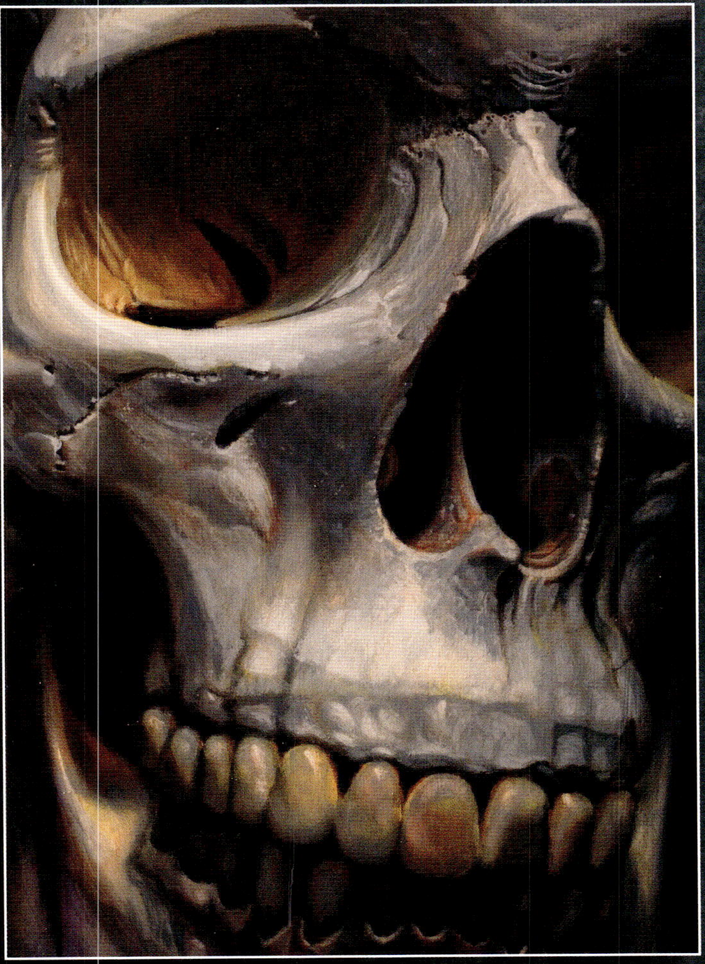

Jeffrey Srsic

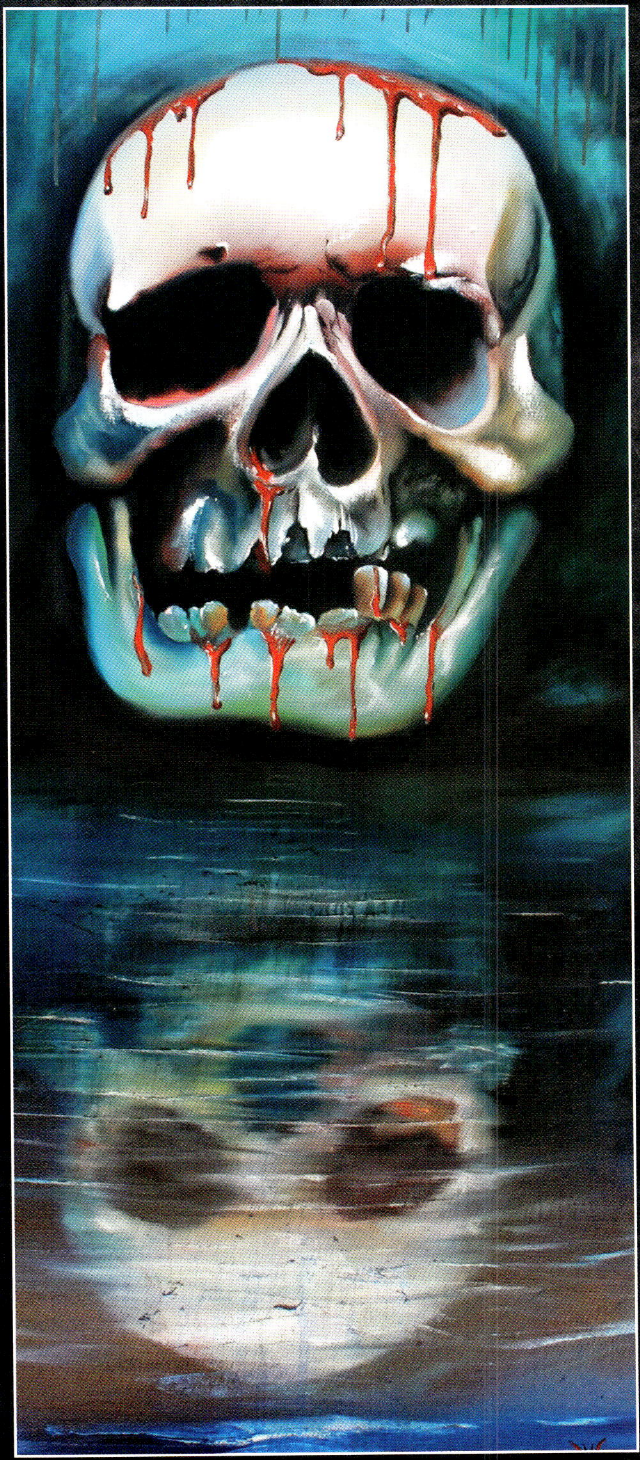

Kyle Cotterman & Nick Harrison

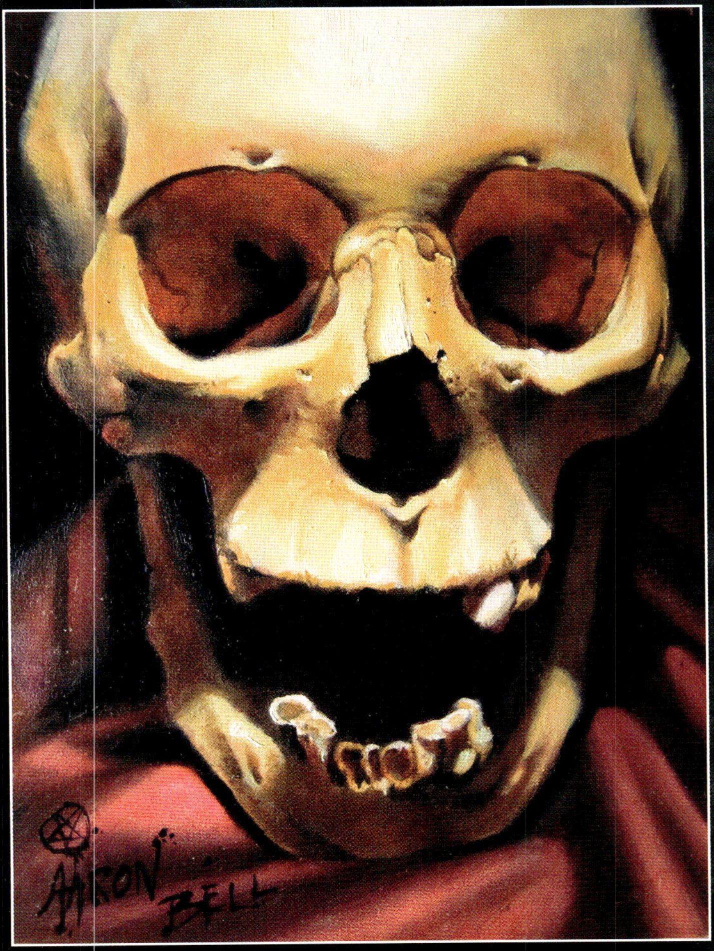

Aaron Bell

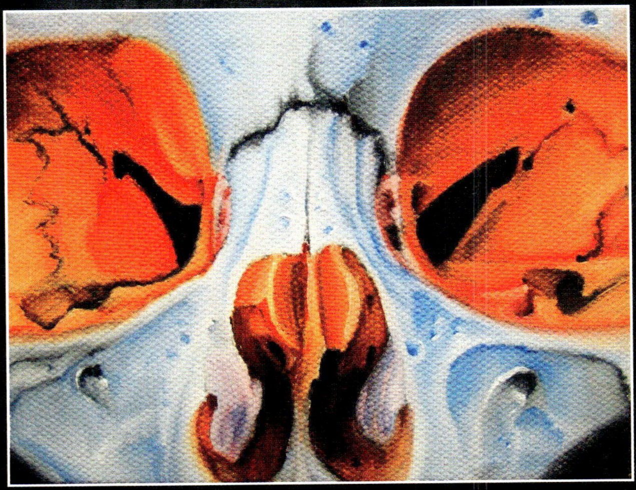

Bobby Deneen

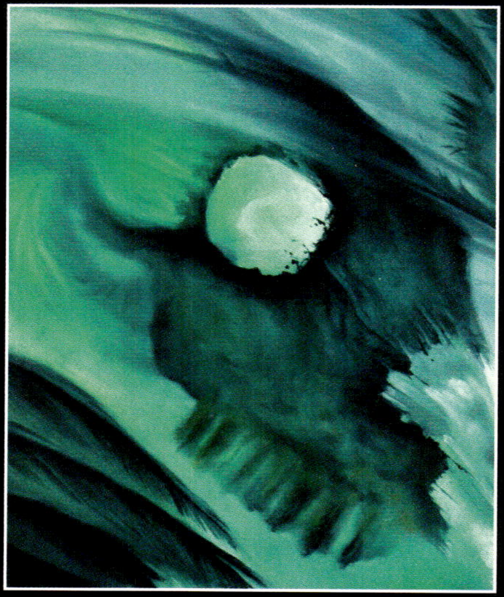

Gifford Kasen

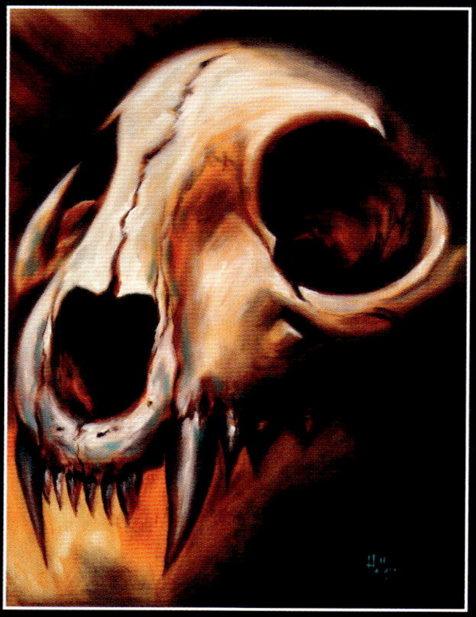

Bobby Holland

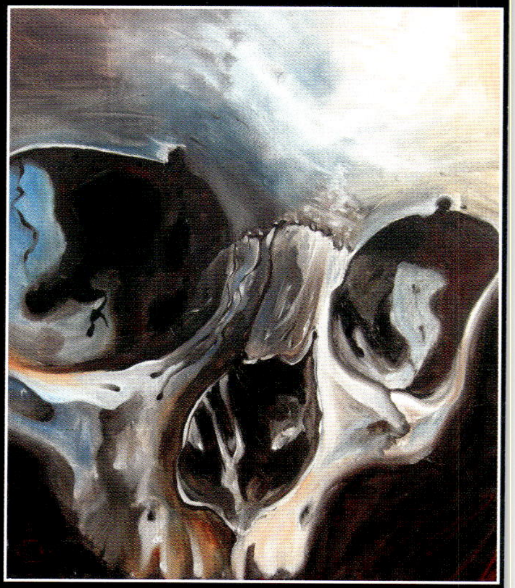

Adam James Chiodo

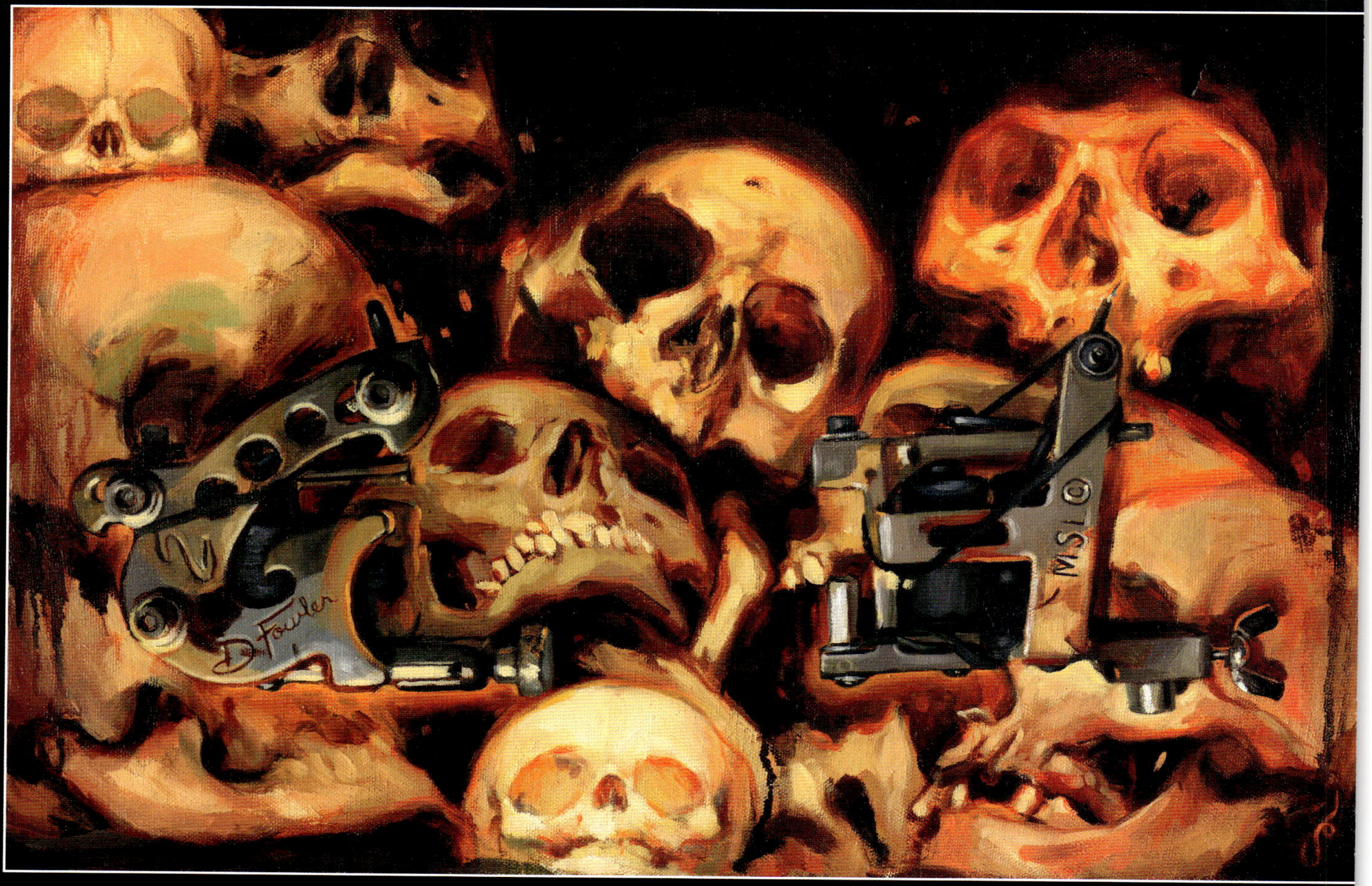

Shawn Barber

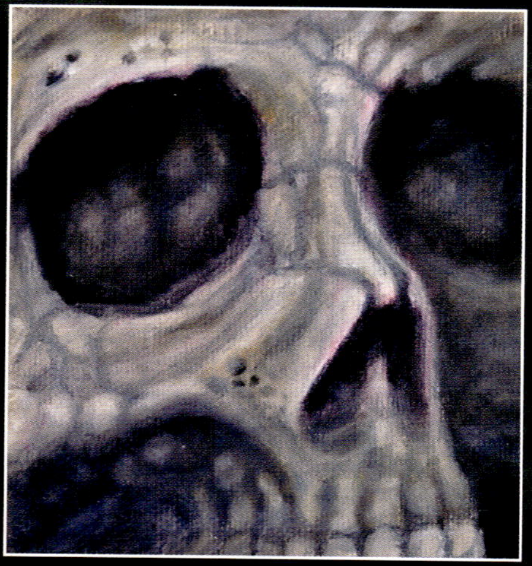

Jeff Johnson

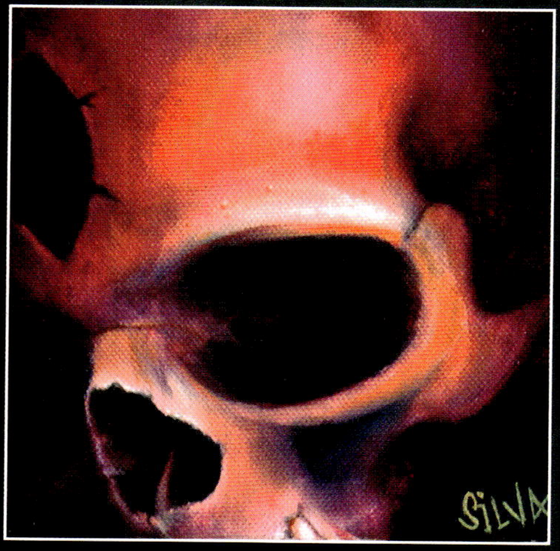

Marvin Silva

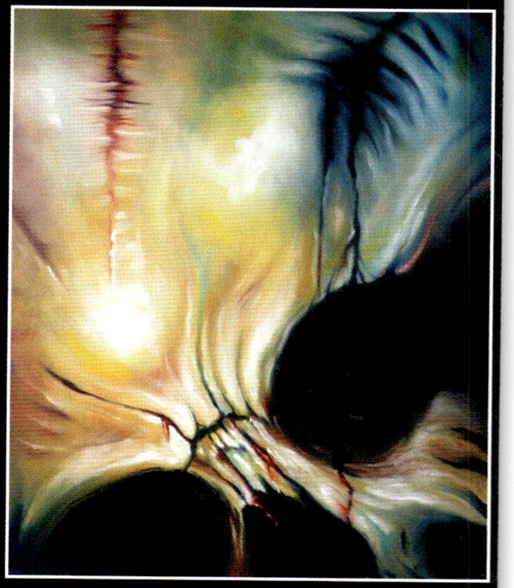

Cory Norris

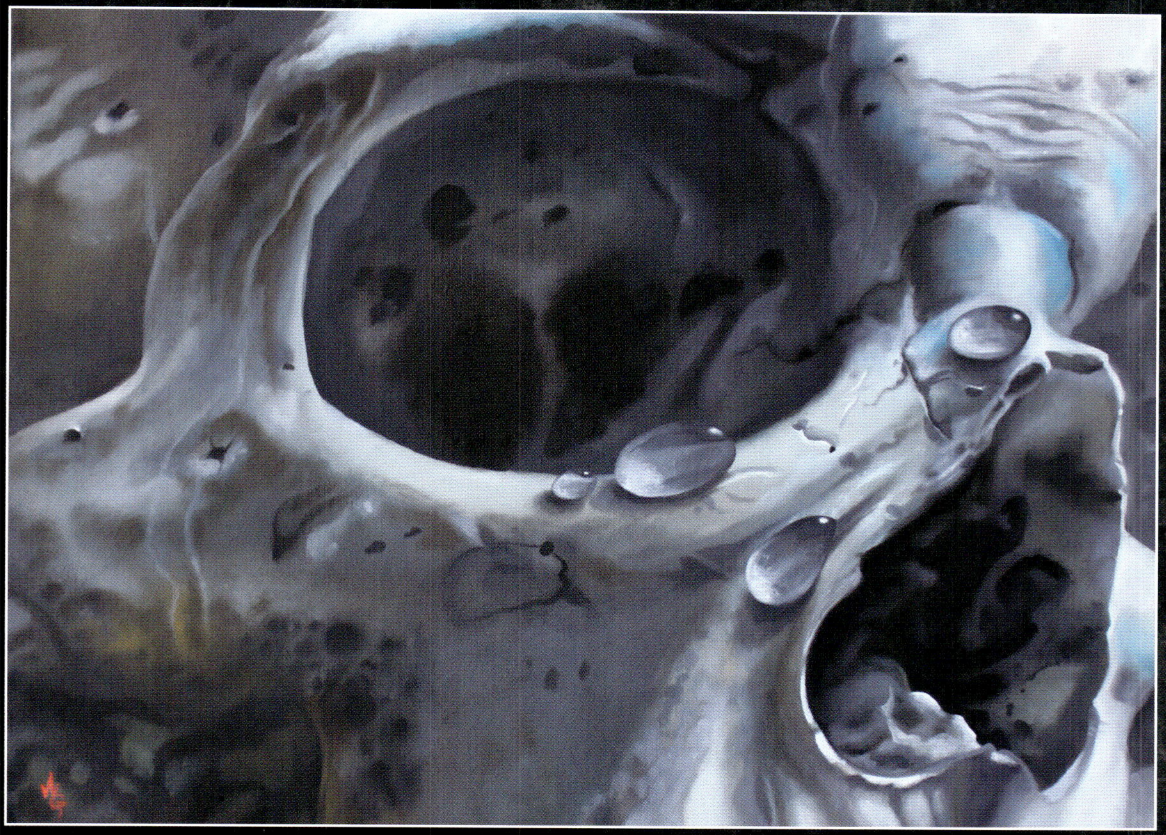

Will Gonzalez

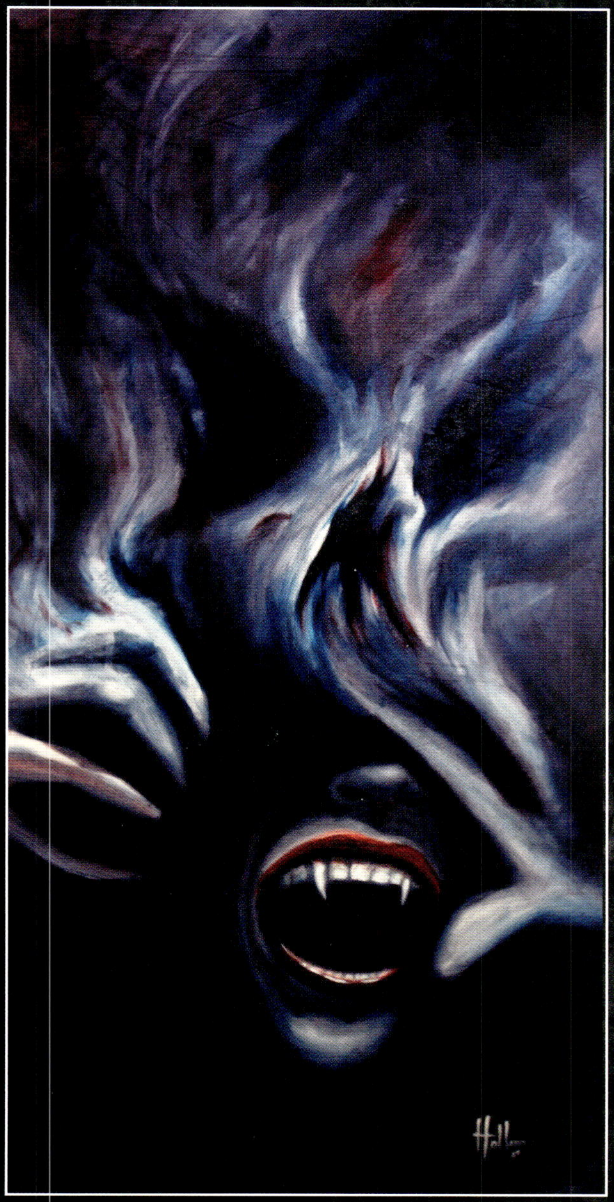

Bobby Holland

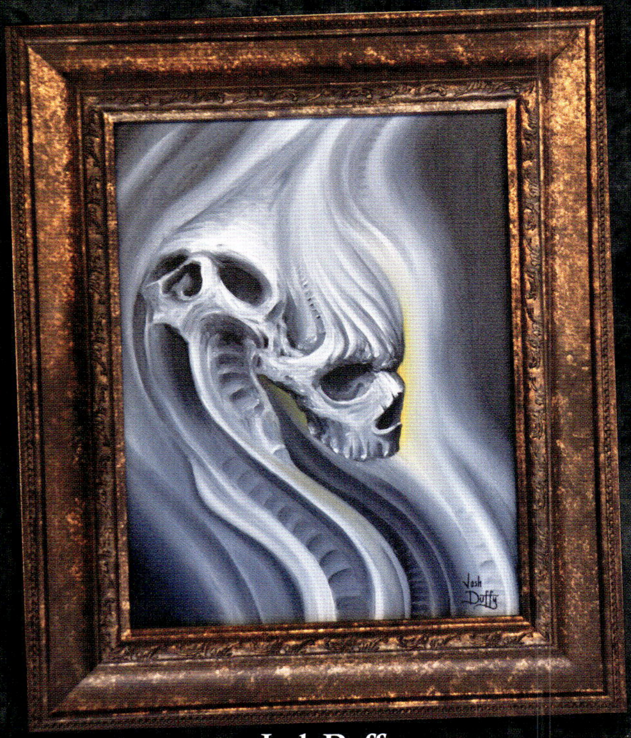

Josh Duffy

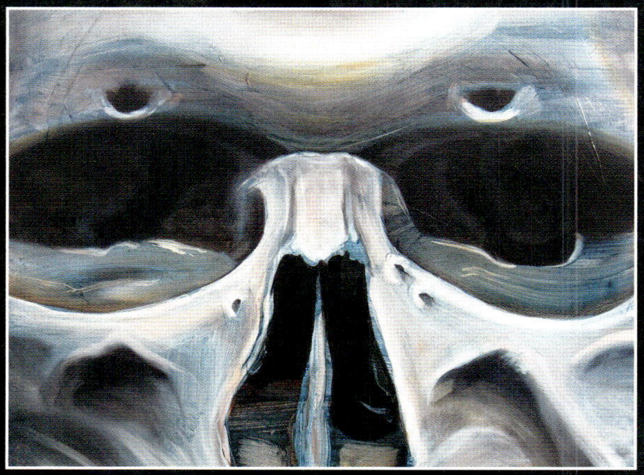

Adam James Chiodo

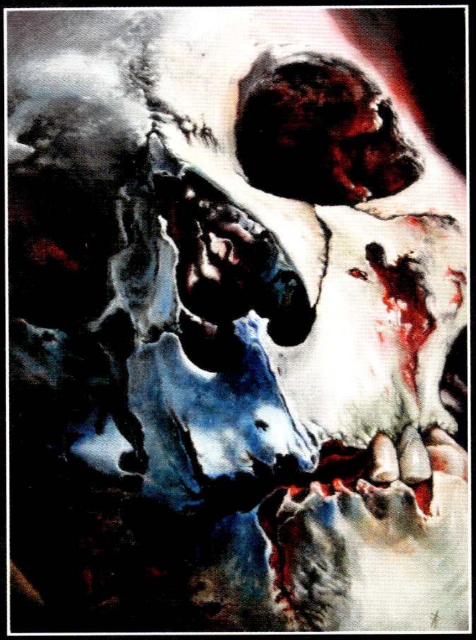
Thomas Kynst

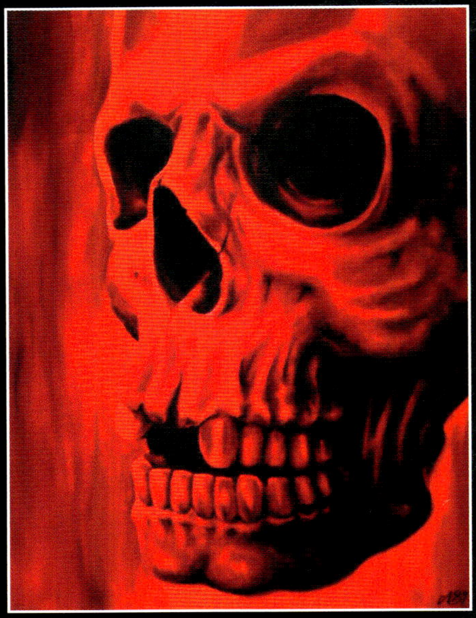
Nadine Bryant

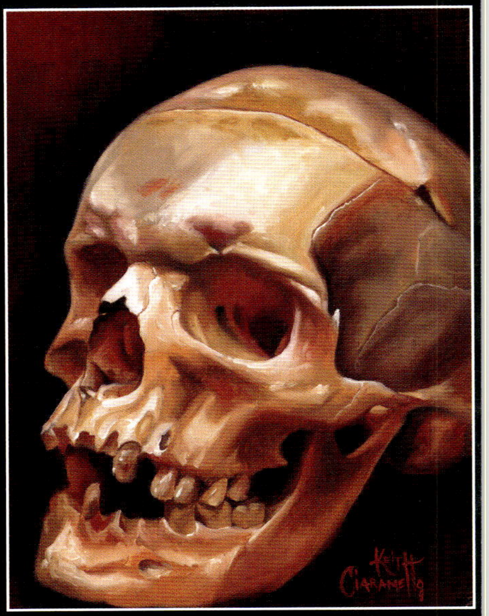
Keith Ciaramello

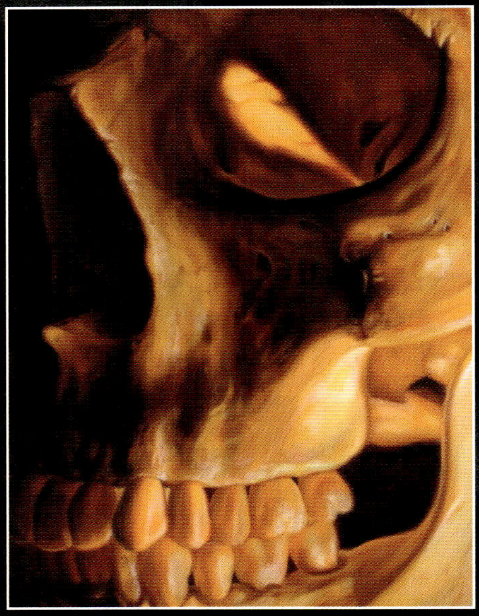
Jeffrey Srsic

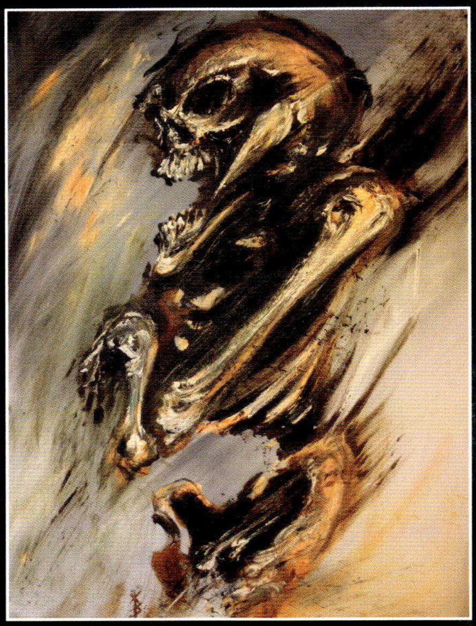
Kari Barba

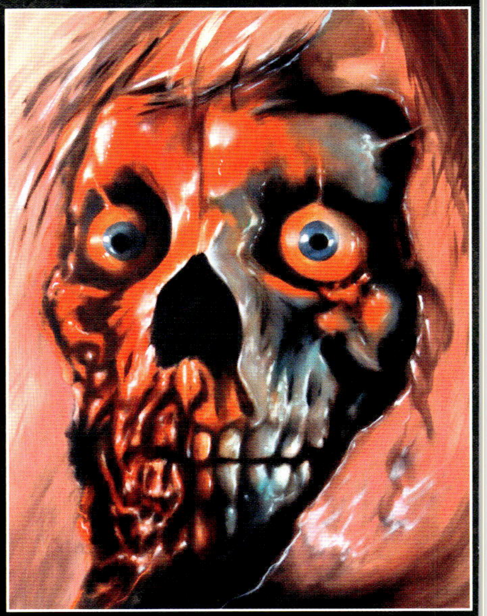
Matthew Lukesh

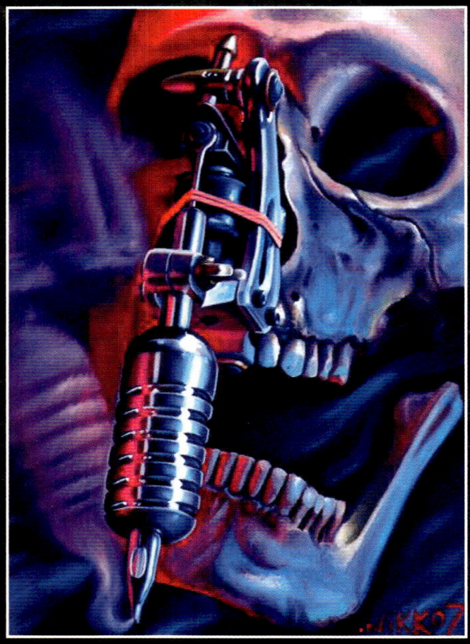
Nikko Hurtado

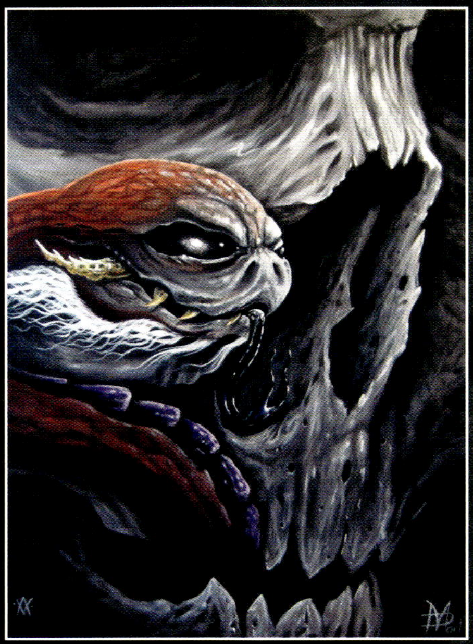
Mathew Delamort & Will Dose Eckies

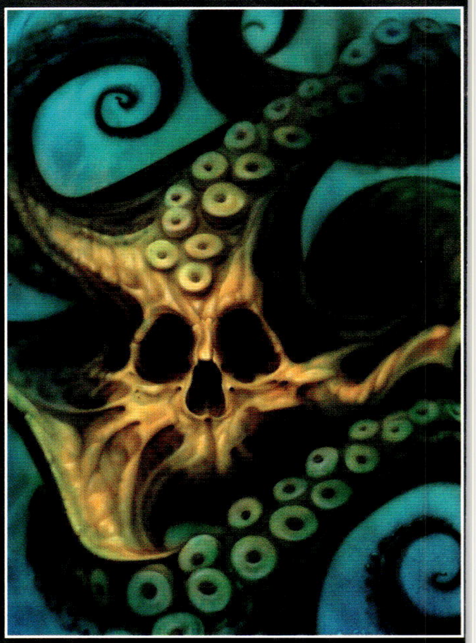
Will Ralston

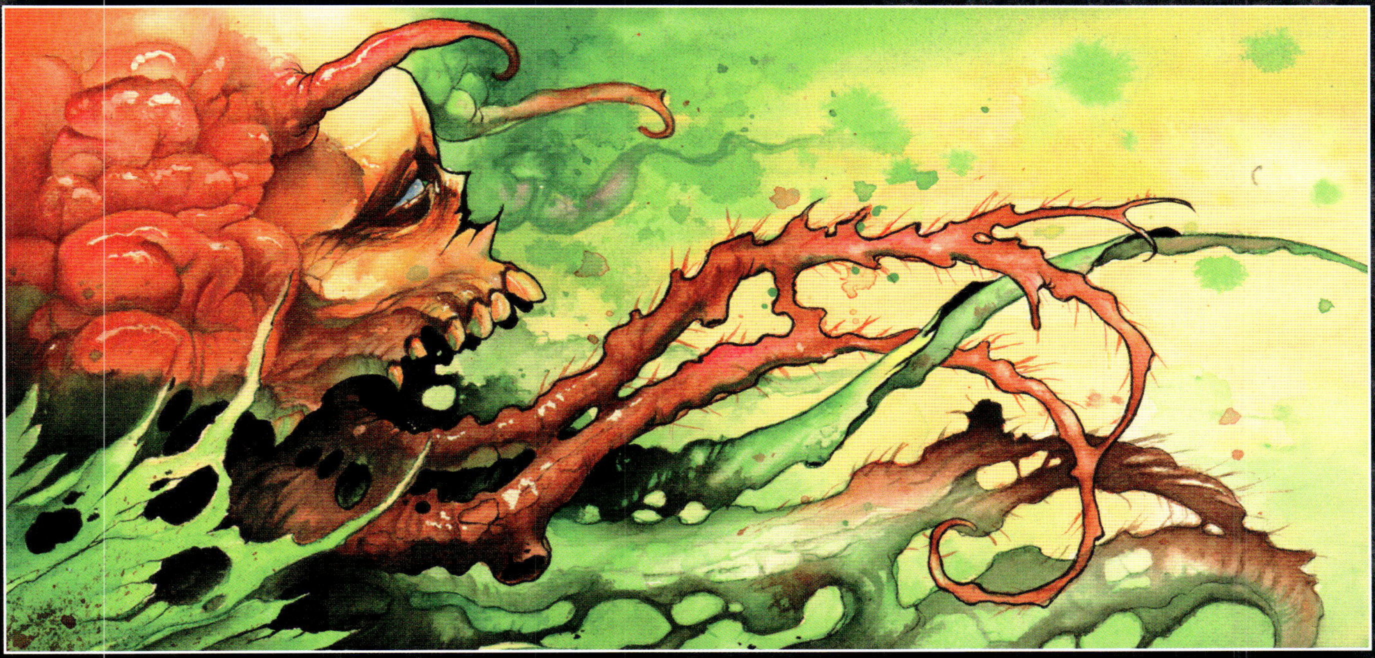

Tom Strom

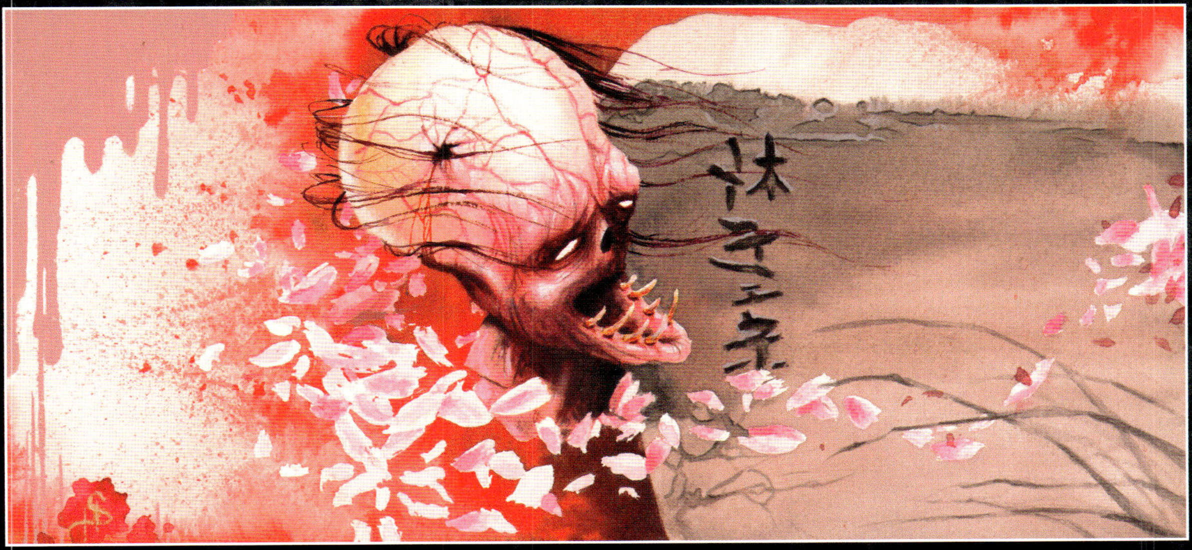

Tom Strom

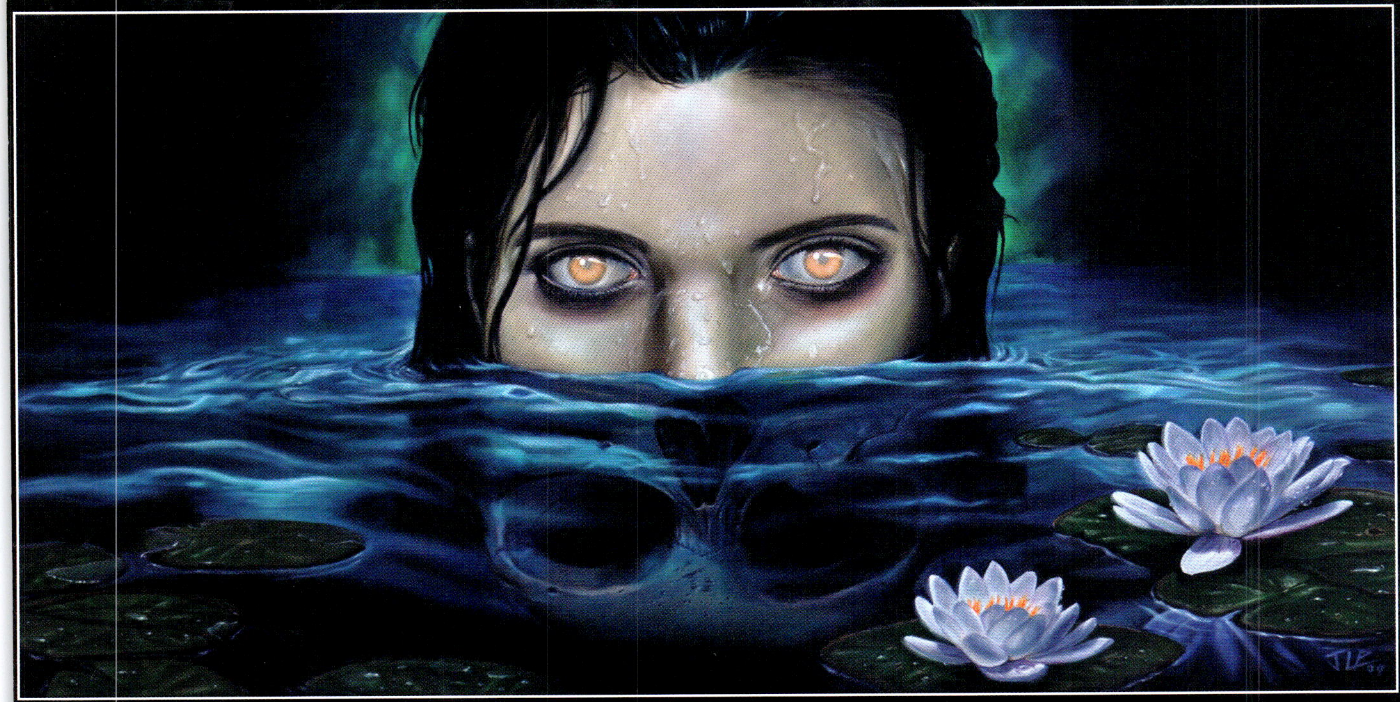

Jamie Lee Parker

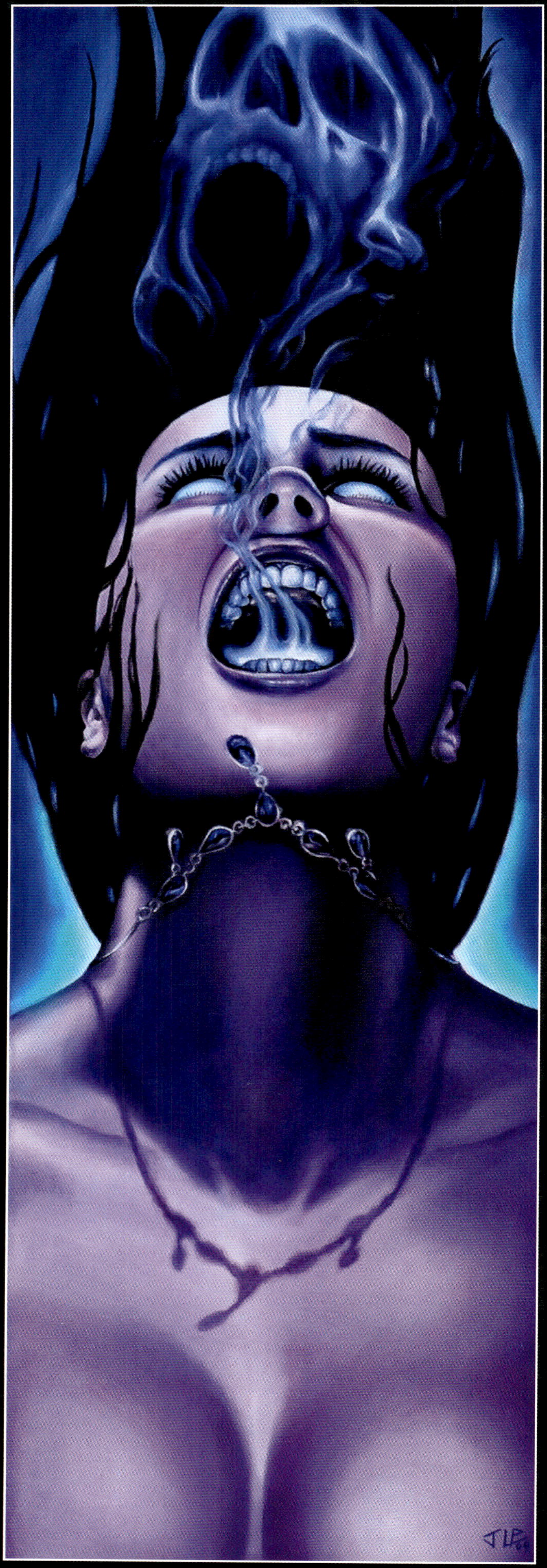

Jamie Lee Parker

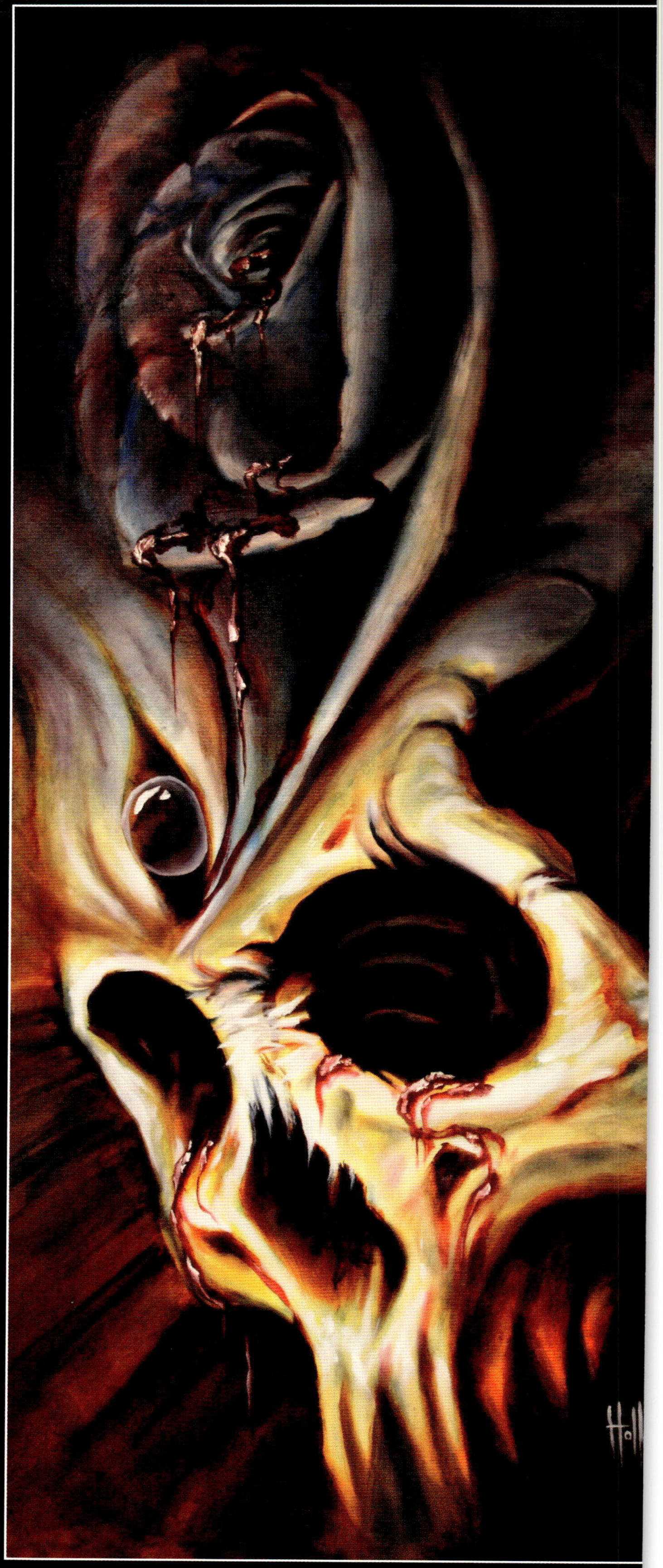

Bobby Holland

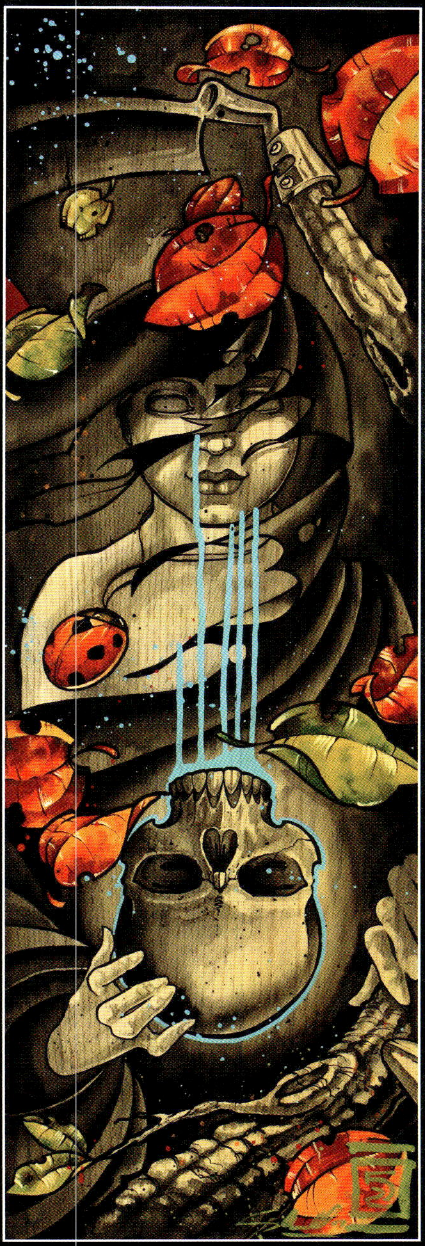

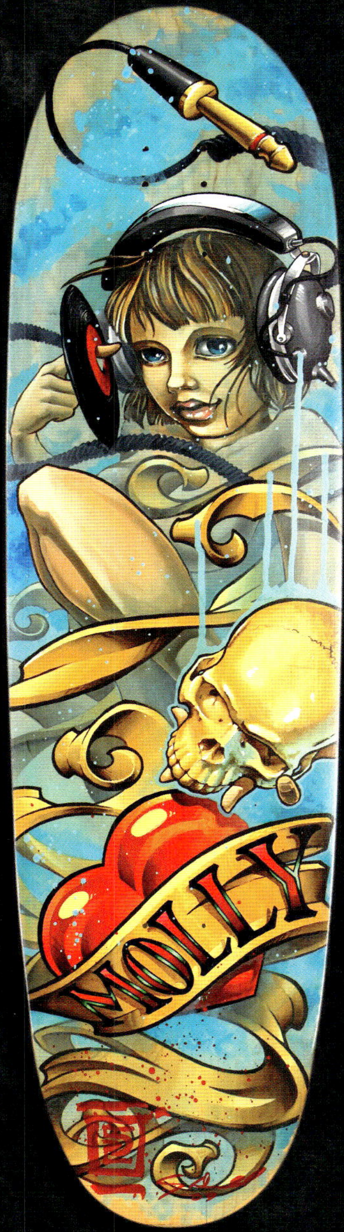

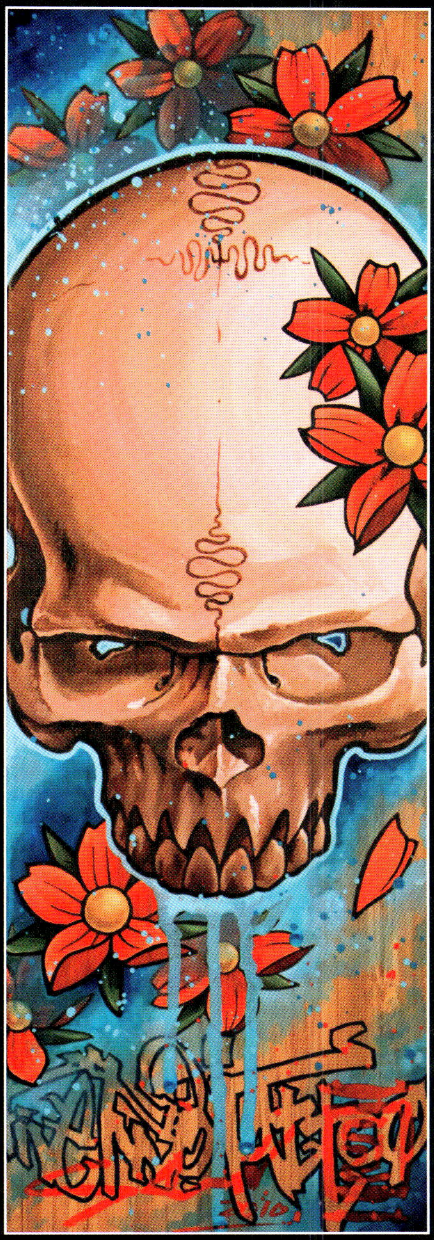

Craig Driscoll

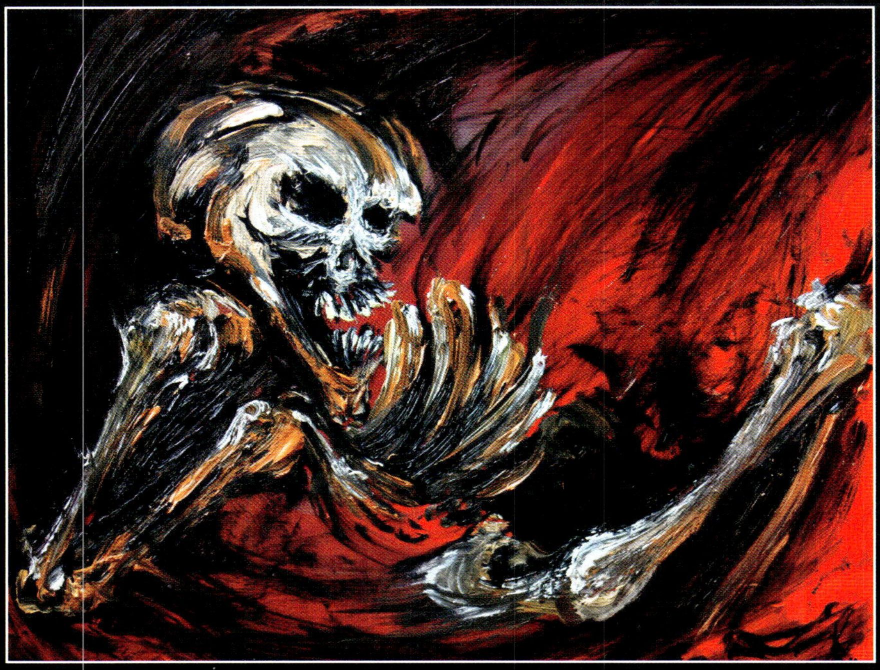

Kari Barba

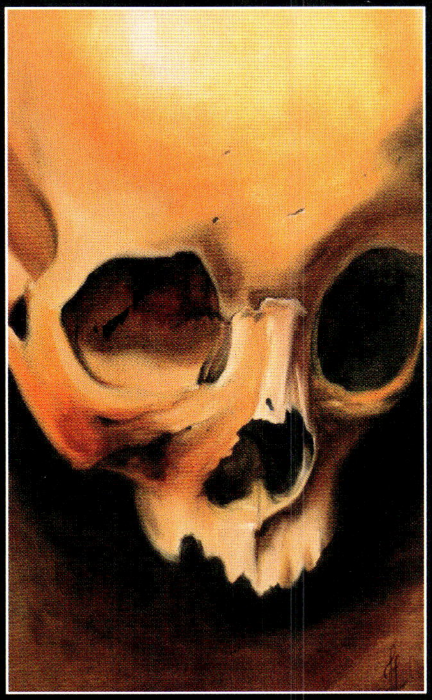

Jesso

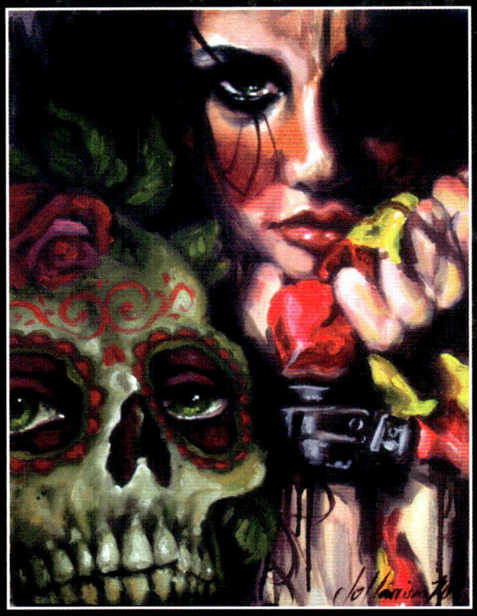

Jo Harrison

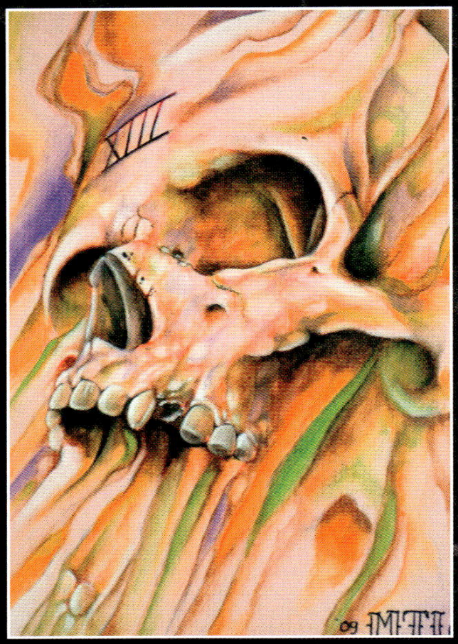

Marcus Lund

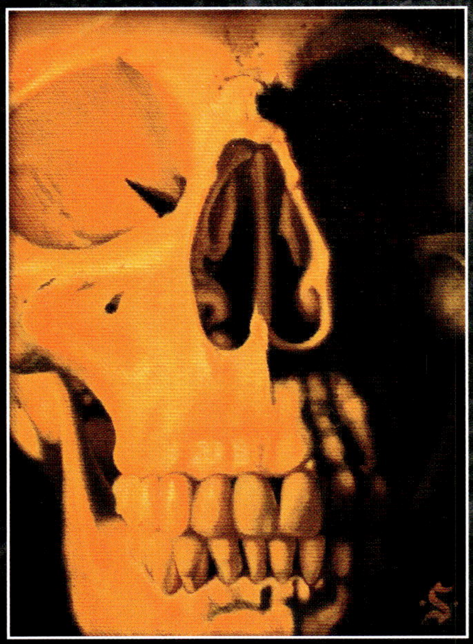

Jeffrey Srsic

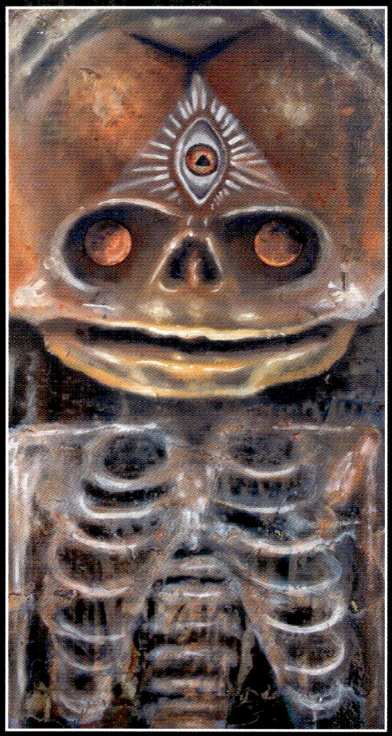

Dave Barton

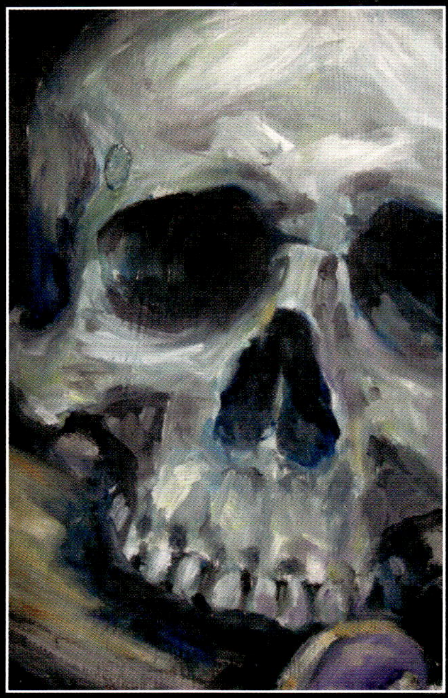

Chuck Day

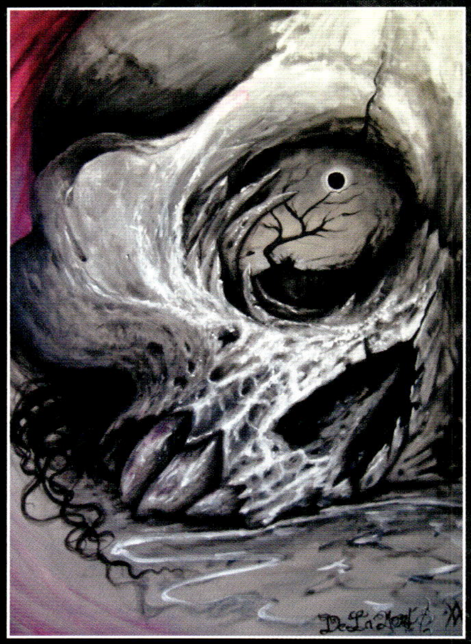

Mathew Delamort &
Will Dose Eckies

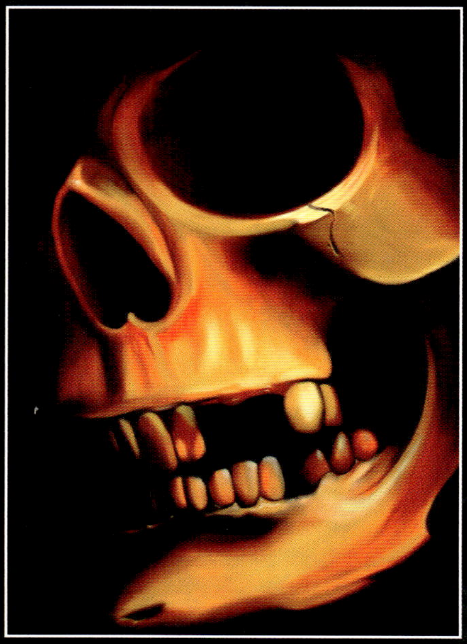

Aric Taylor

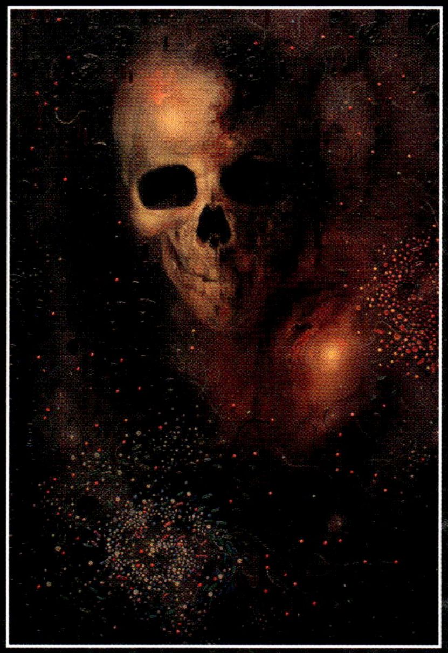

Ewelina Ferruso

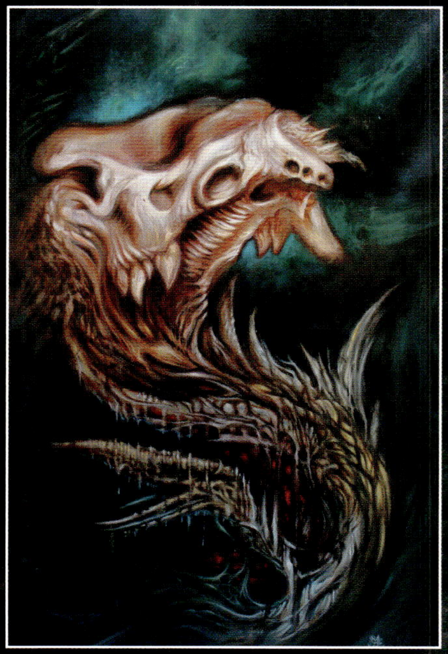

Matt Zumberge

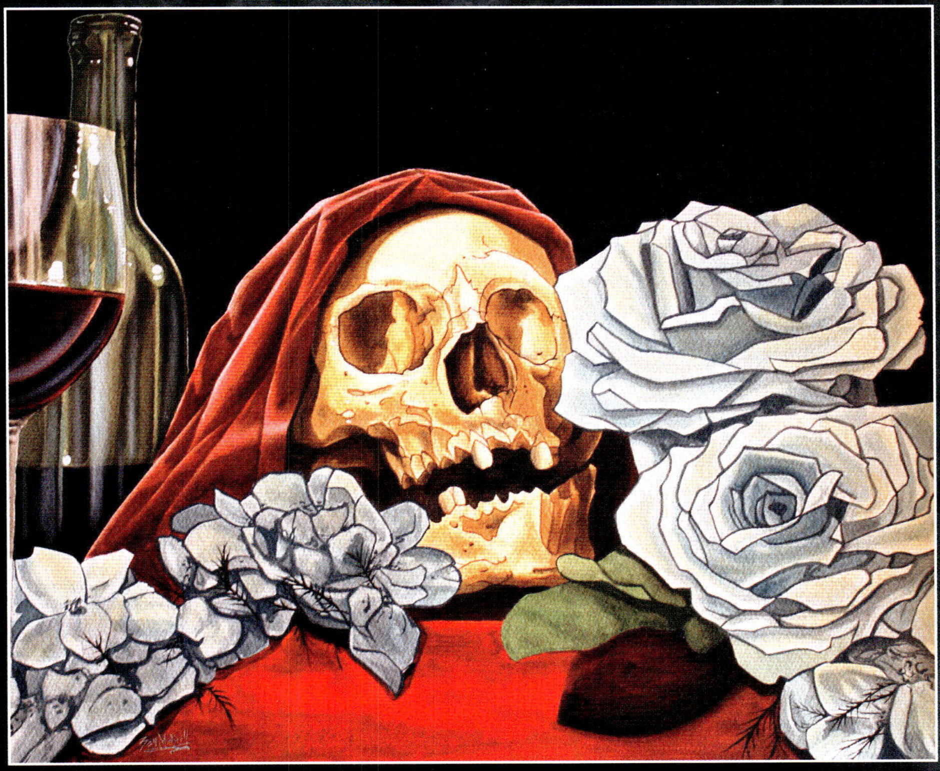

Ben Merrel

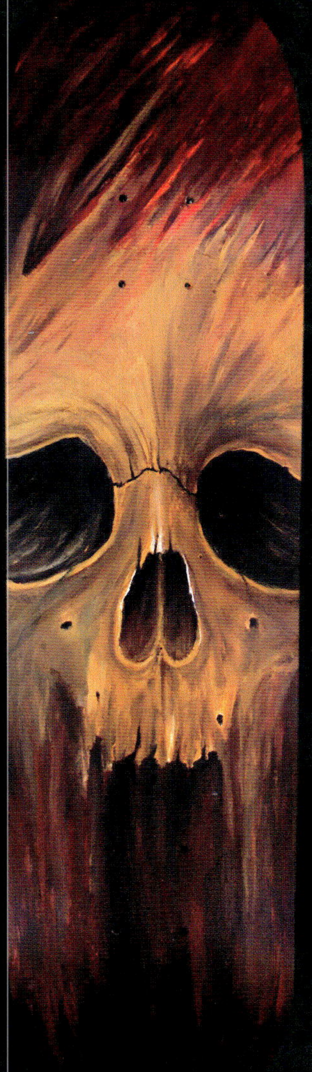

John Graefe

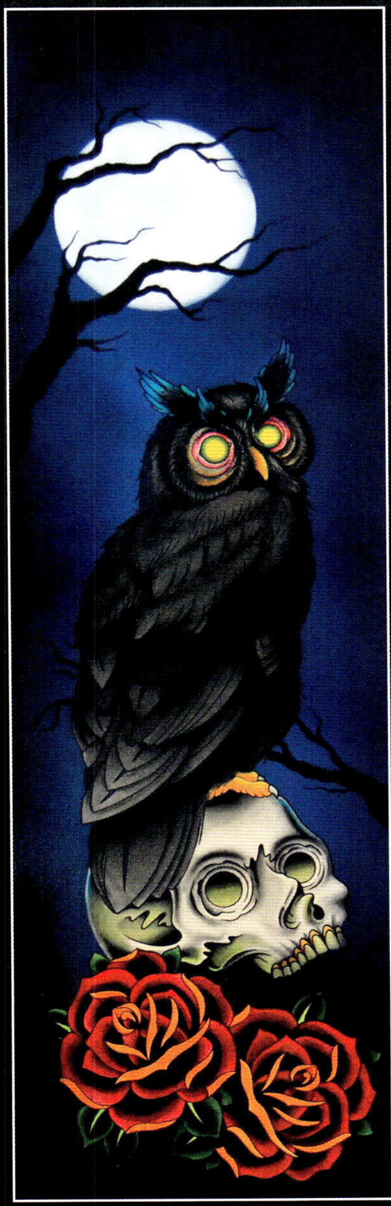

Jim Sylvia

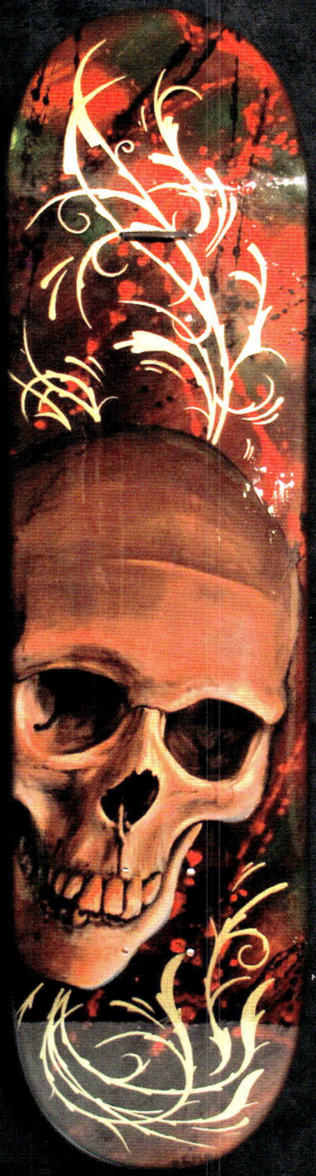

Chris Blinston

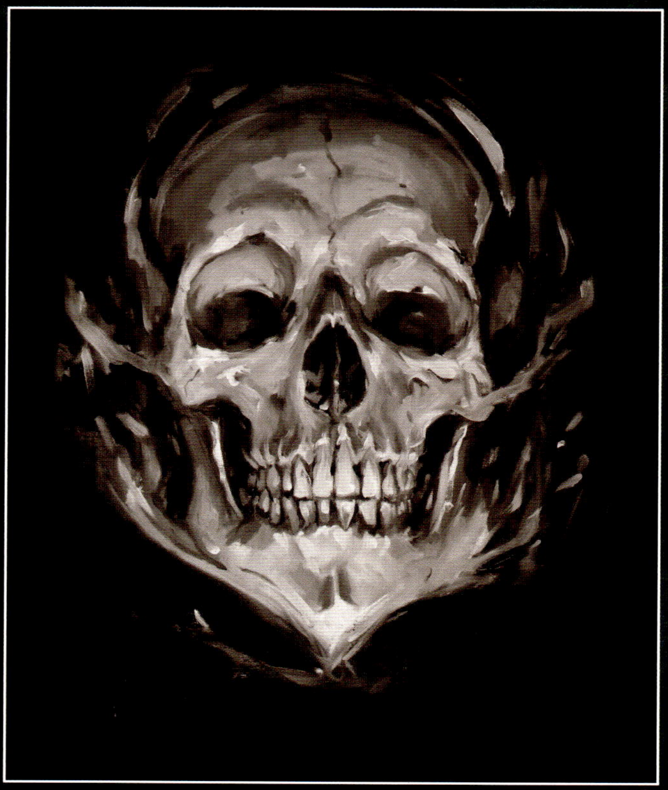

Ryan Smith & Shawn Barber

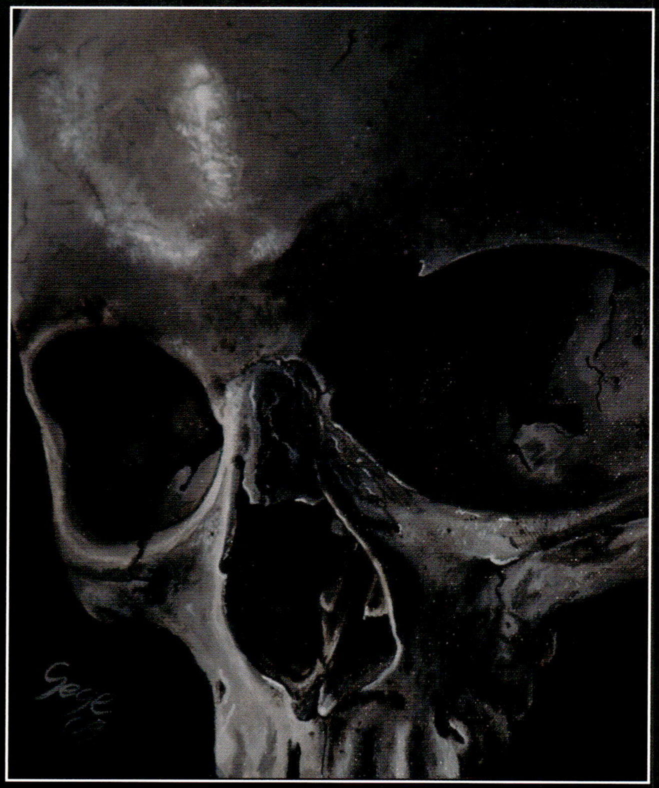

Gege

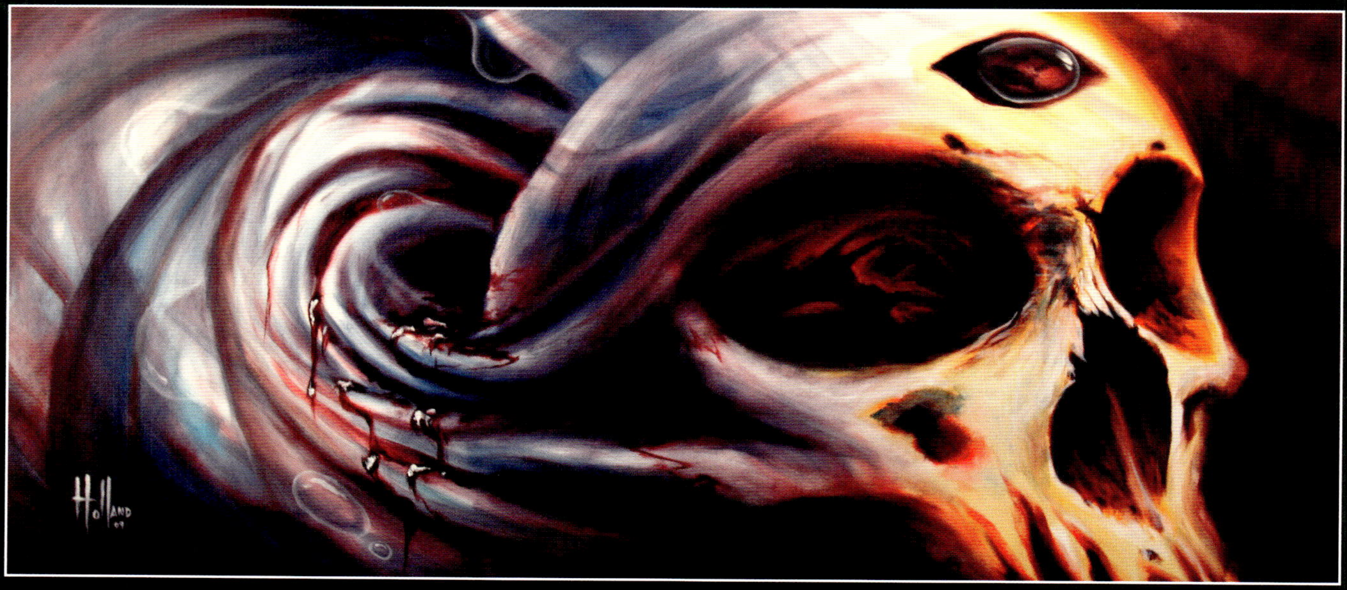

Bobby Holland

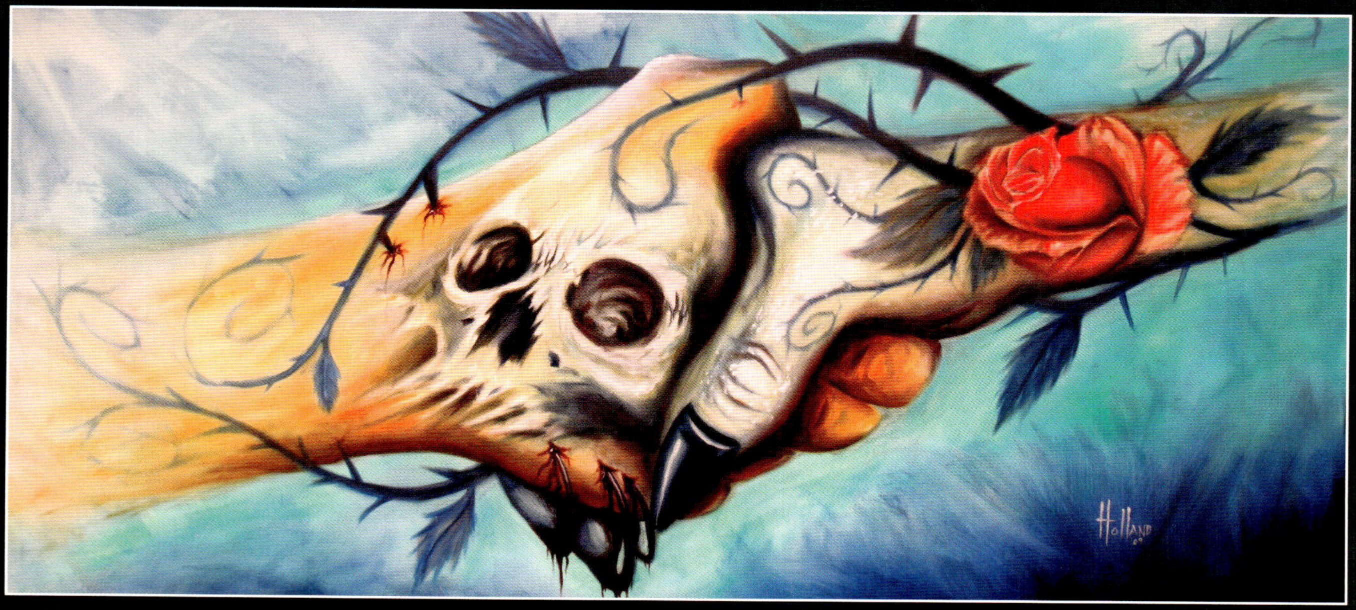

Bobby Holland

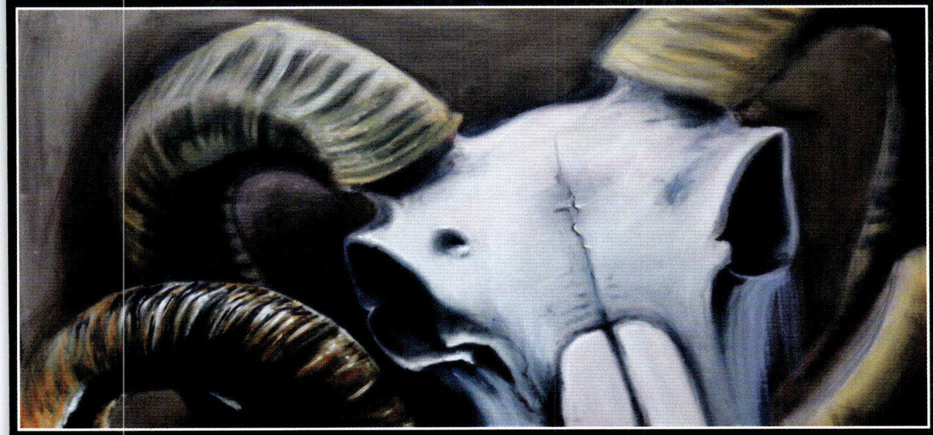

Ty McEwen

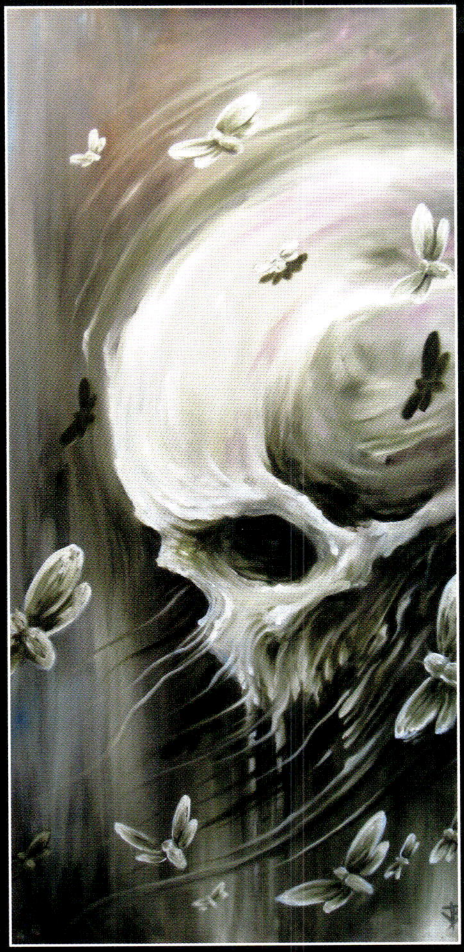

Jeremiah Barba

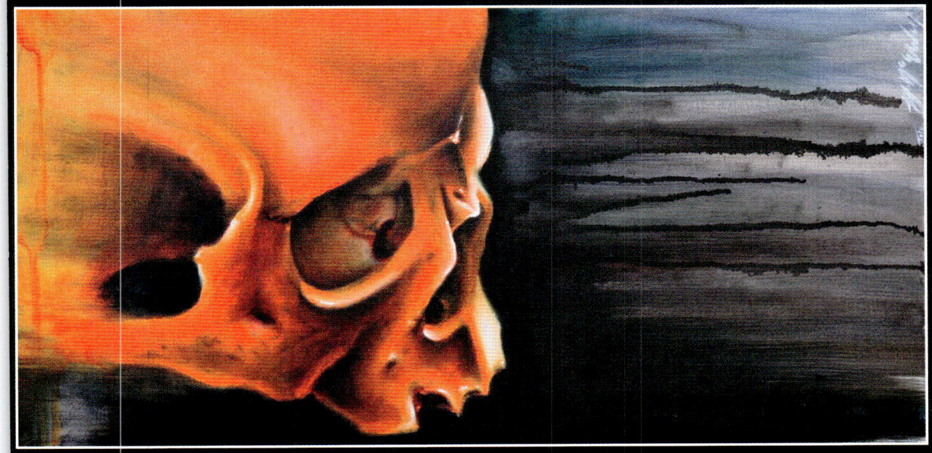

Ty McEwen

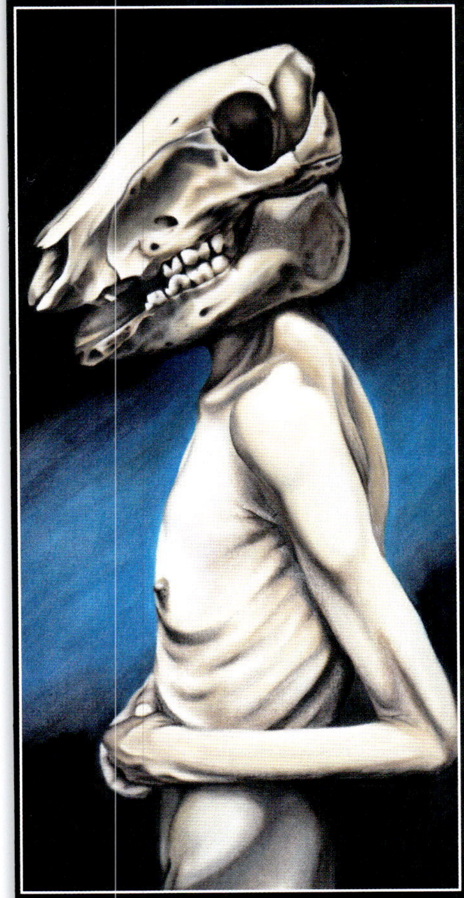

Matthew Lukesh

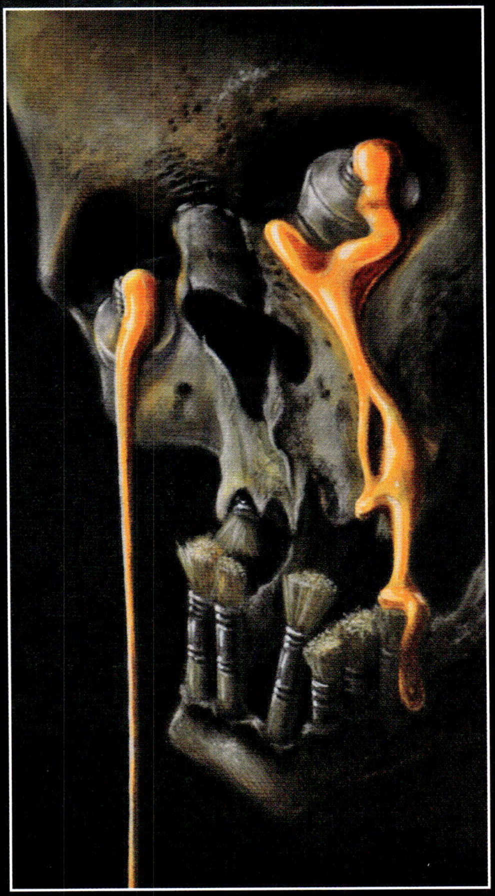

Alexis Vaatete

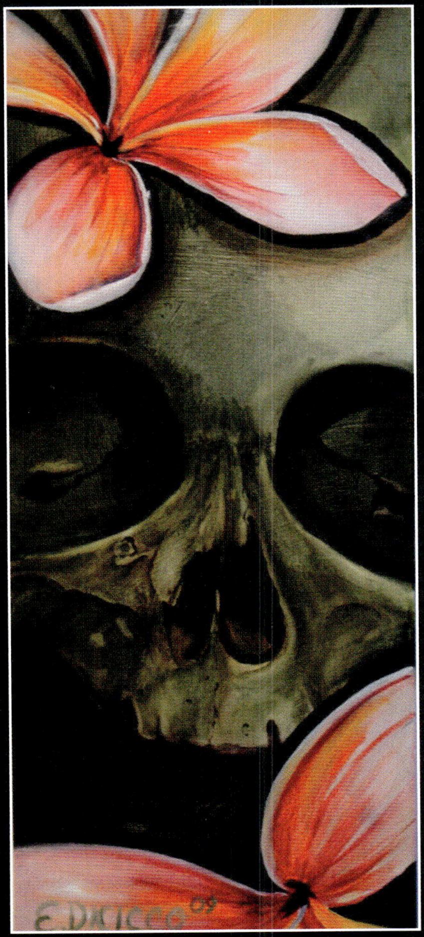

Eric DiCicco

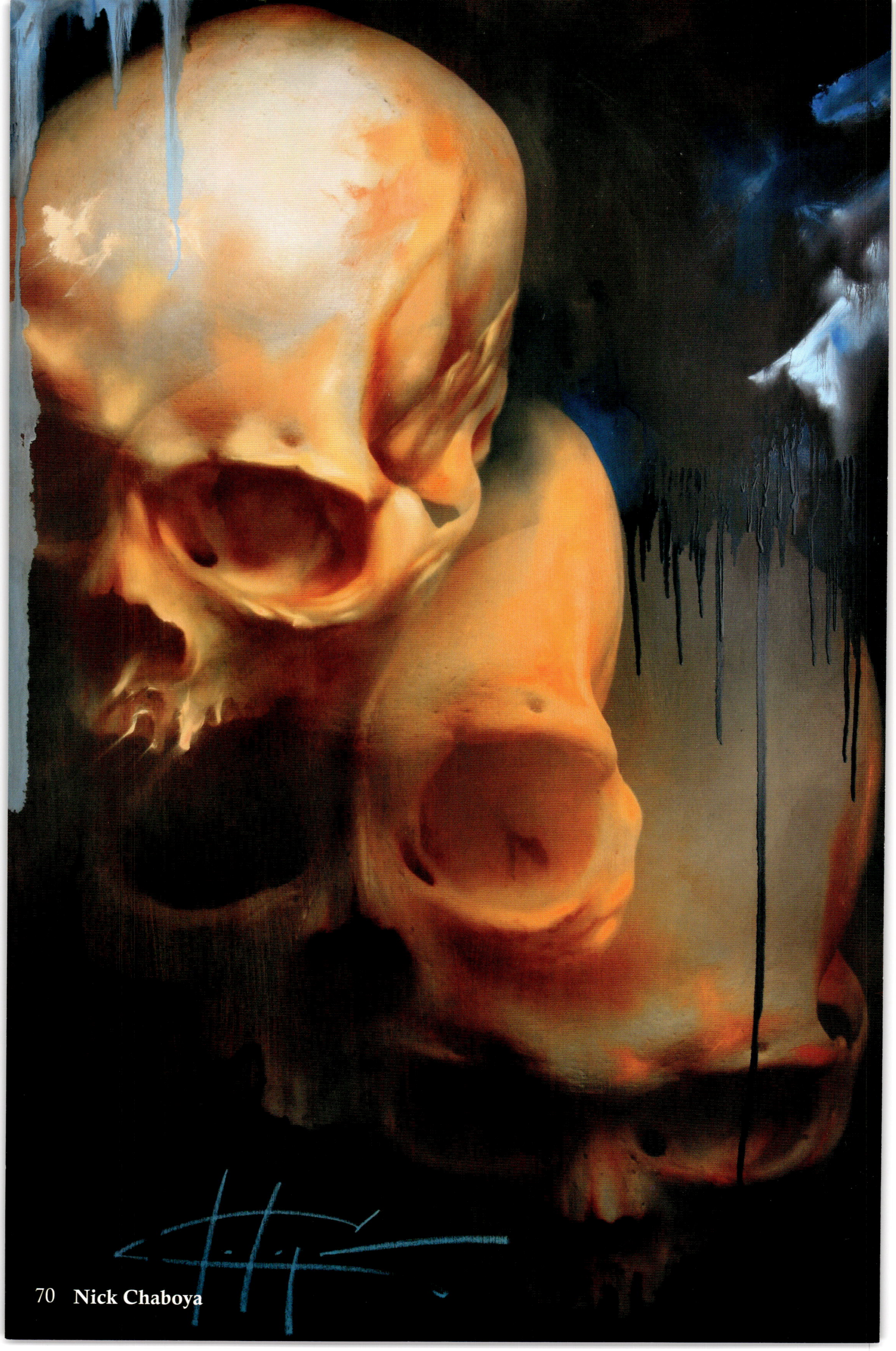

 Nick Chaboya

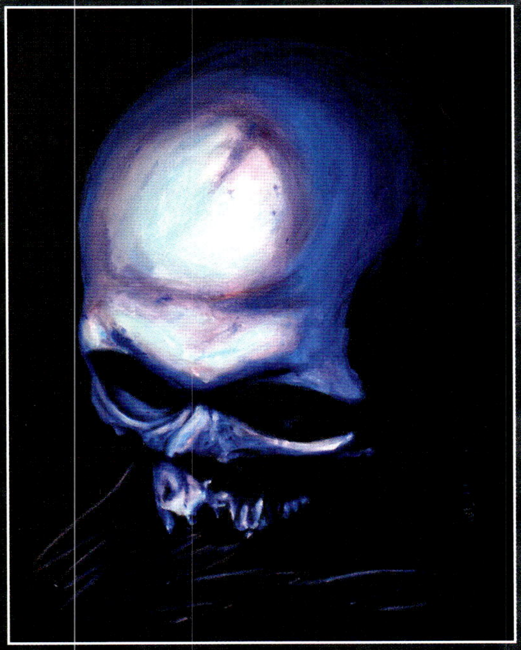

Tommy Coon

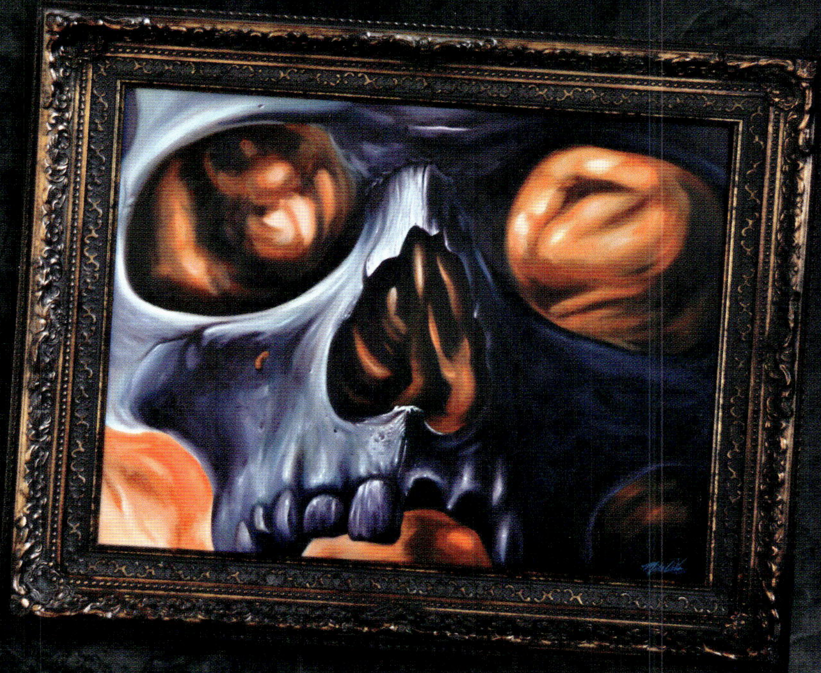

Mike DeVries

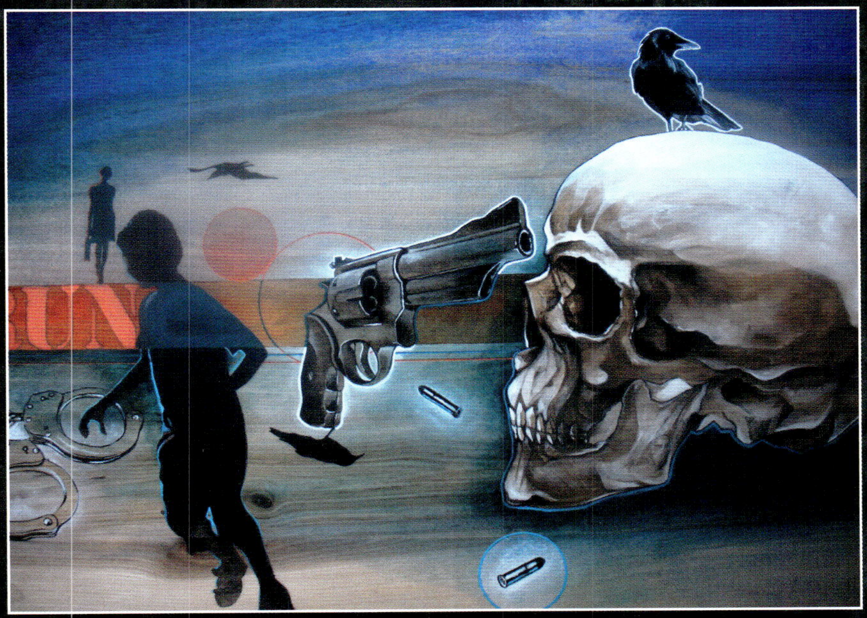

Alvin Chong

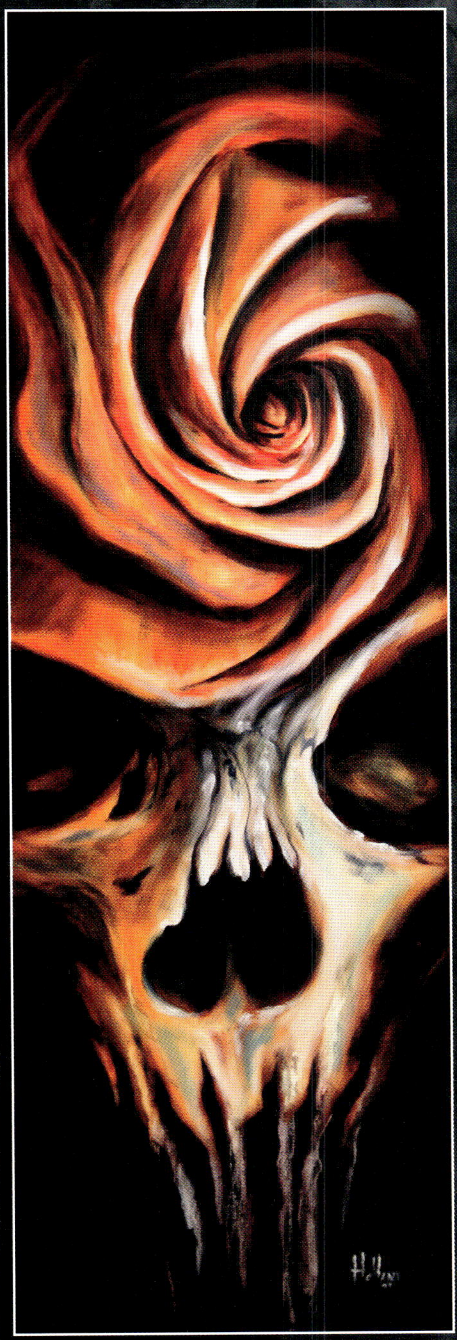

Bobby Holland

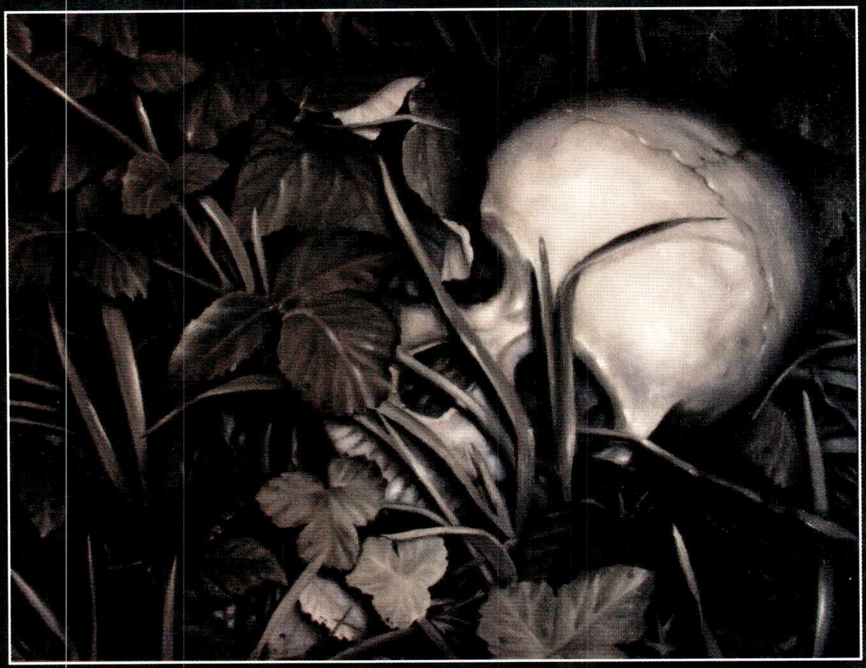

Cory Norris

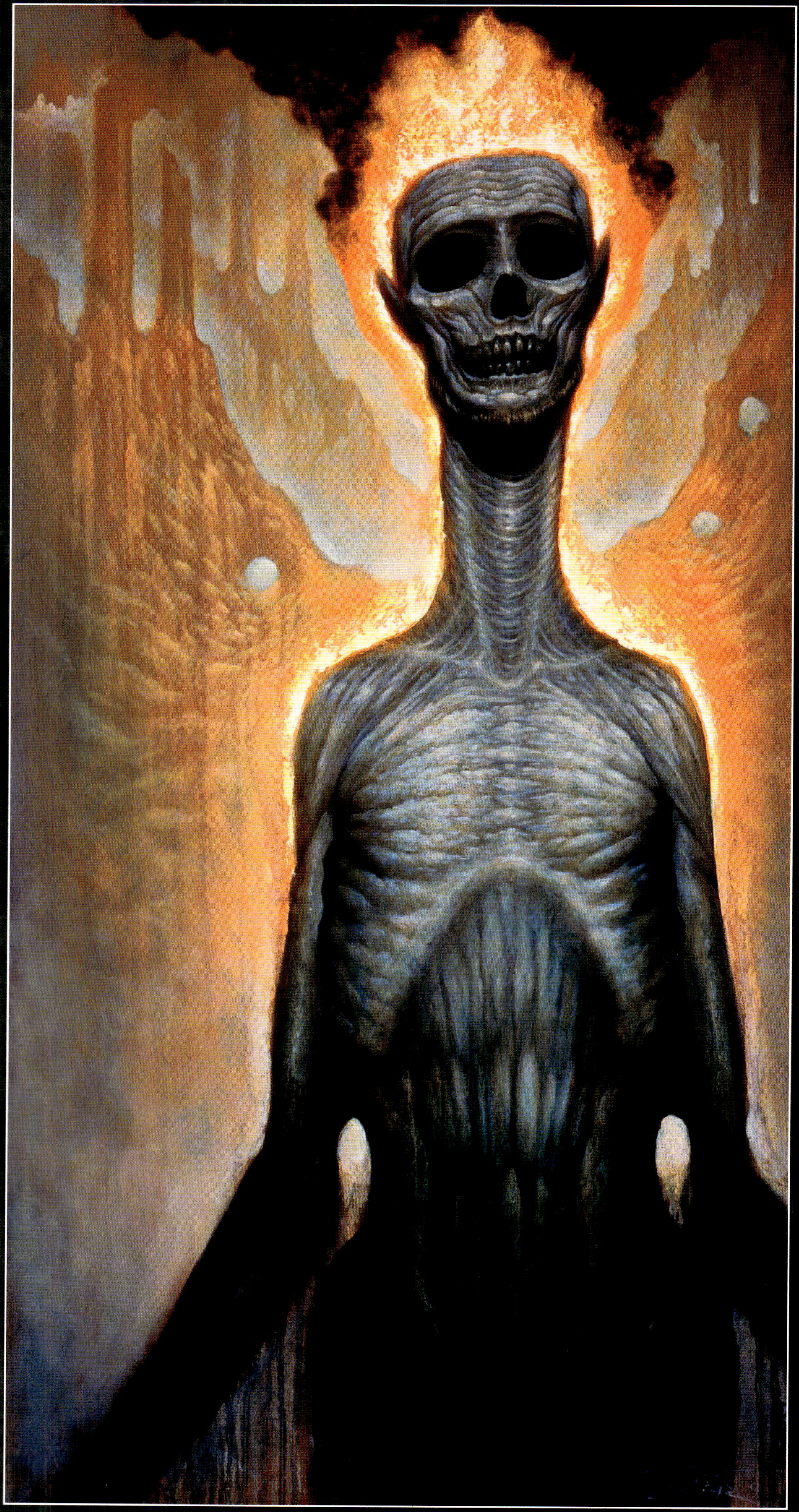

Chet Zar

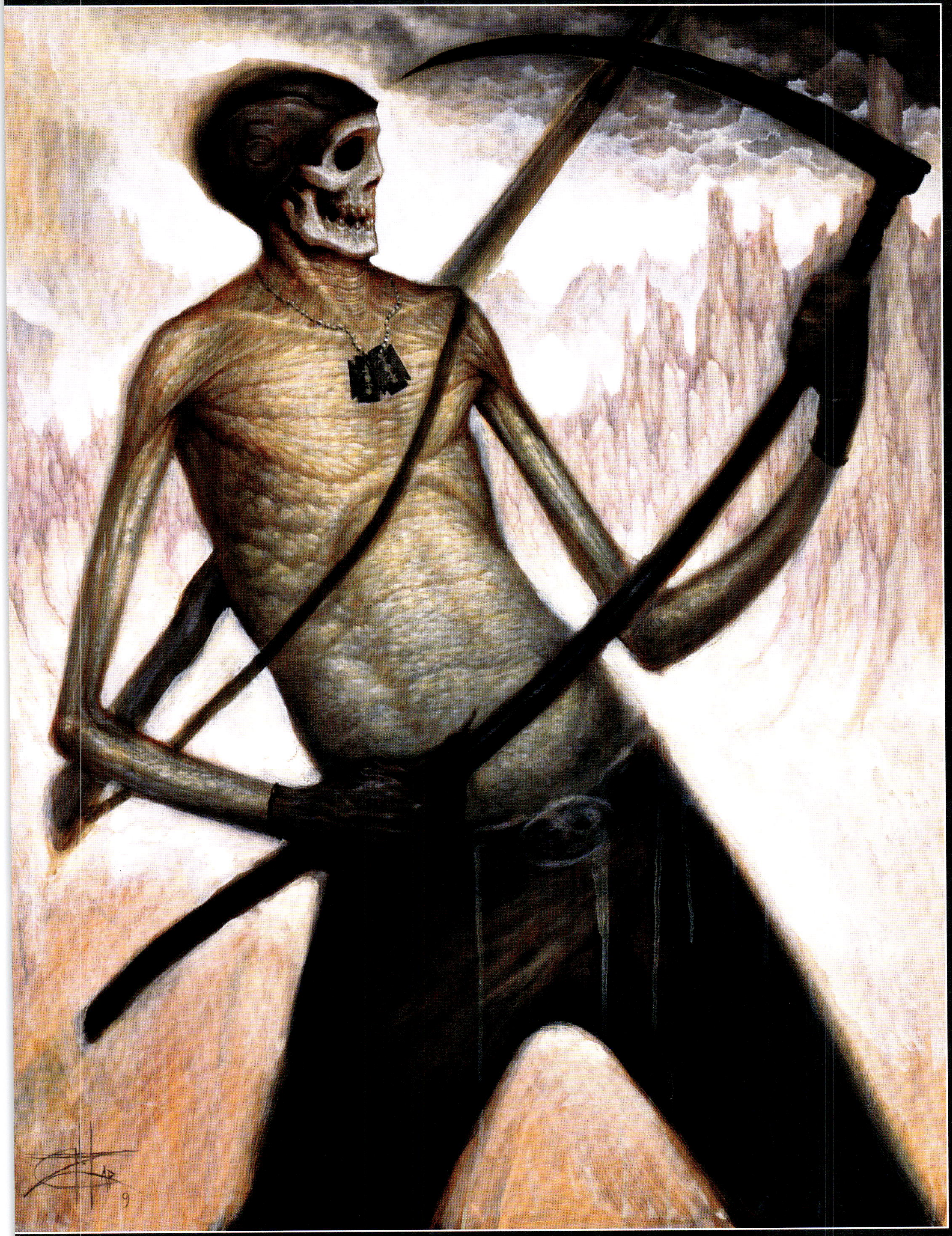

Chet Zar

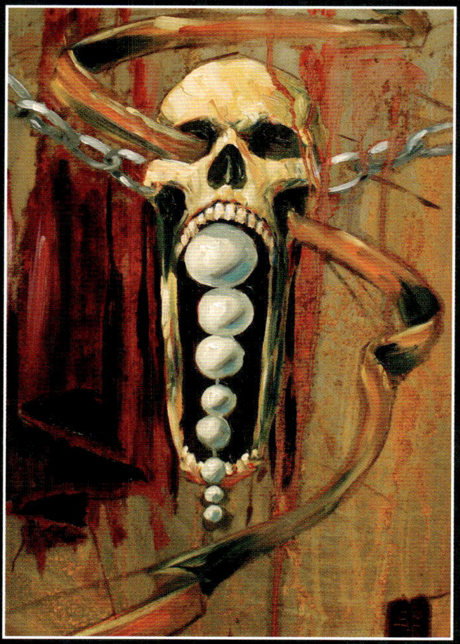

Scott Glazier

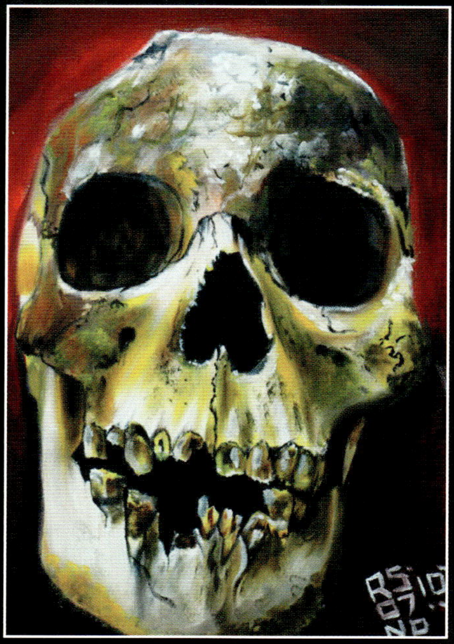

Ron Russo

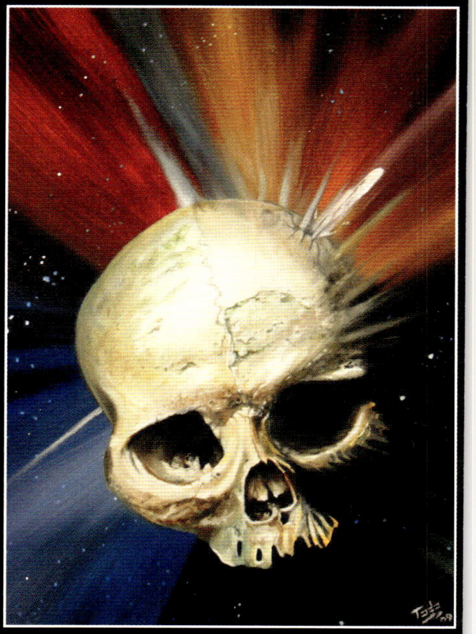

Todo

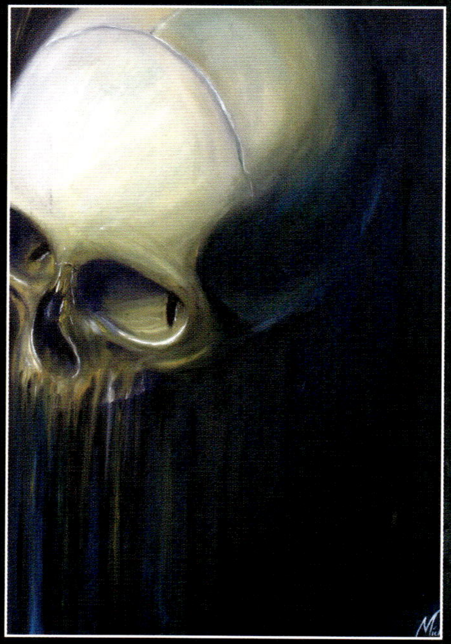

Mick Squires

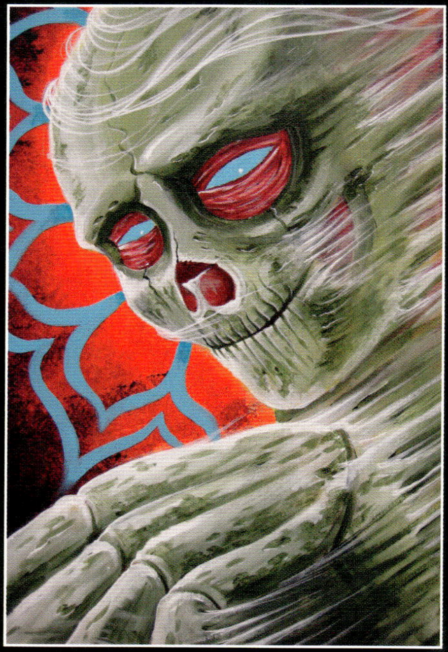

Pete Vaca

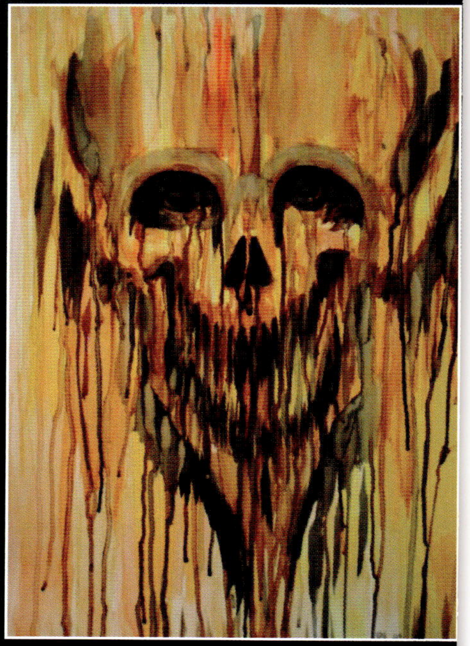

Robert Adams

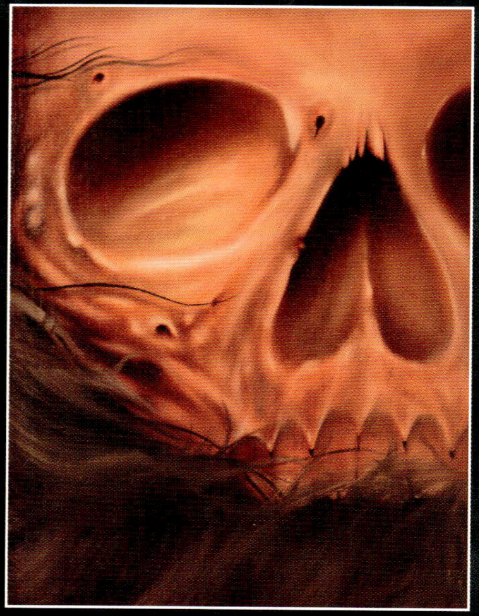

Ryan Hadley

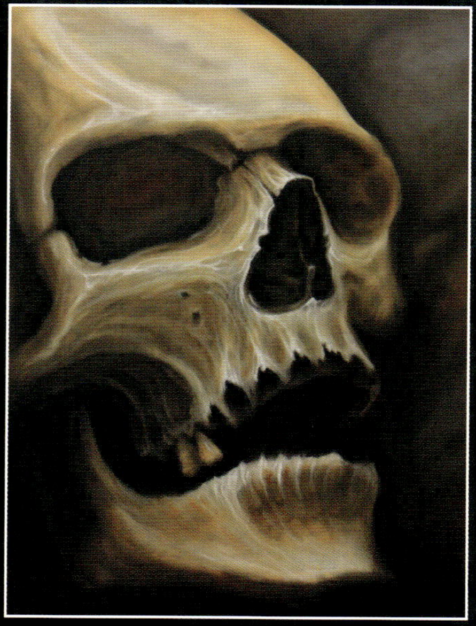

Nathan Petz

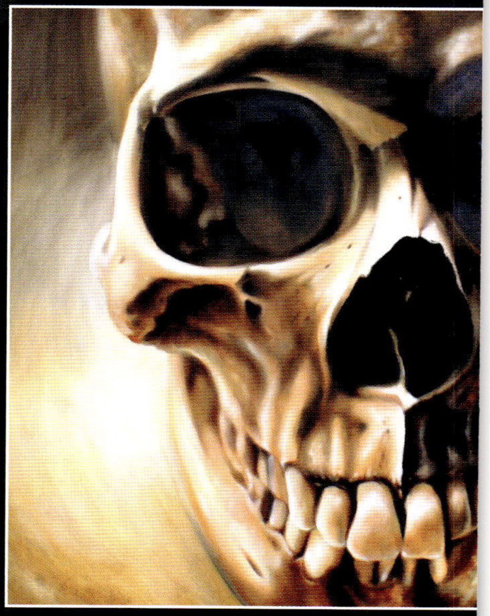

Matthew Lukesh

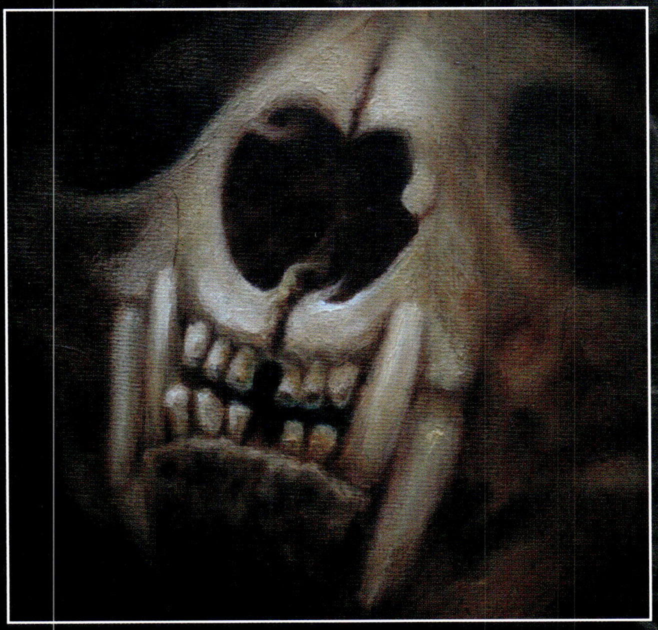

Nathan Kostechko

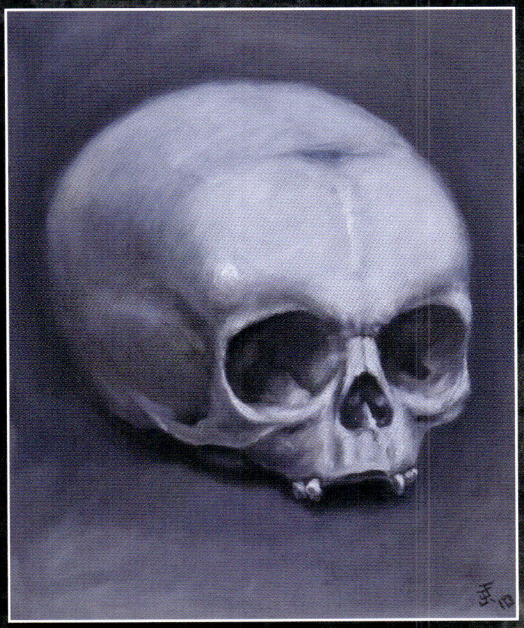

Jeff Johnson

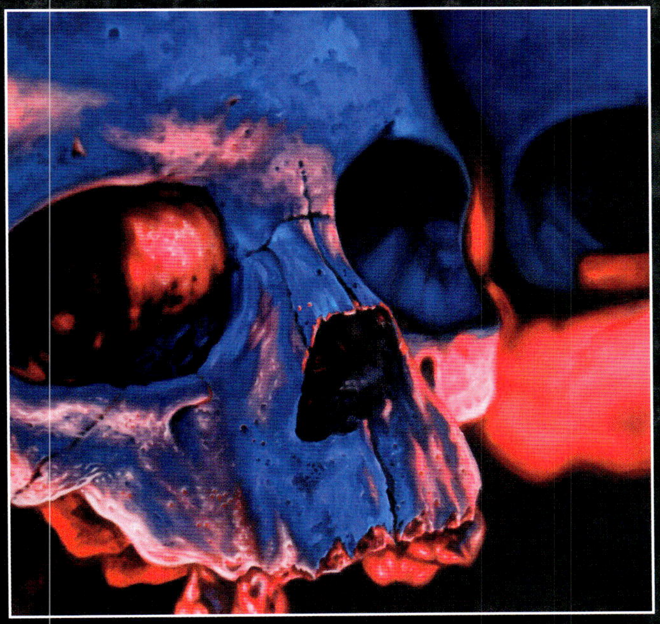

Sean Zee

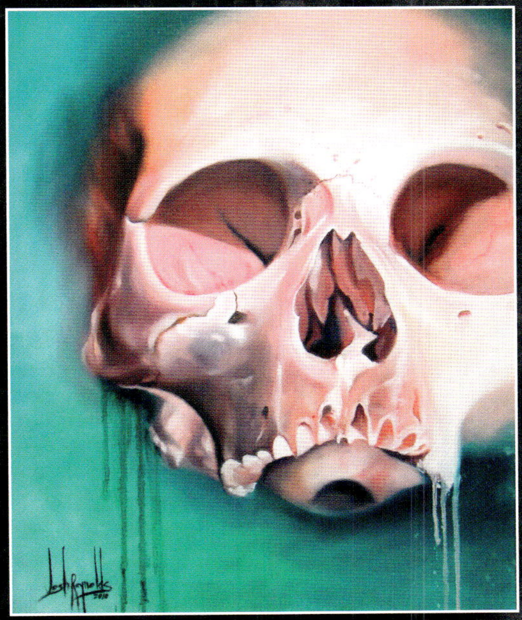

Josh Reynolds

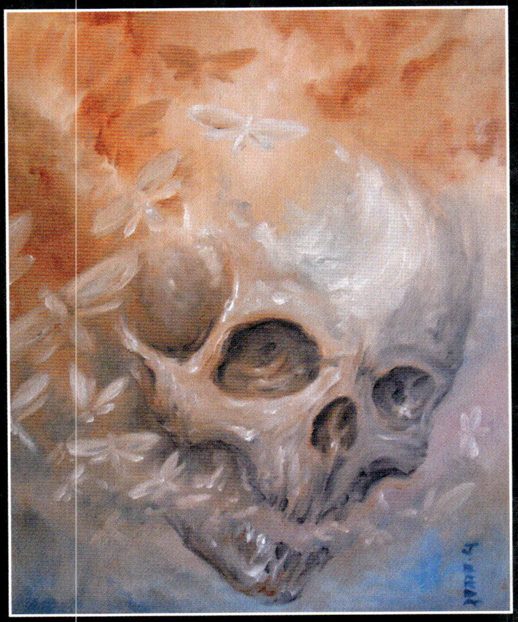

Jeremiah Barba

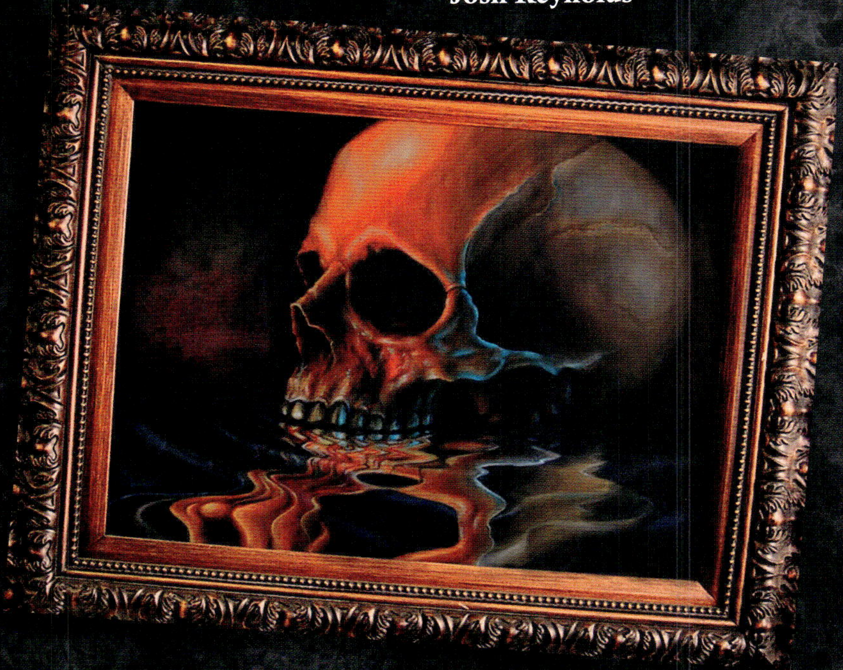

Vince Antony

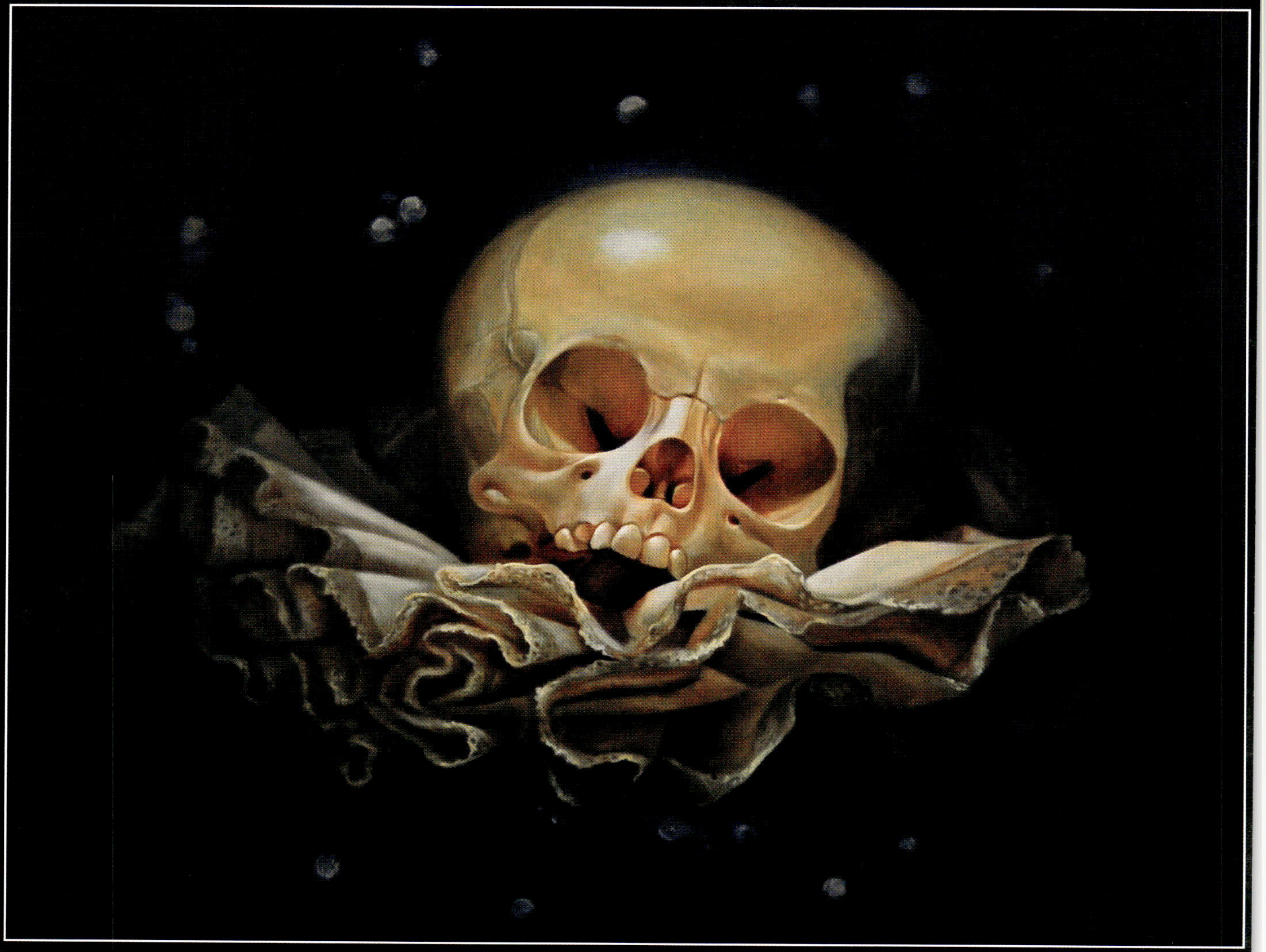

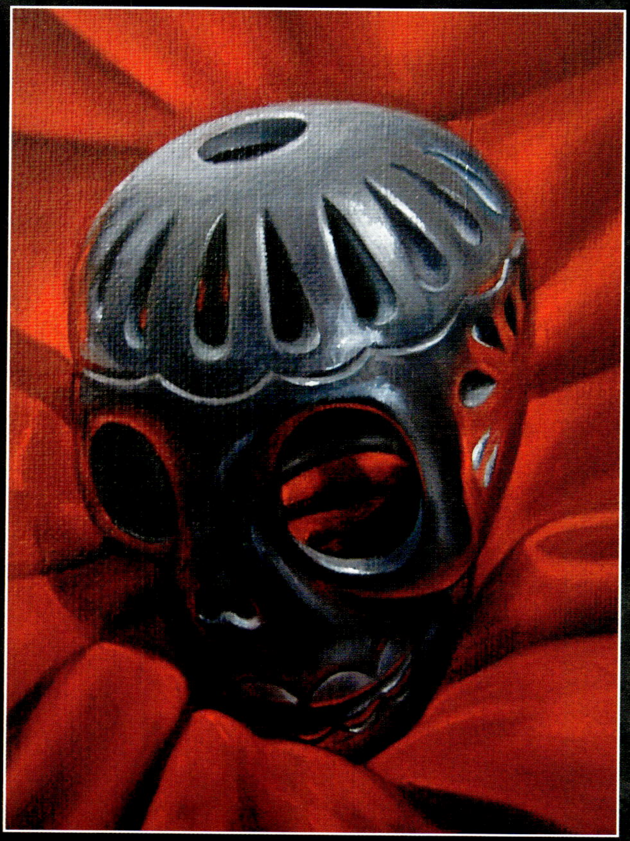

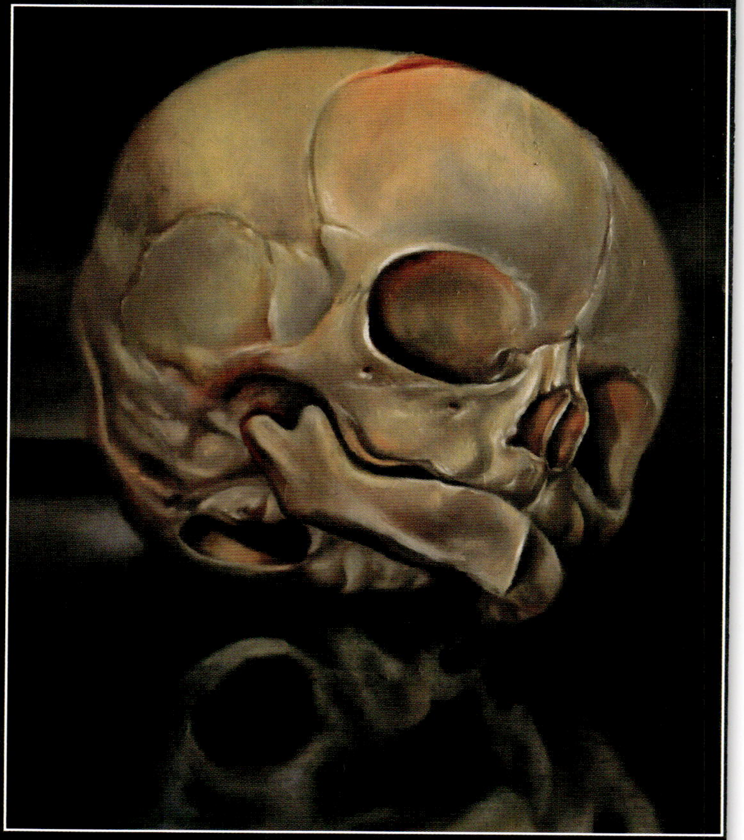

Carlos Rojas

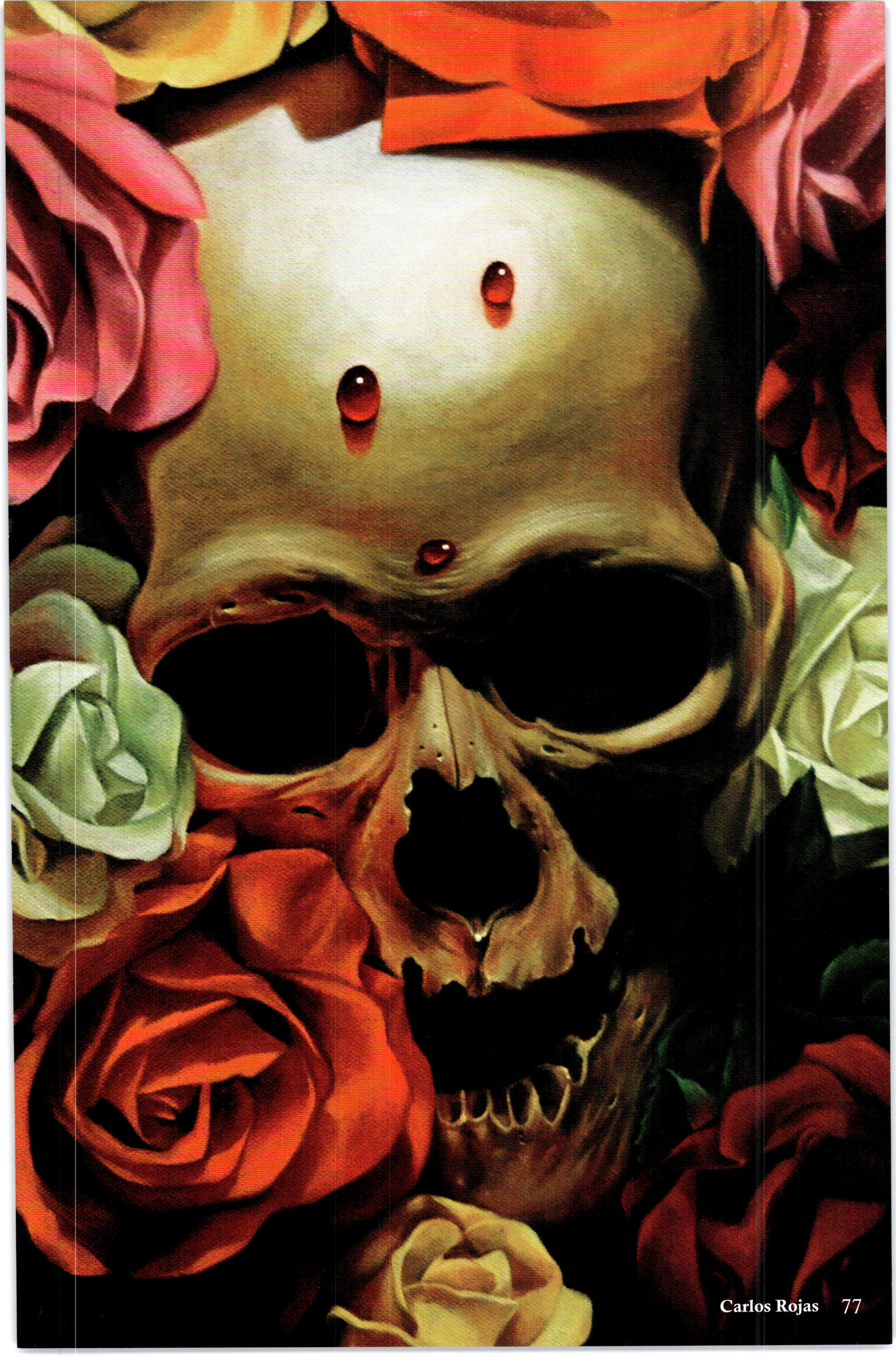

Carlos Rojas

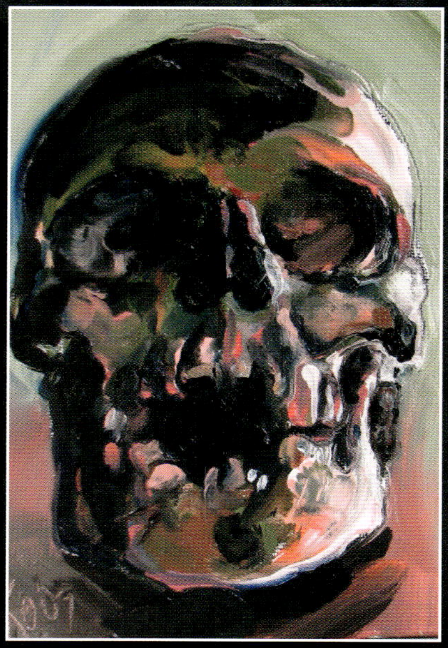

Kim Reed

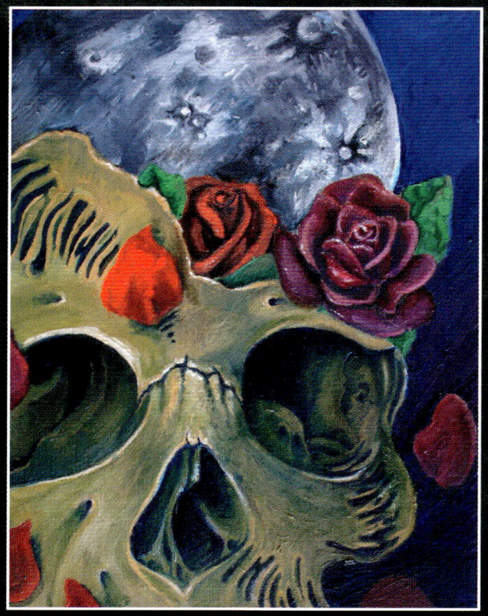

Katelyn Crane

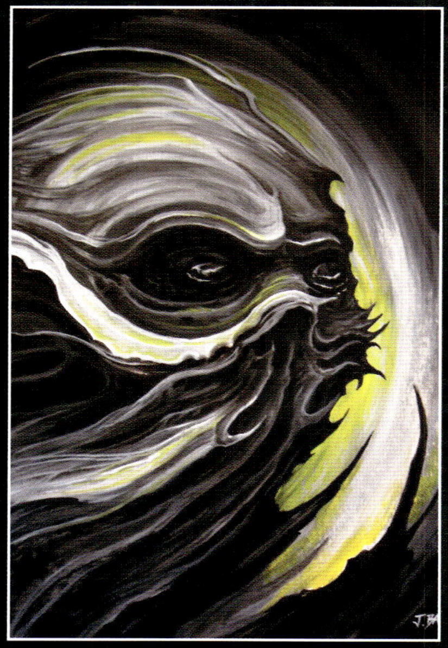

Jeremiah Barba

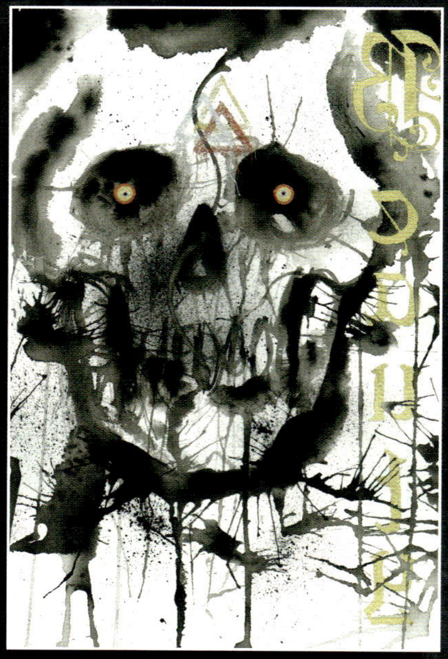

Alex Schultz

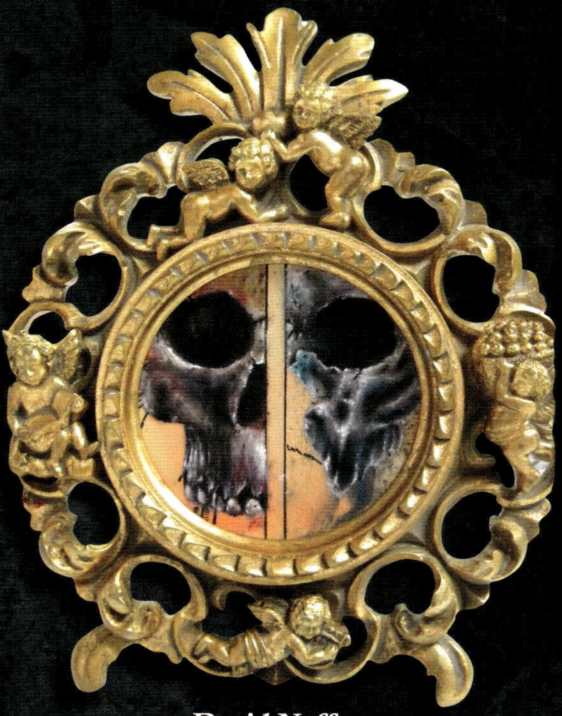

David Neff

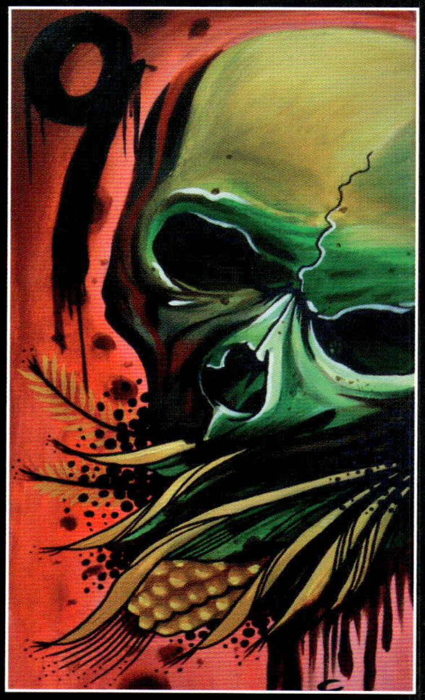

Craig Christianson

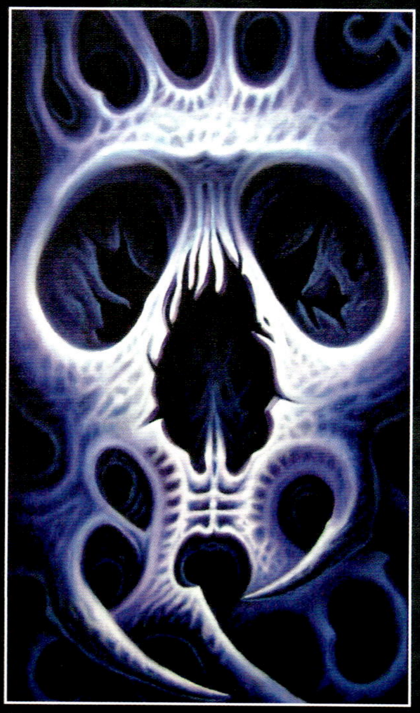

Josh Ruff

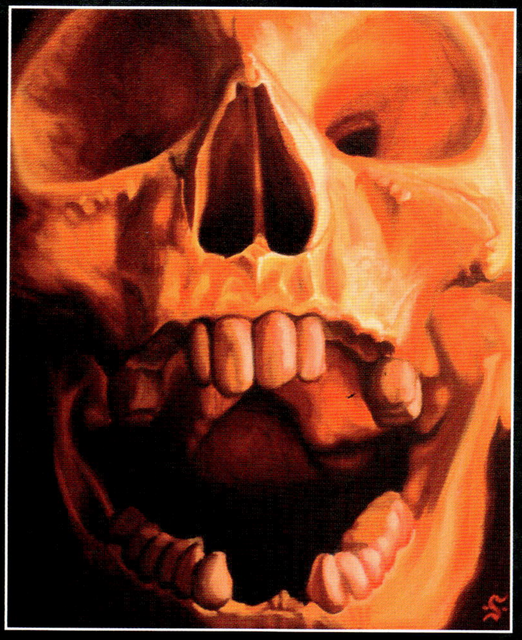

Jeffrey Srsic

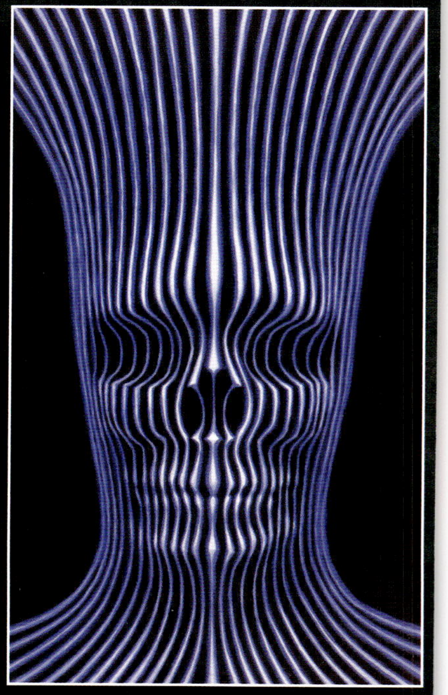

Alex Grey

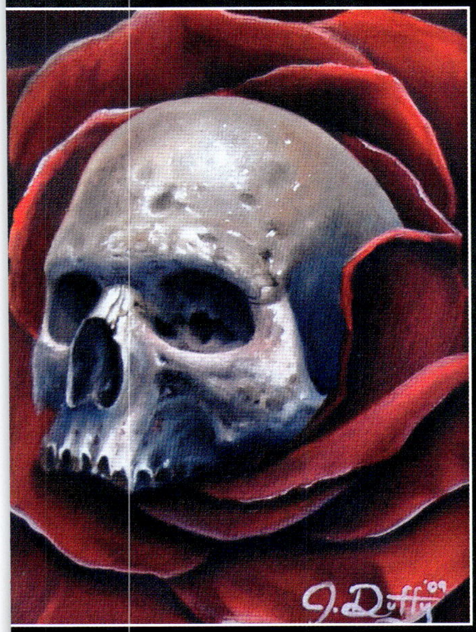

Josh Duffy

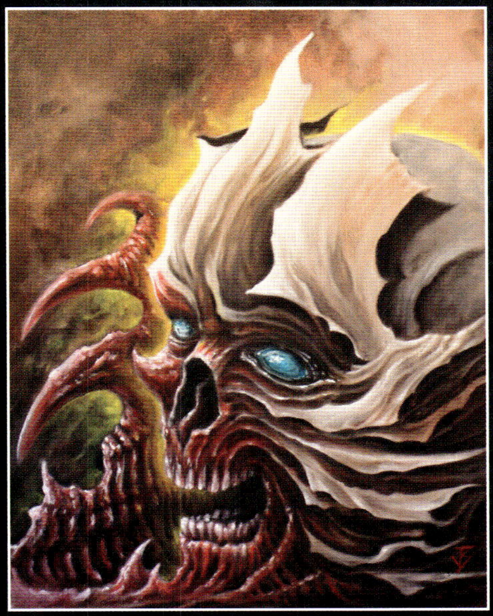

Jeremiah Barba

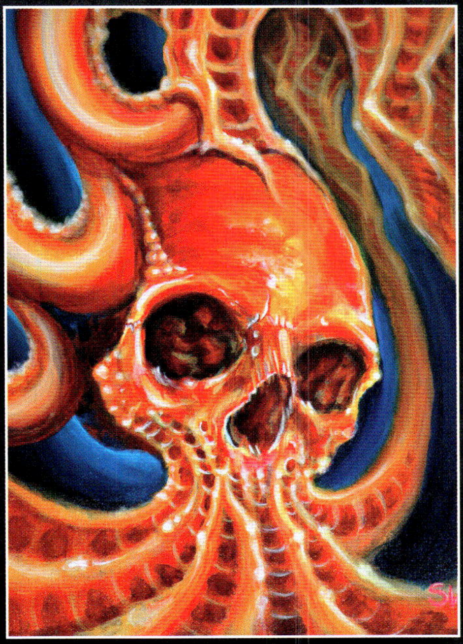

Shawn Will

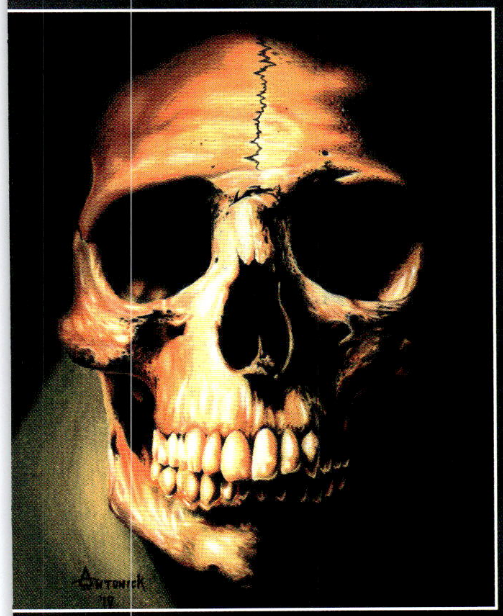

Ron Antonick

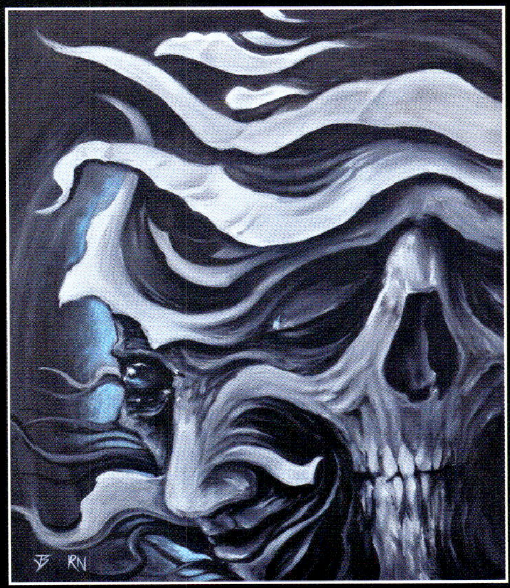

Ryan Needles & Jeremiah Barba

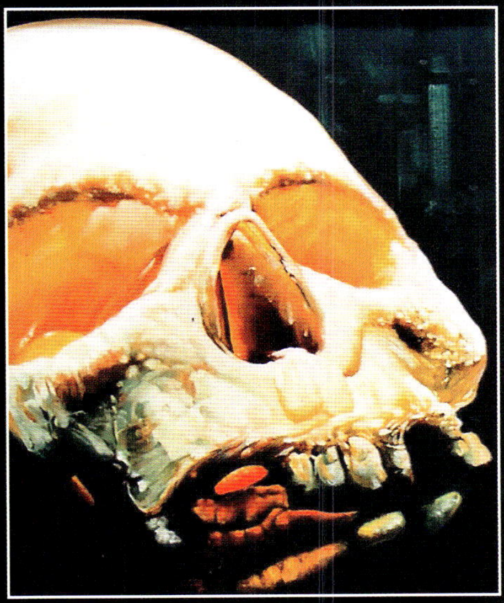

Monte

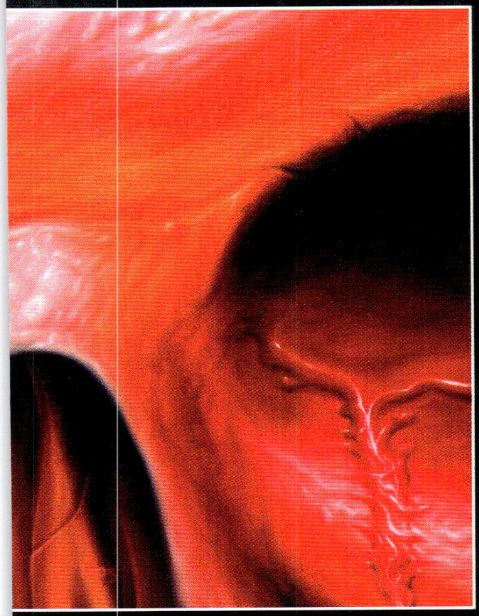

Nick Baxter

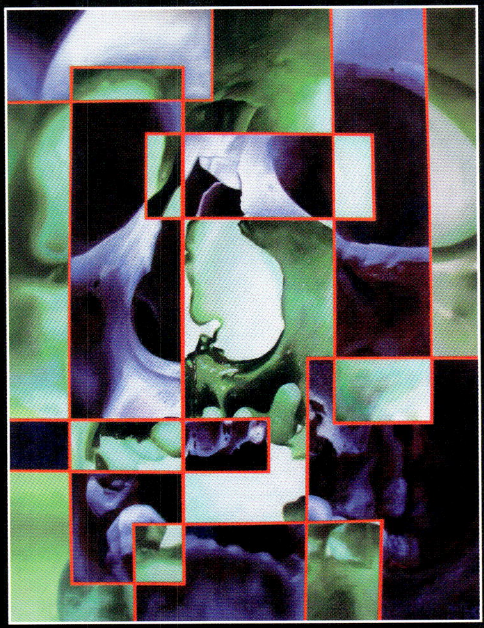

Nate Heim

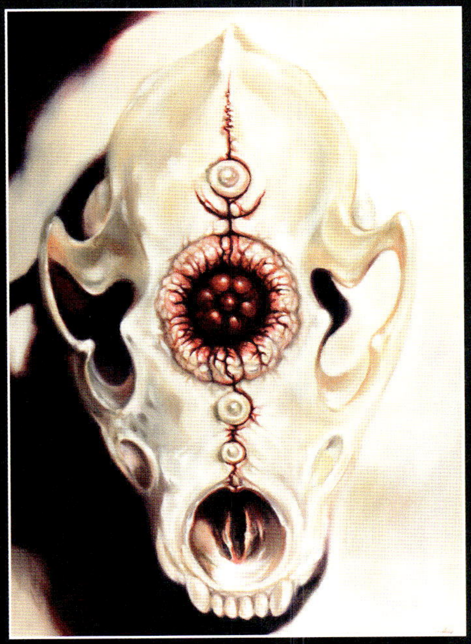

Muriel Zao

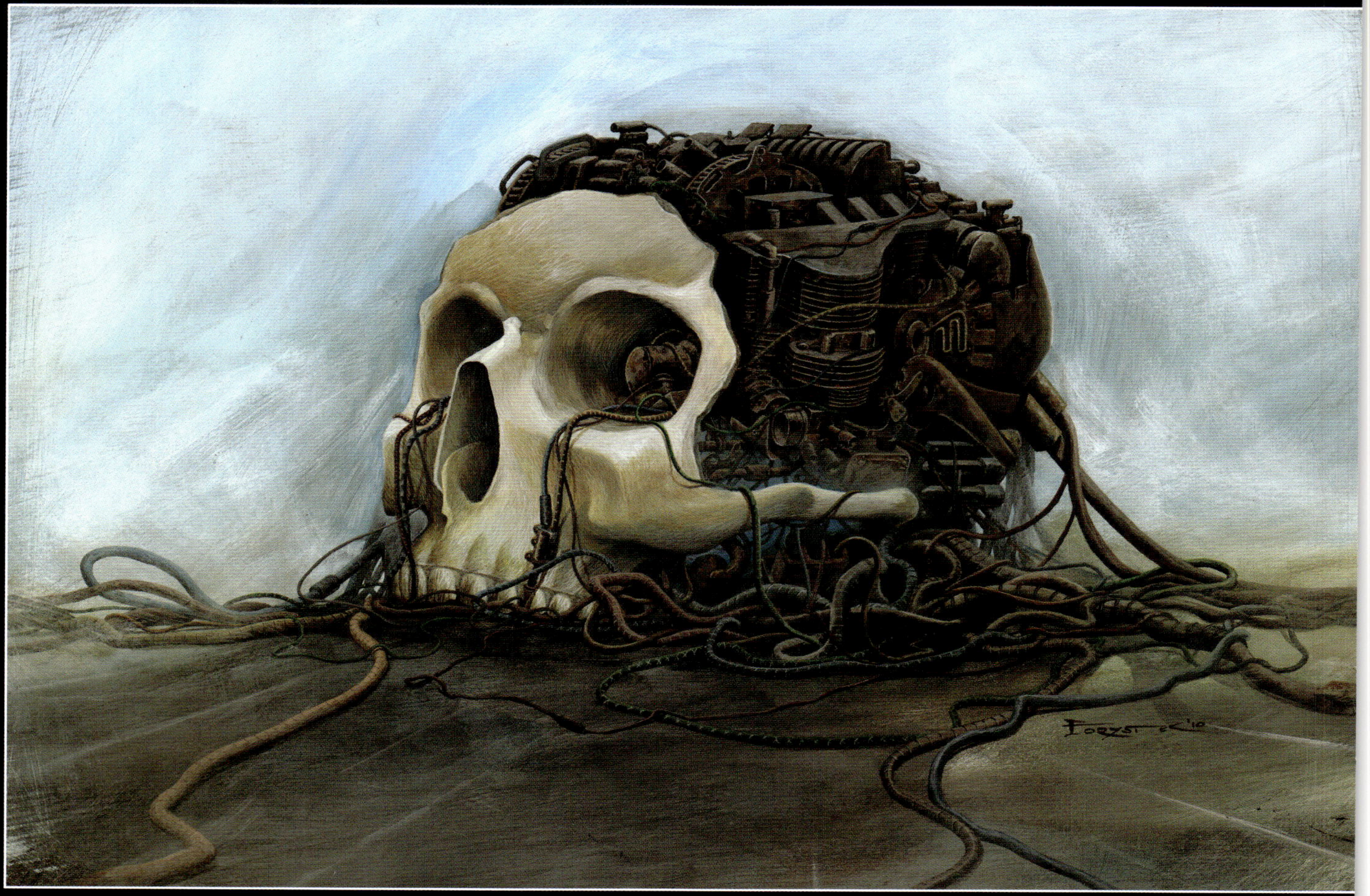

Peter Forystek

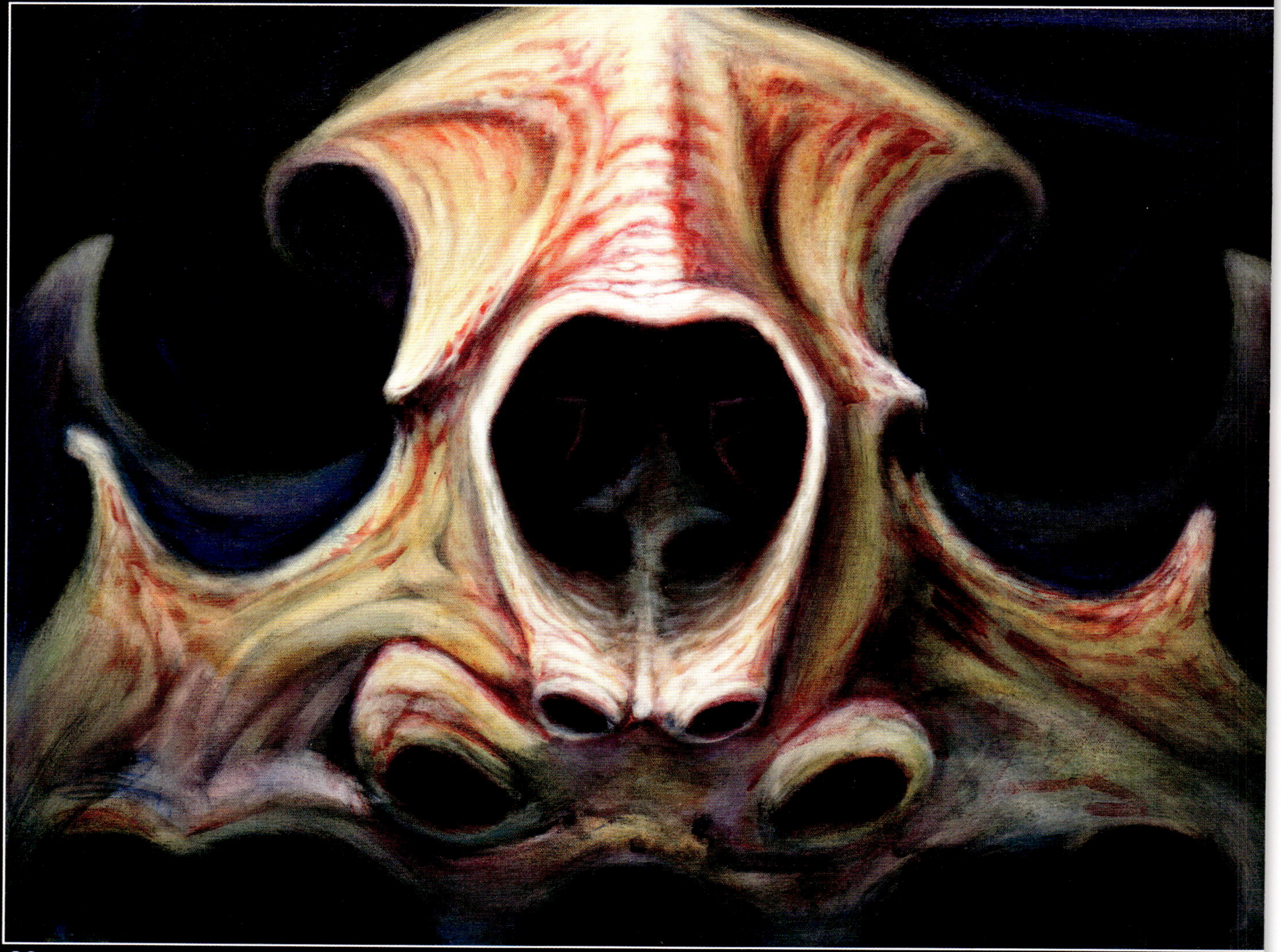

Carson Hill

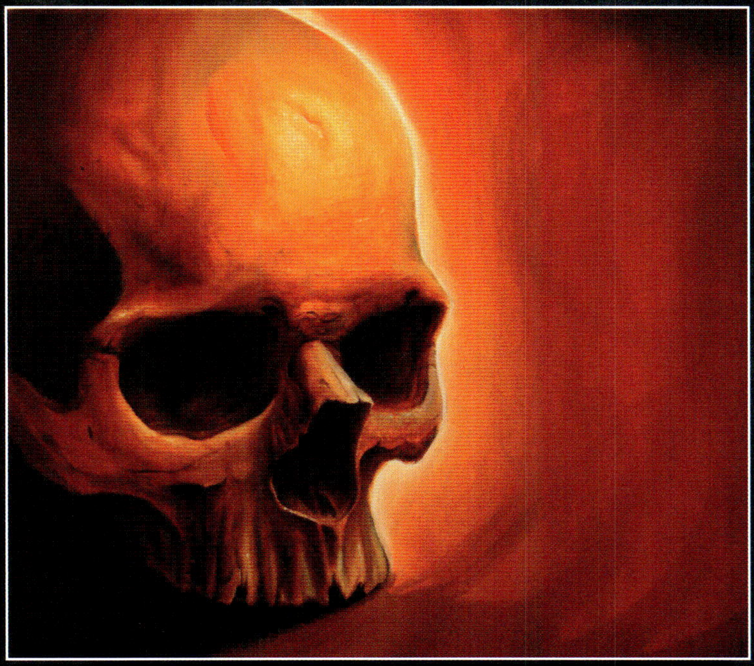
Nate Euvrard

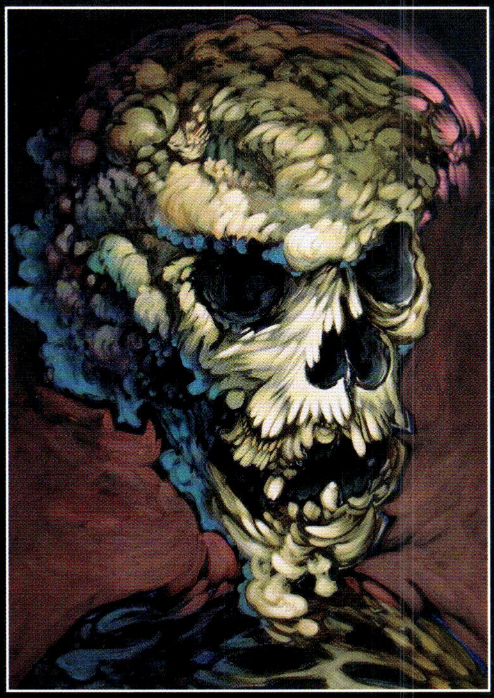
Chris Dingwell

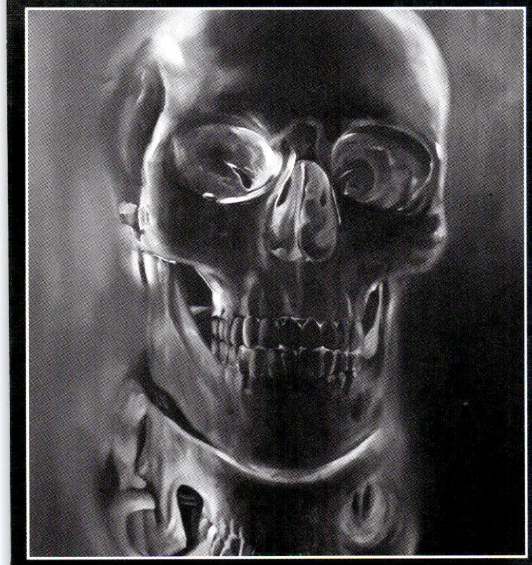
Shay Bredimus

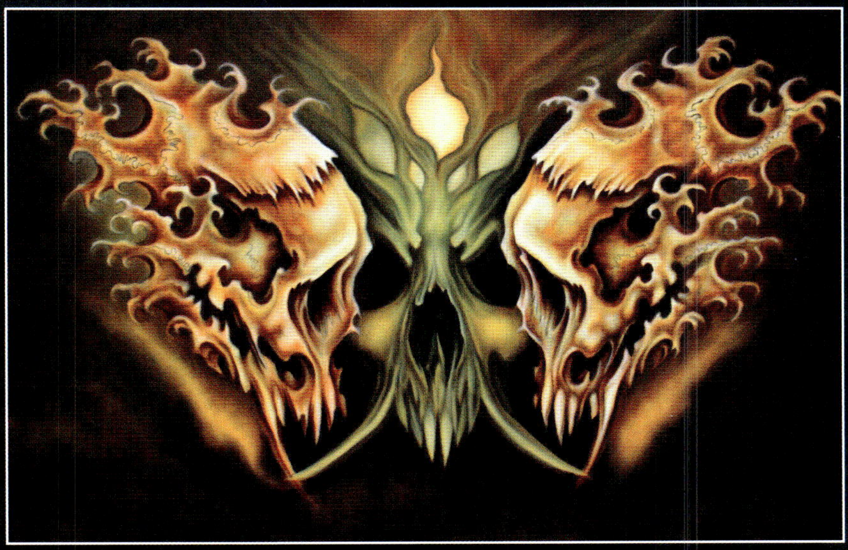
Matthew Lukesh

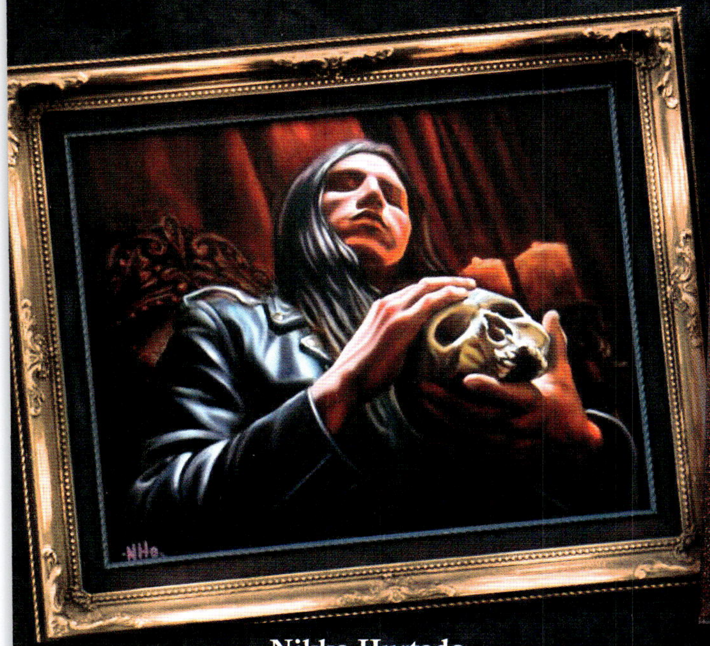
Nikko Hurtado

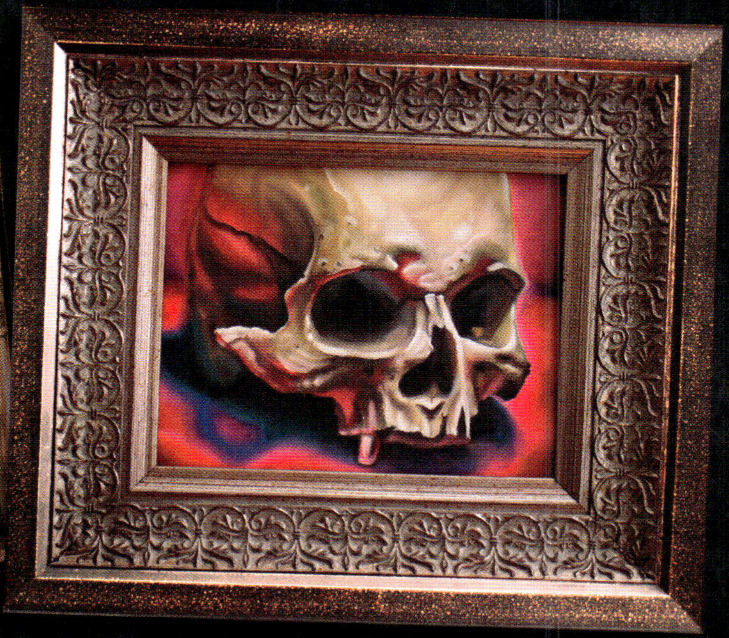
Sean Herman

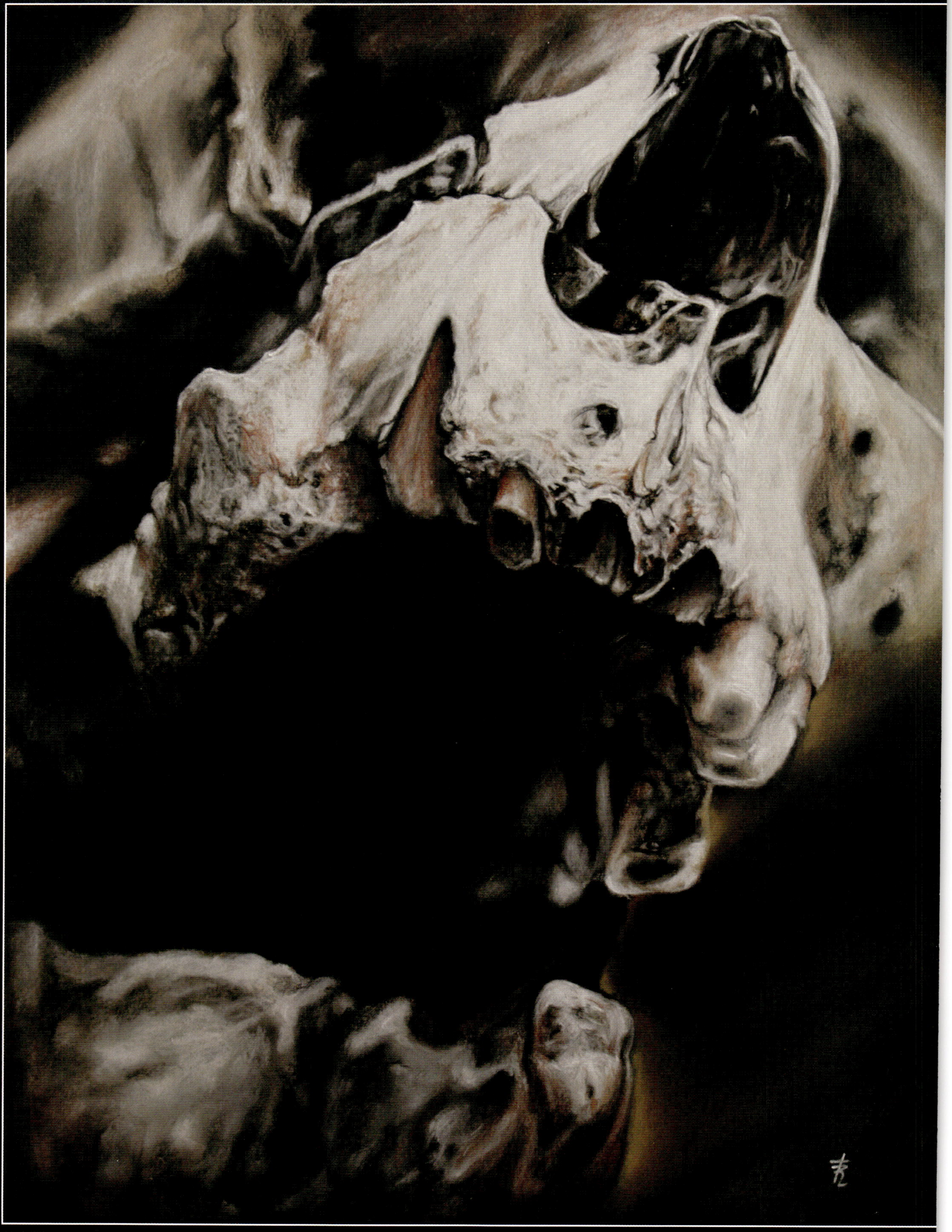

Thomas Kynst

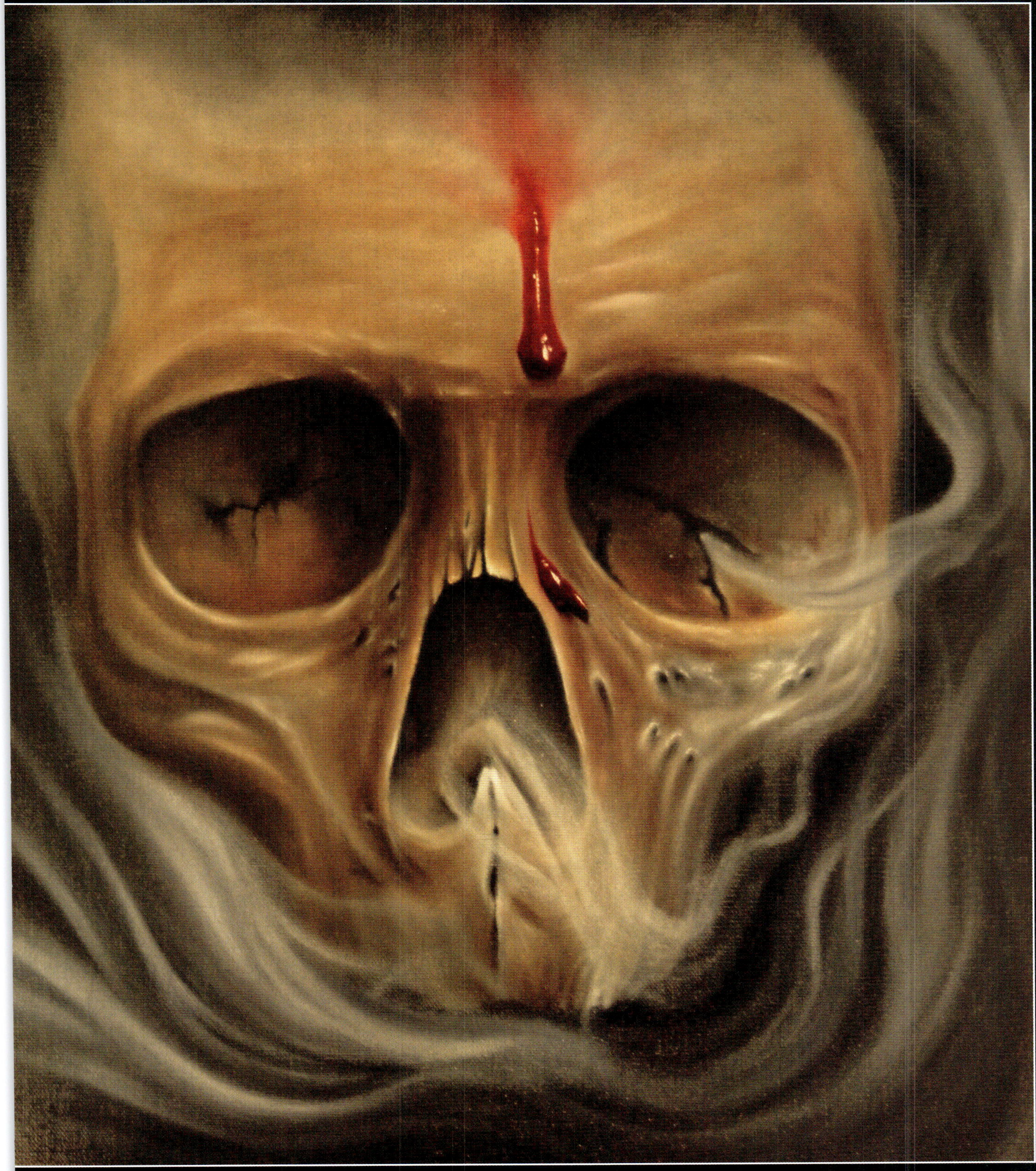

Ryan Hadley

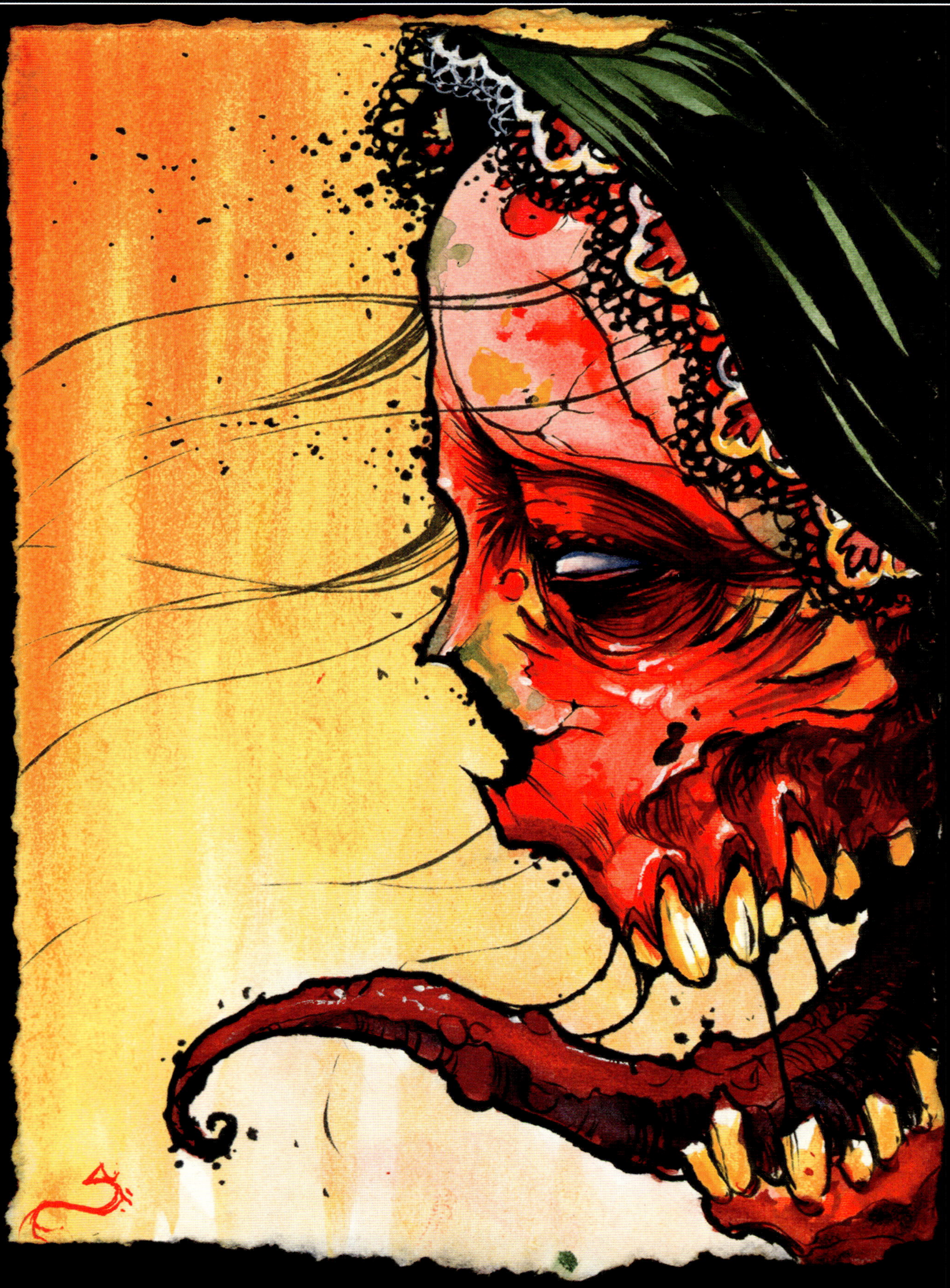

Tom Strom

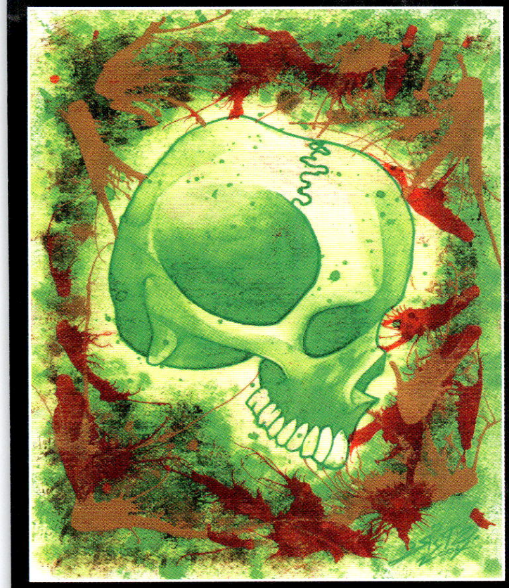
Spiro Kambitsis

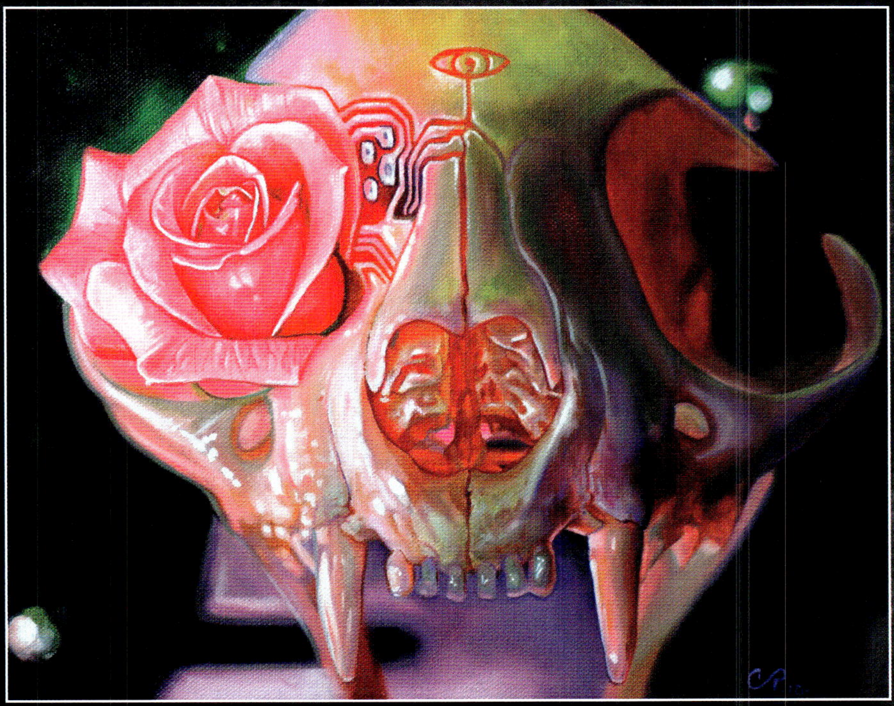
Christian Perez

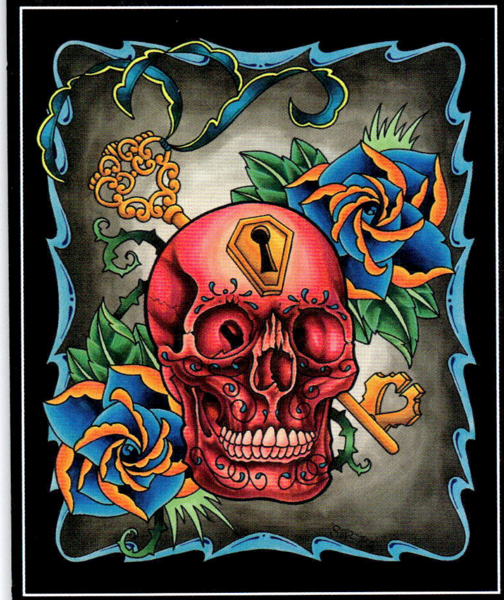
Stretch

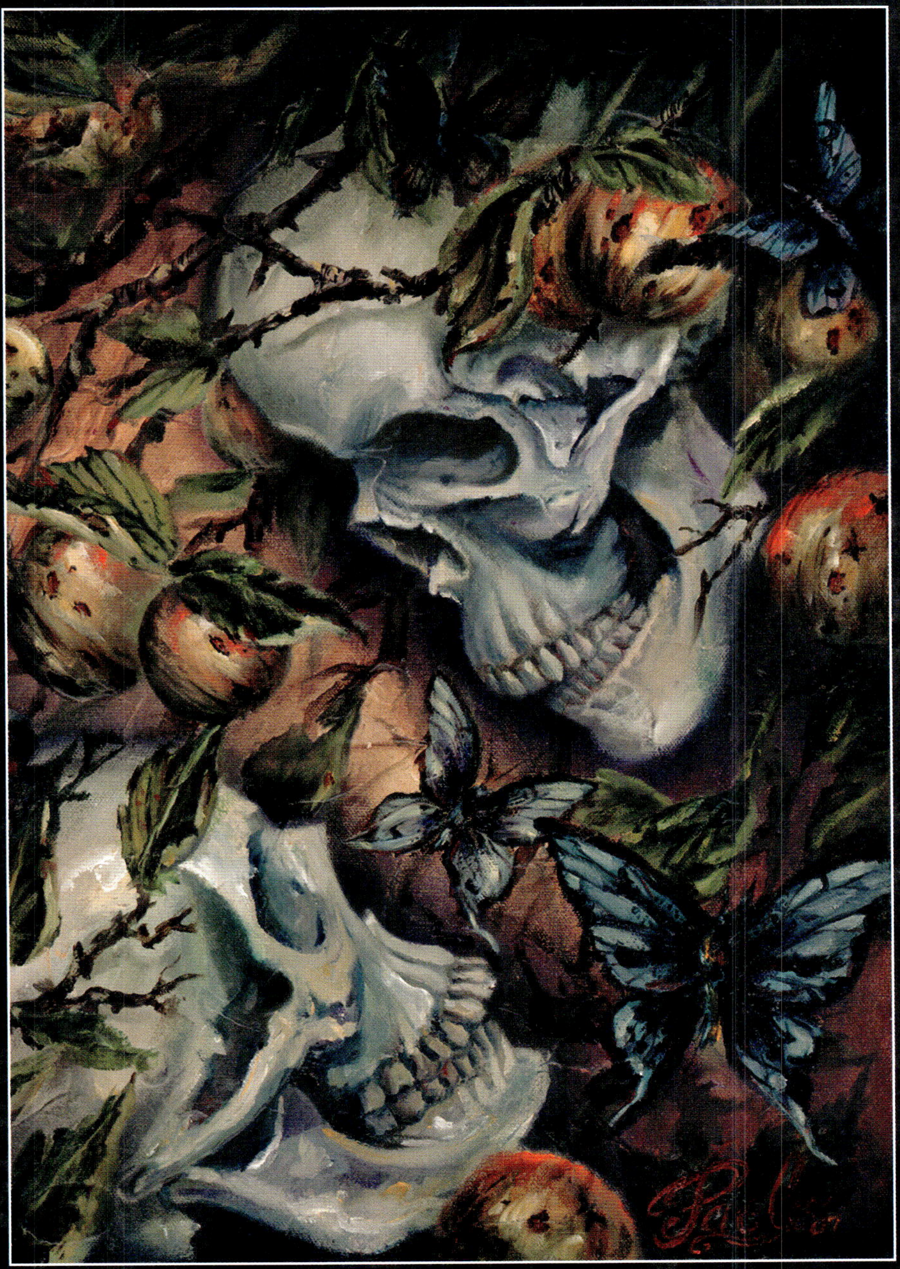
Paolo Acuna

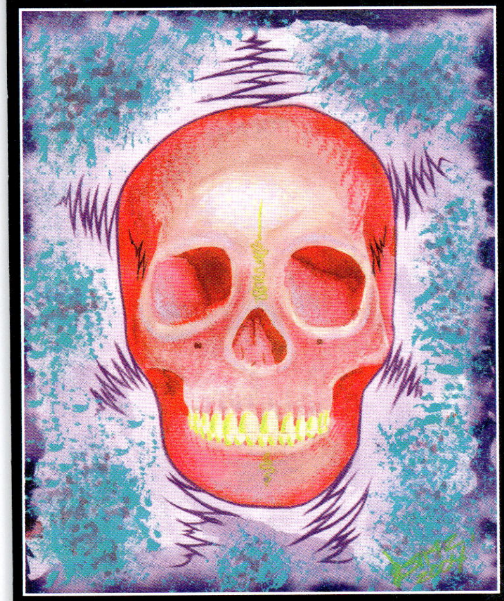
Spiro Kambitsis

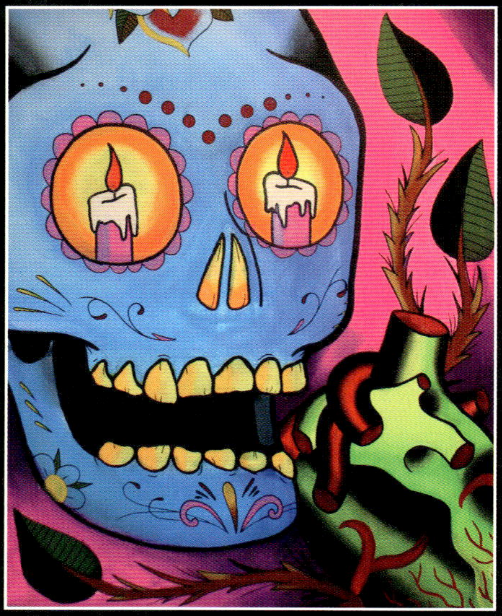

Benjamin Duarte

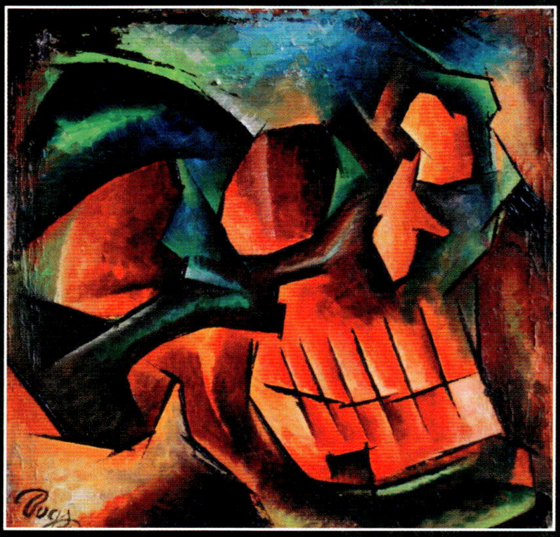

Bugs

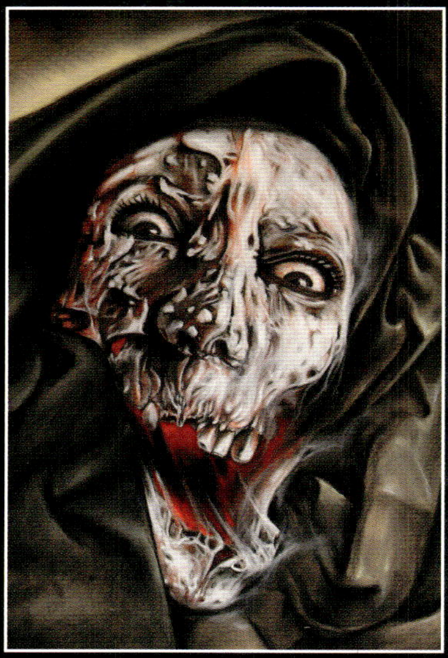

Thomas Kynst

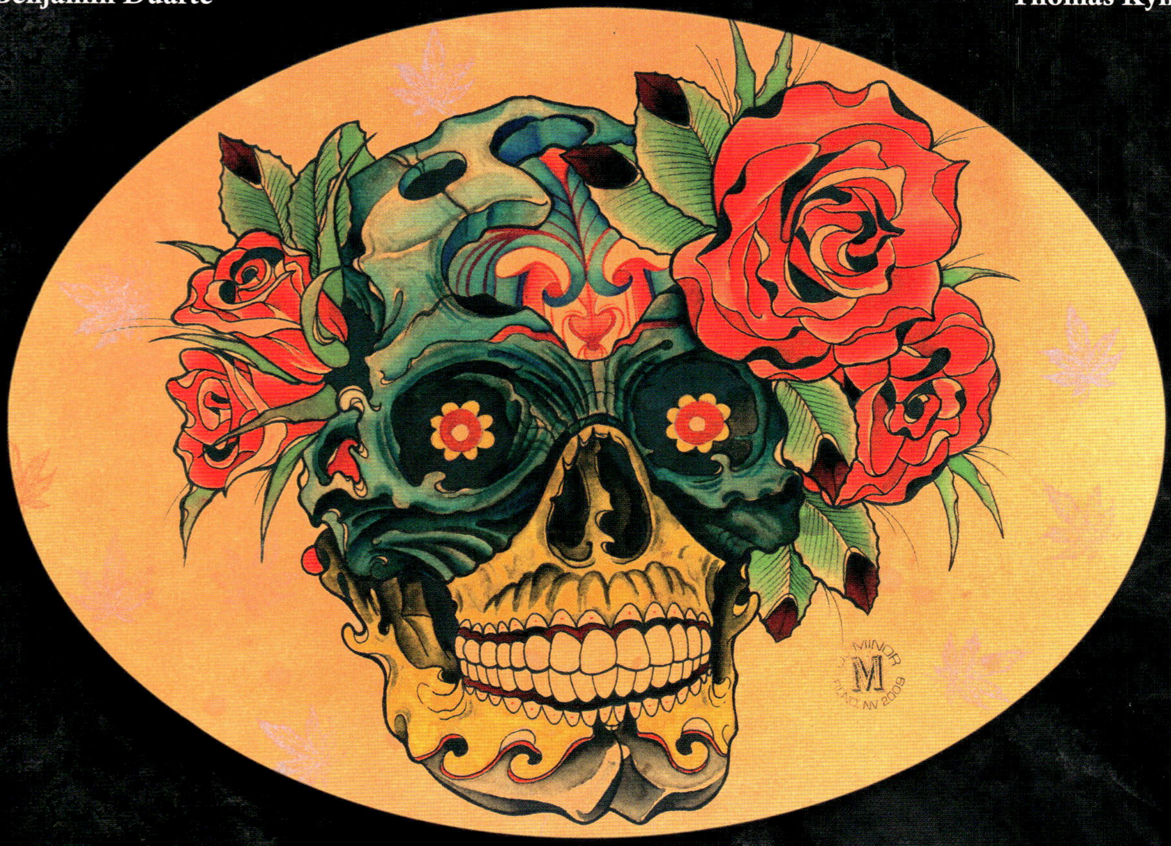

Dj Minor

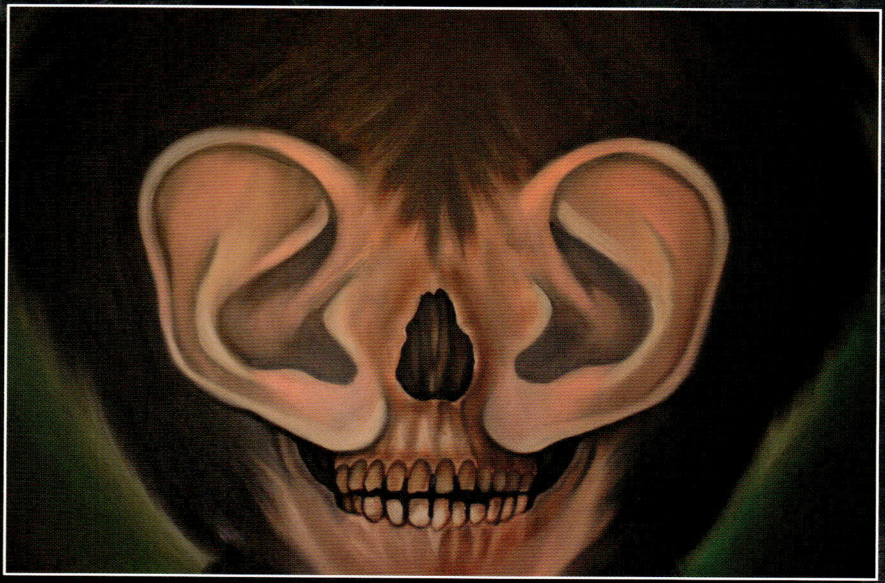

Benjamin Duarte

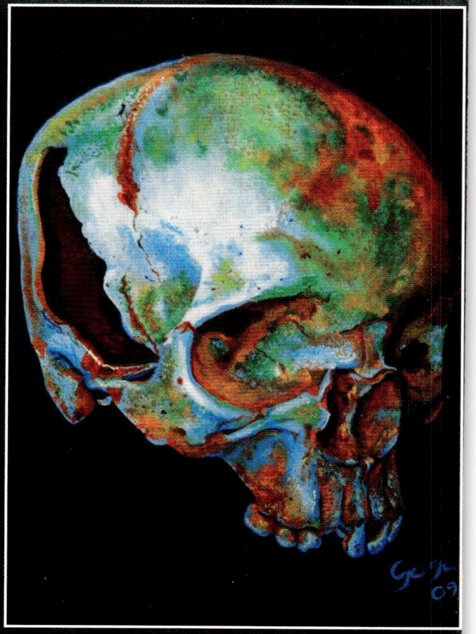

Gege

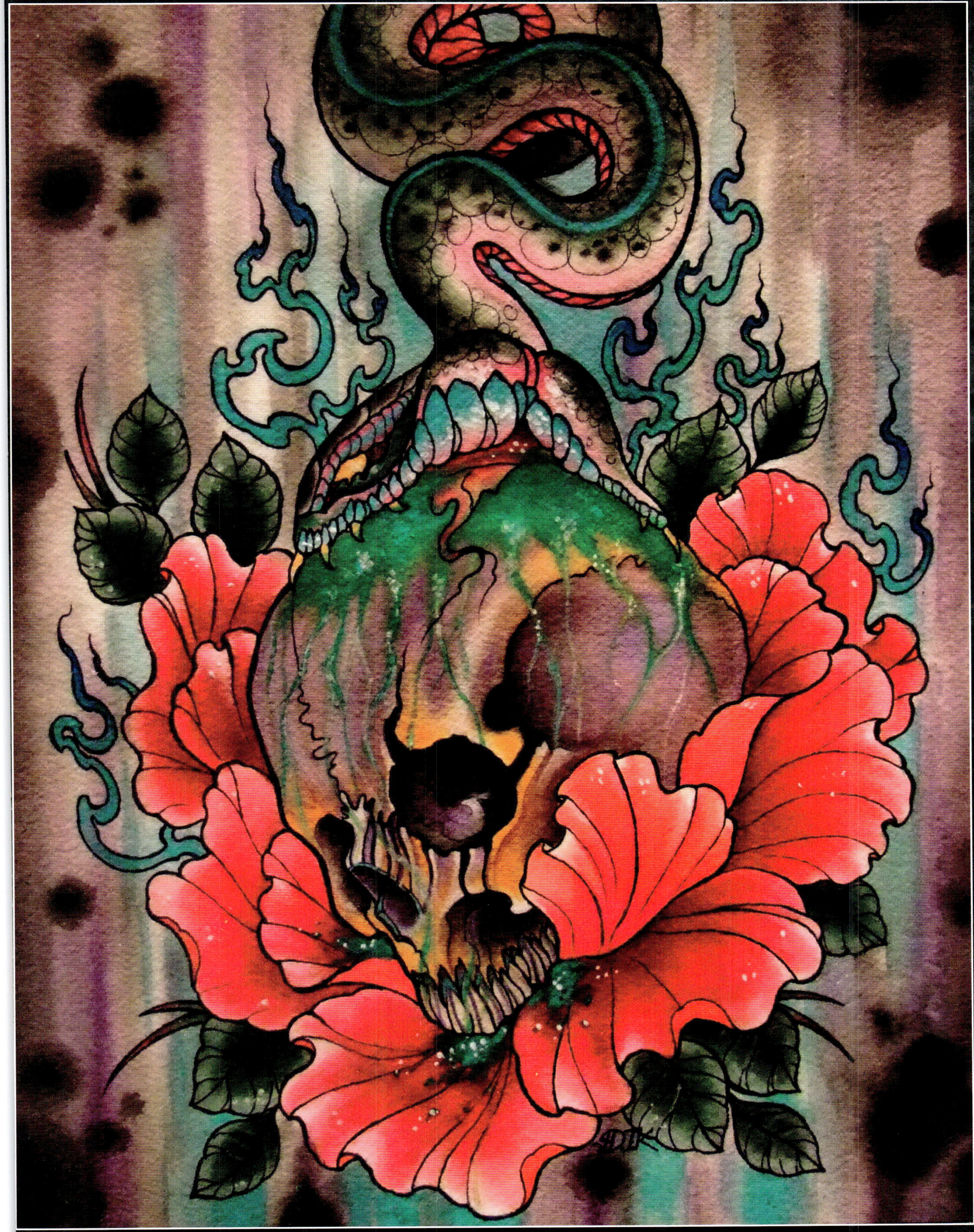

Dean Kalcoff

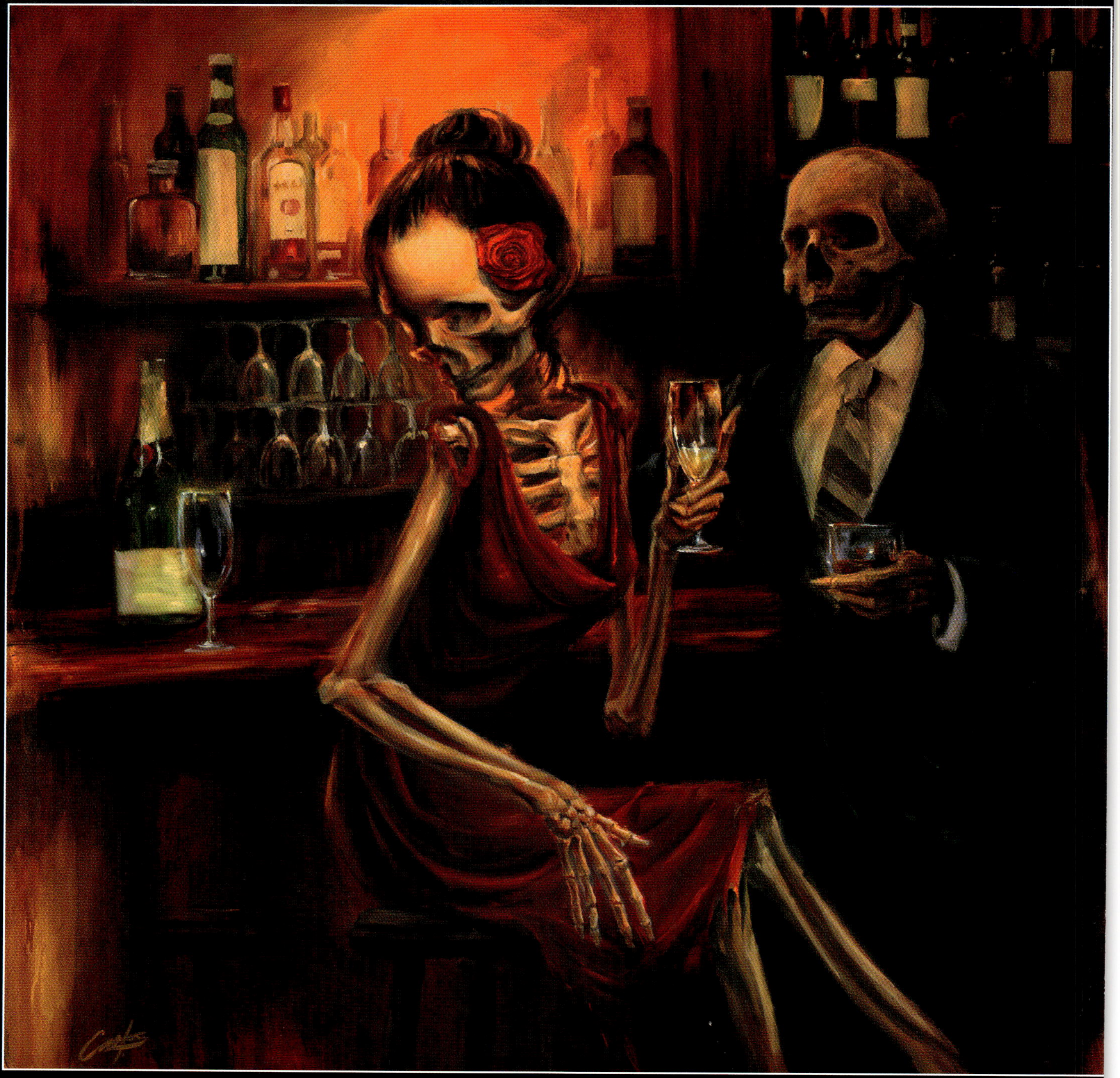

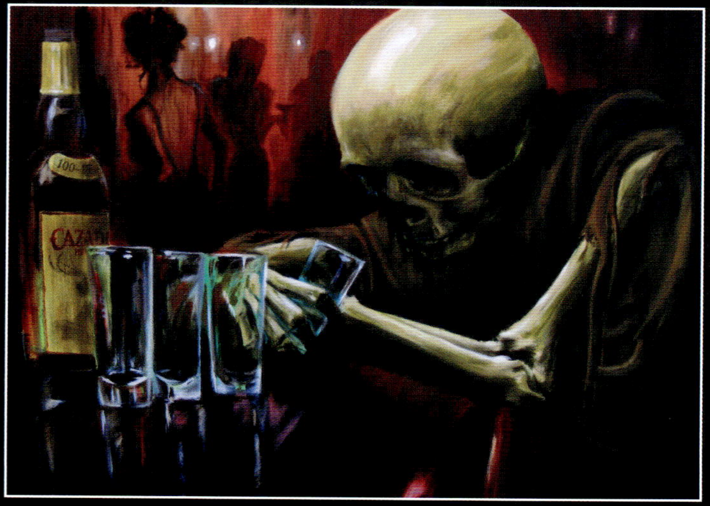

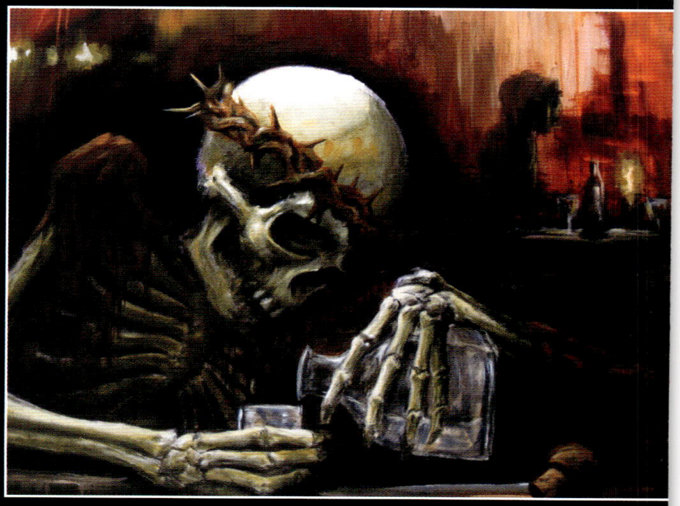

Carlos Torres

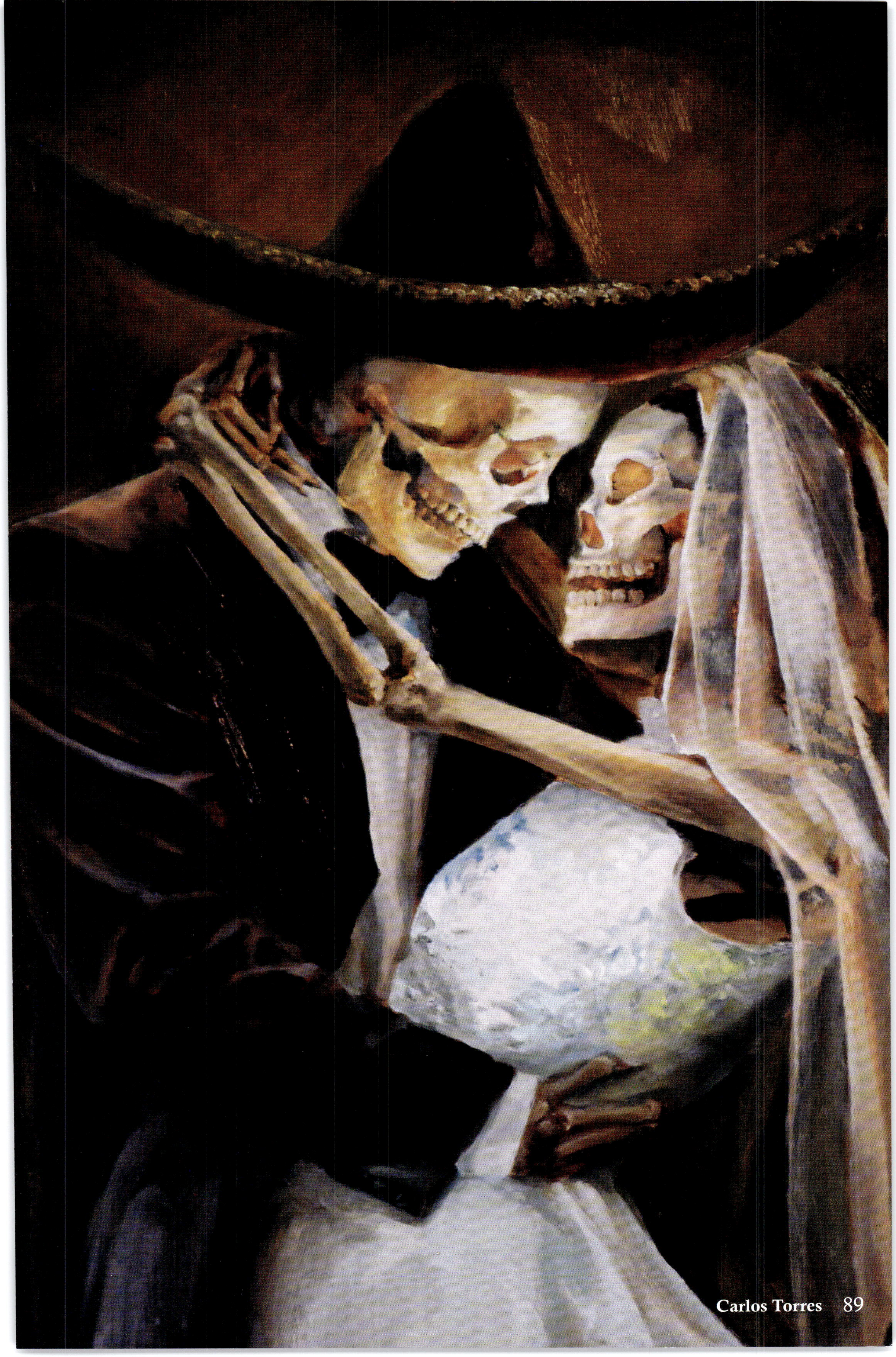

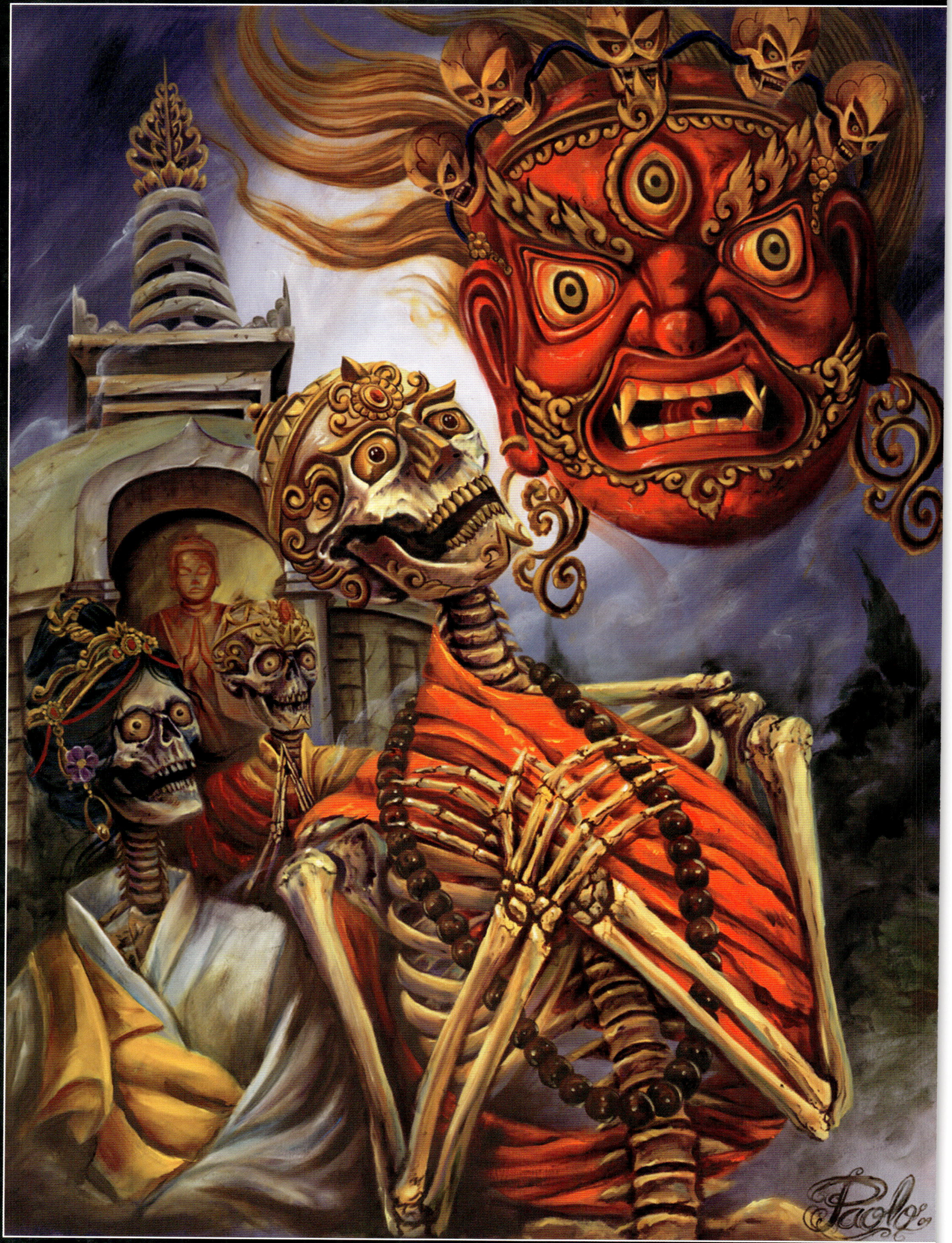

Paolo Acuna

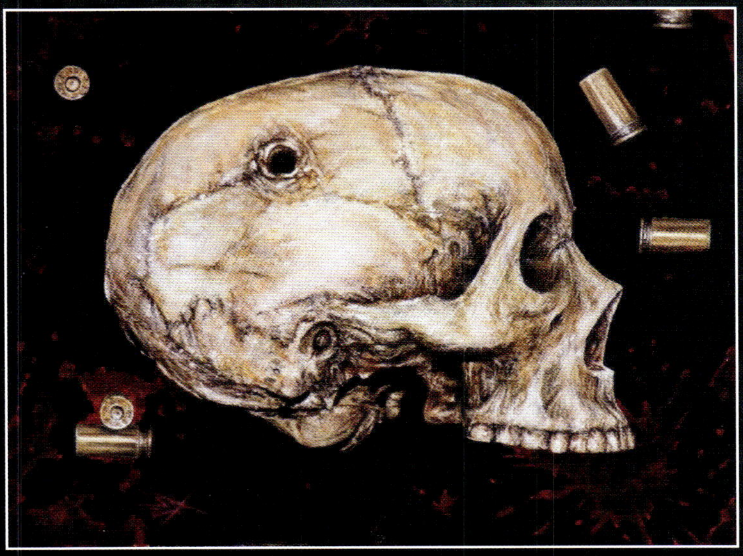

Dan Henk

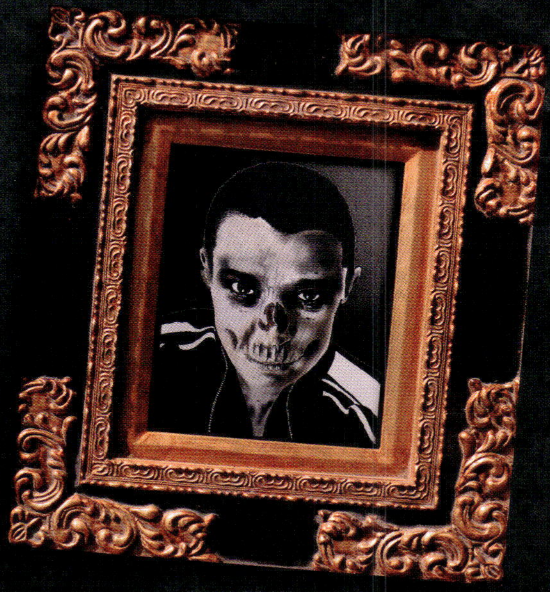

Volko Merschky

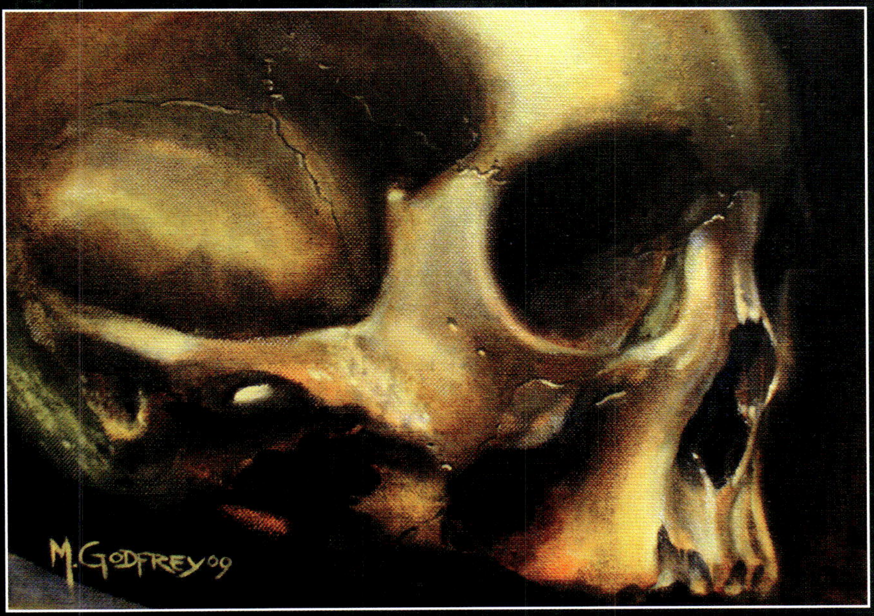

Mike Godfrey

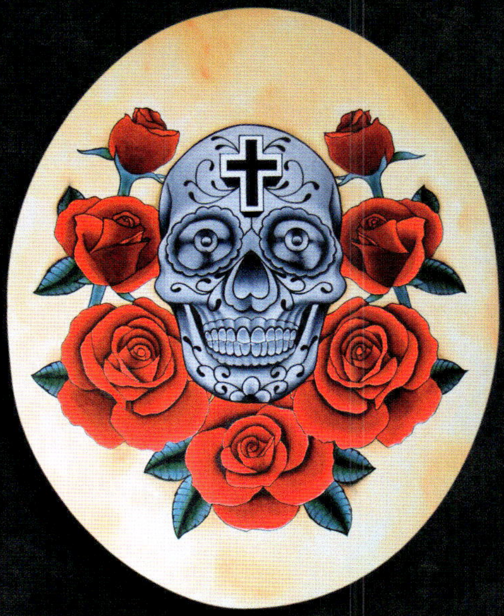

John Graefe

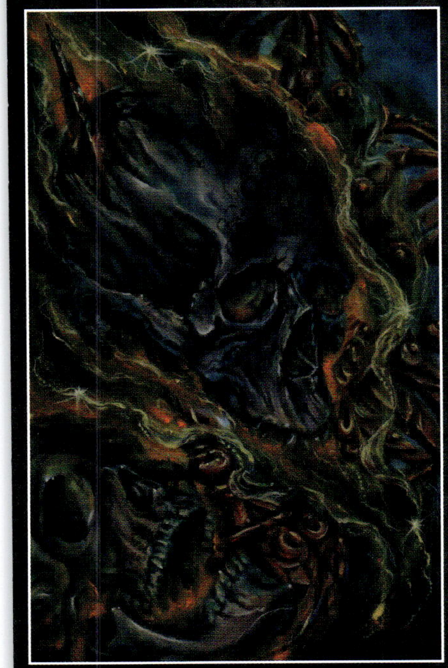

Paolo Acuna

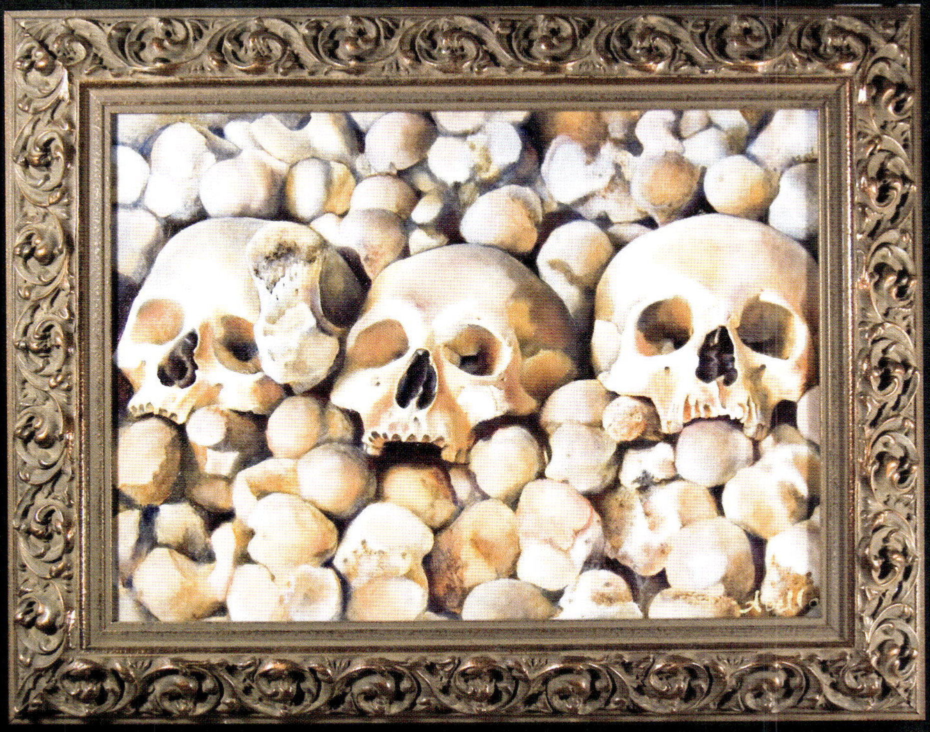

Aaron Bell

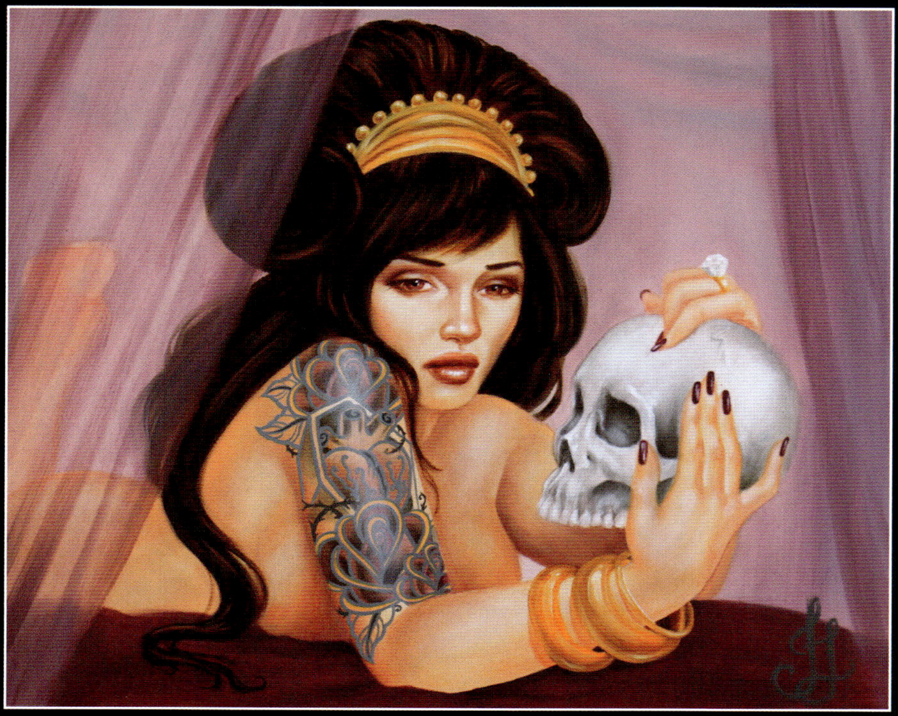

Jesso

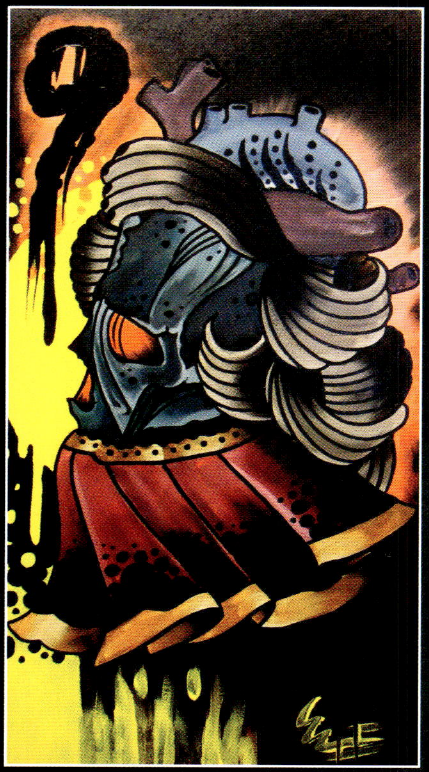

Craig Christianson

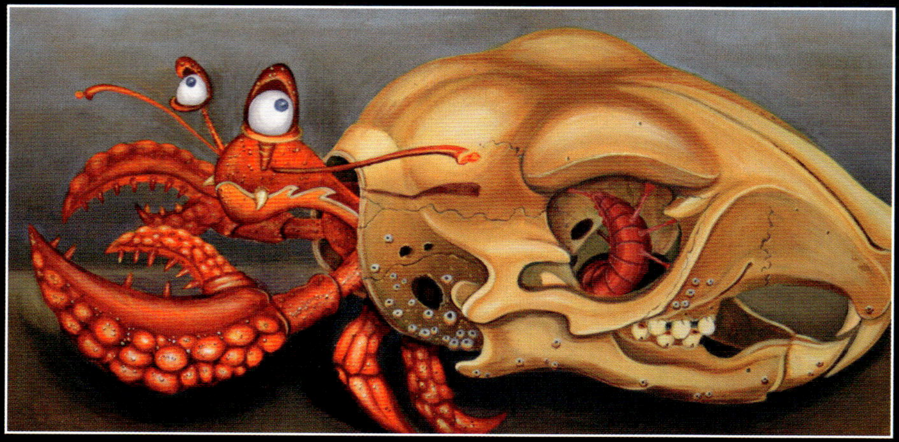

Jason Stephan

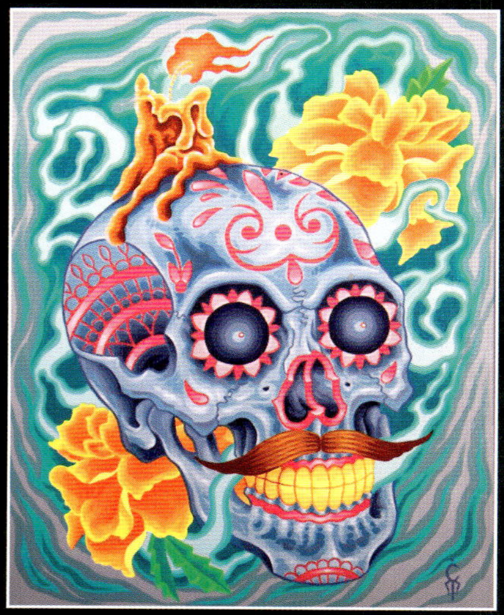

Canman

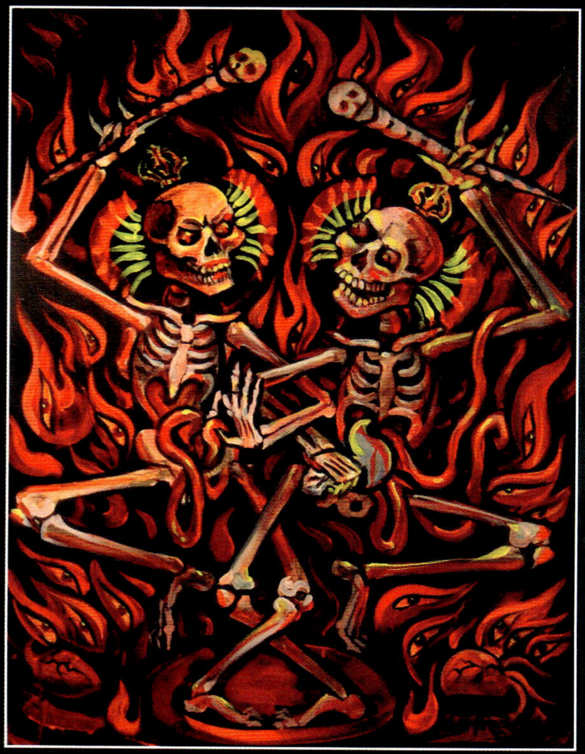

Alex Grey

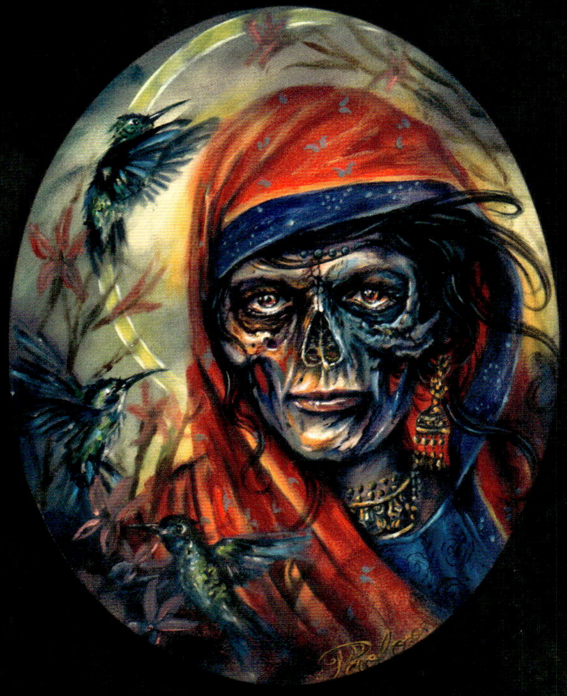

Paolo Acuna

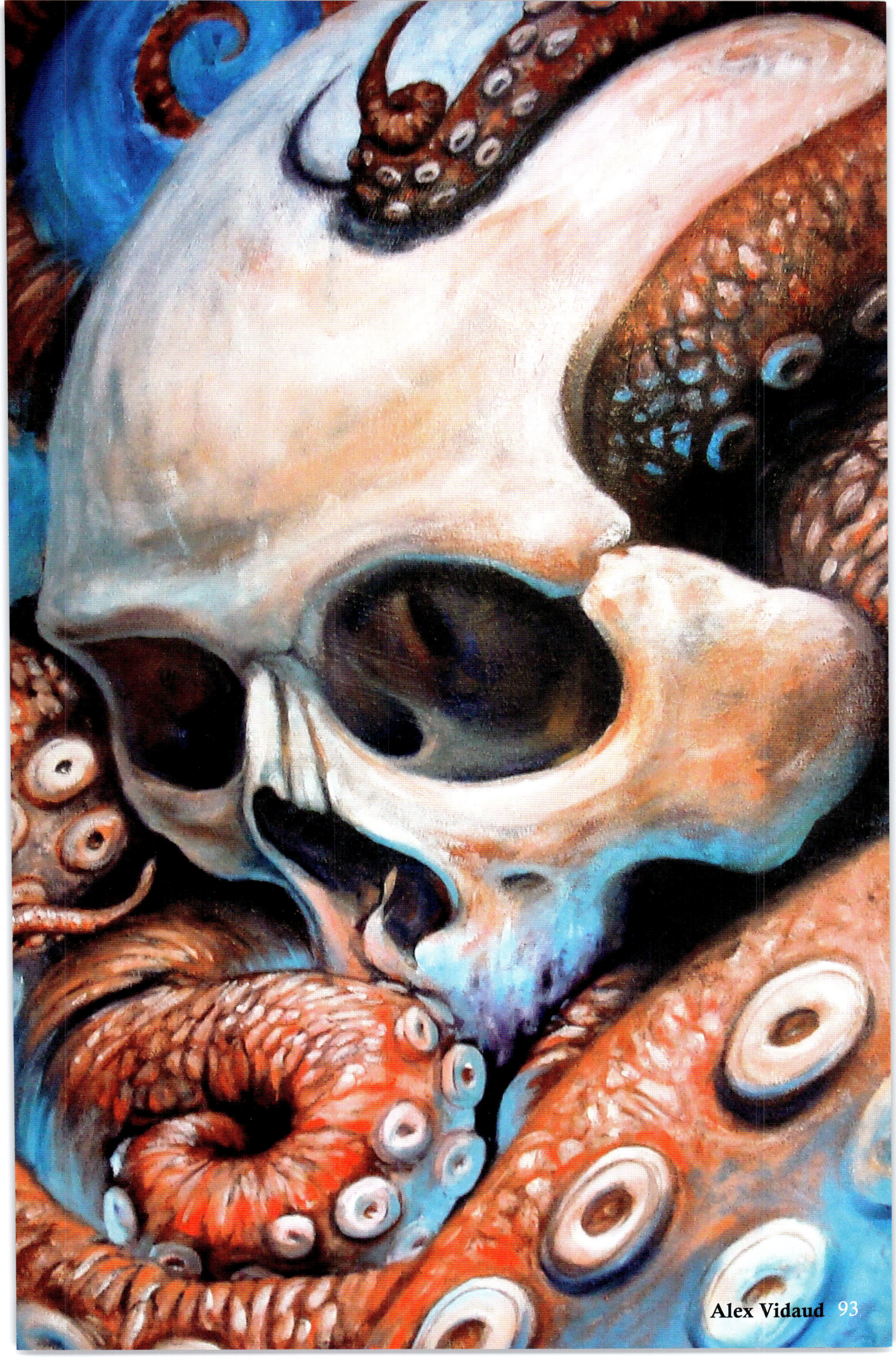

Alex Vidaud

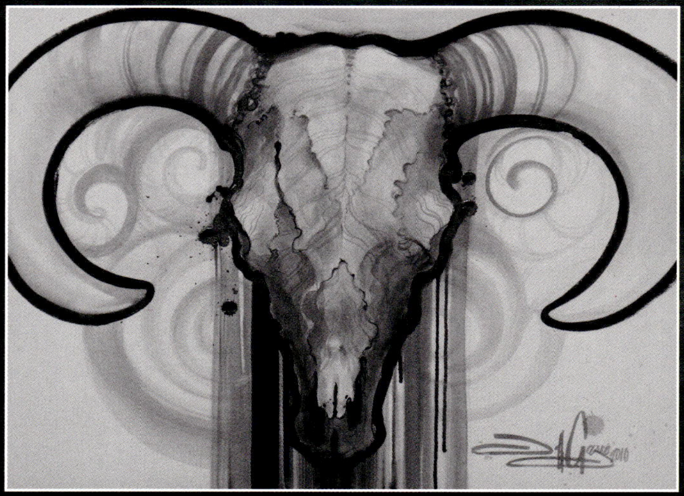

Jeff Gogué

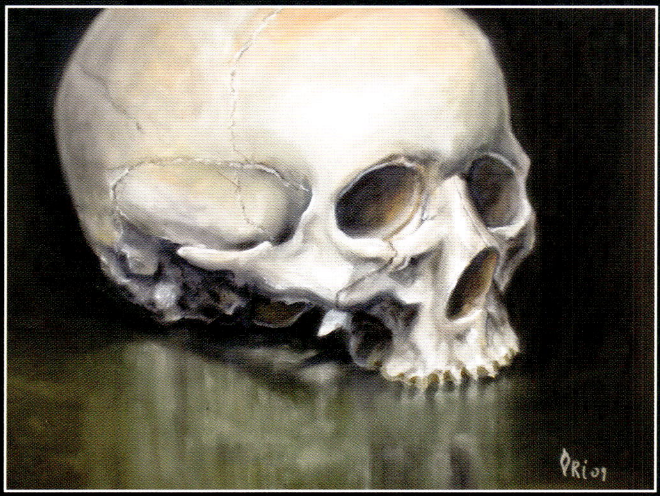

Oriol Zambudio

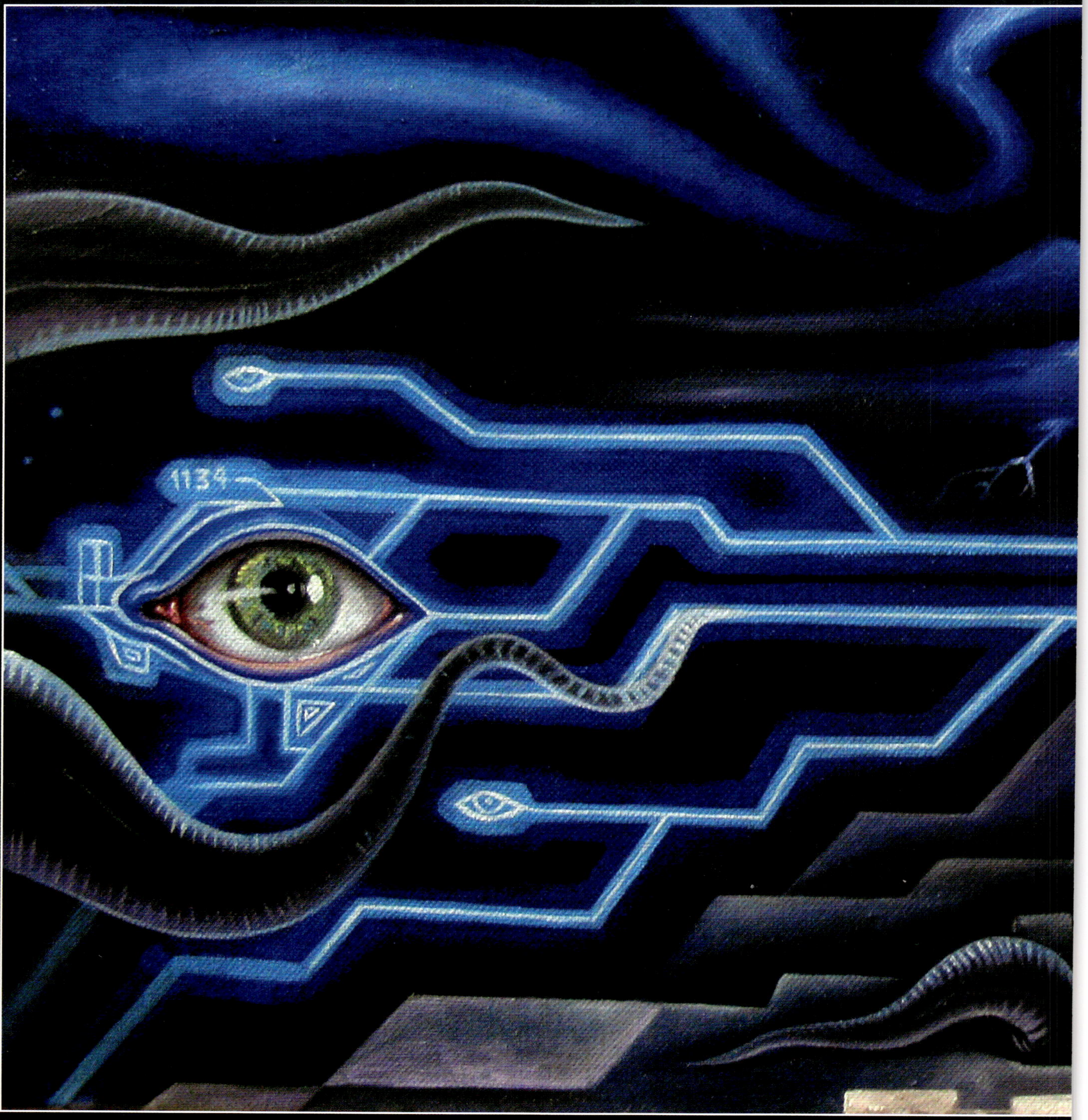

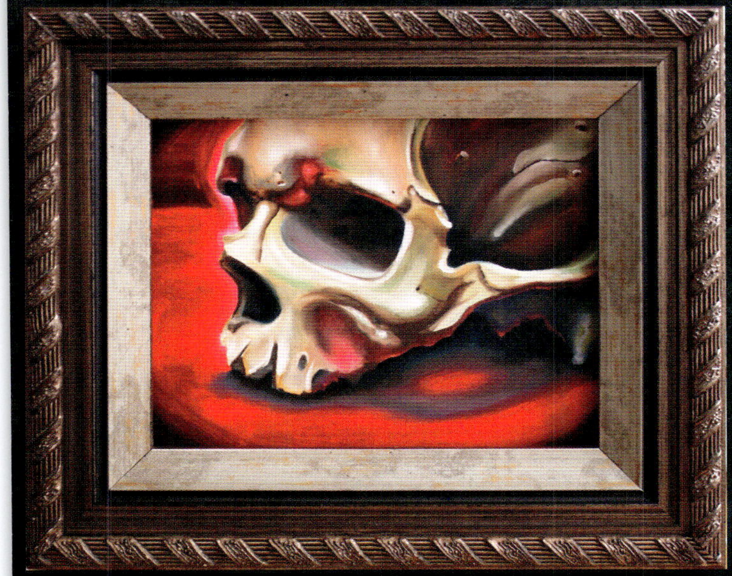

Sean Herman

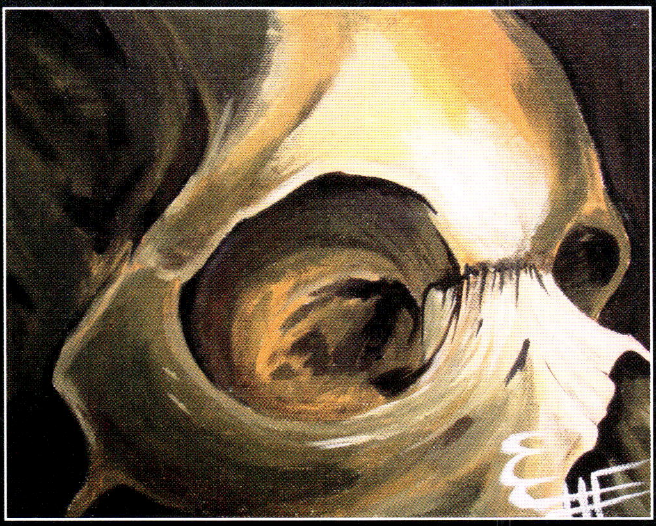

Craig Christianson

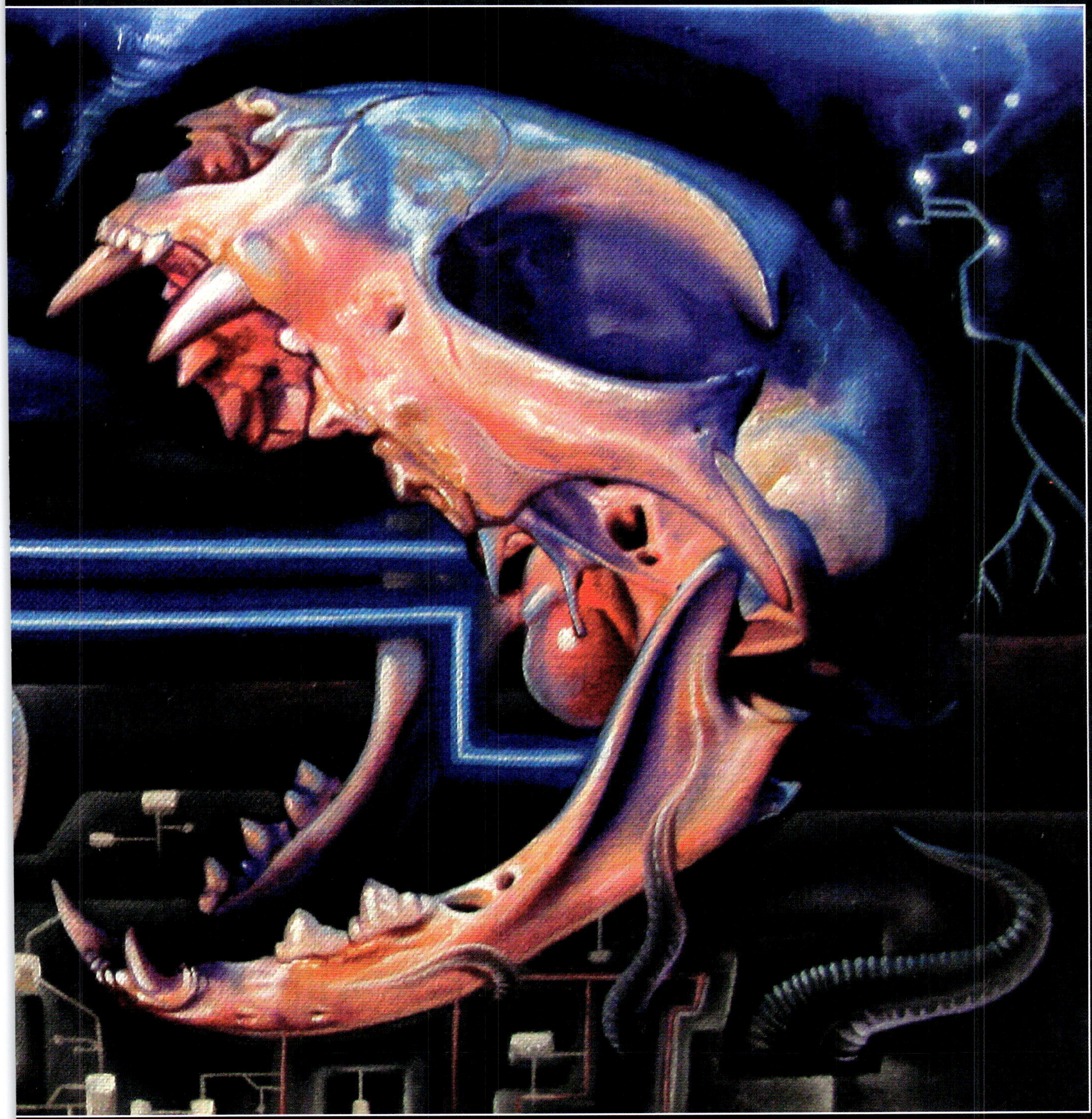

Christian Perez

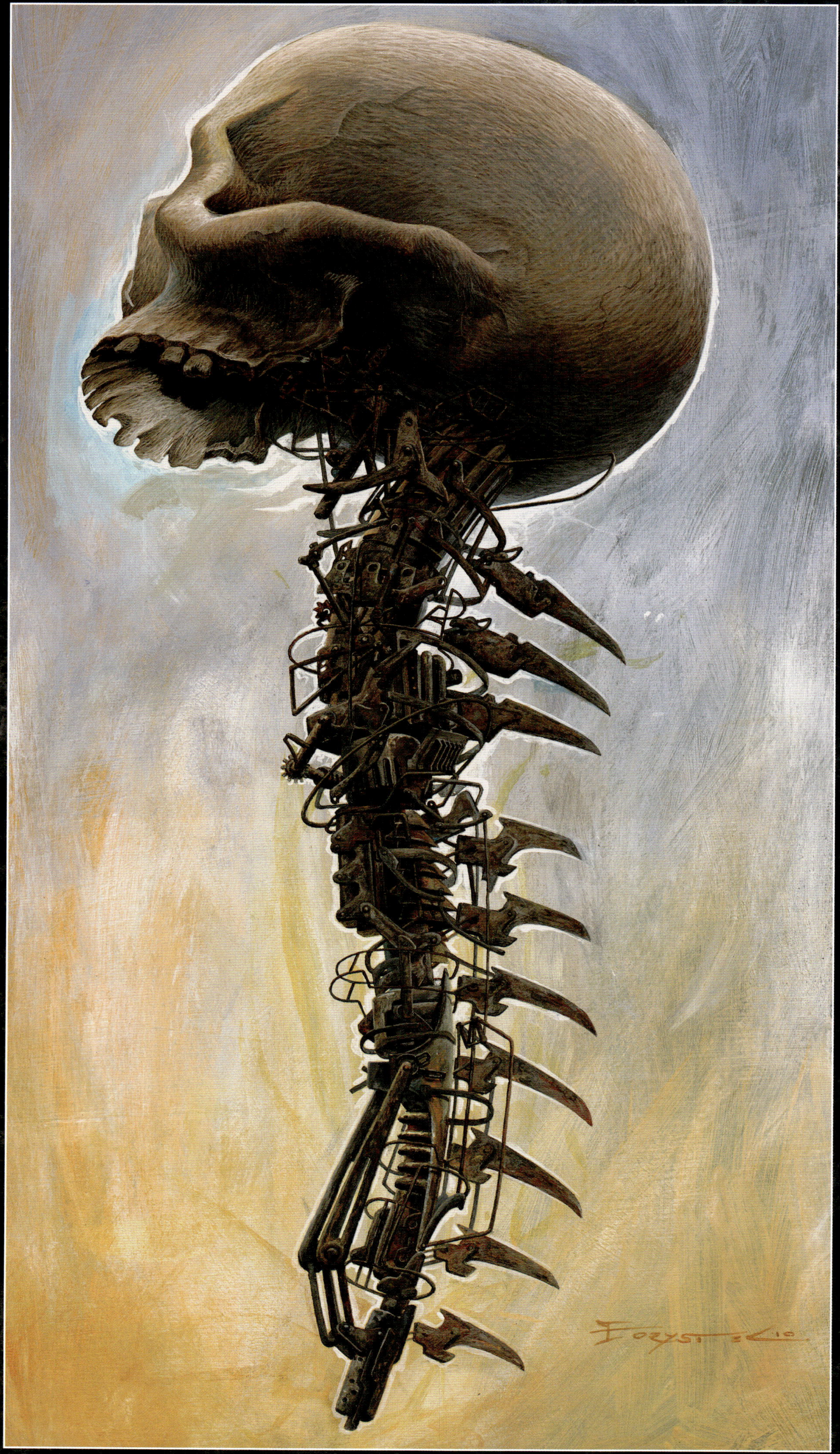

Peter Forystek

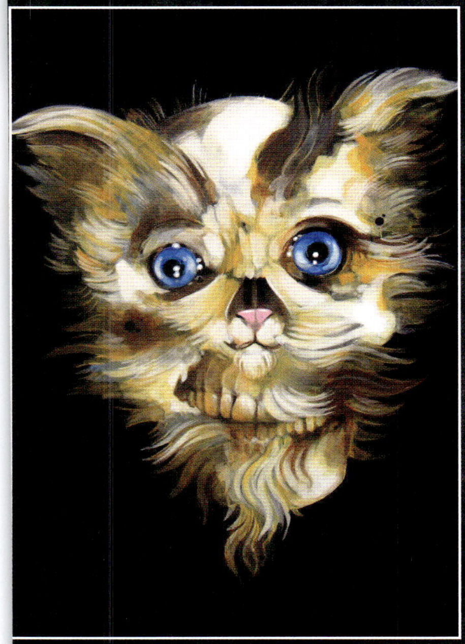

Christopher Gay

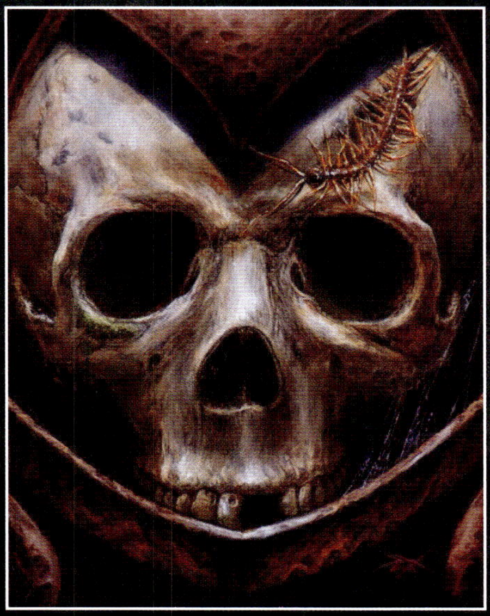

Dan Henk

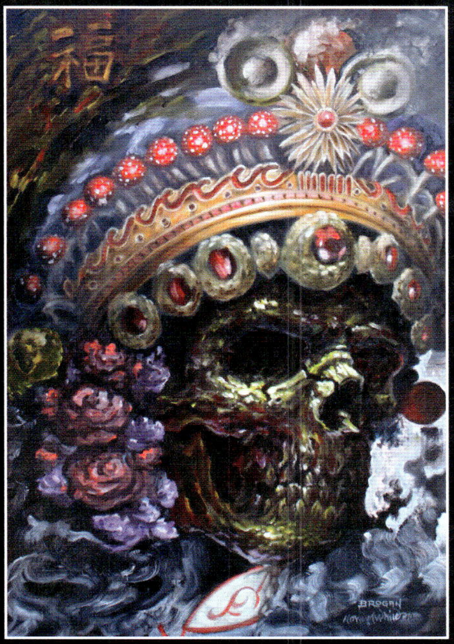

Larry Brogan & Darrin White

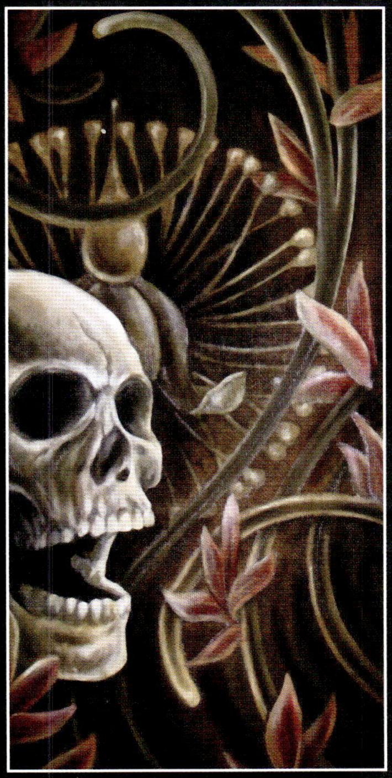

Jessica Rocha

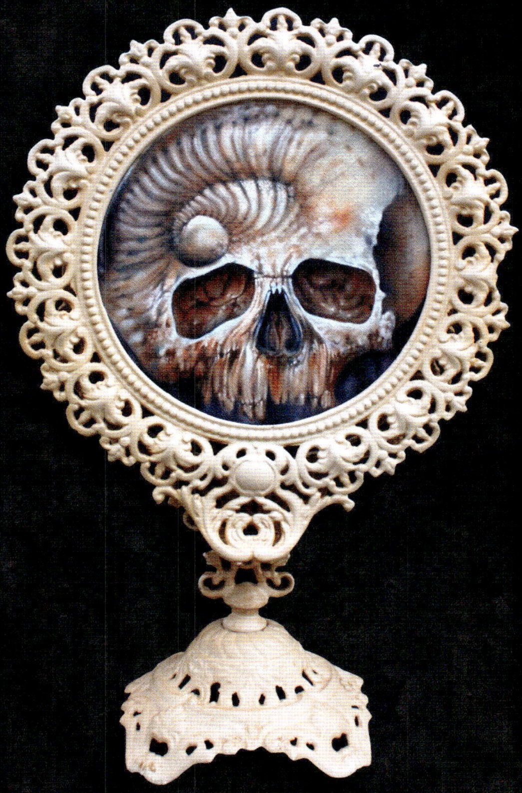

David Page

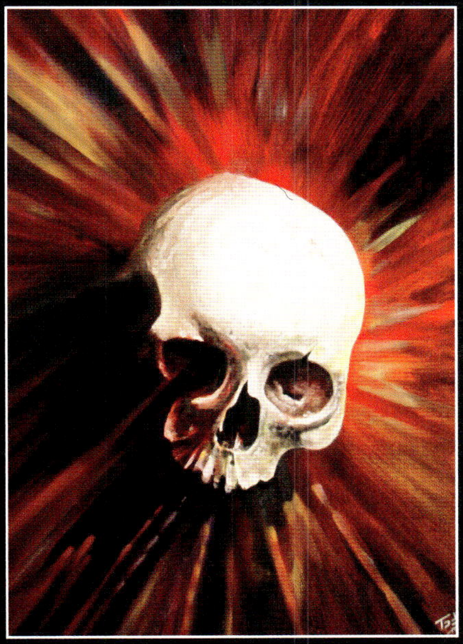

Todo

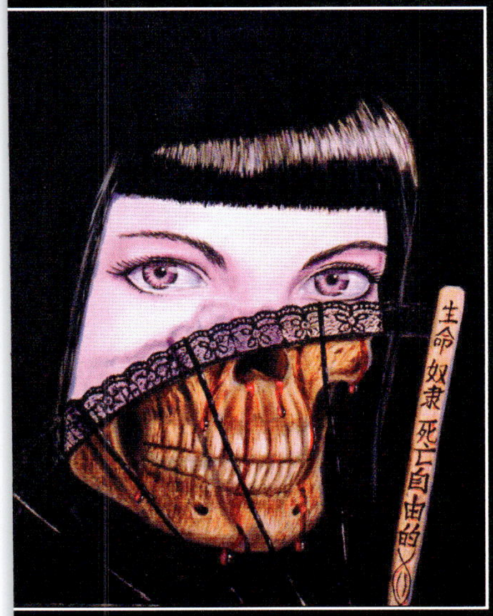

Mike Lucus

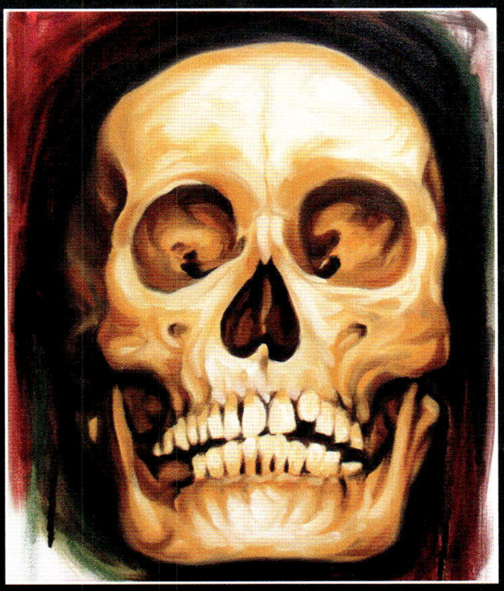

Canman

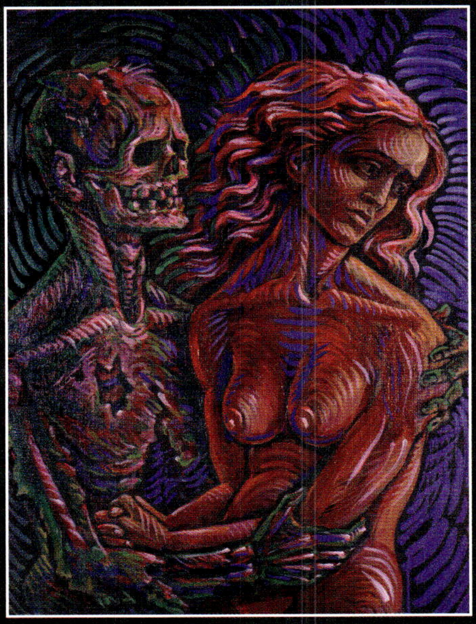

Alex Grey

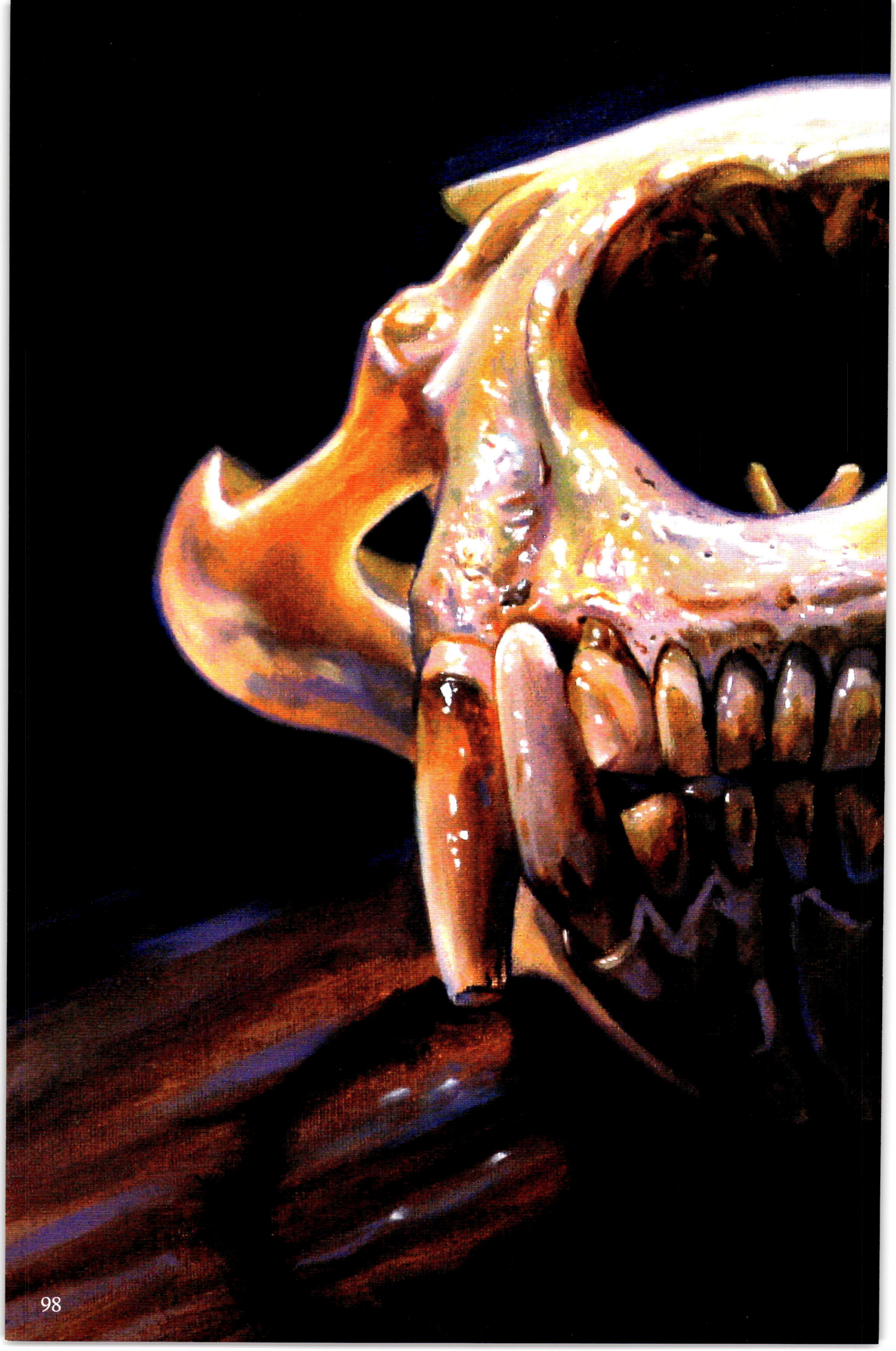

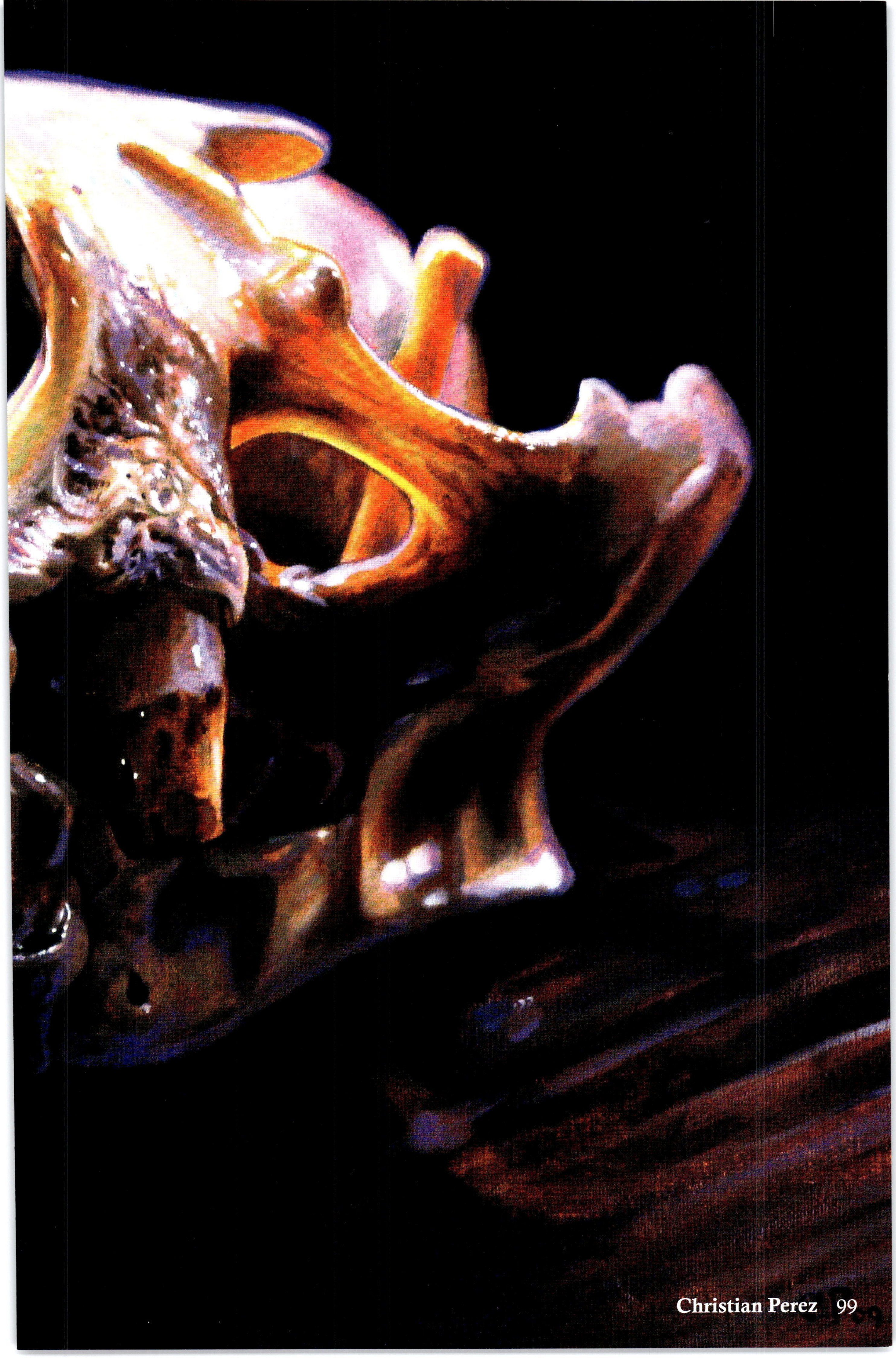

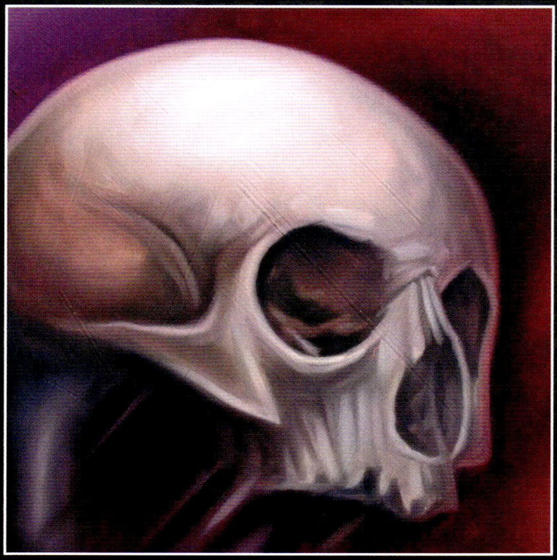

Taylor Cort

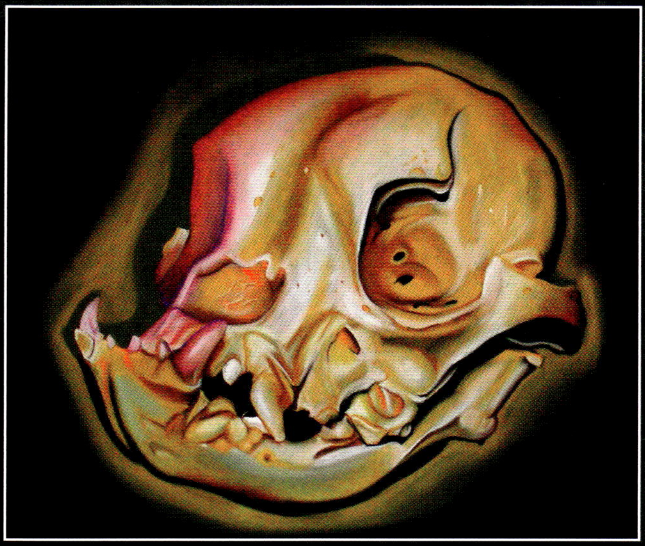

John Freyvogel

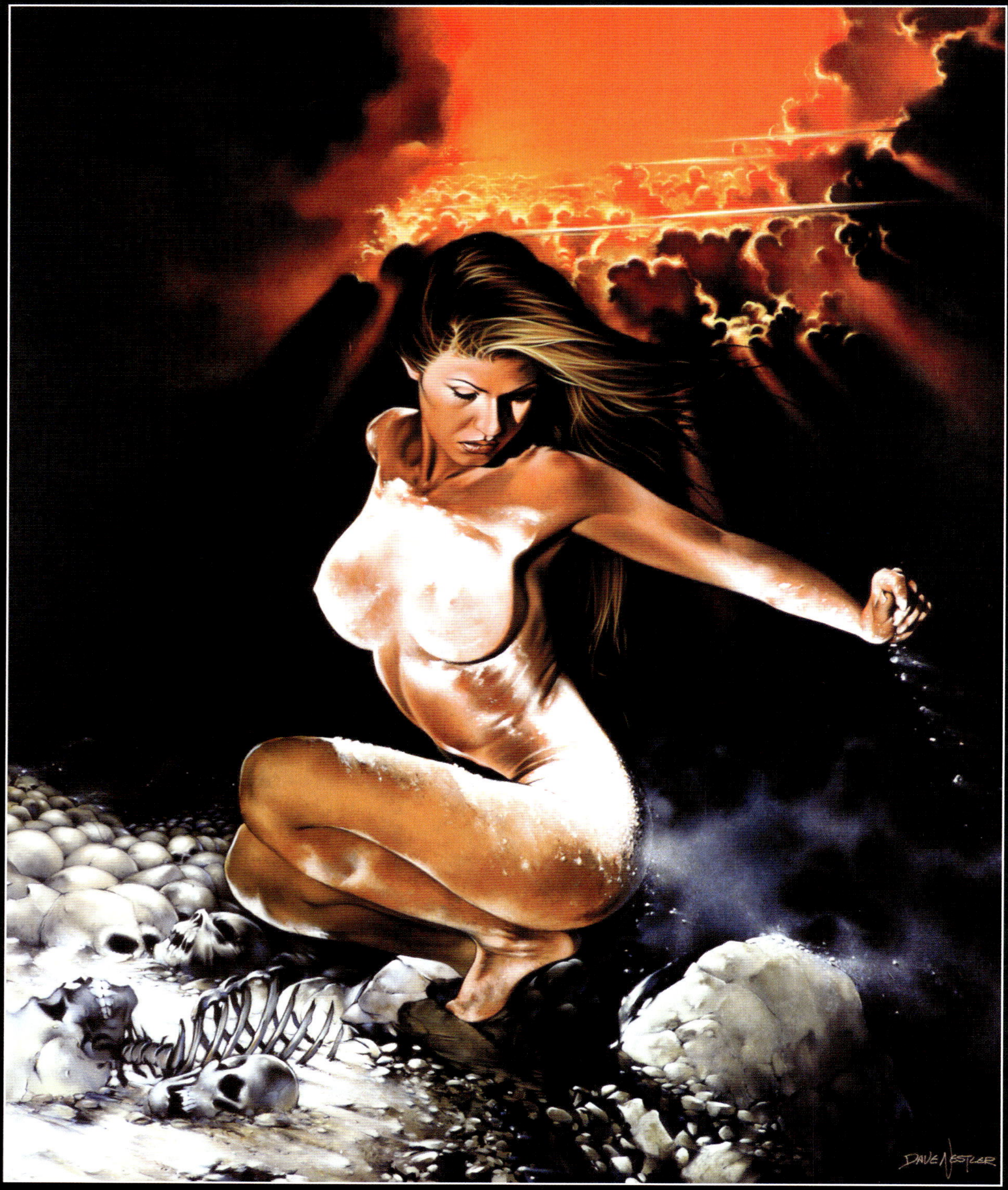

Dave Nestler

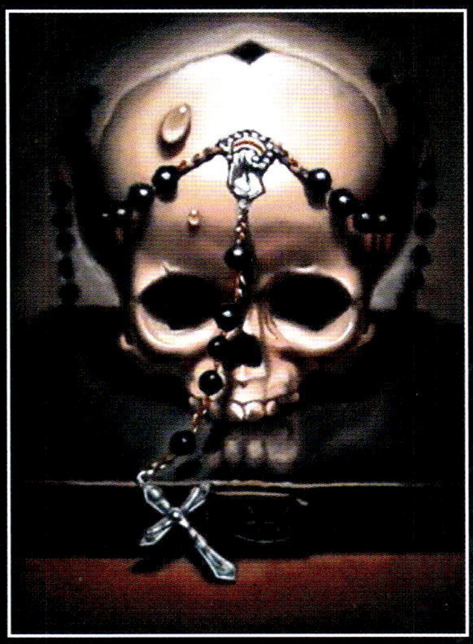
Nikko Hurtado

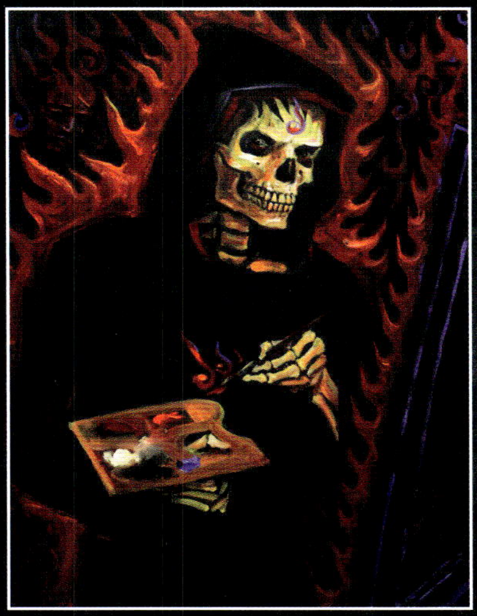
Alex Grey

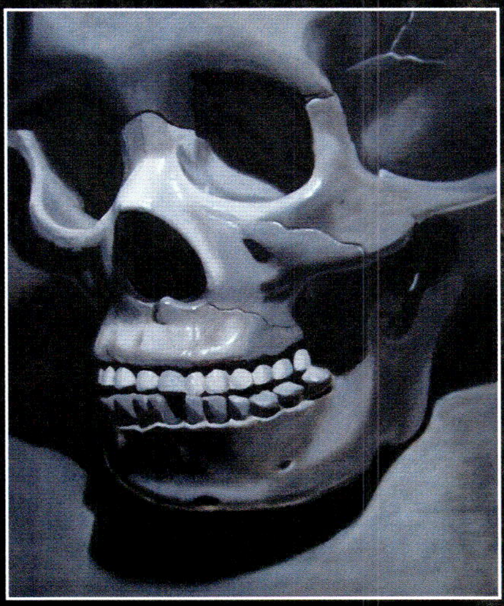
Alex Rios

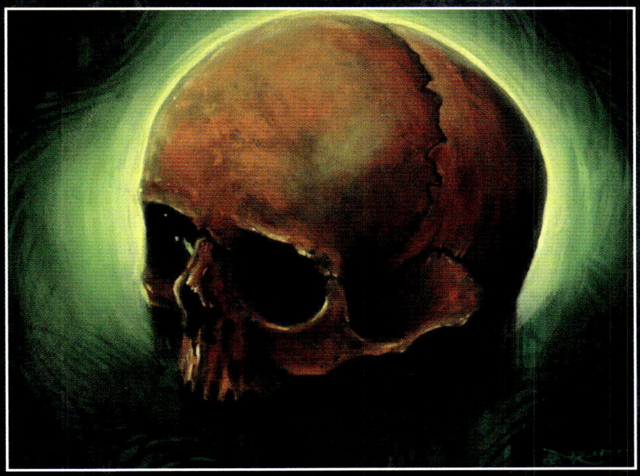
Robert Bennett

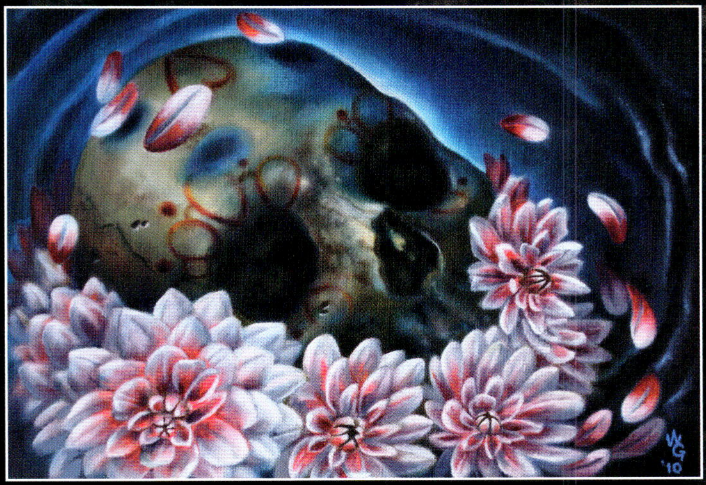
Will Gonzalez

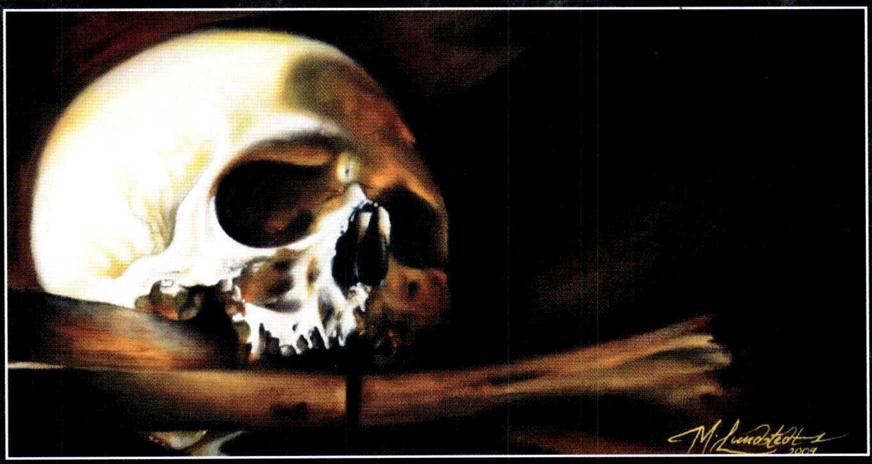
Marcus Lund

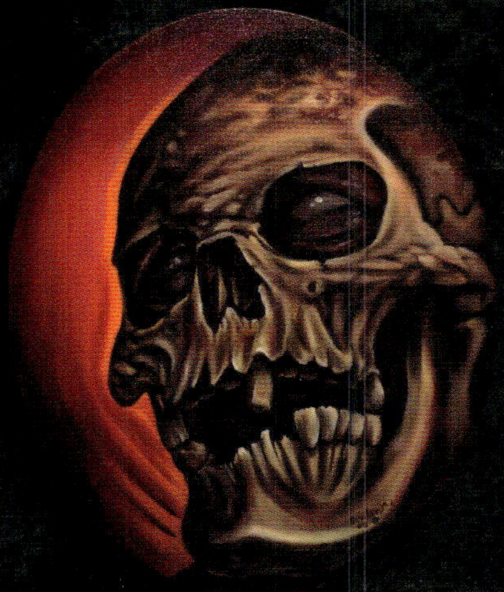
Trevor Bonkers

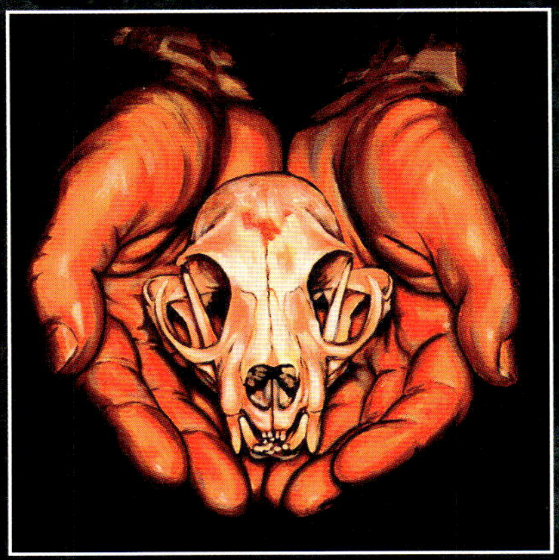
Robert Mankowski

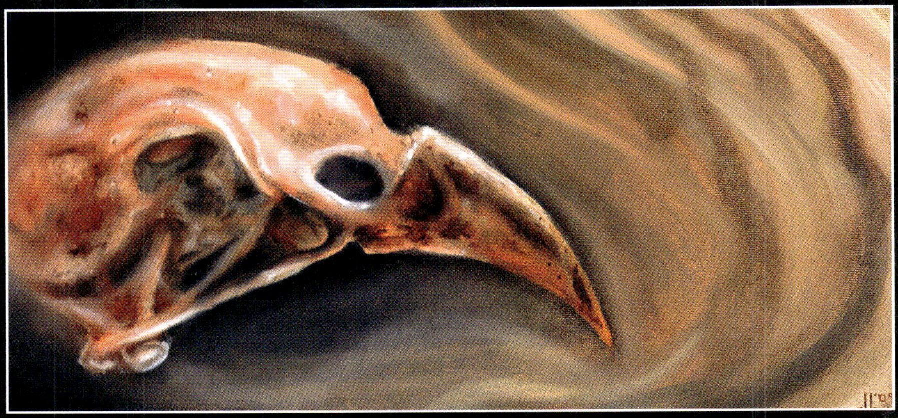
Dave Barton

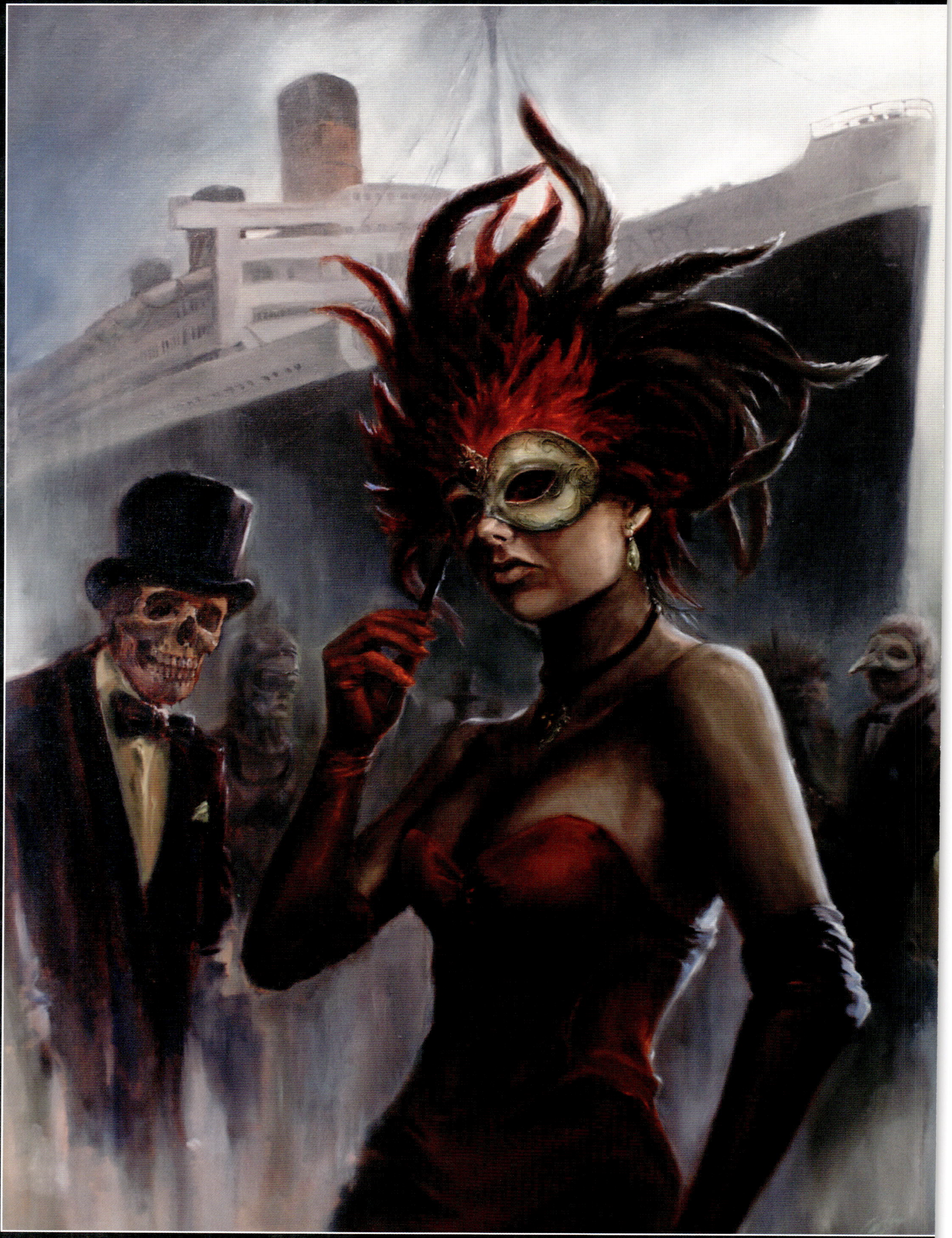

Carlos Torres

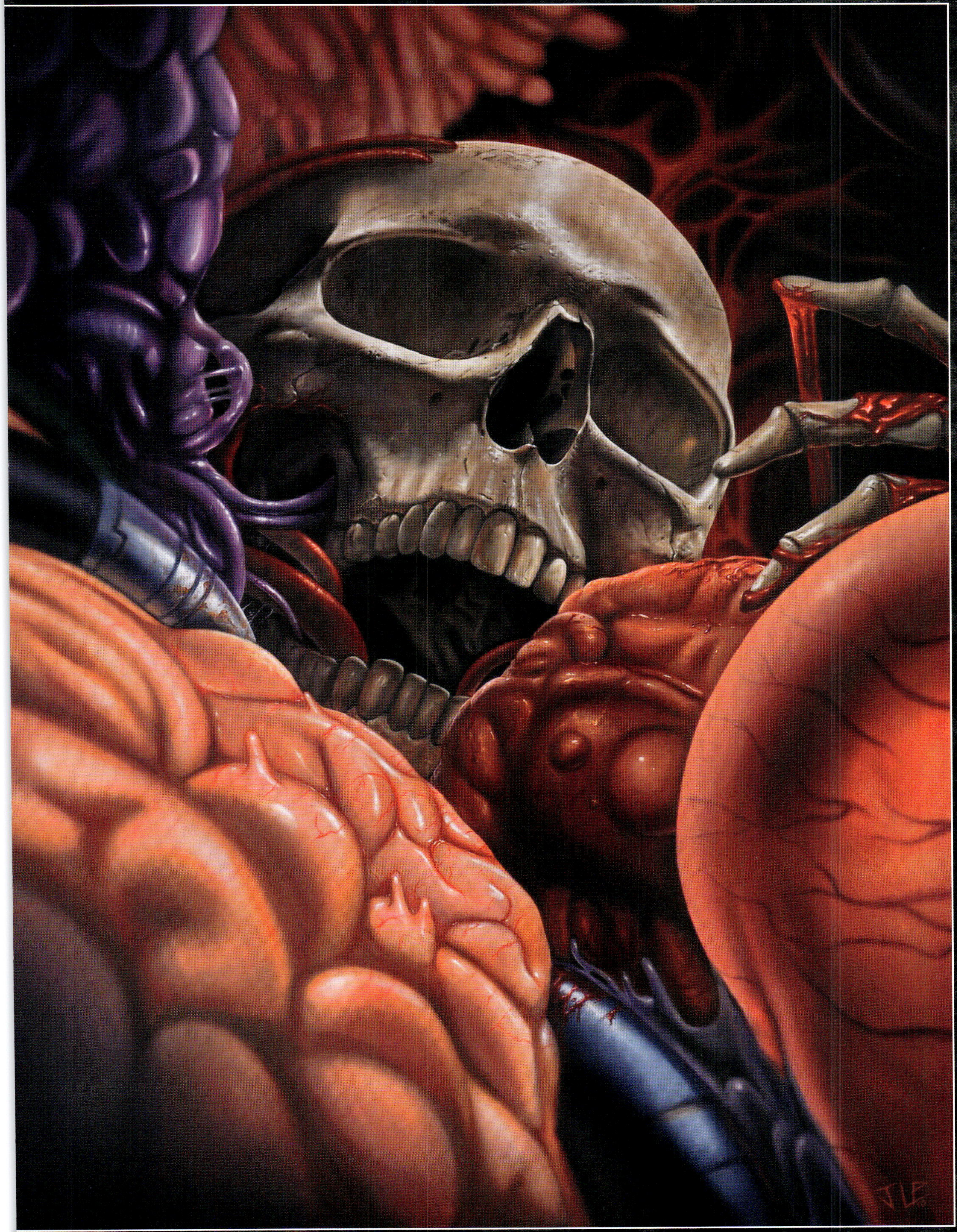

Jamie Lee Parker

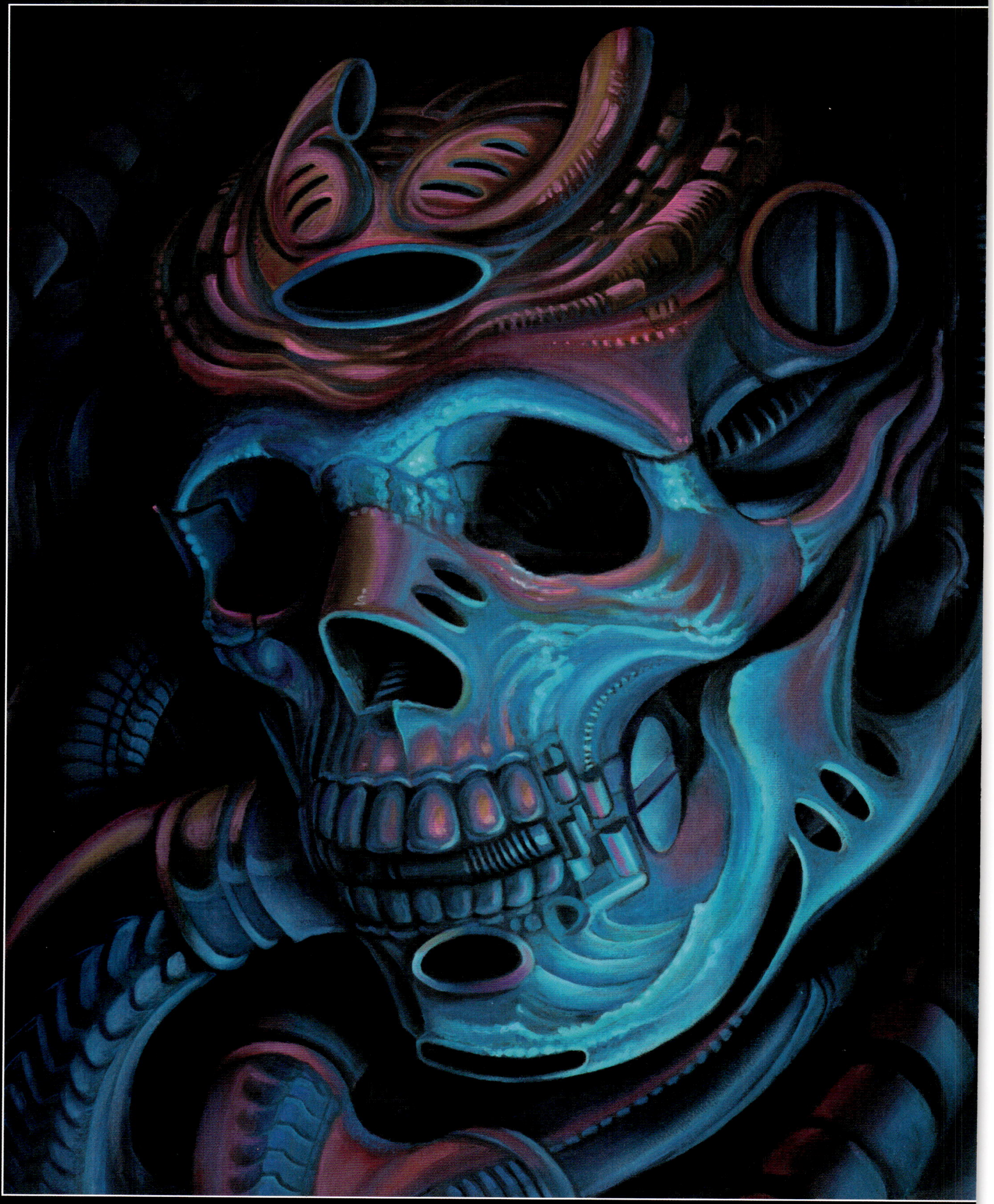

Don Mcdonald

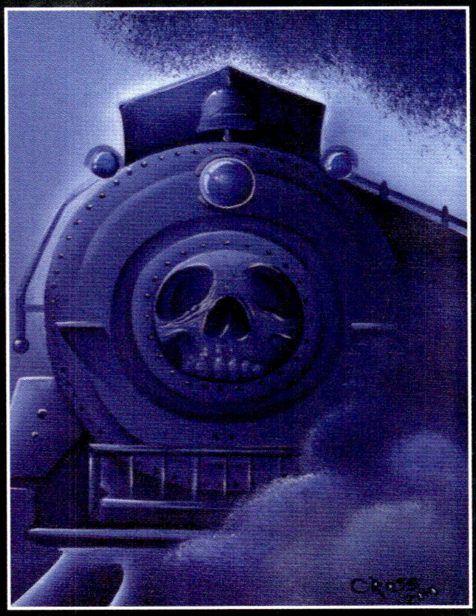
Jamie Cross

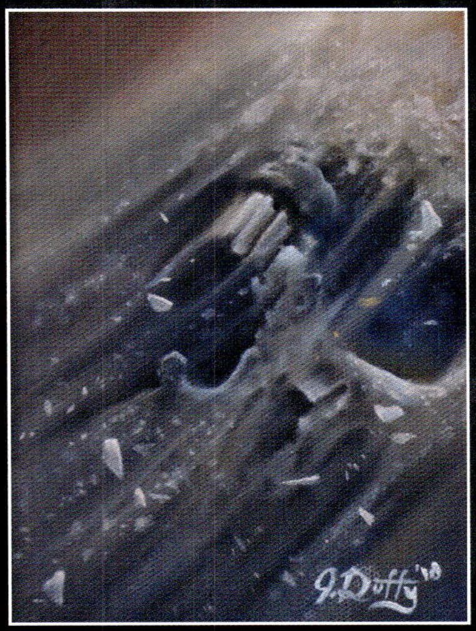
Josh Duffy

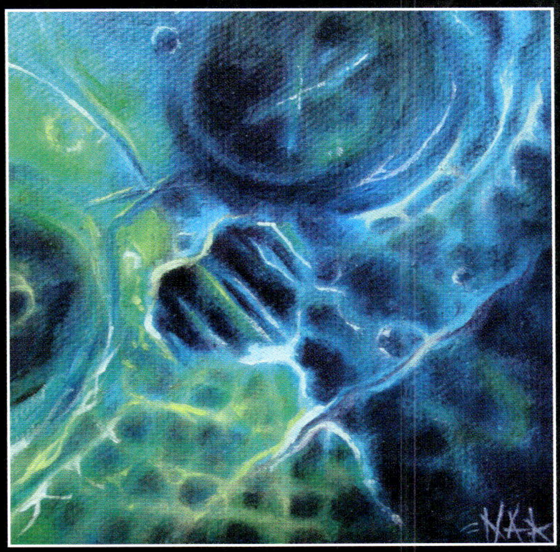
Nathan Kostechko

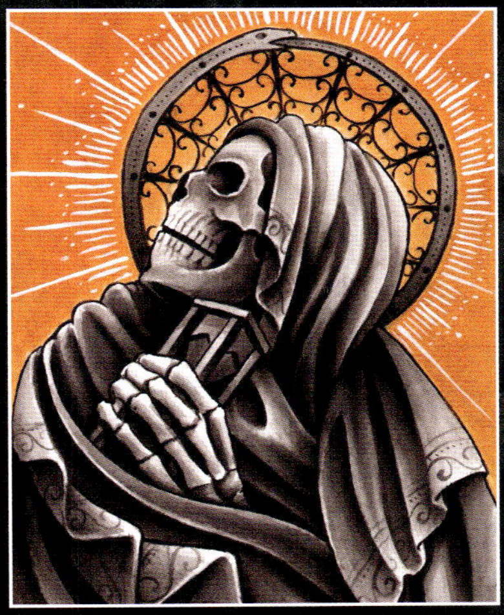
Xris Kukiel

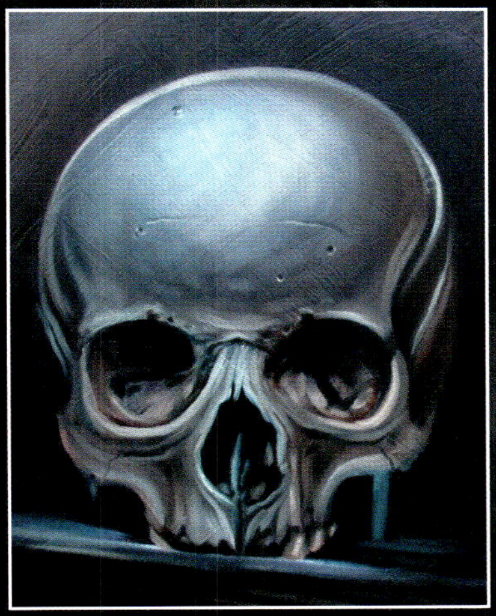
Taylor Cort

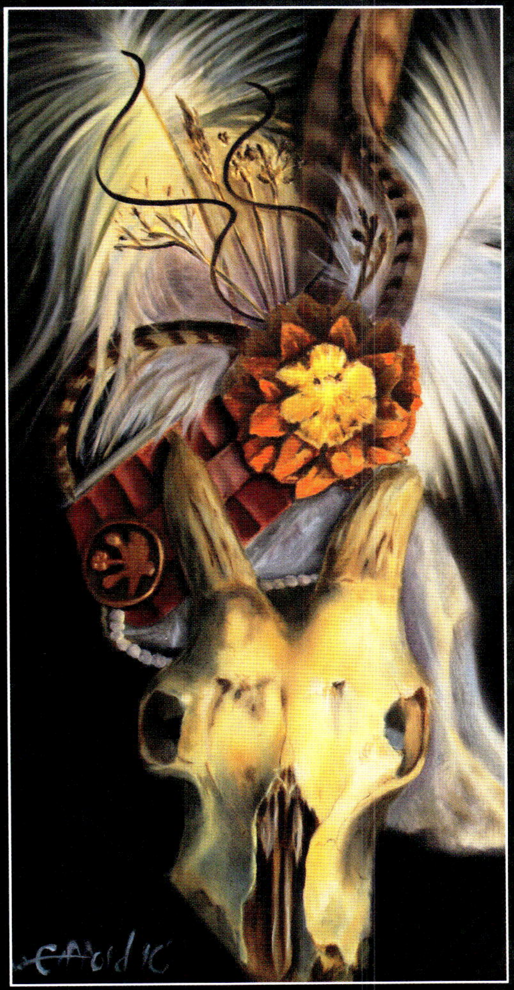
Gifford Kasen

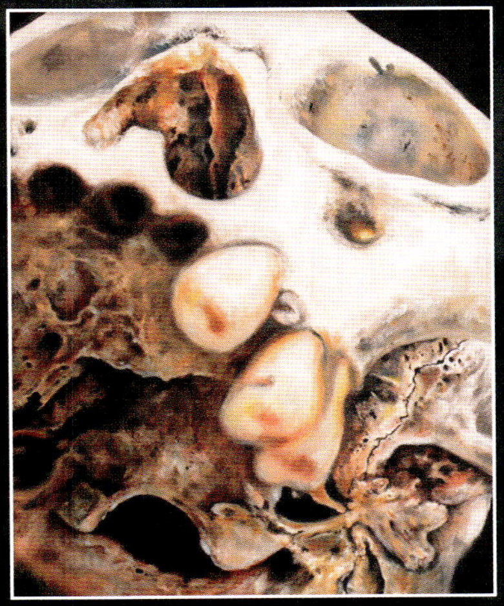
Alex Vidaud

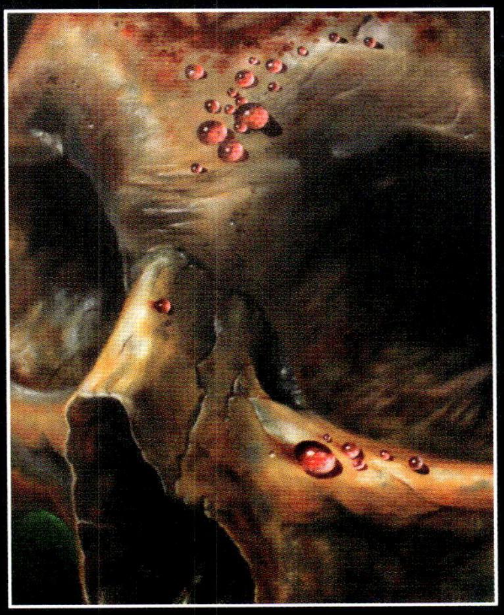
Will Gonzalez

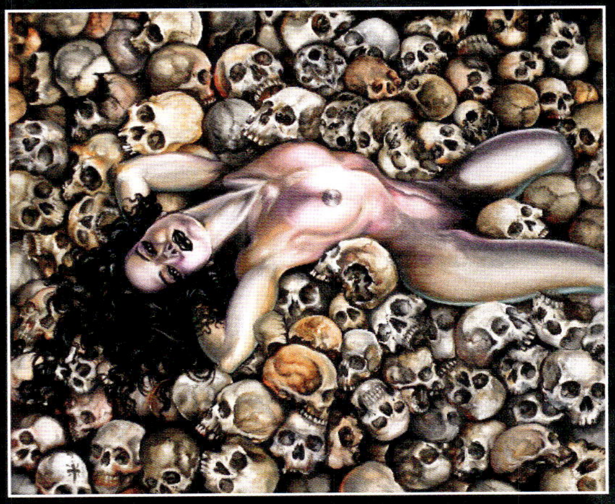
Johnny Thief

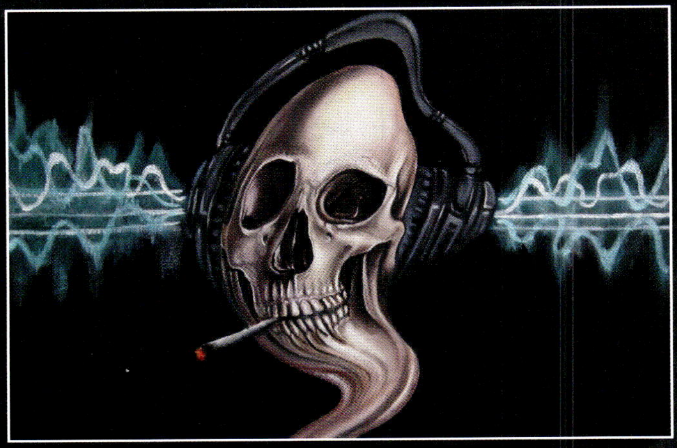
Daniel Chashoudian

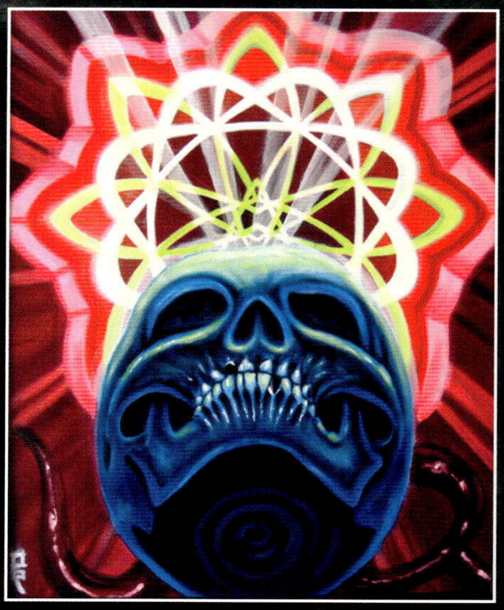
Phil Robertson

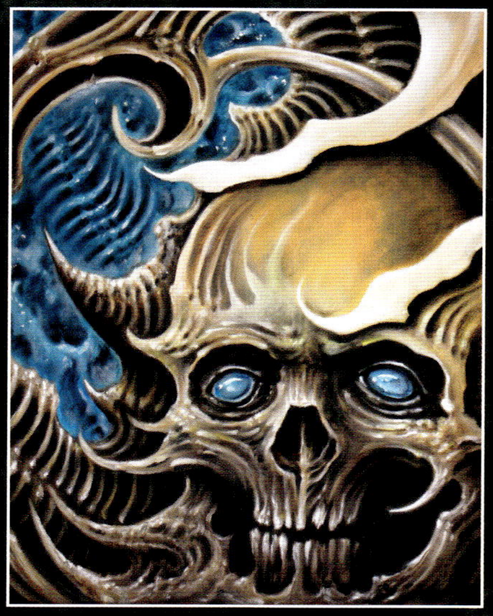
Jeremiah Barba

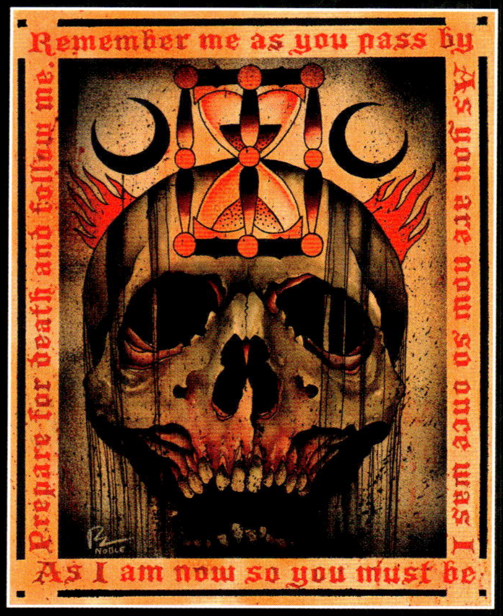

Derek Noble

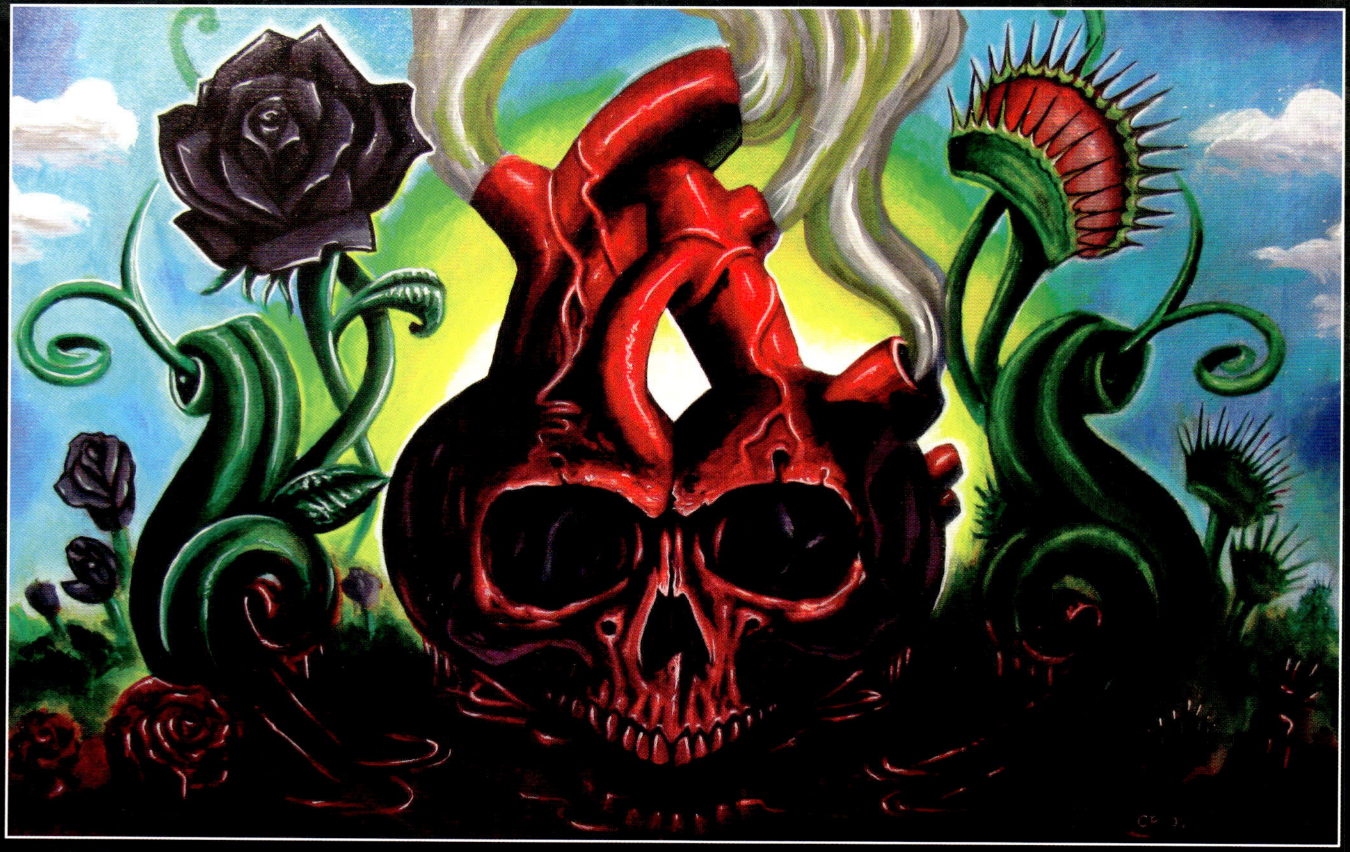
Christian Perez

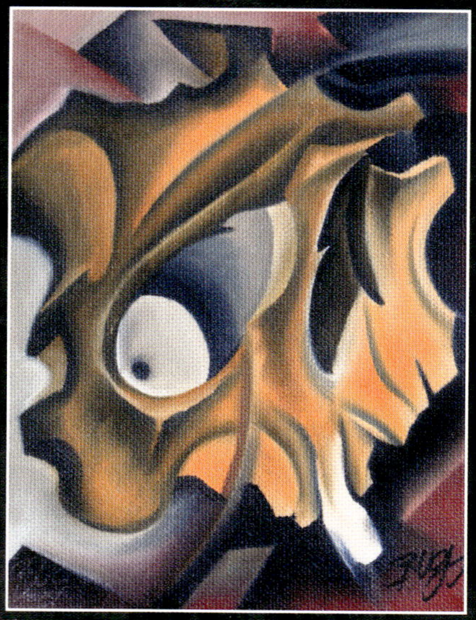

Bugs

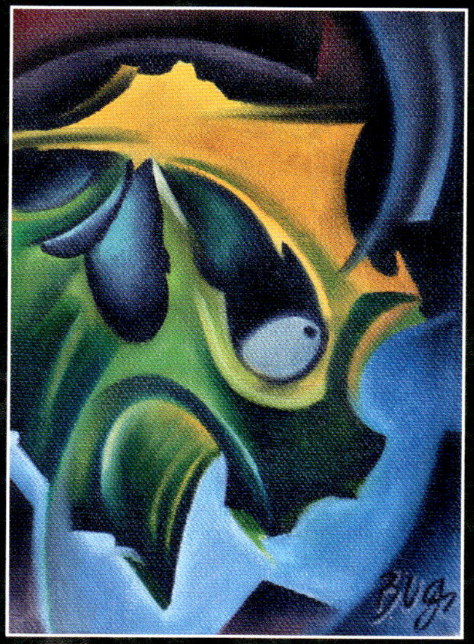

Bugs

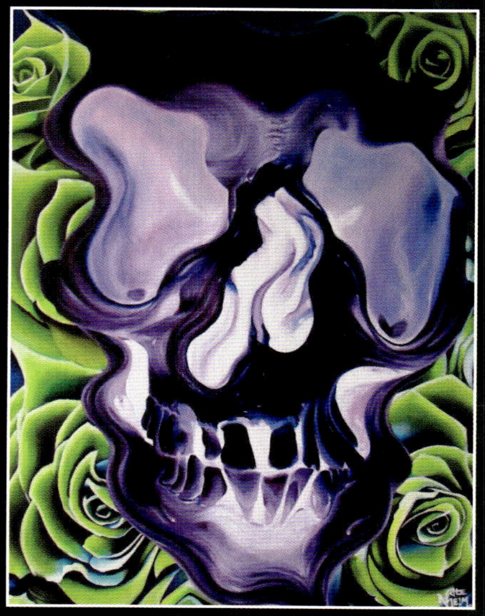
Nate Heim

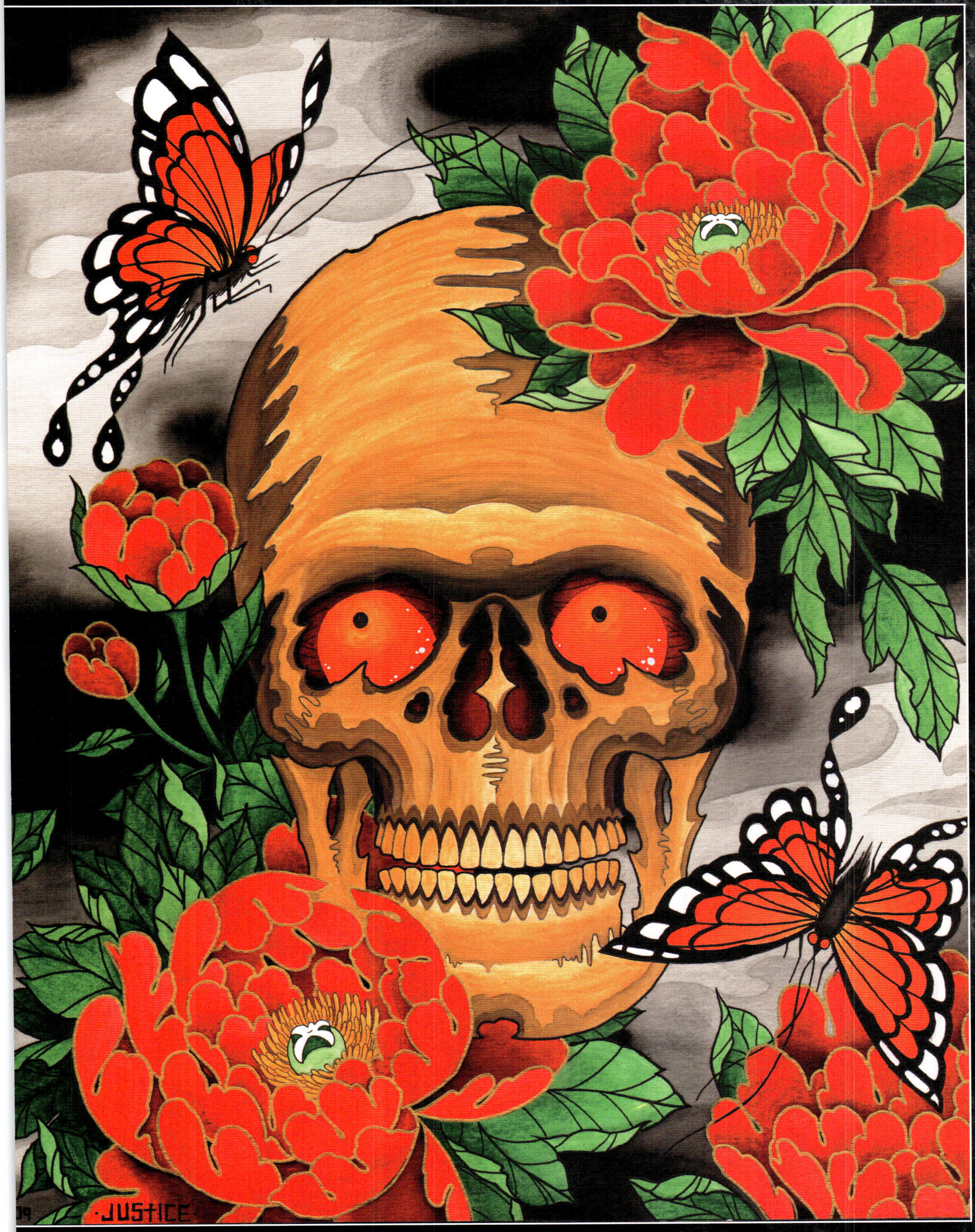

Jeremy Justice

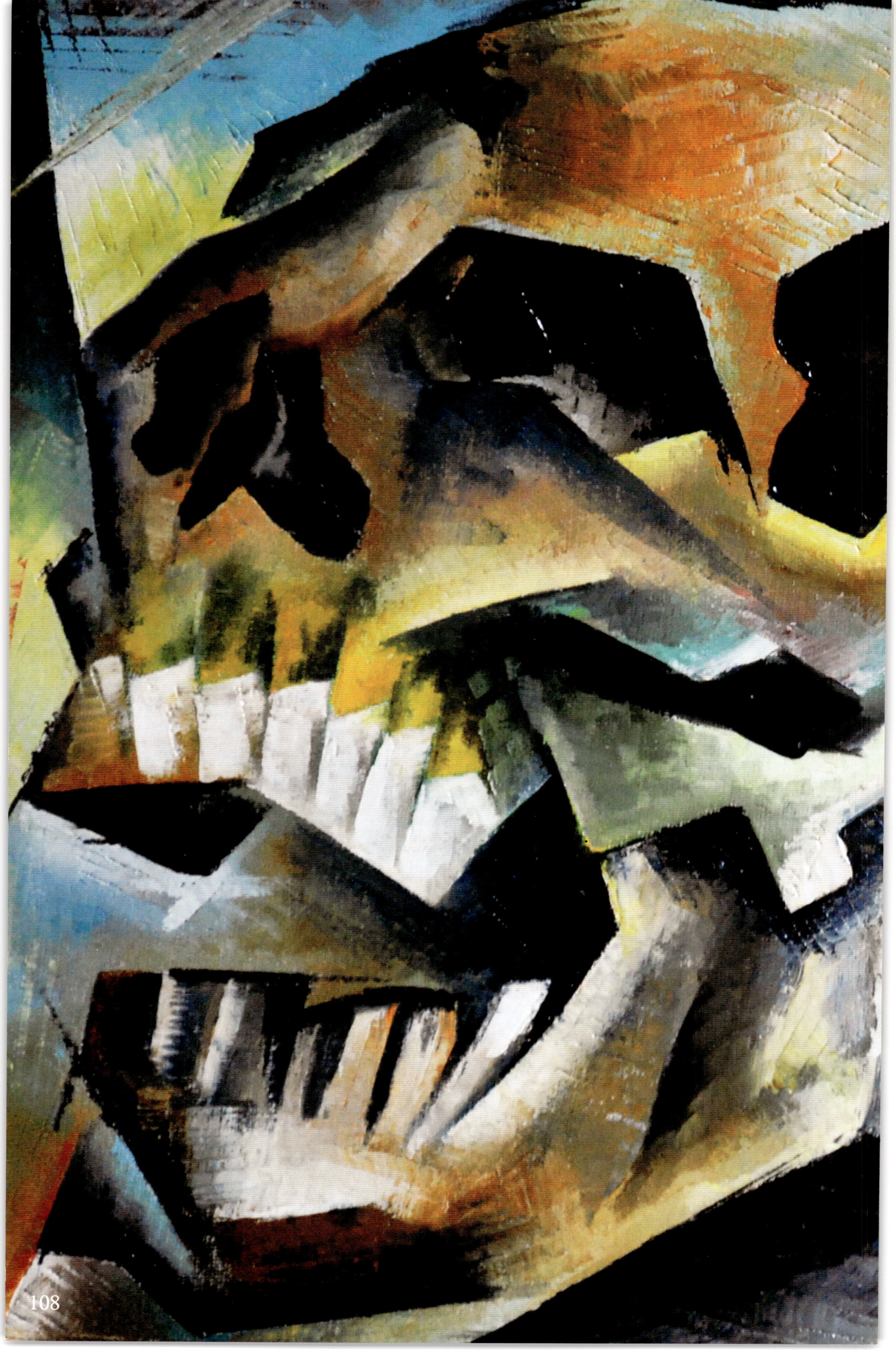

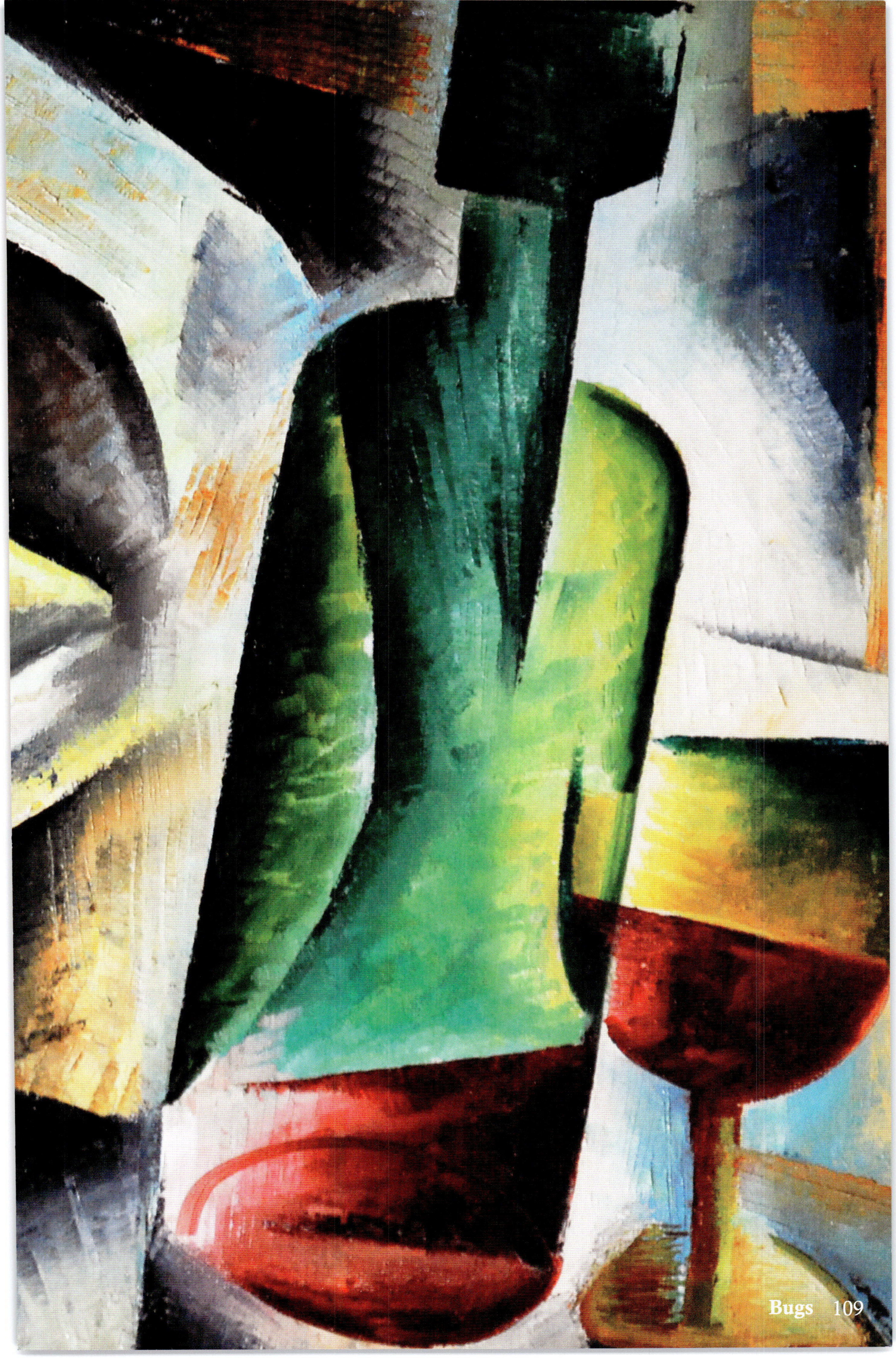

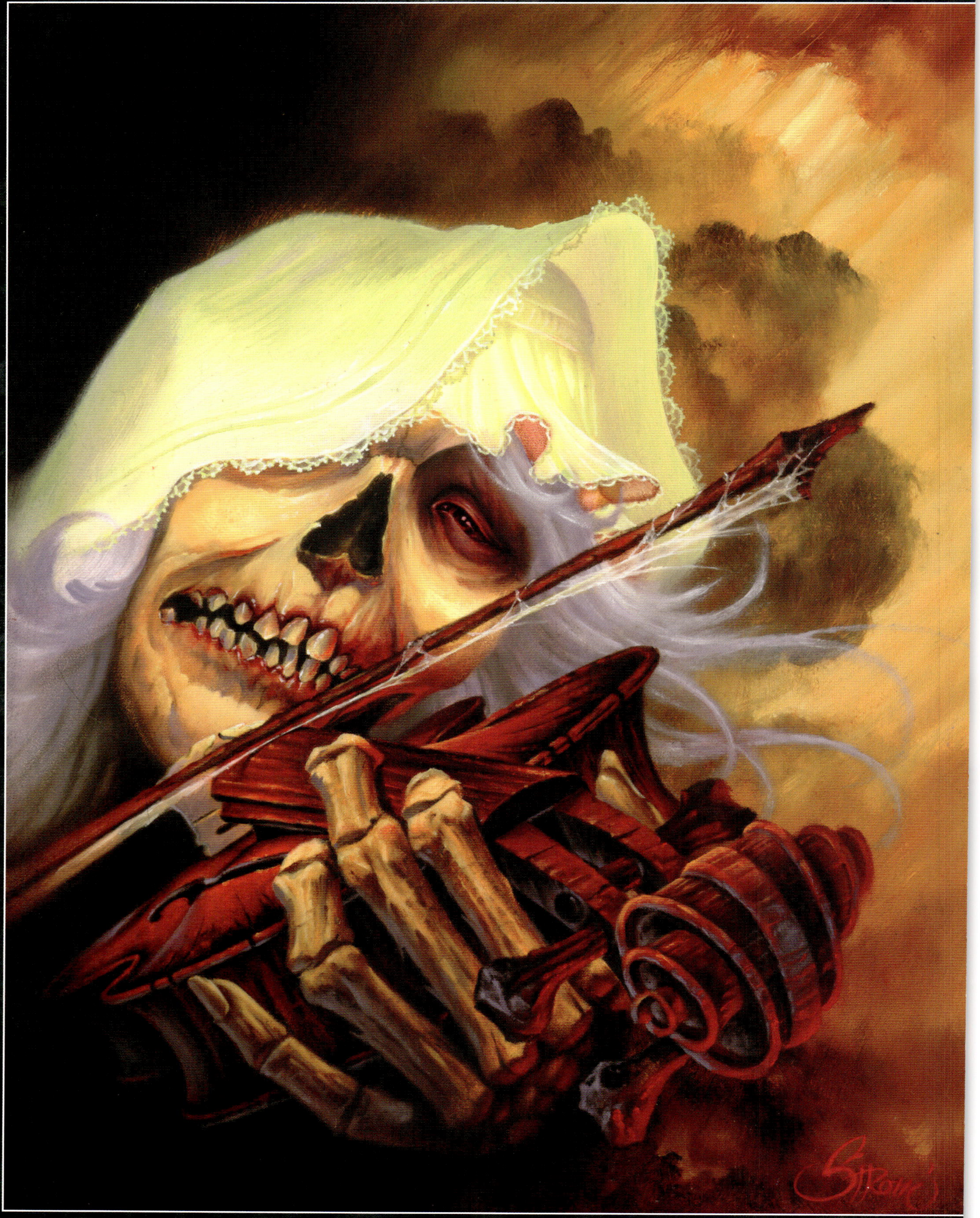

Tom Strom

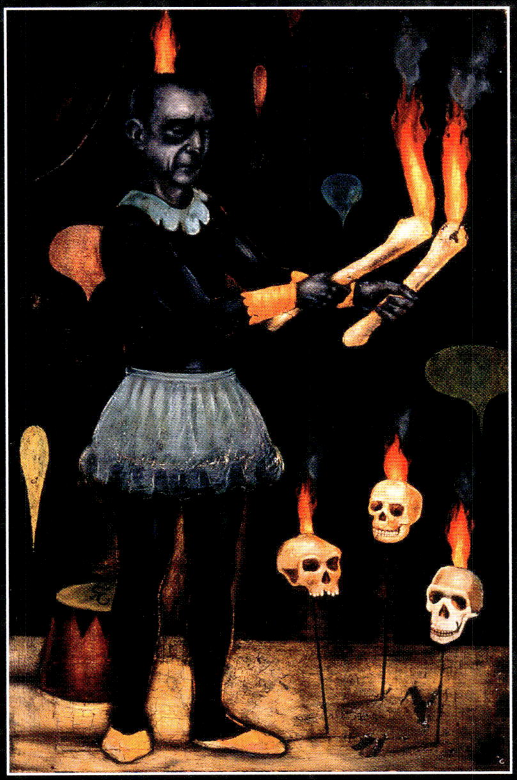

Craig LaRotonda

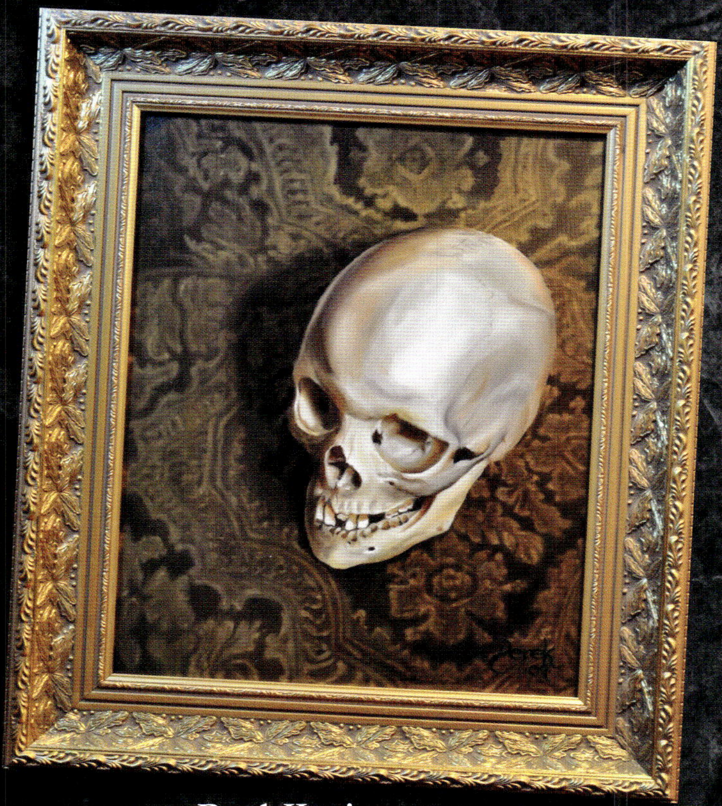

Derek Harrison

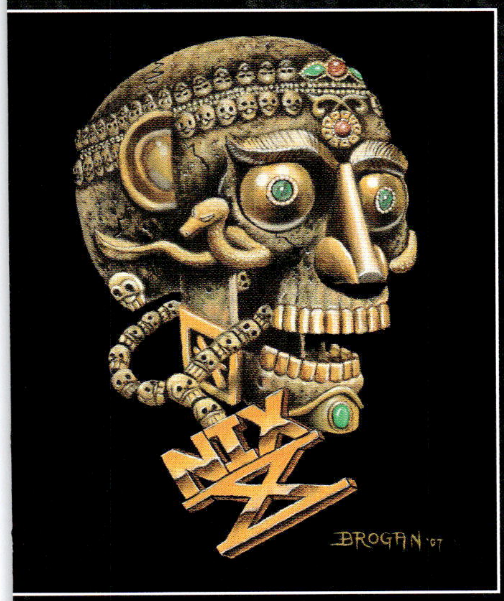

Larry Brogan

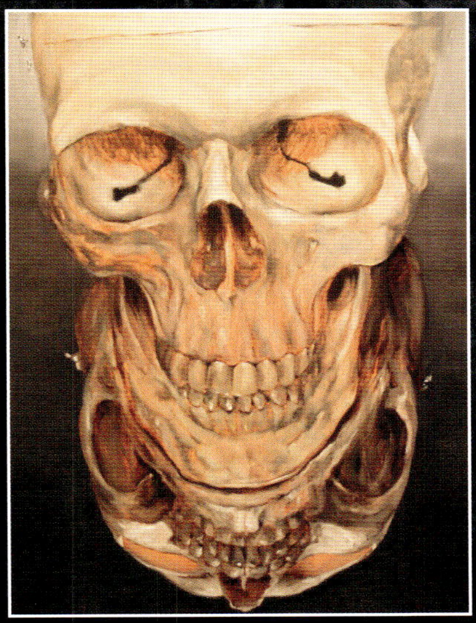

Shay Bredimus

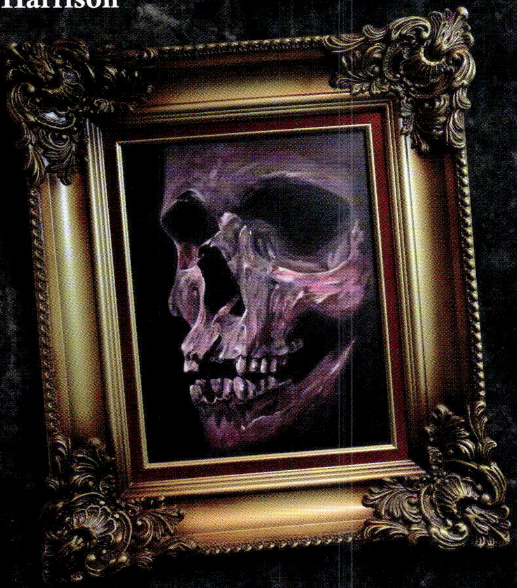

Chris Macdonald

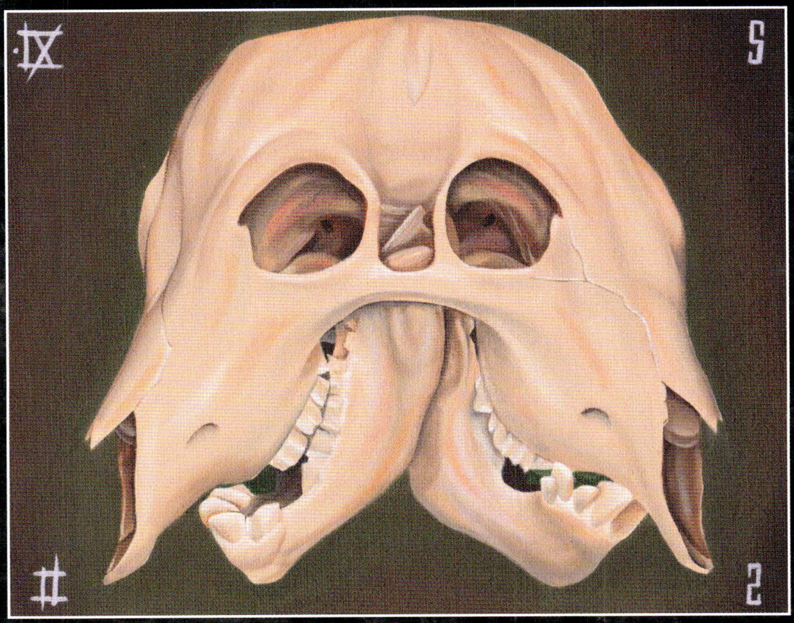

Steven Roberts

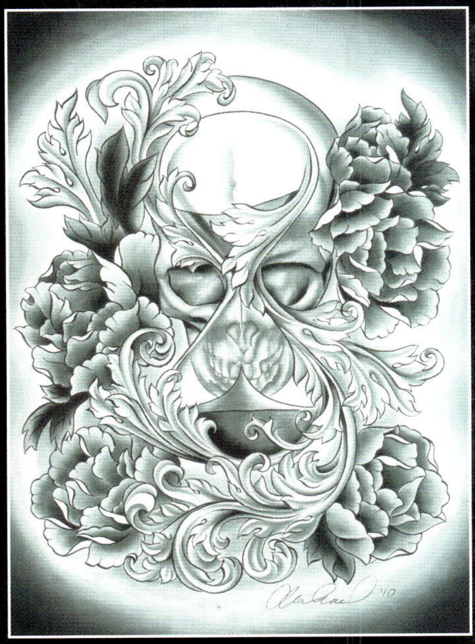

Megan Hoogland

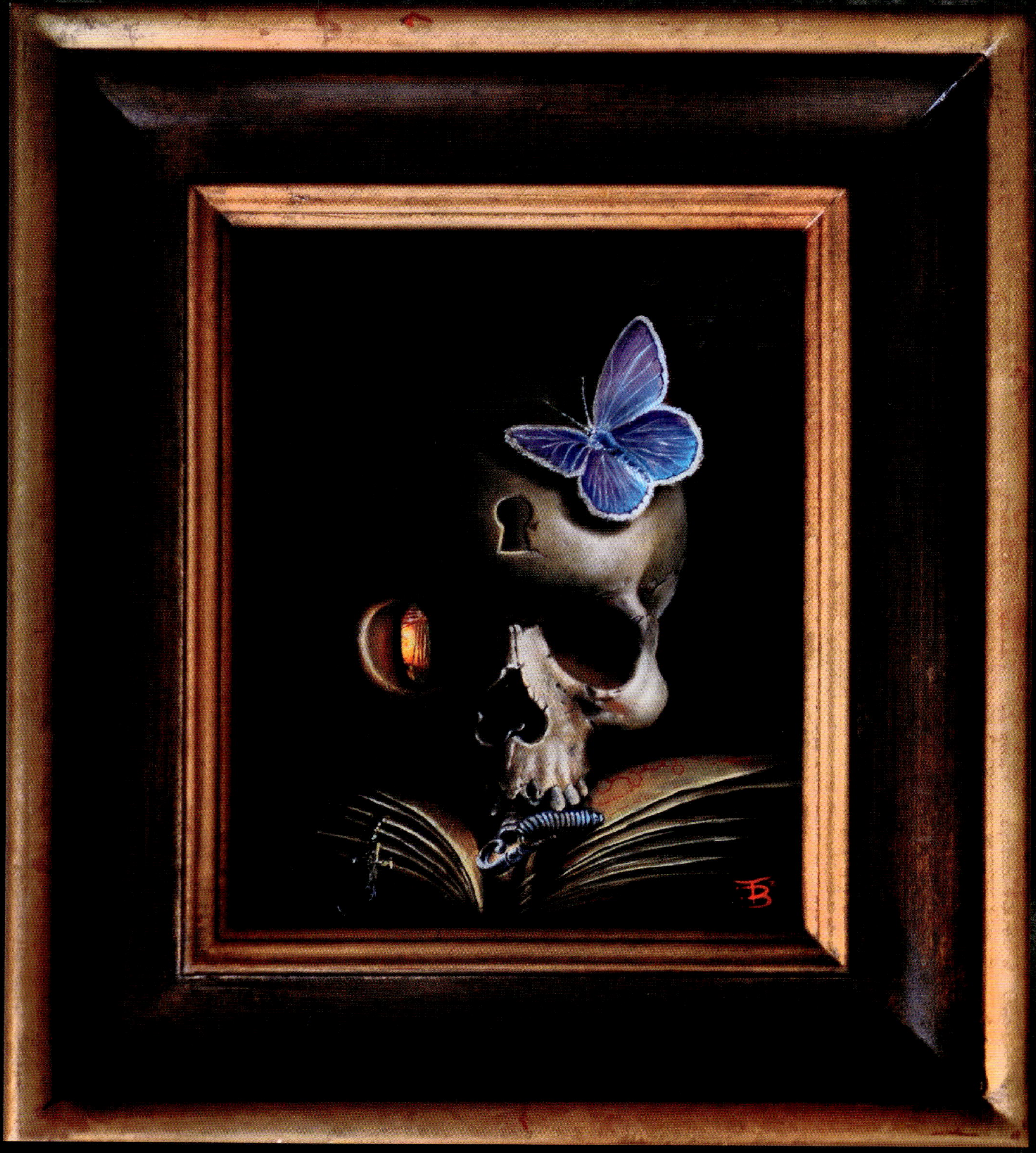

Timothy Boor

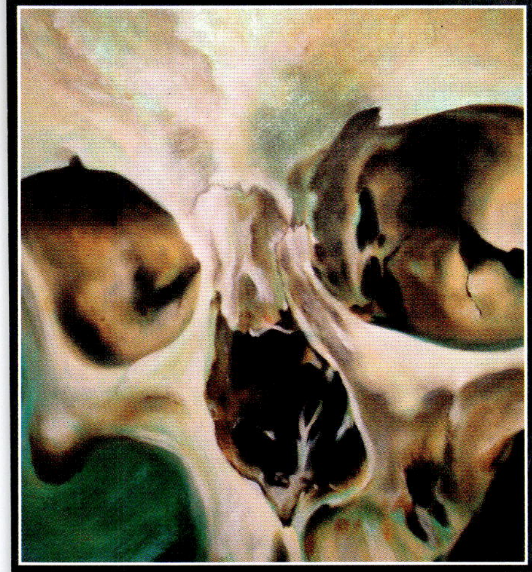

Matthew Lukesh

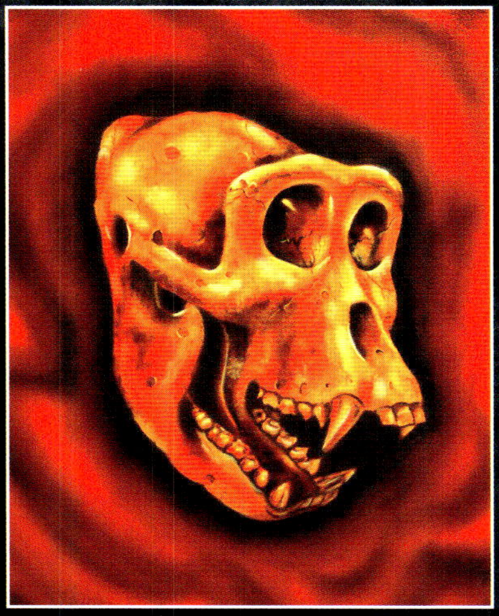

Charity Oetgen

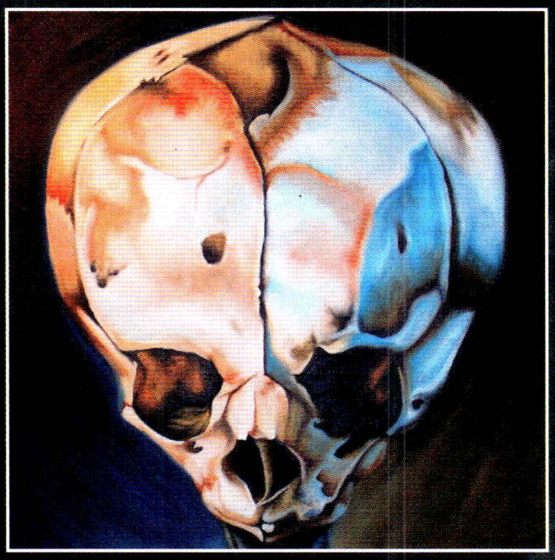

Matthew Lukesh

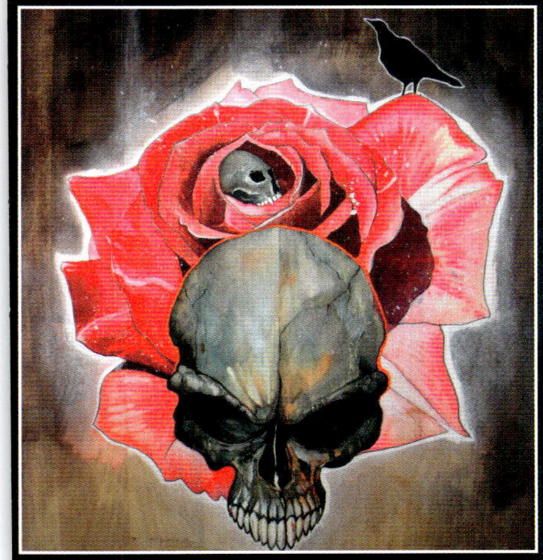

Alvin Chong

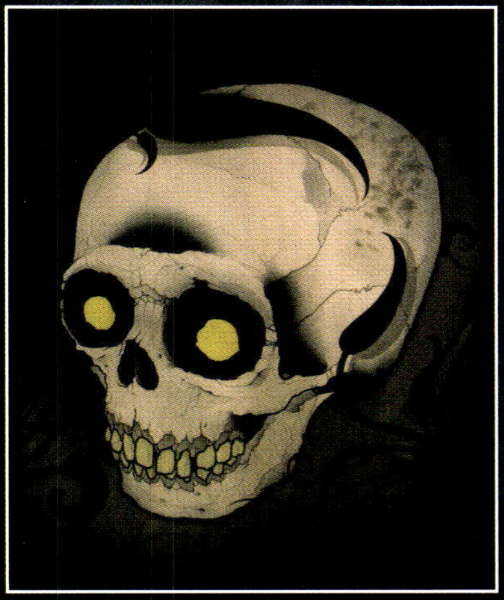

Adam Turk

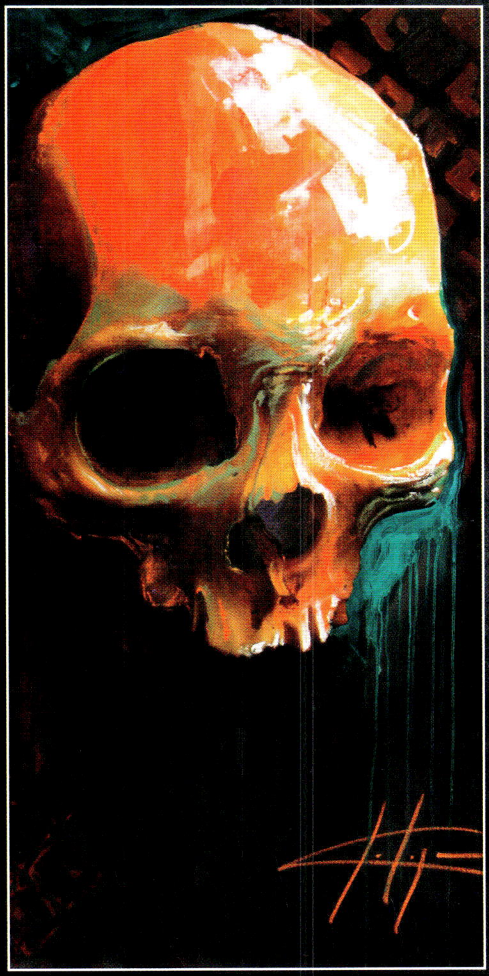

Nick Chaboya

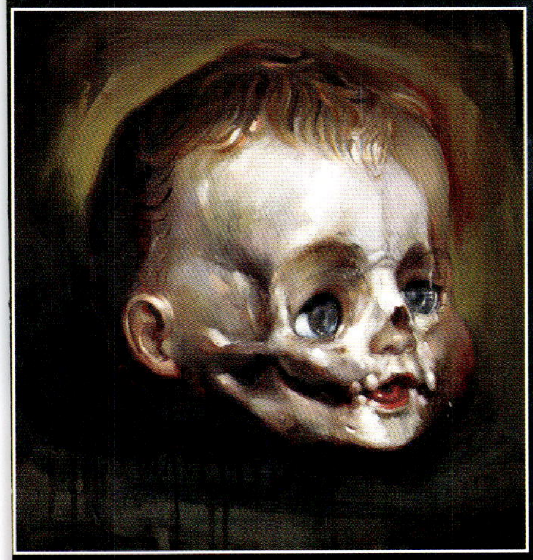

Shawn Barber

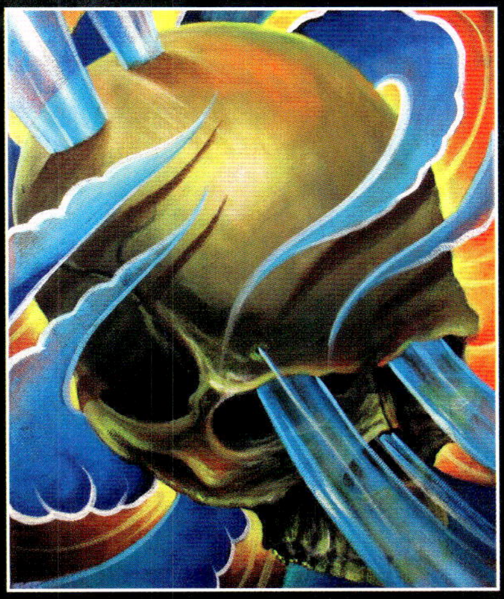

Gabe Tucker

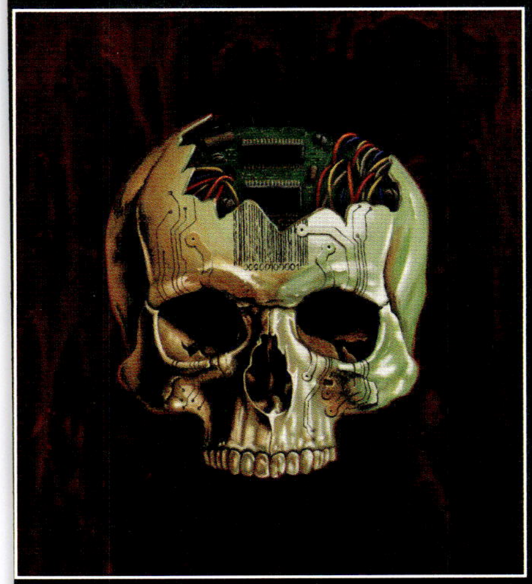

Robert Mankowski

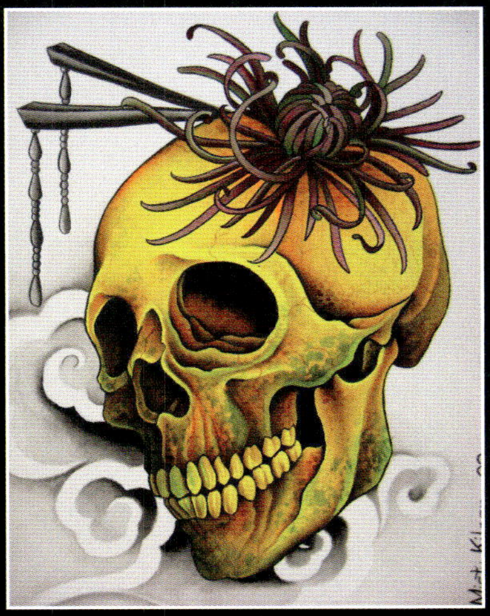

Misty Kilgore

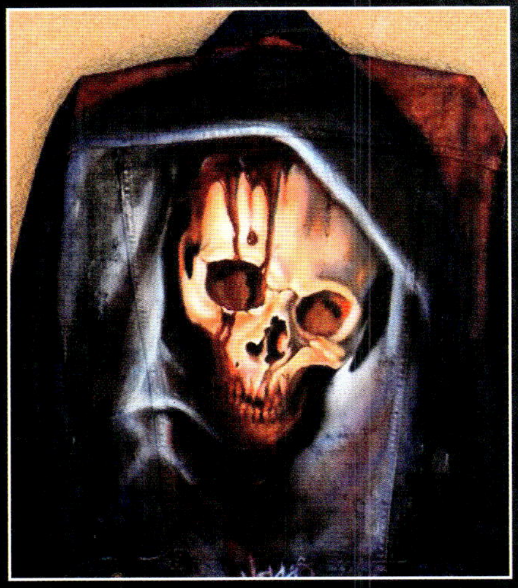

Nikko Hurtado

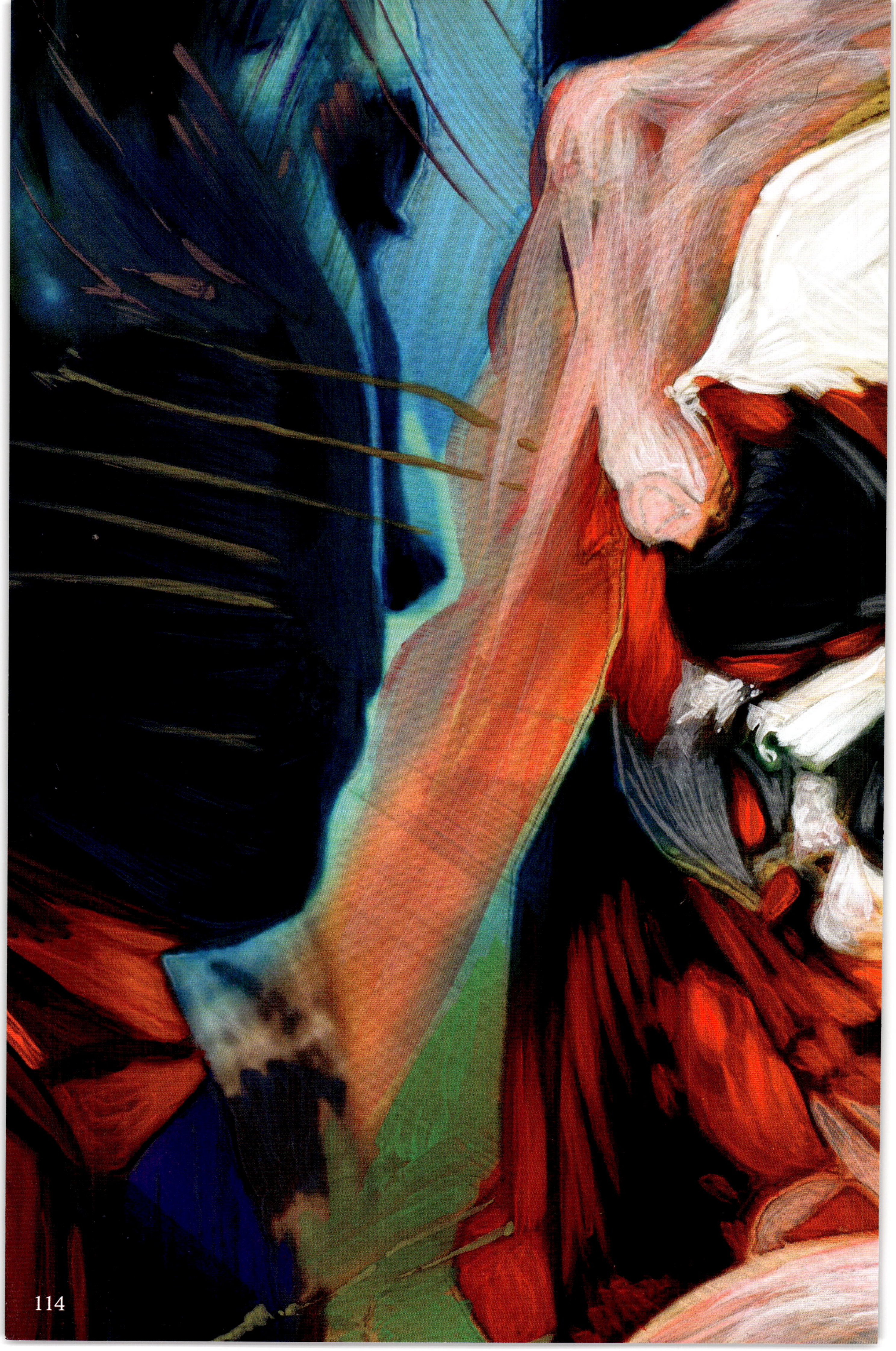

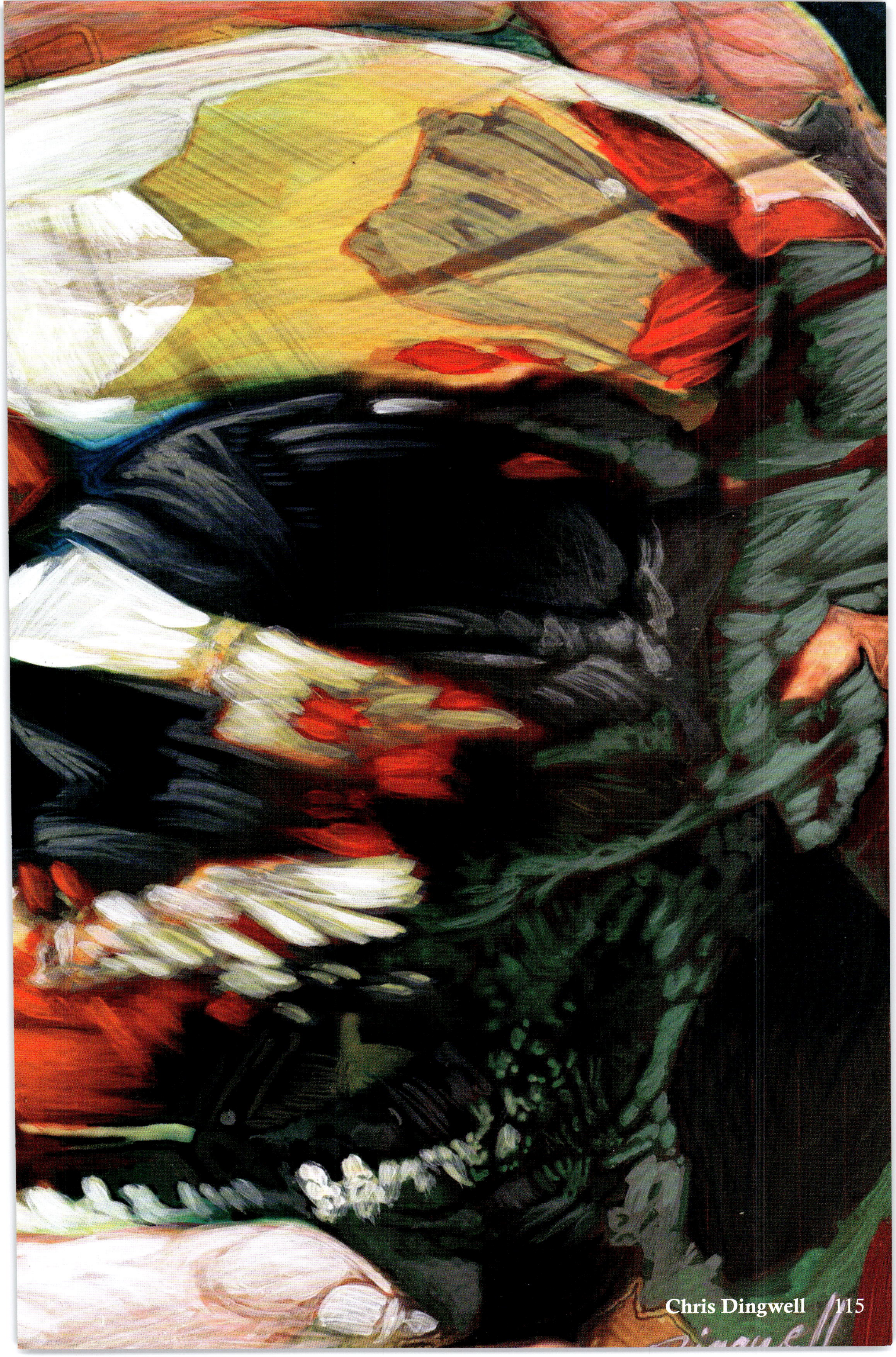

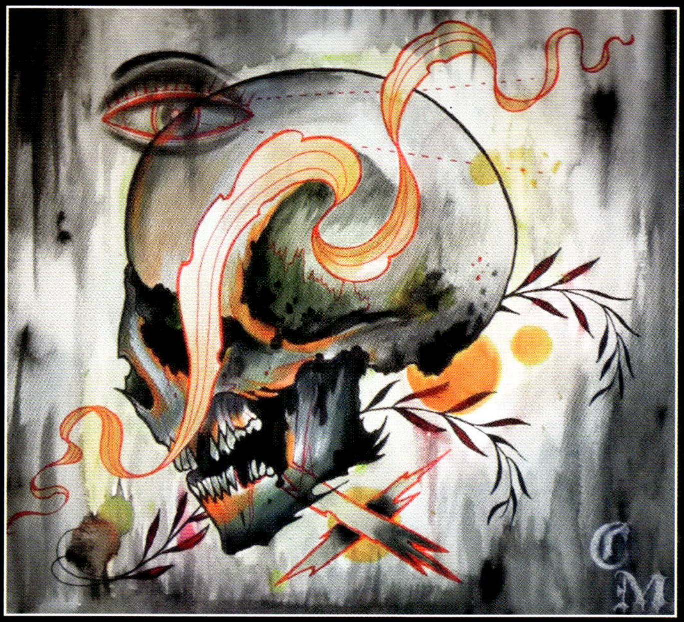

Chris Macdonald

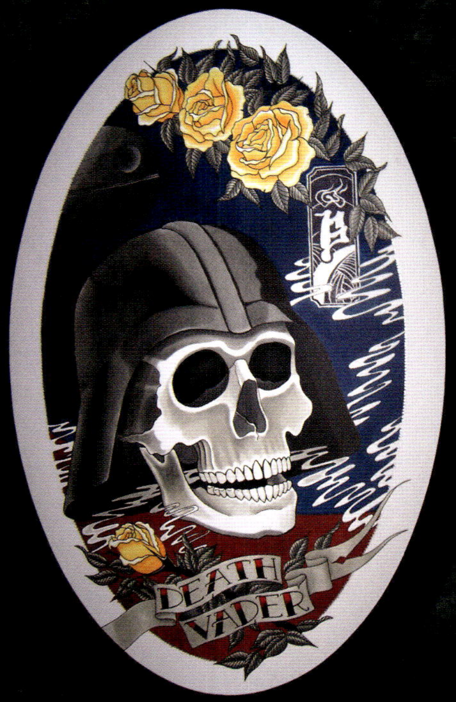

Reuben Briggs

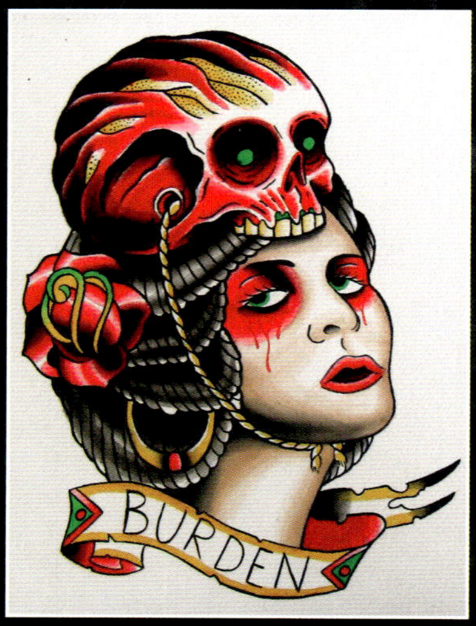

Nick Phillips

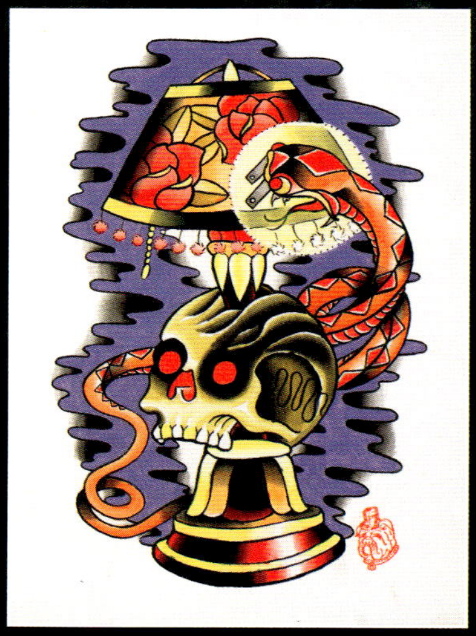

Nick Phillips

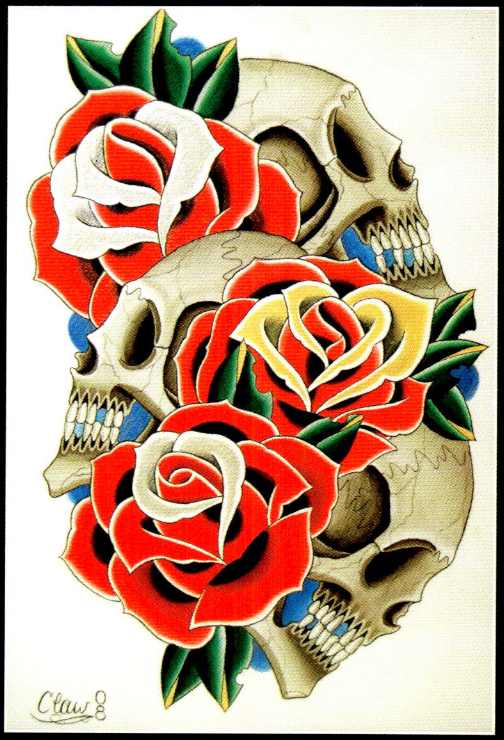

Matthew Verseput

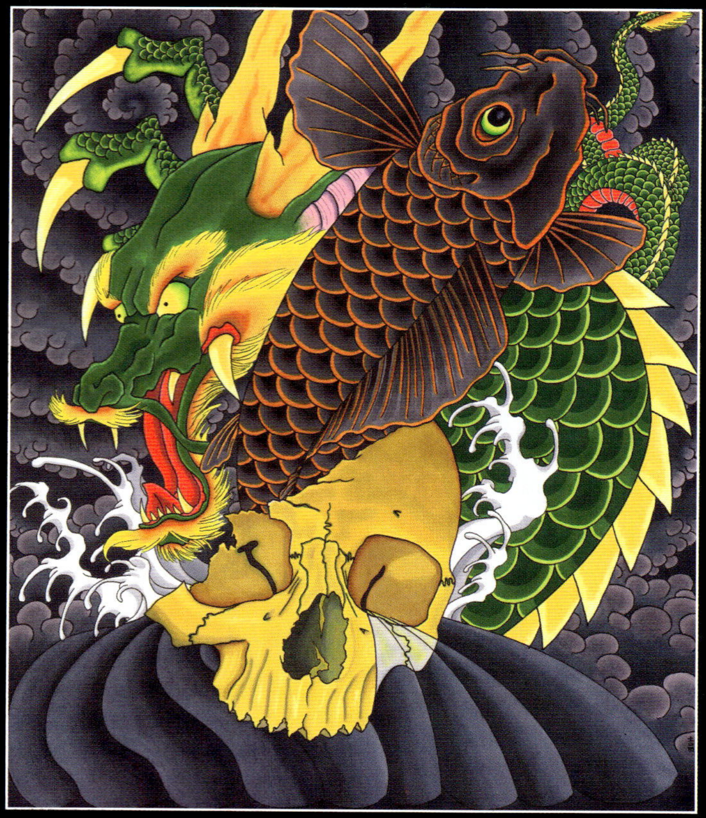

Lucas Yoshida

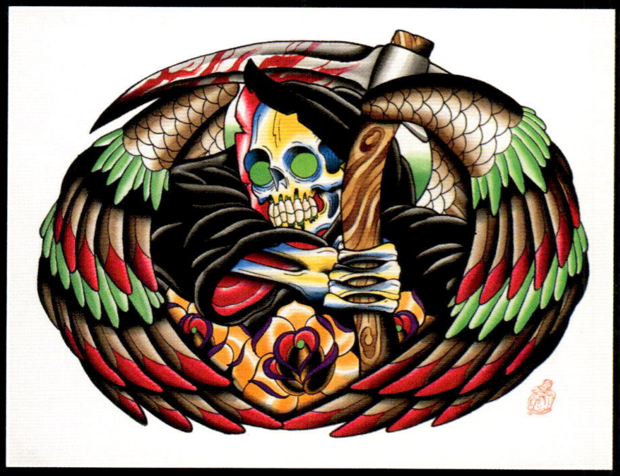

Nick Phillips

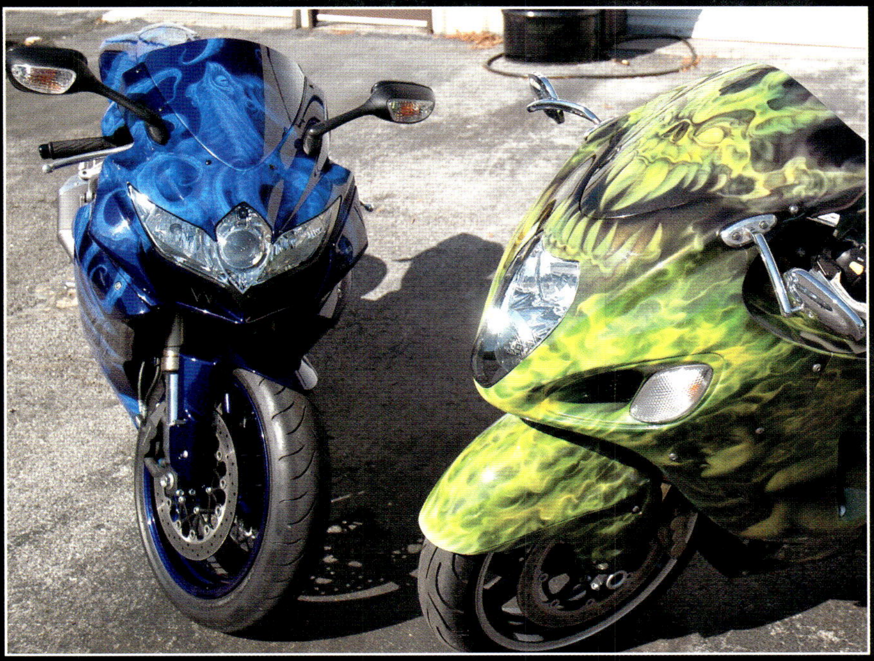
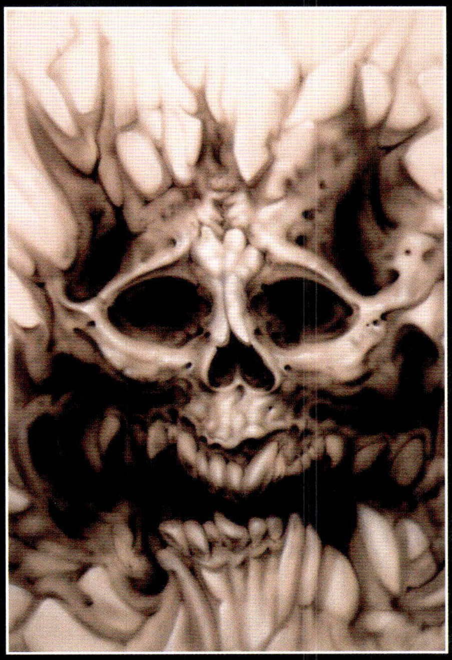
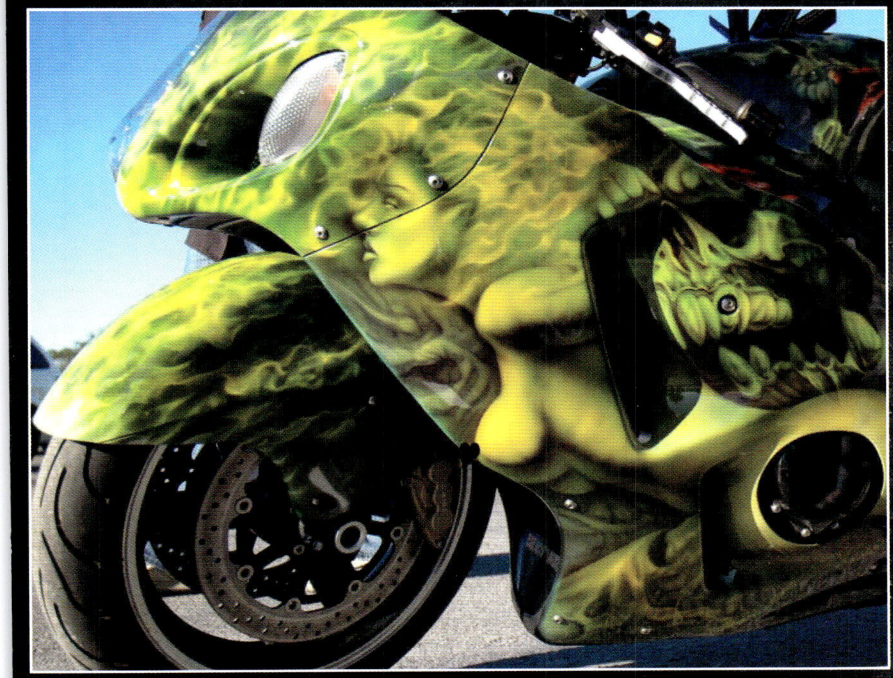
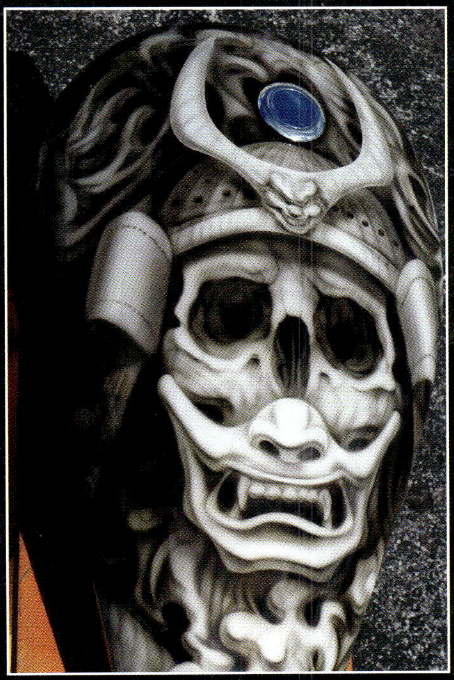
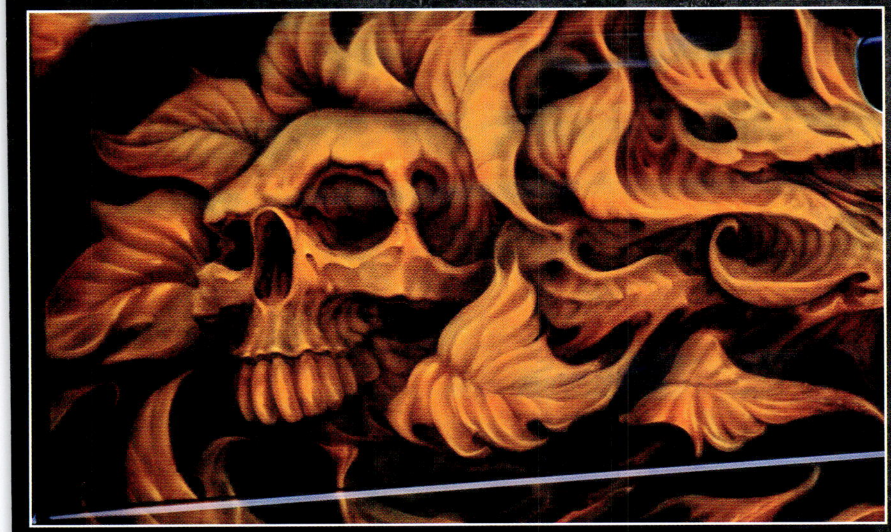
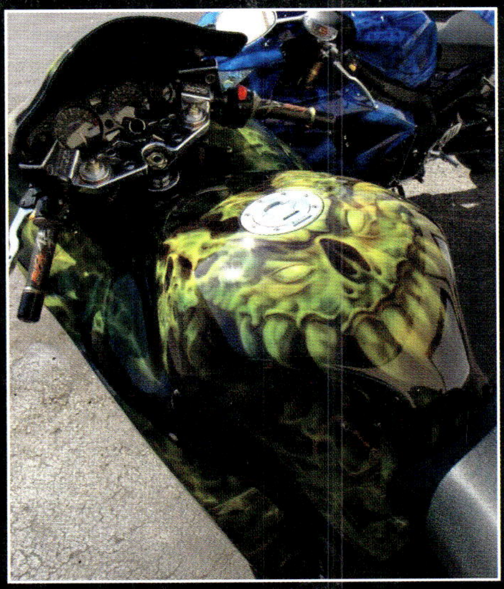

Will Ralston

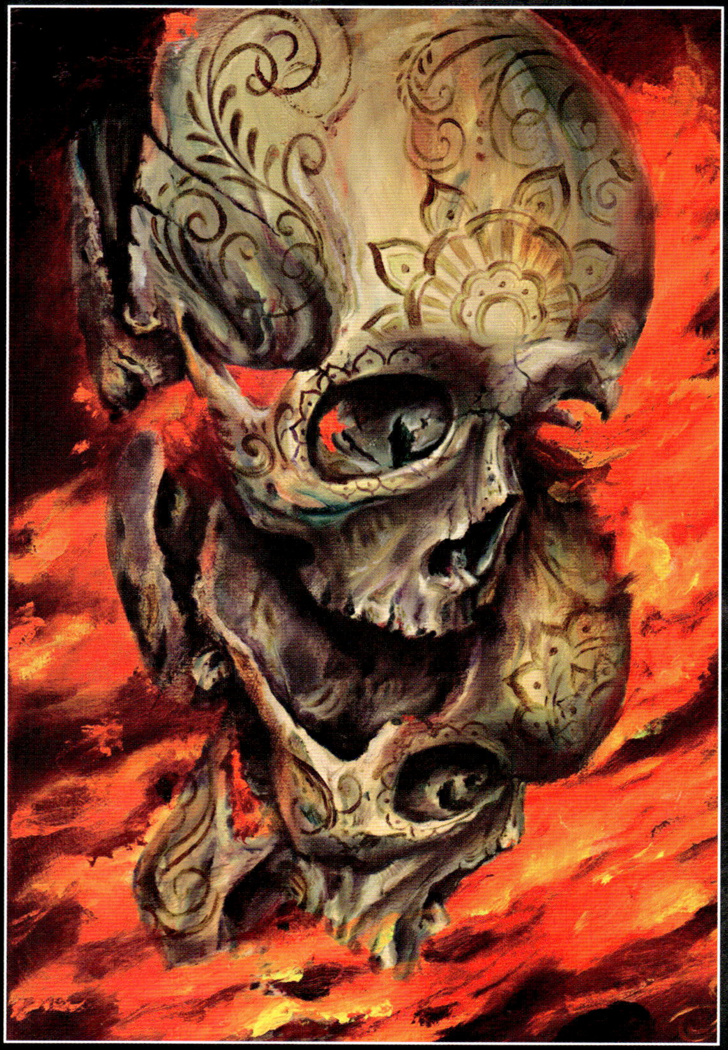

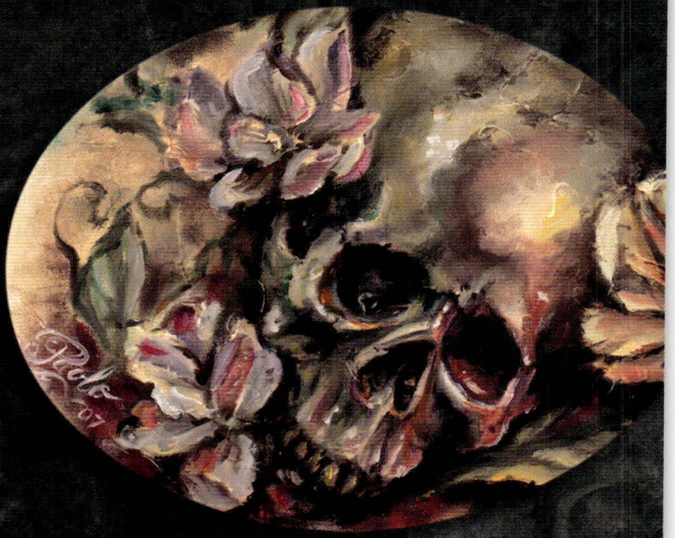

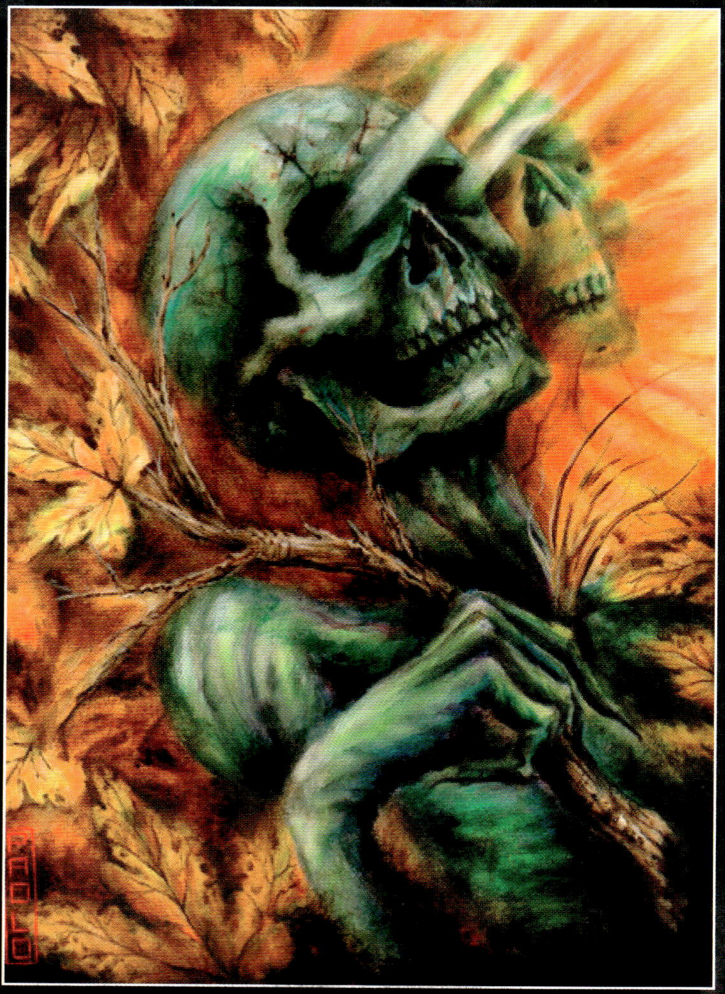

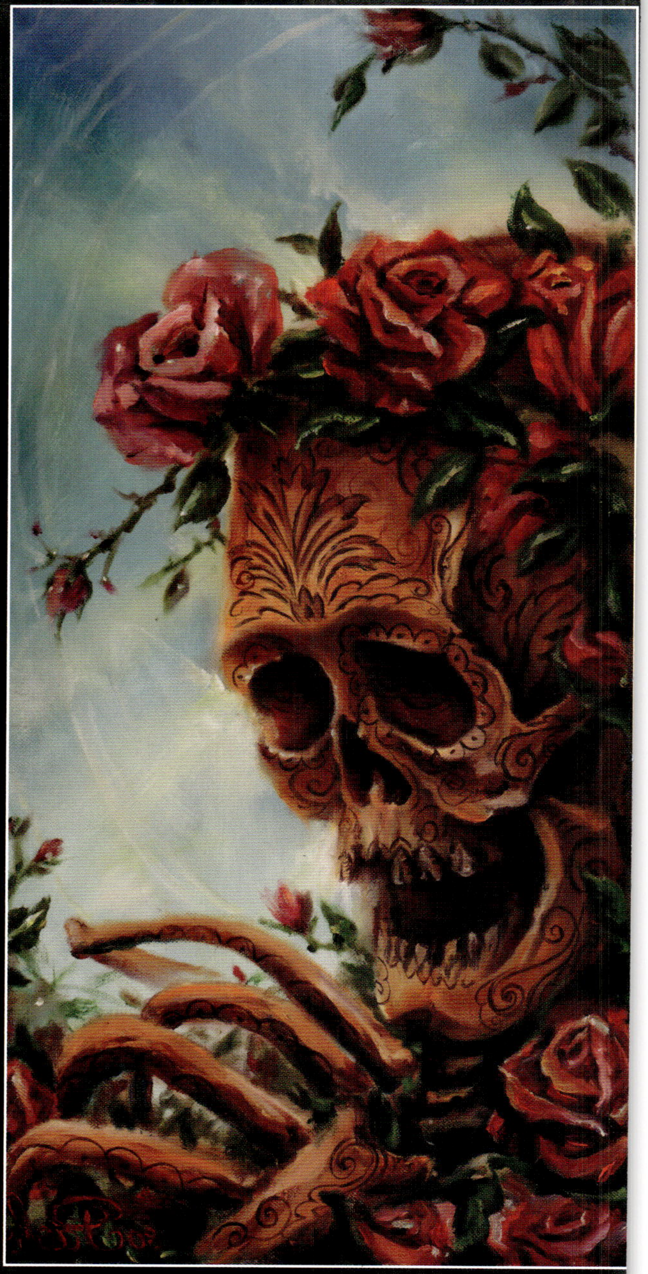

Paolo Acuna

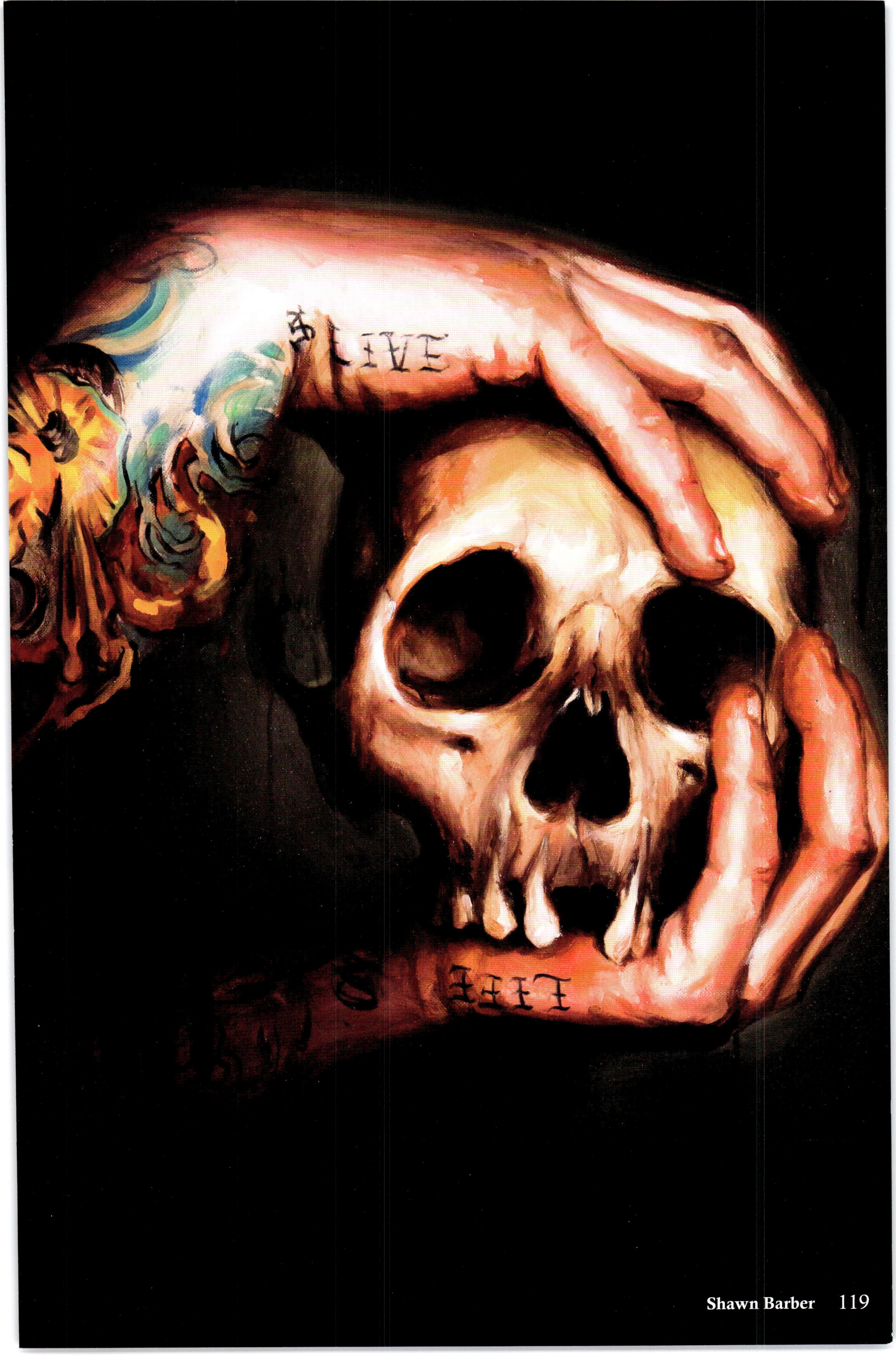
LIVE

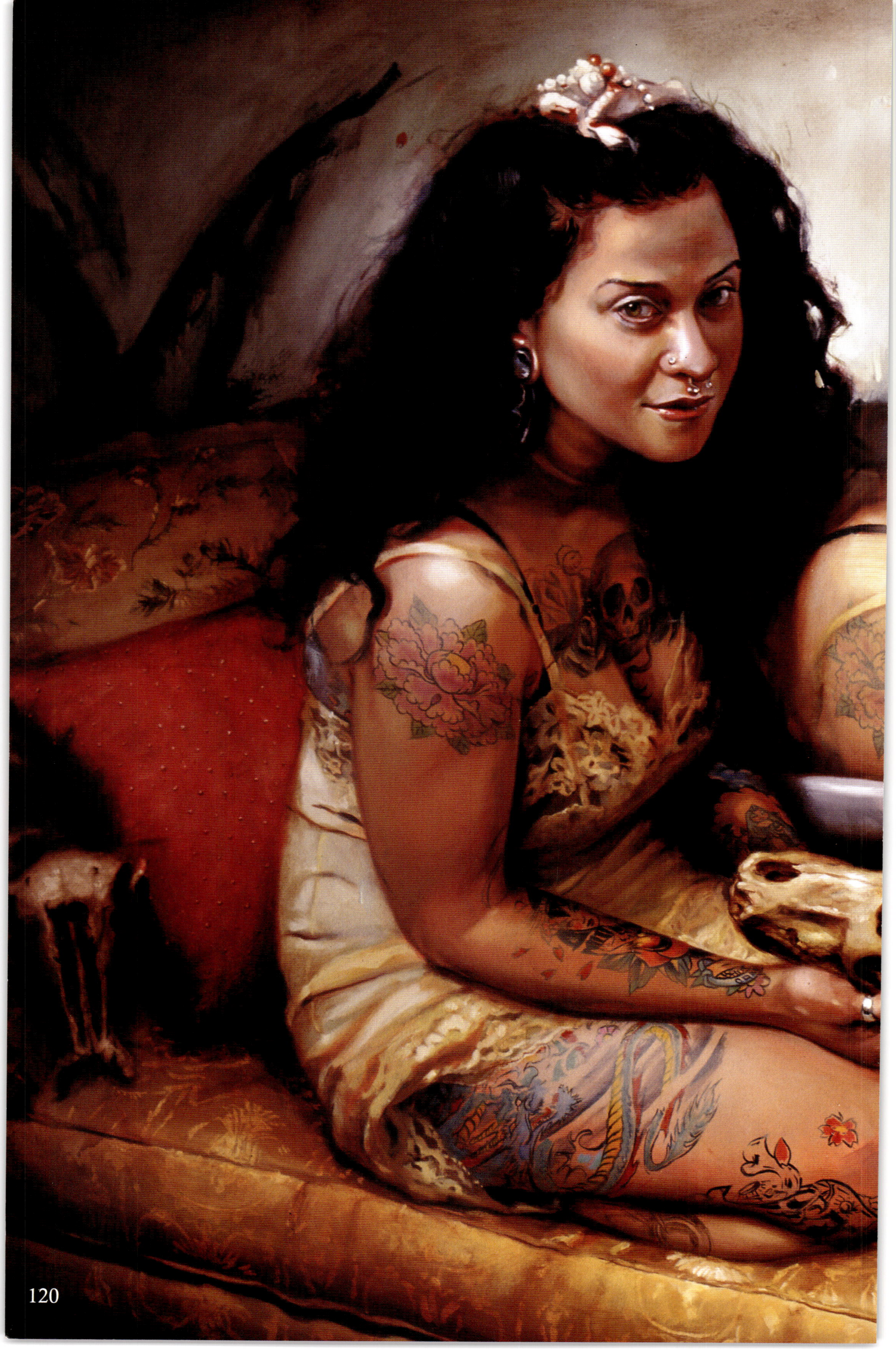

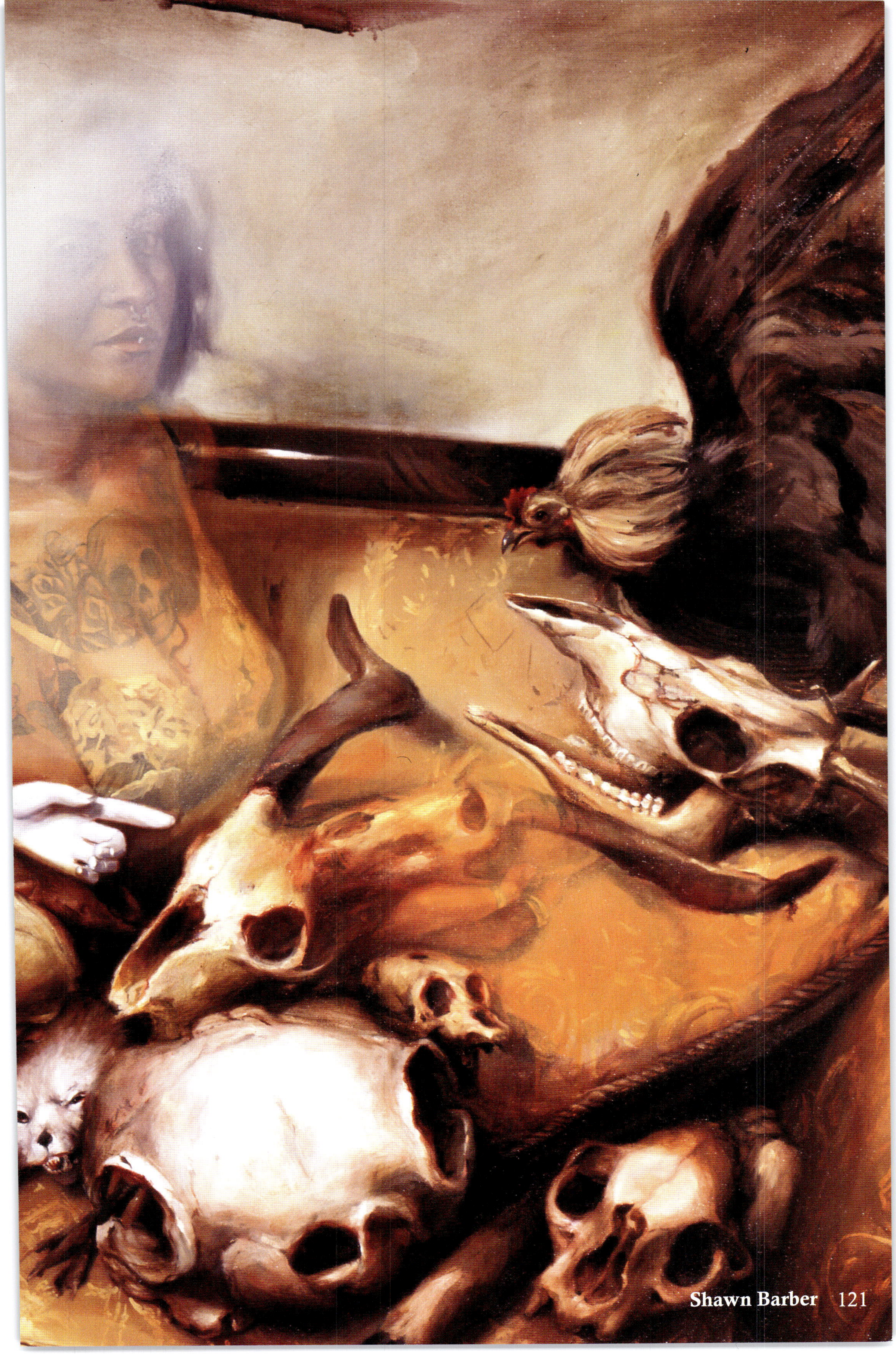

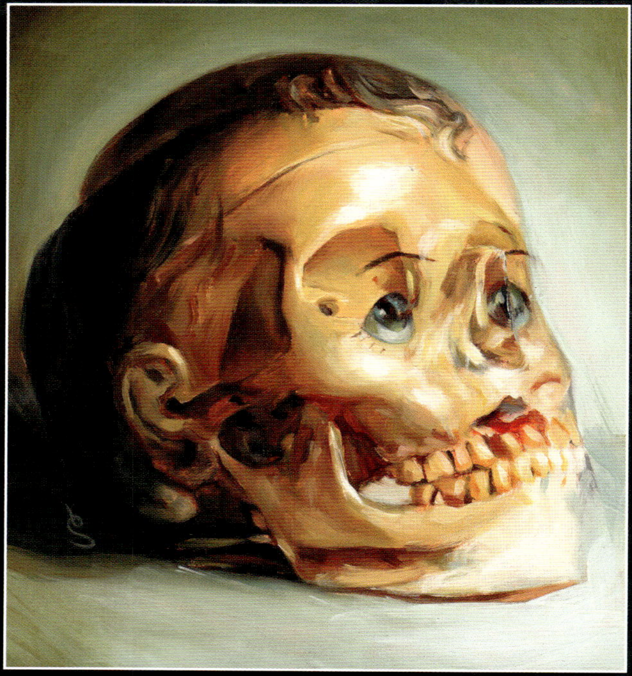

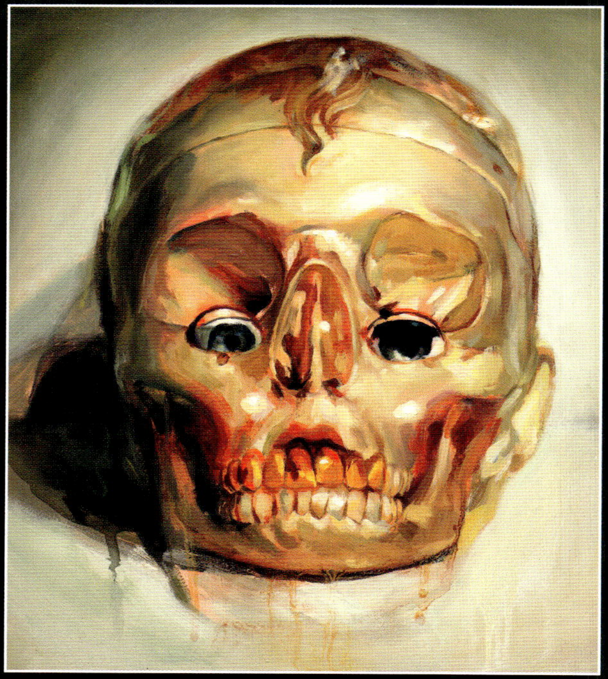

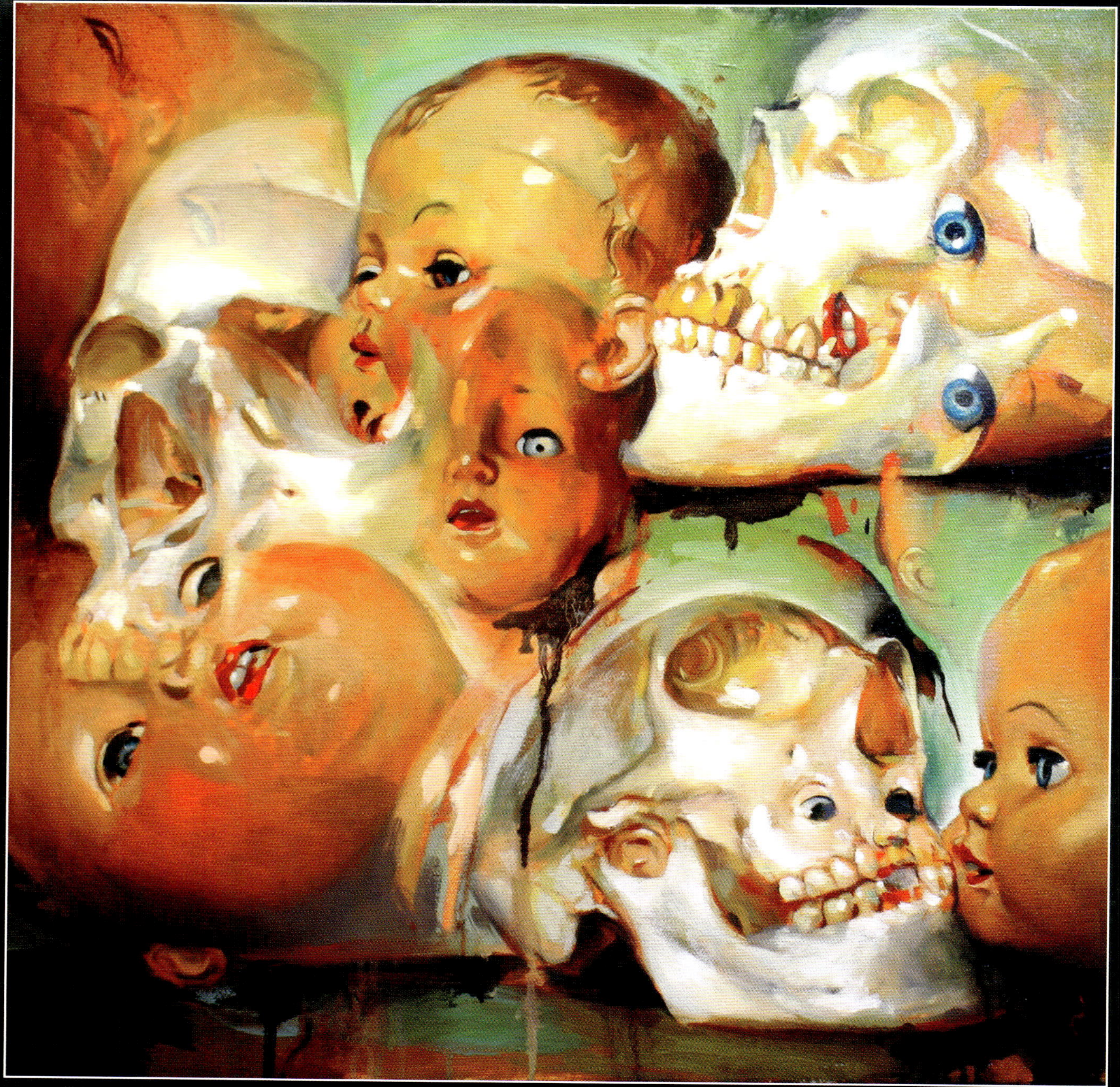

Shawn Barber

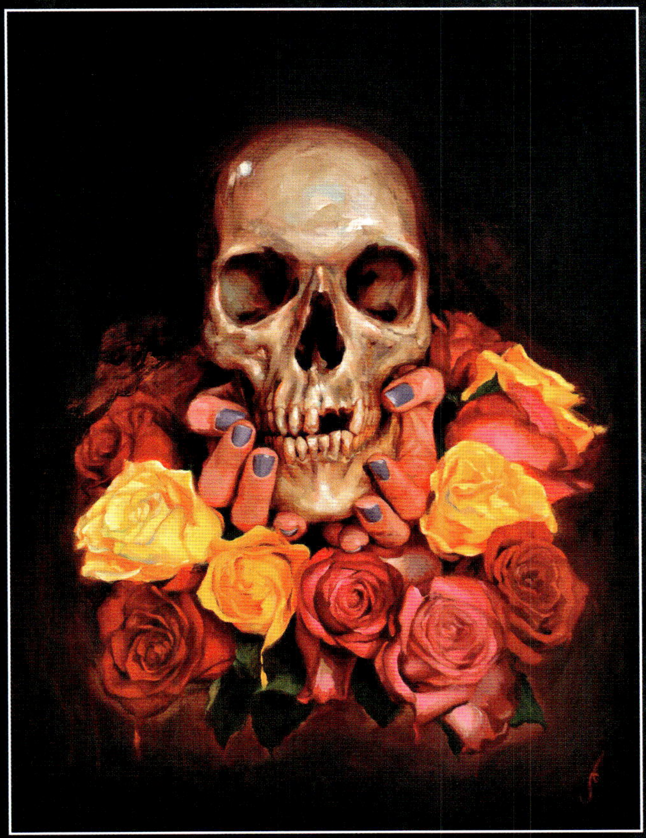

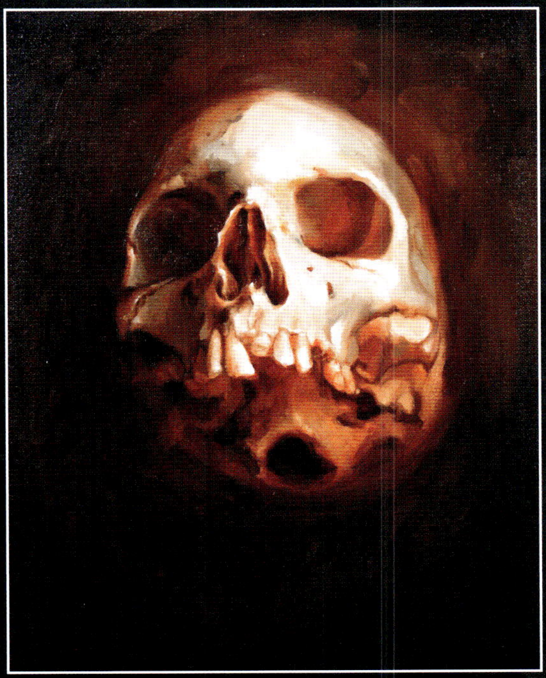

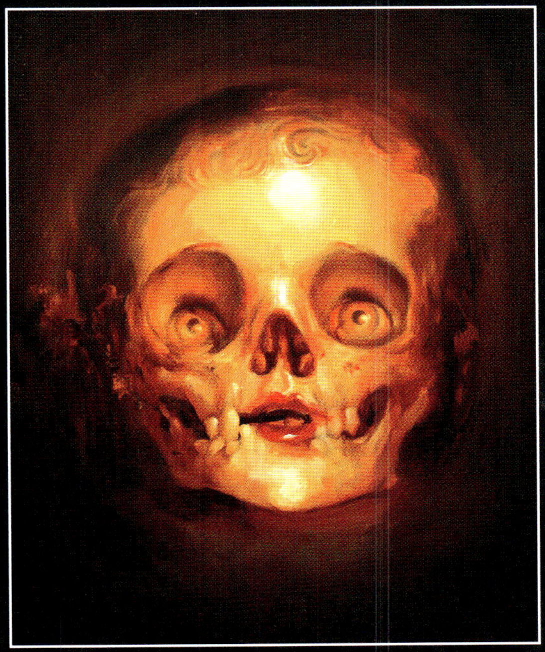

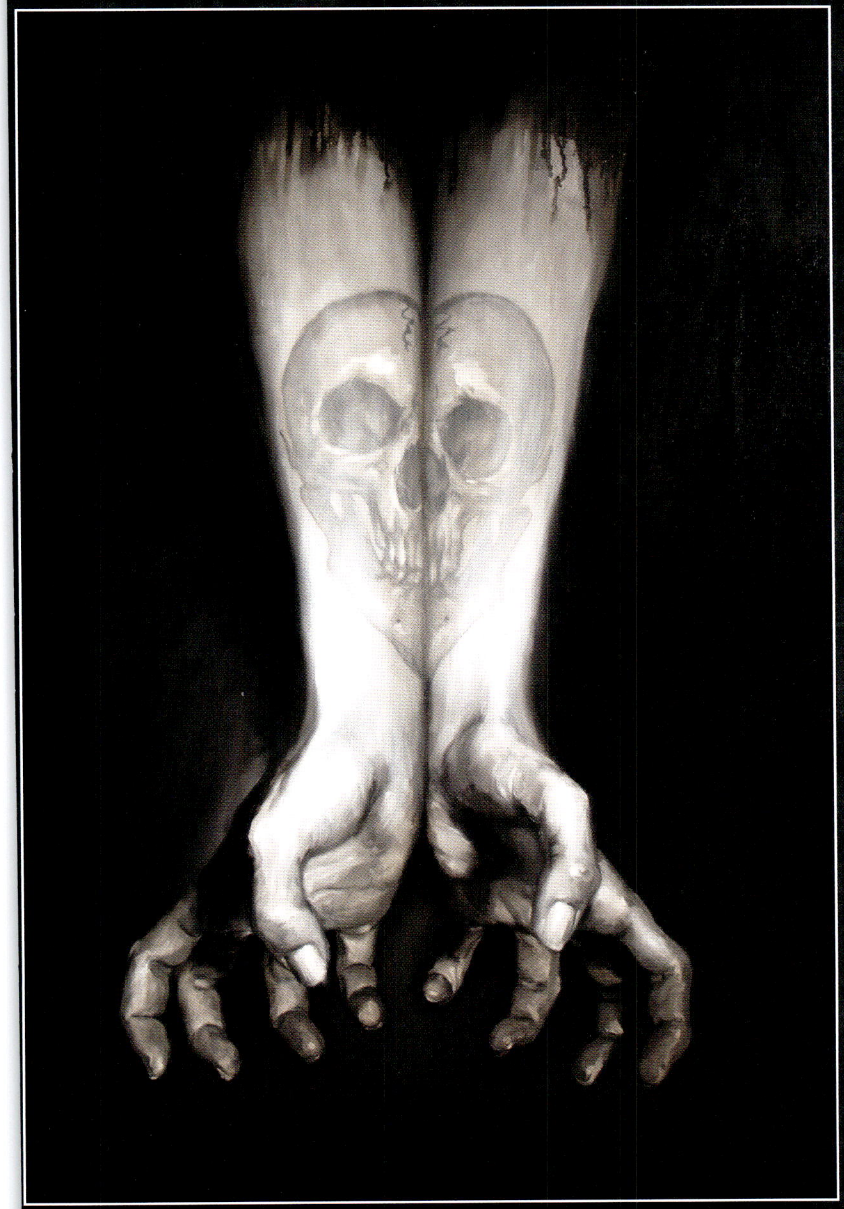

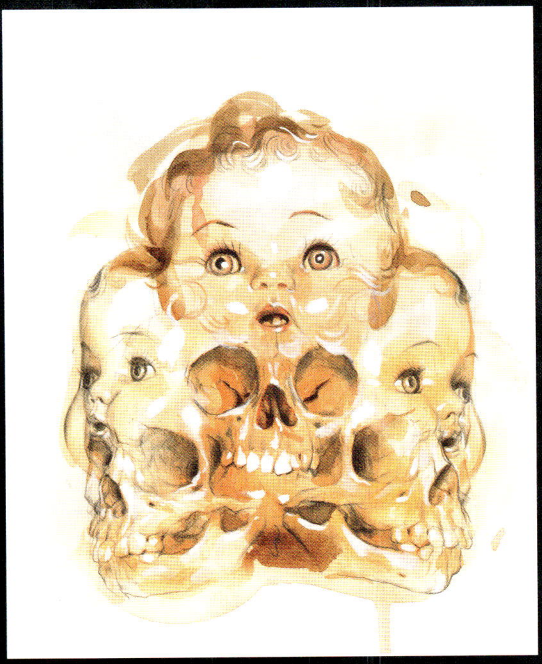

Shawn Barber

DRAWINGS & SKETCHES

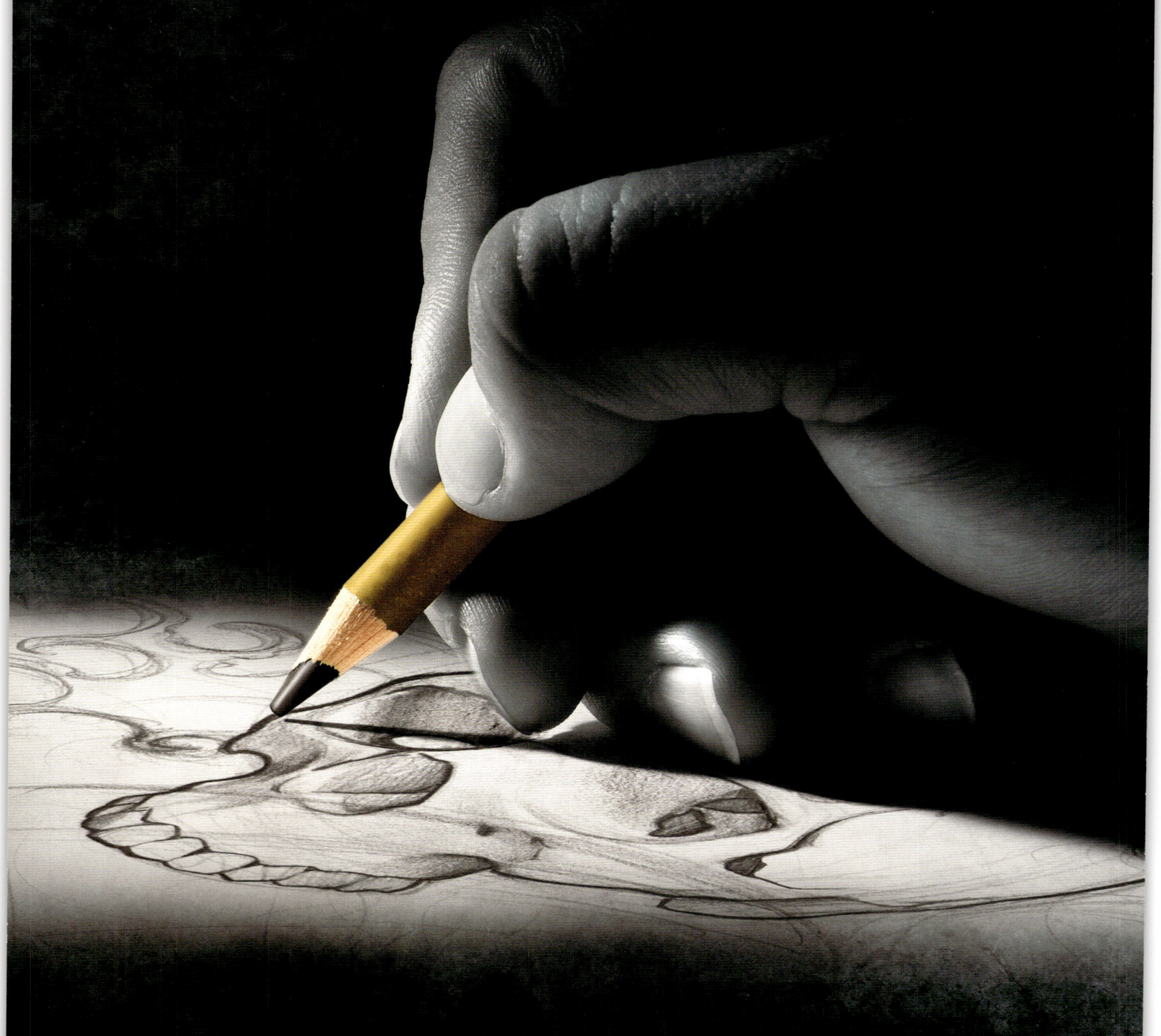

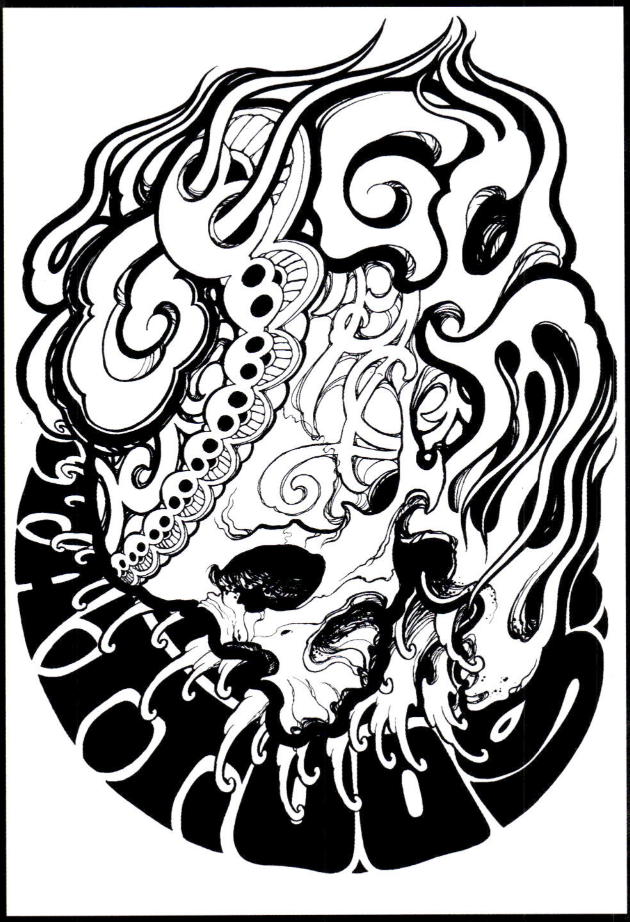

Jeff Gogué

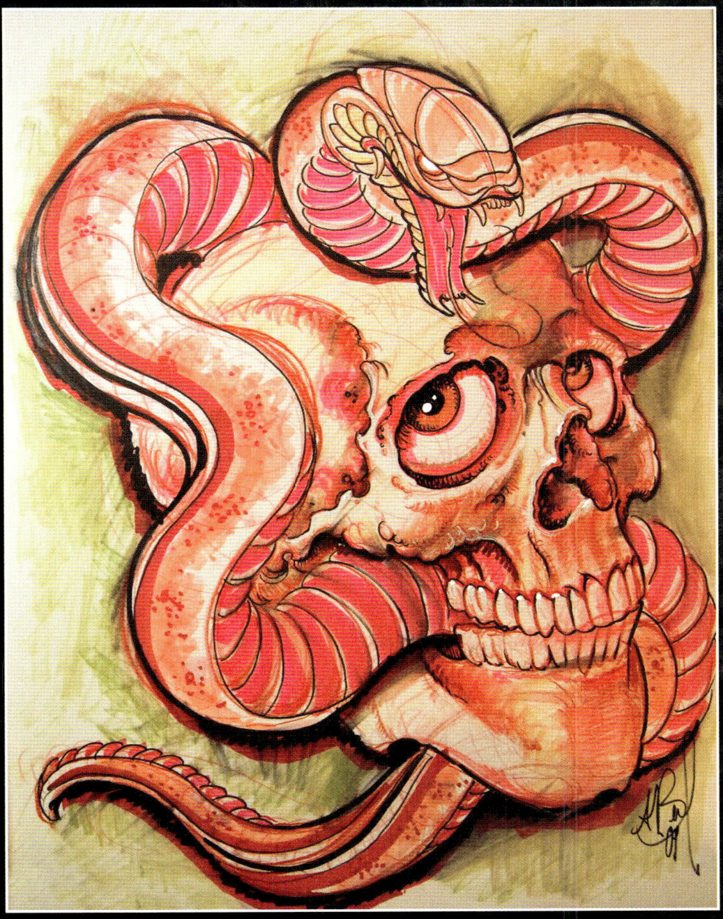

Aaron Bell

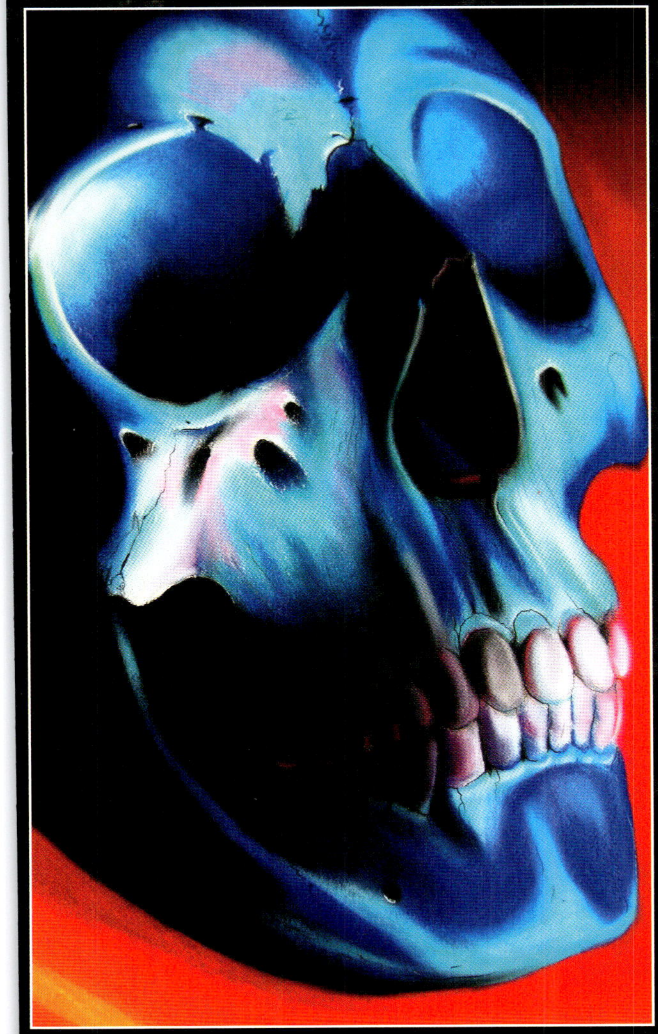

Aric Taylor

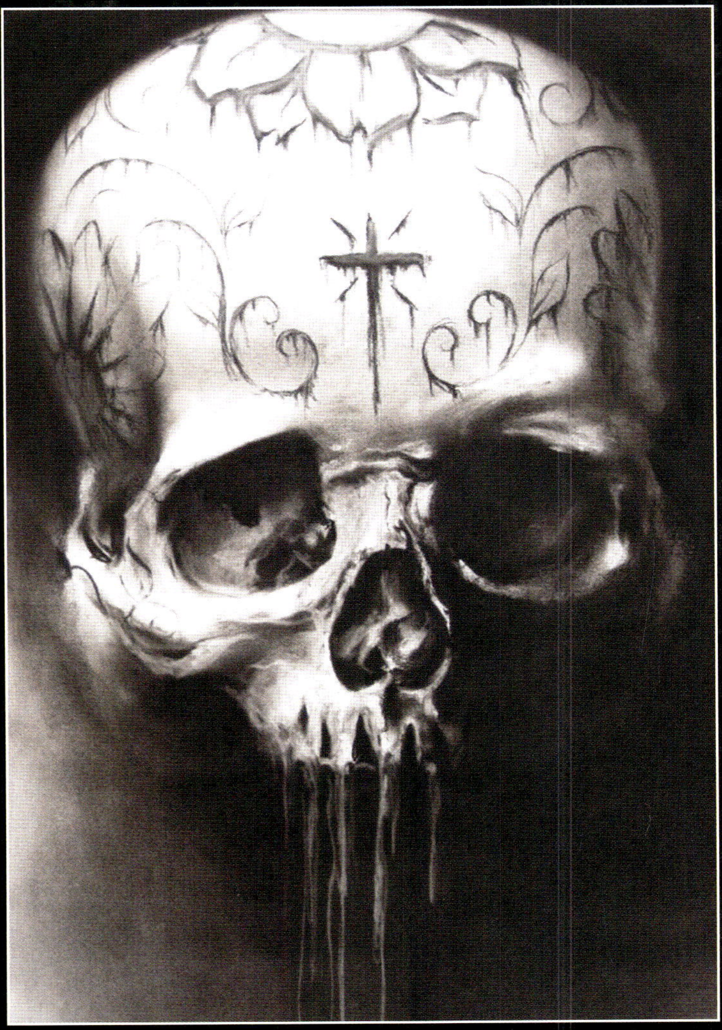

Carlos Torres

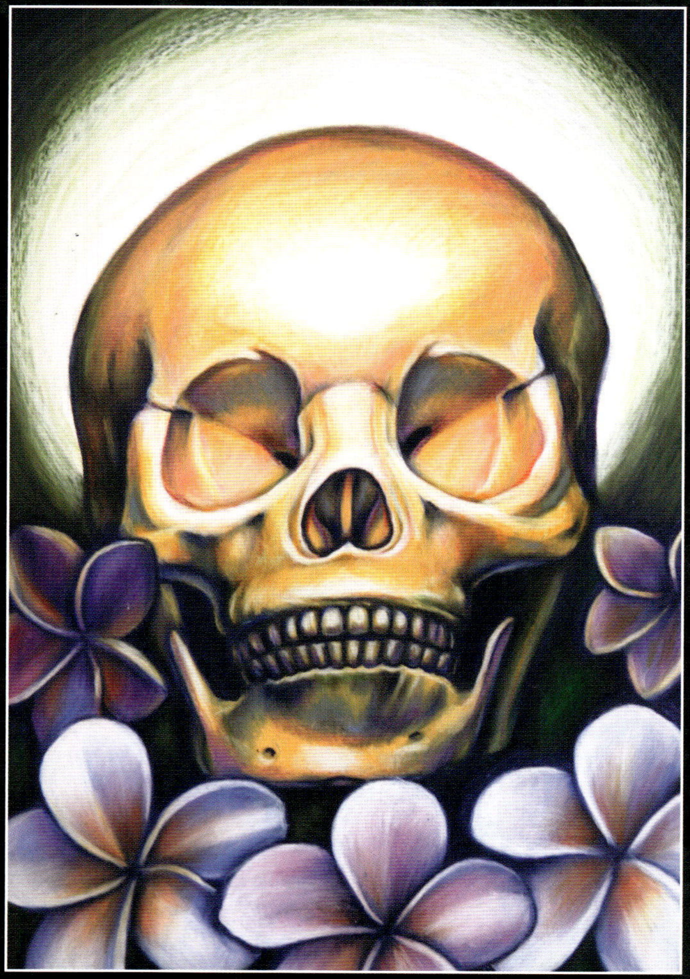

Anna O Donnel

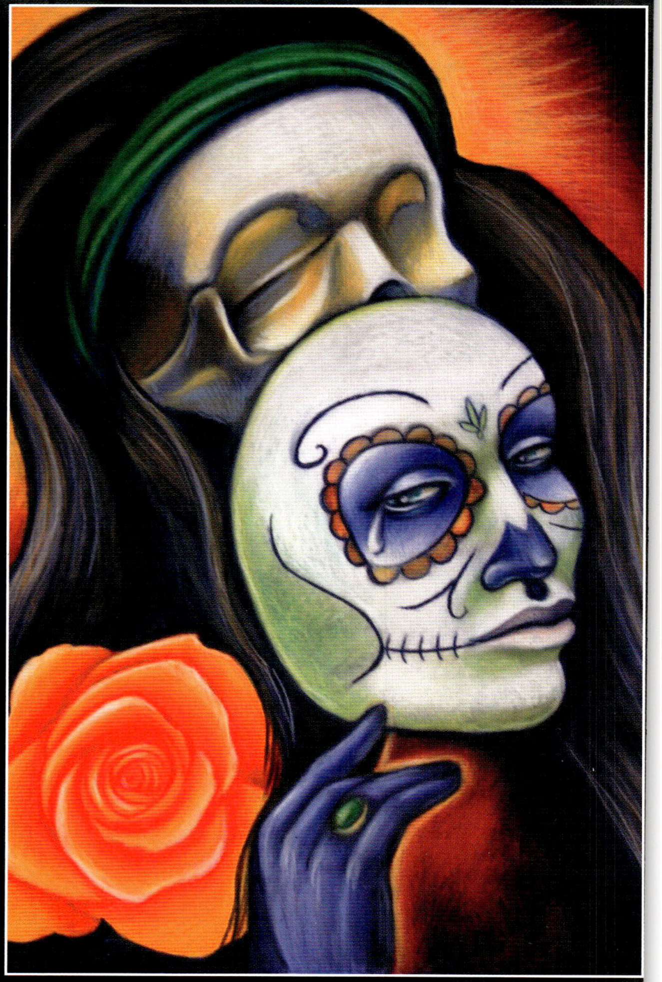

Anna O Donnel

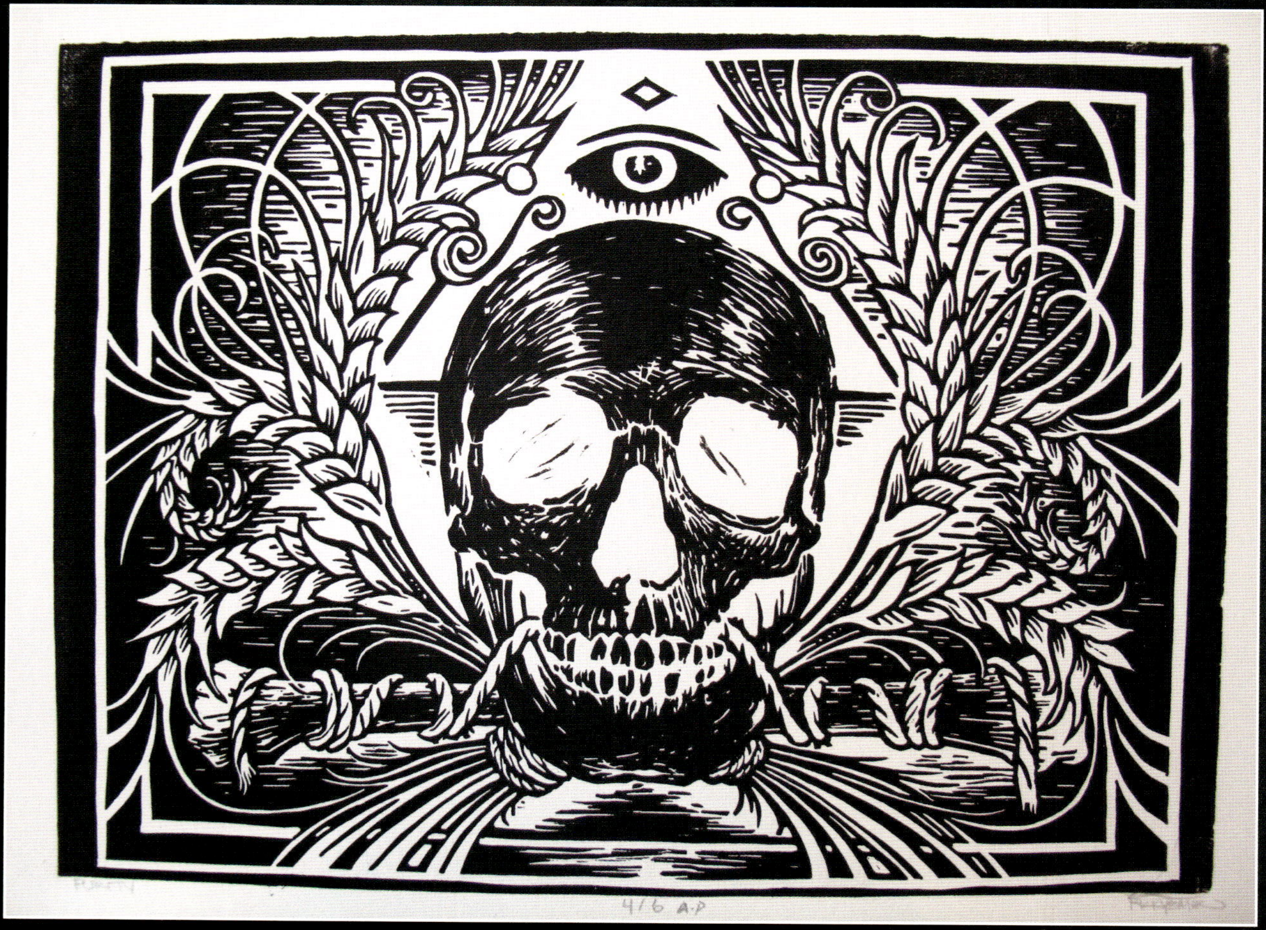

Richard Mcgee

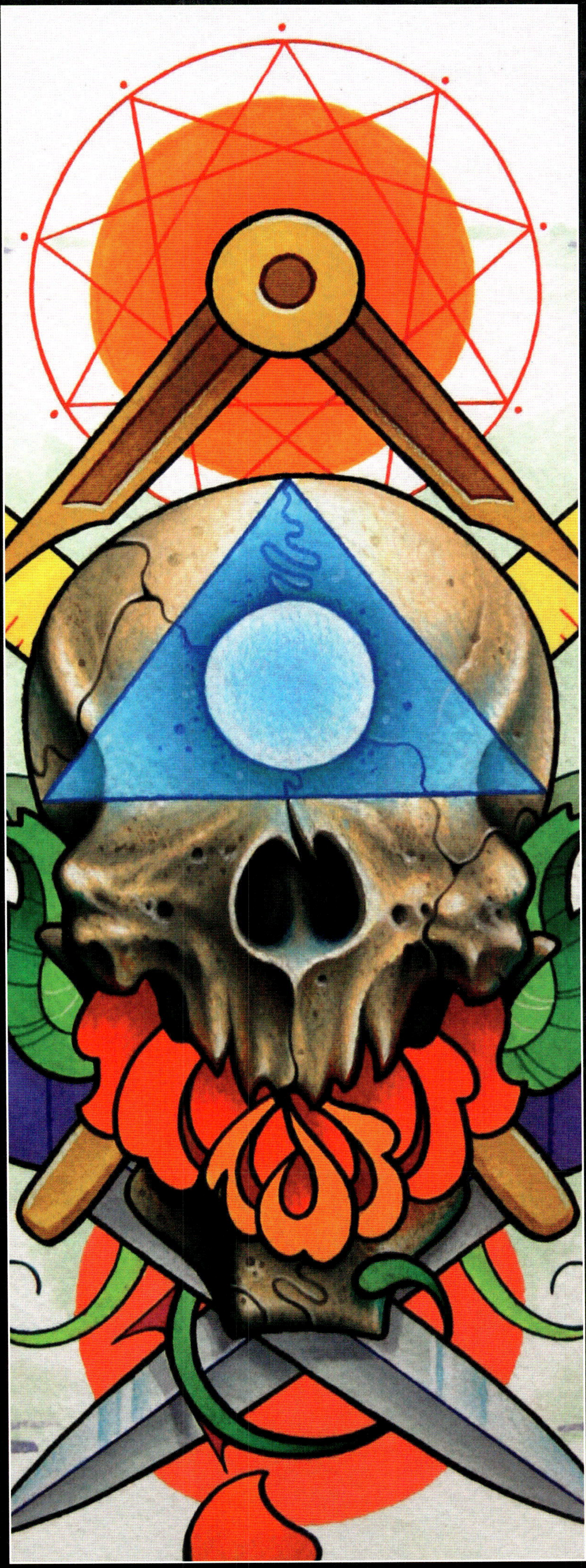

Jeff Ensminger

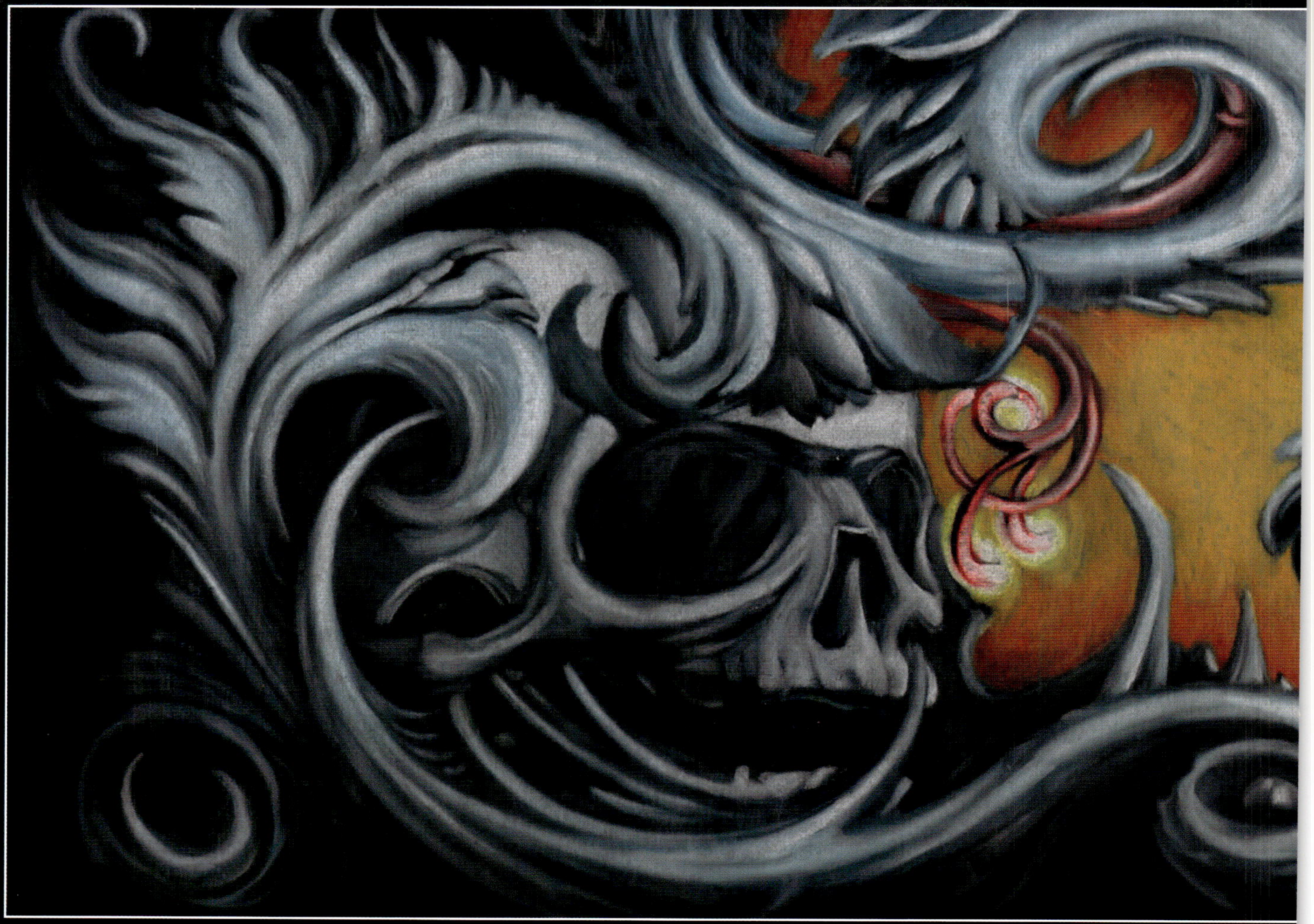

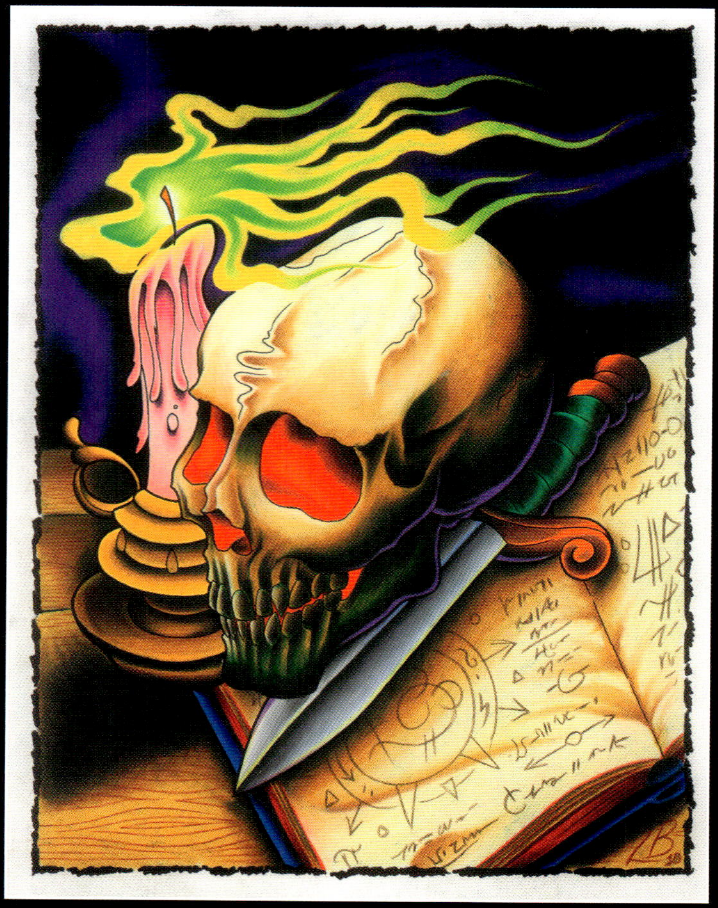

Larry Bone

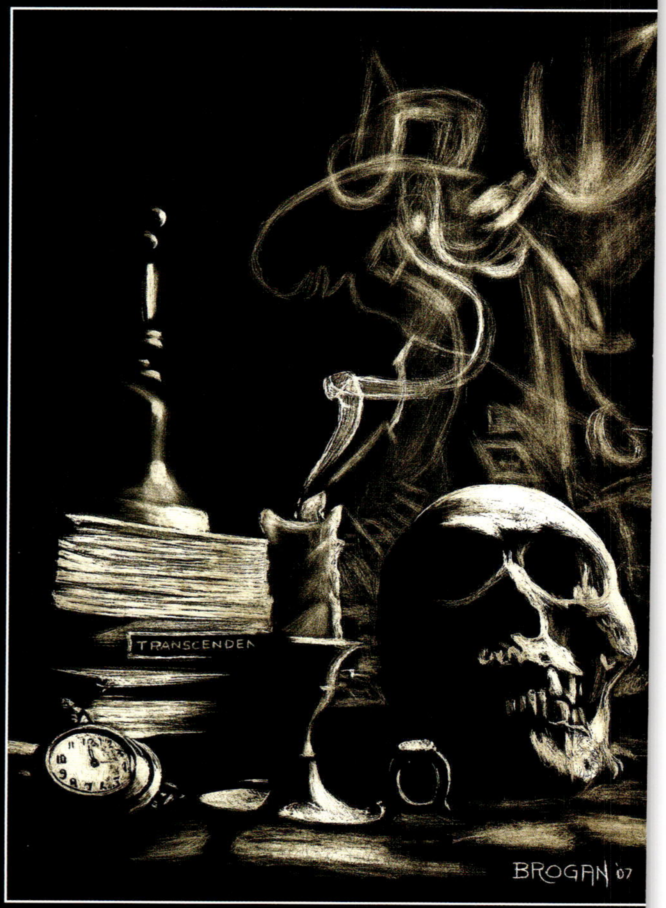

Larry Brogan

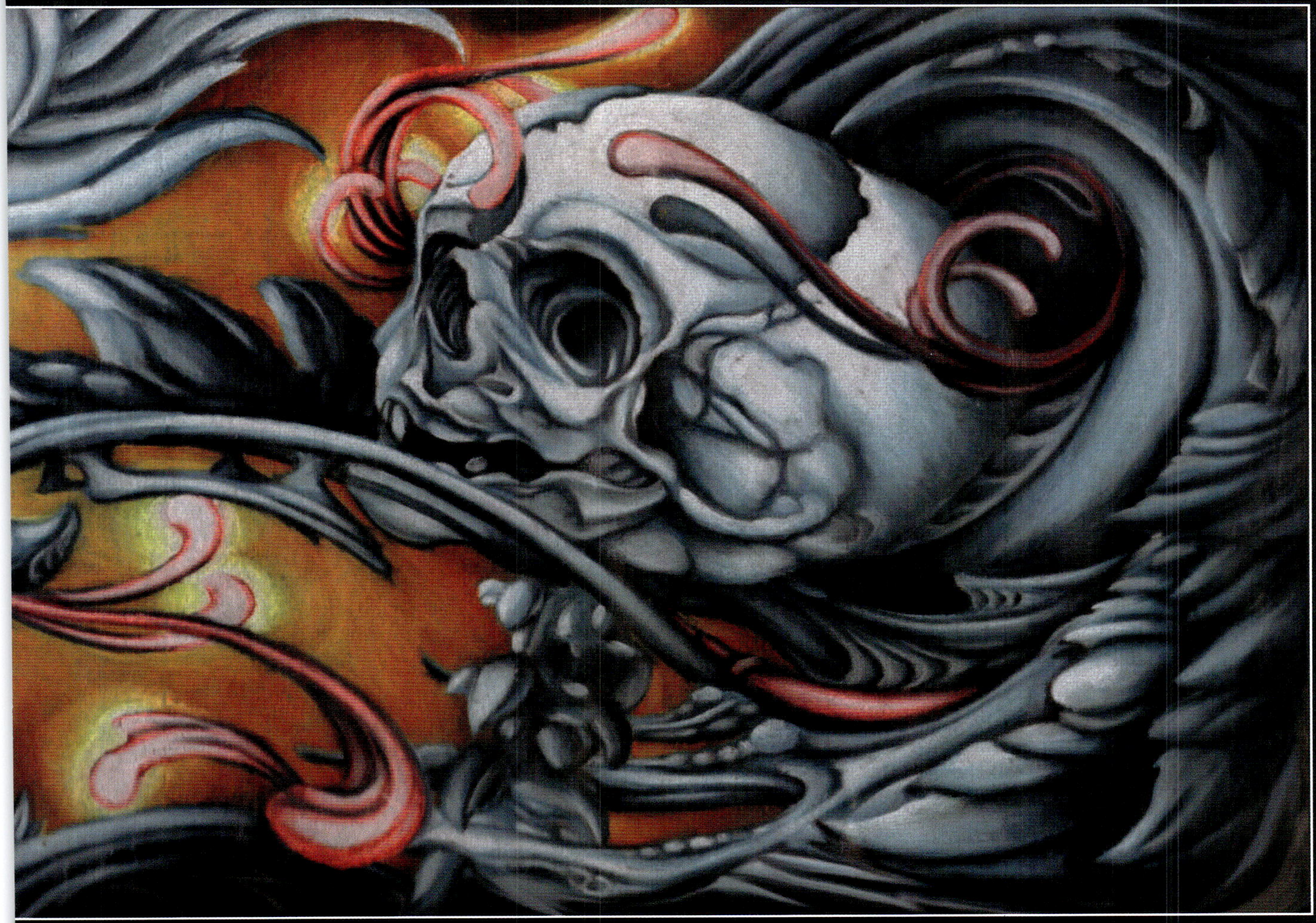

Bobby Deneen & Waylon Rodgers

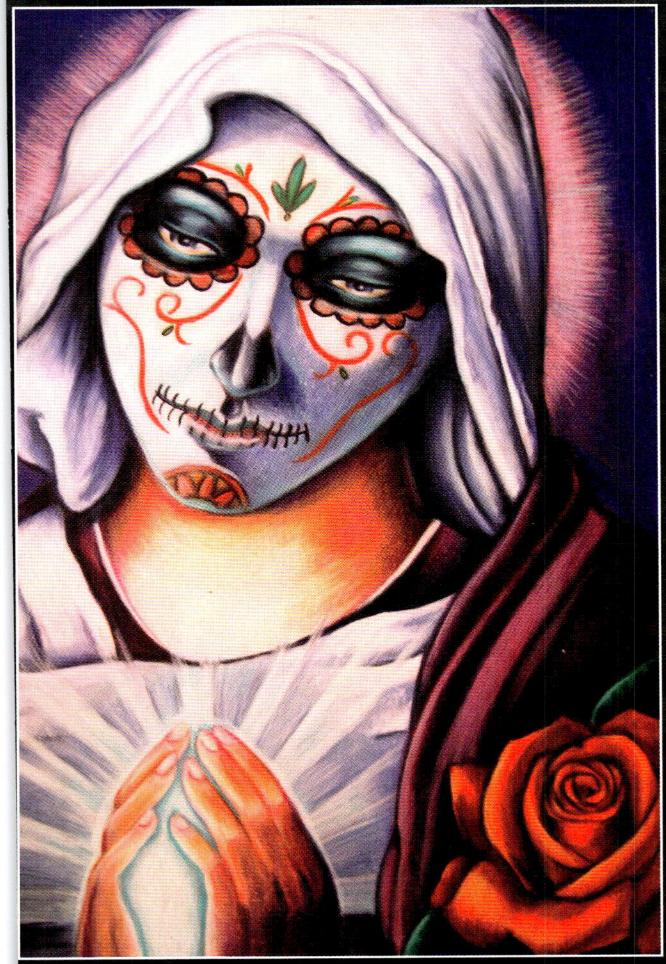

Anna O Donnel

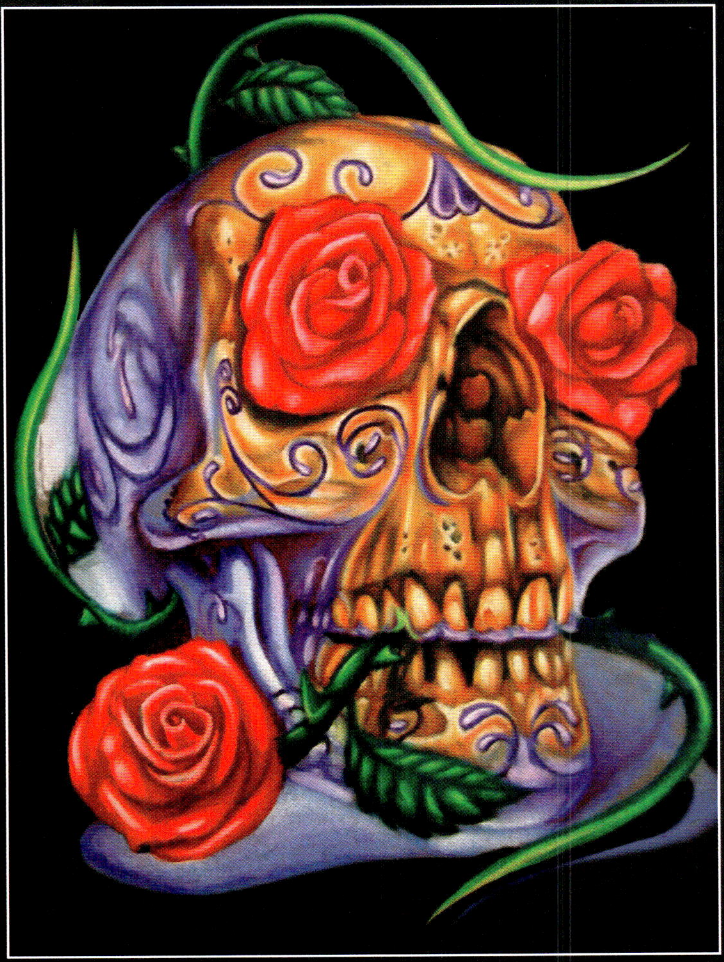

Bobby Deneen

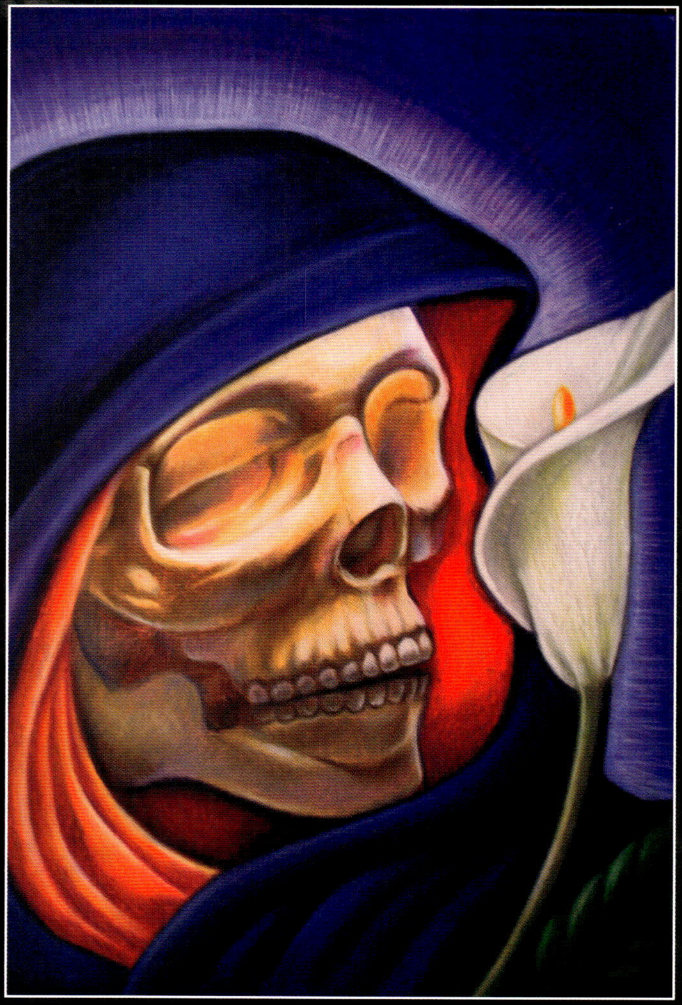
Anna O Donnel

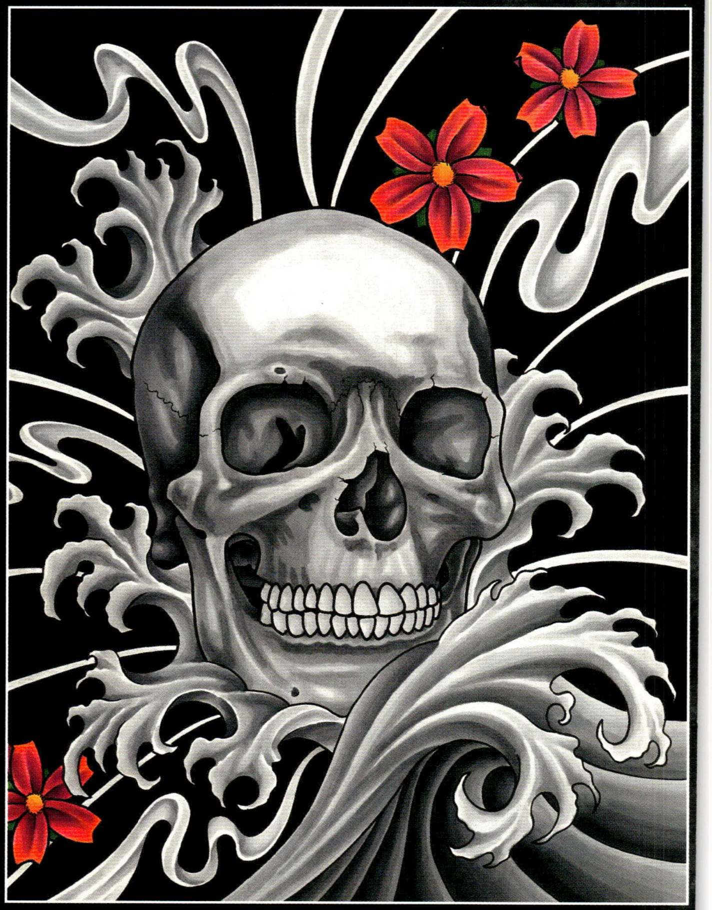
Stretch

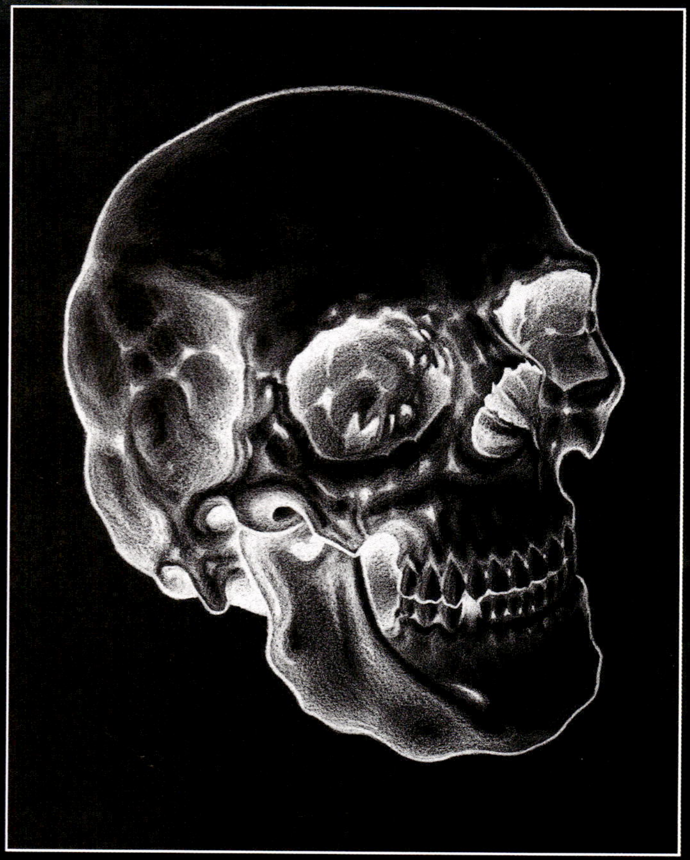
Sean Zee

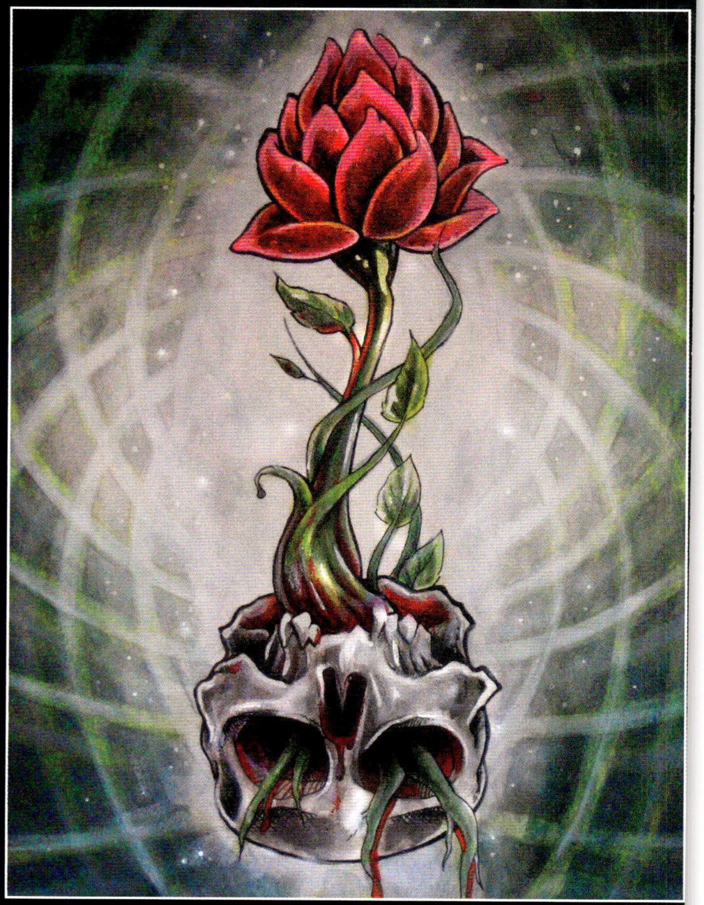
Jessica Rocha

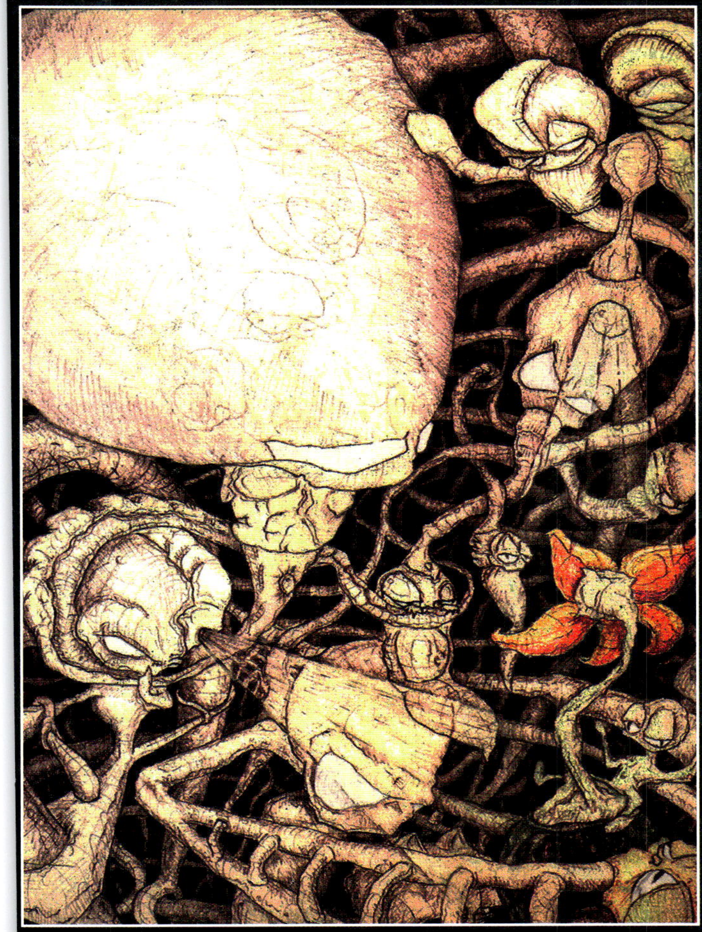

Michael Sexauer

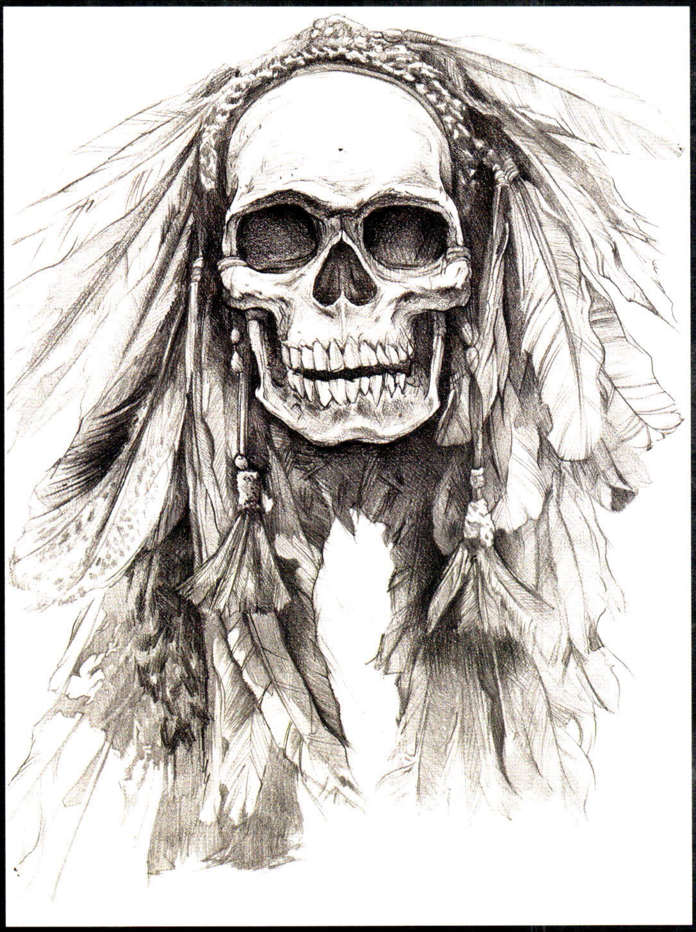

Ryan Smith

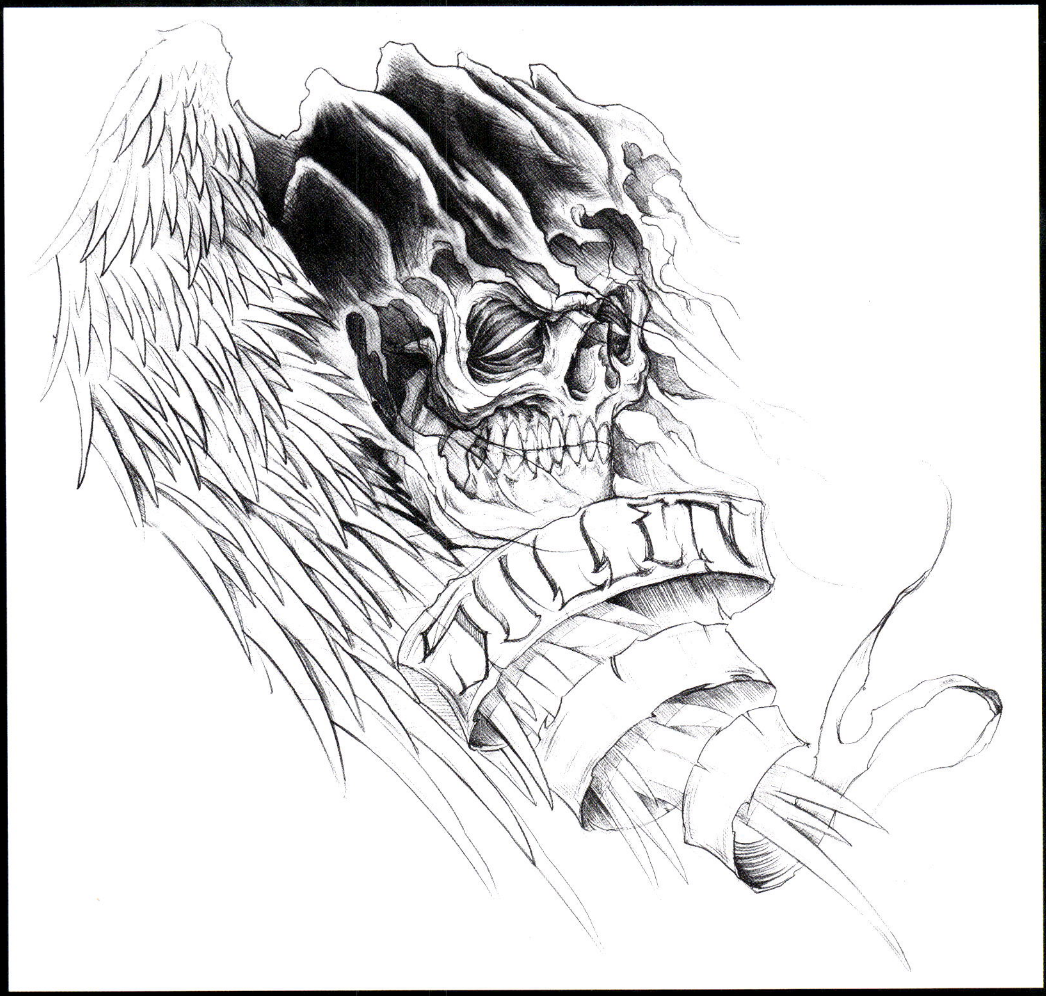

Ryan Smith

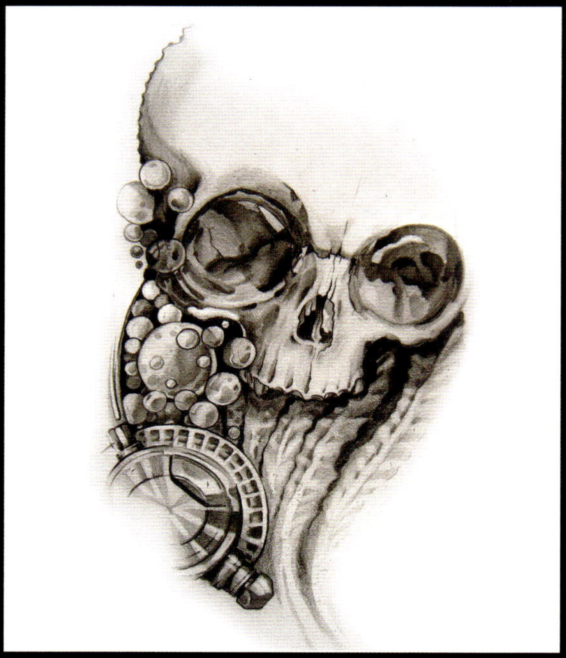

Andrey Barkov

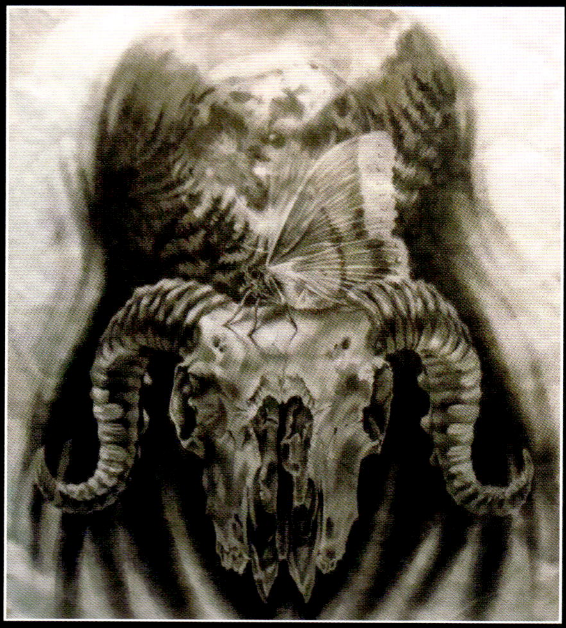

Jack Matisse

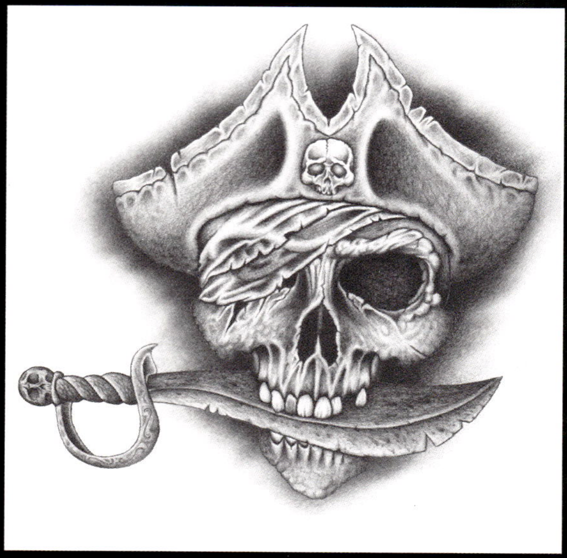

Mogie

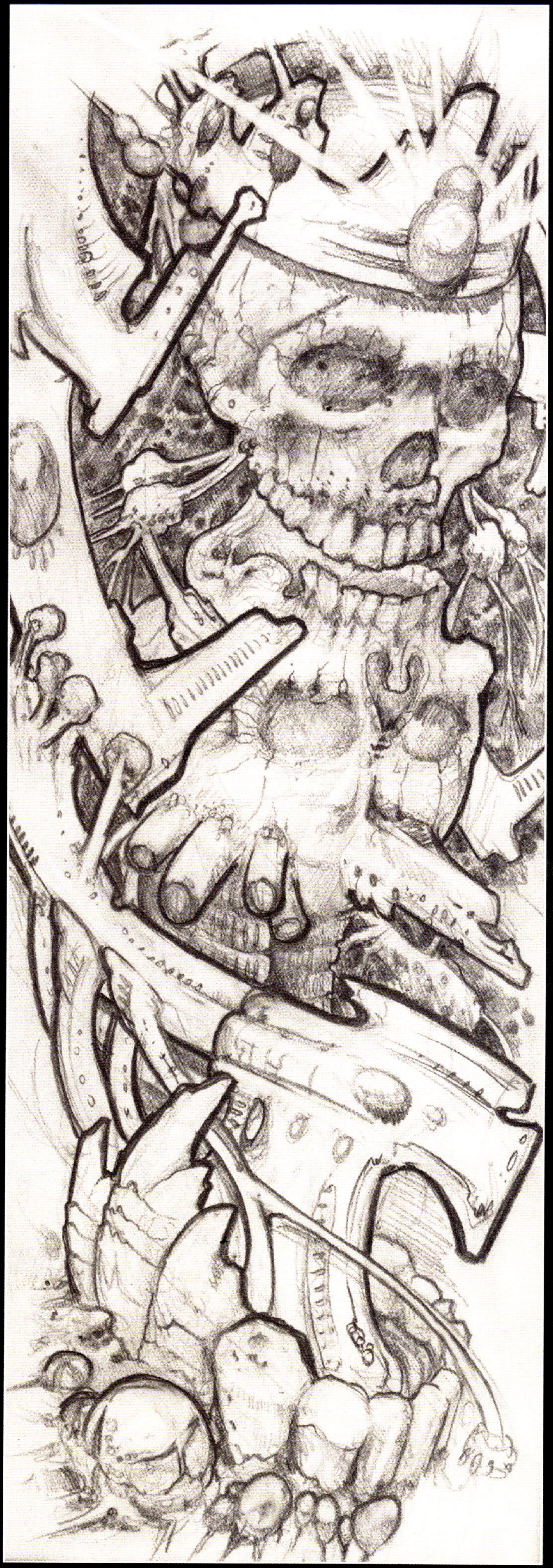

Gifford Kasen

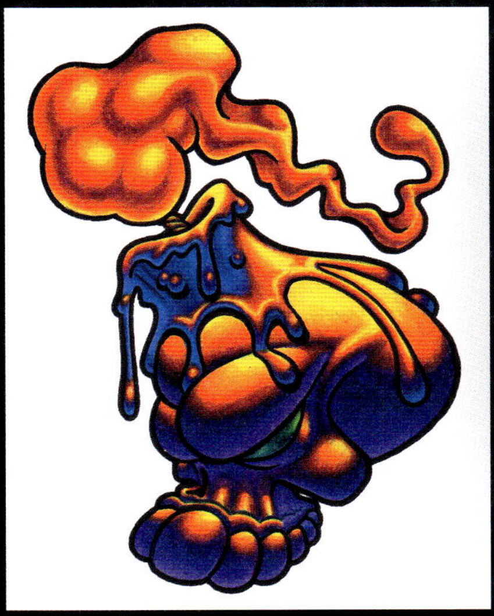
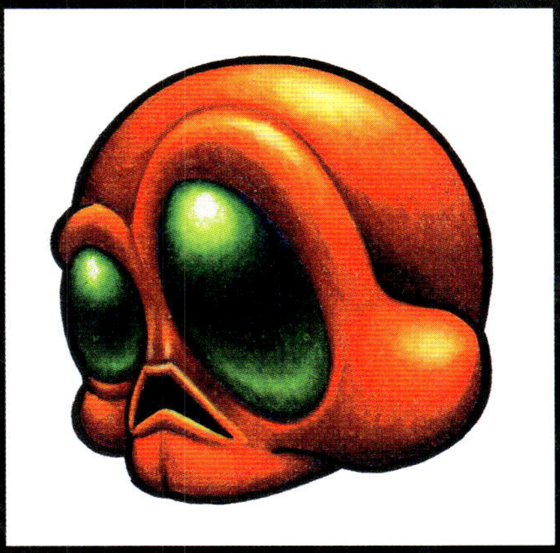
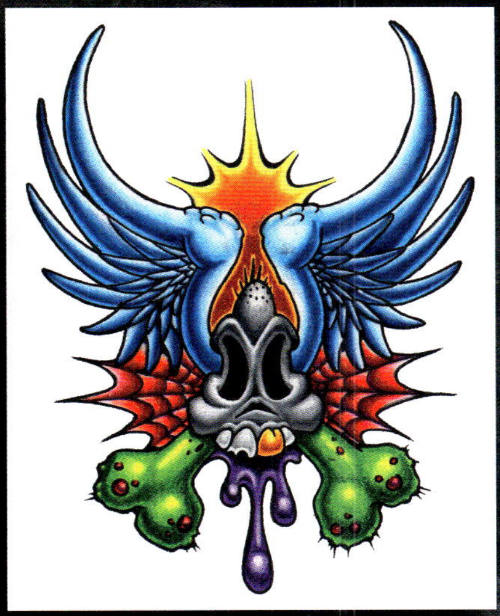
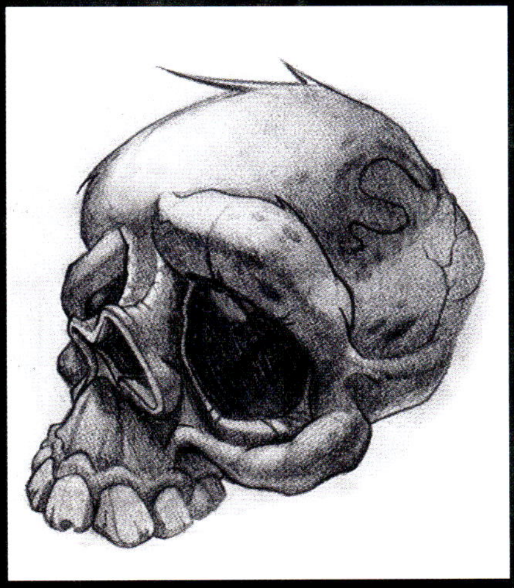
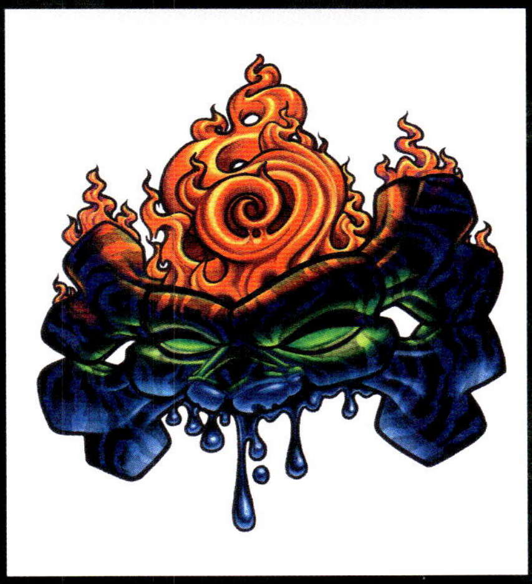
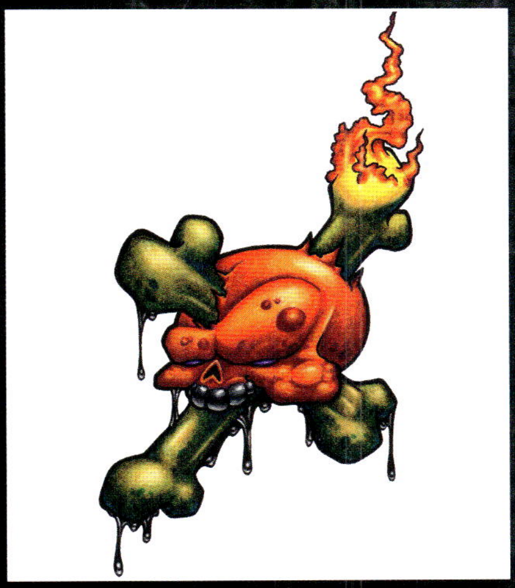
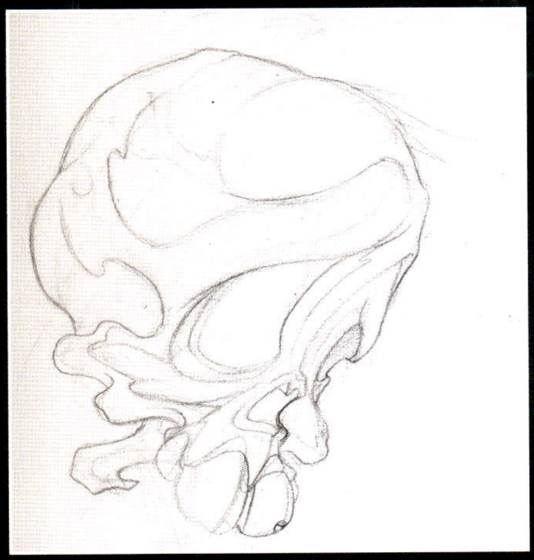
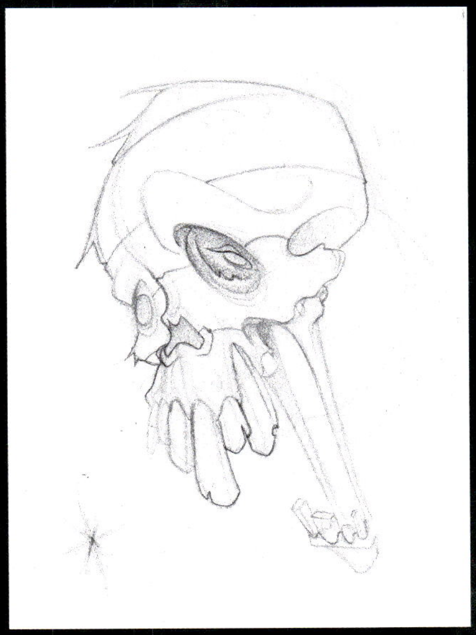
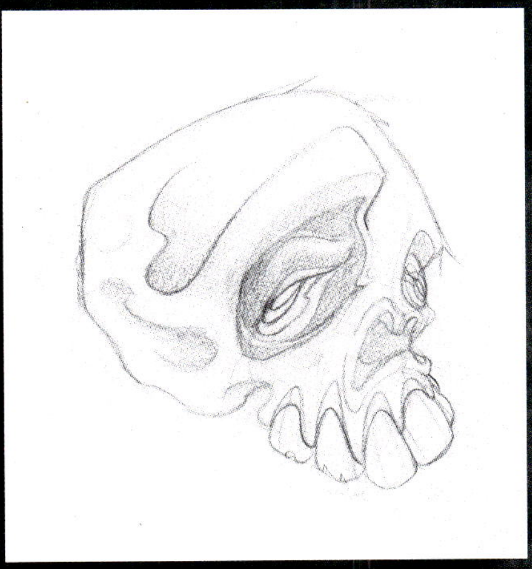
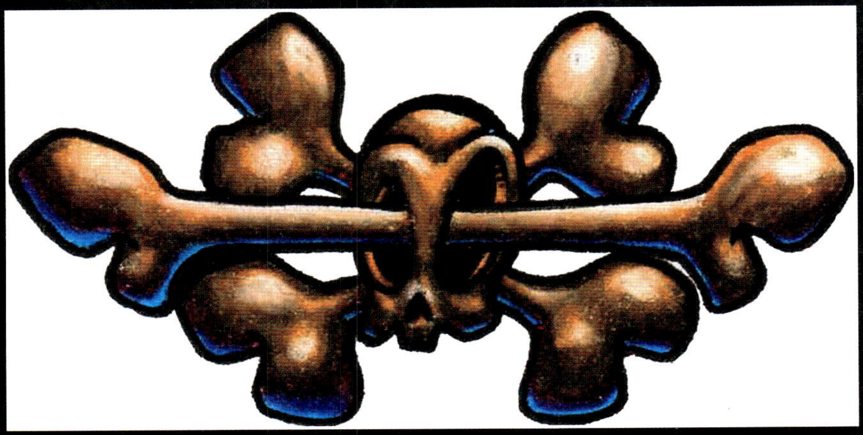

Jesse Smith

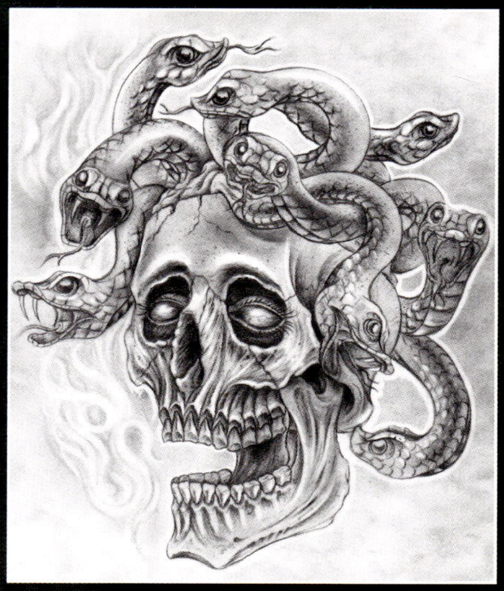
Edwin Marin

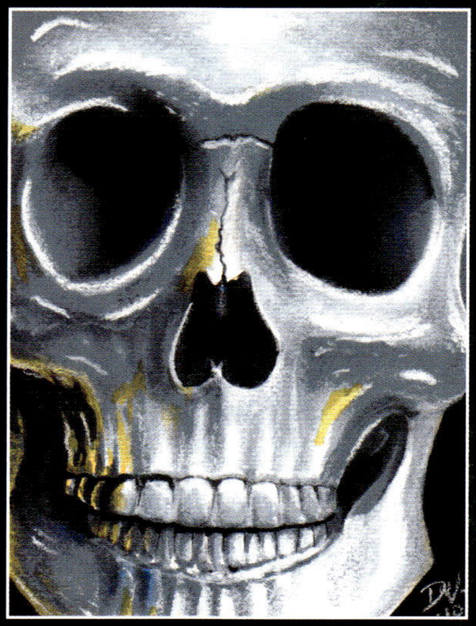
David Meek

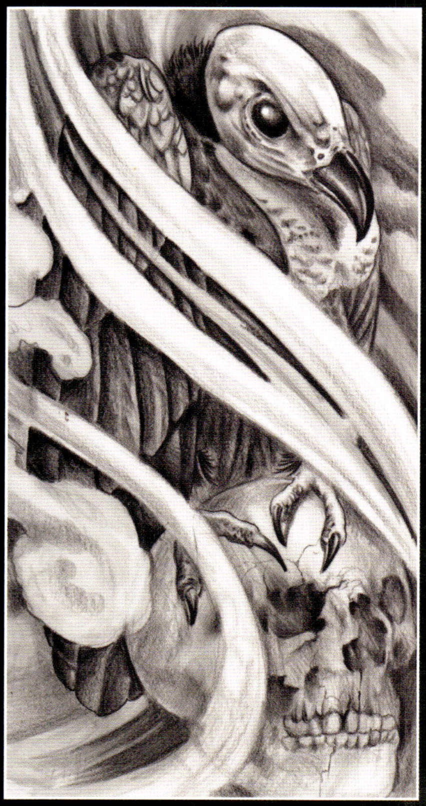
Gifford Kasen

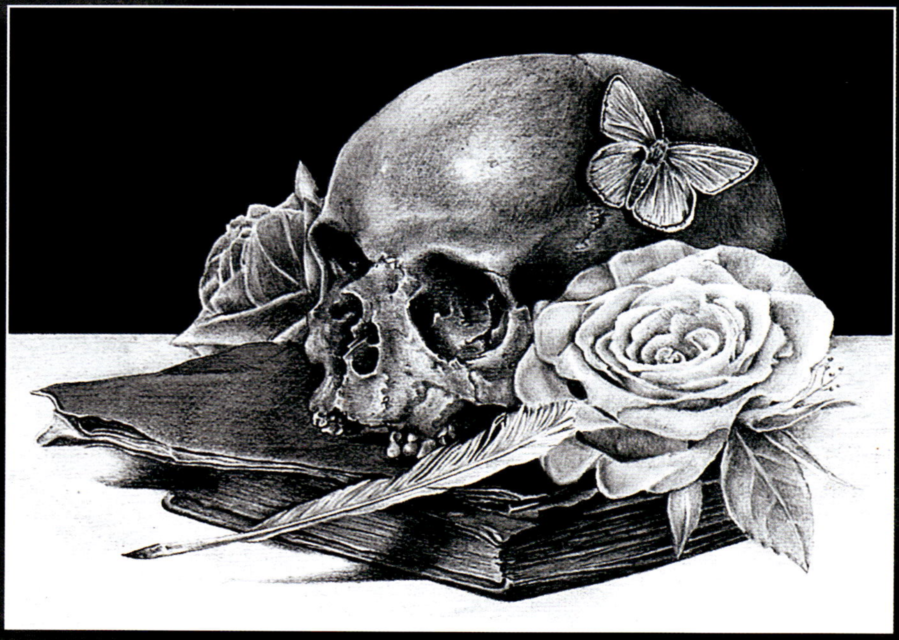
Chris Winterson

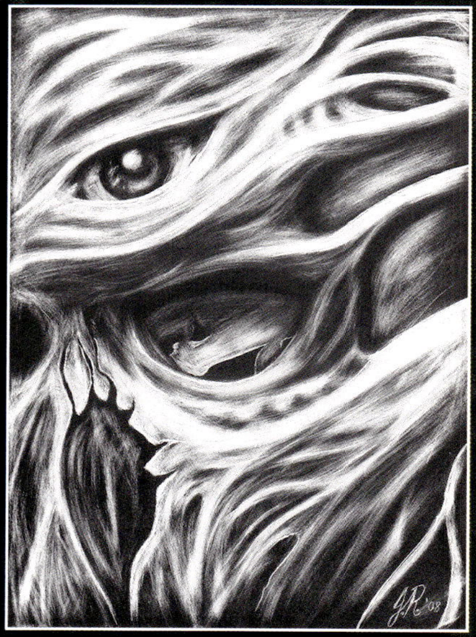
Josh Ruff

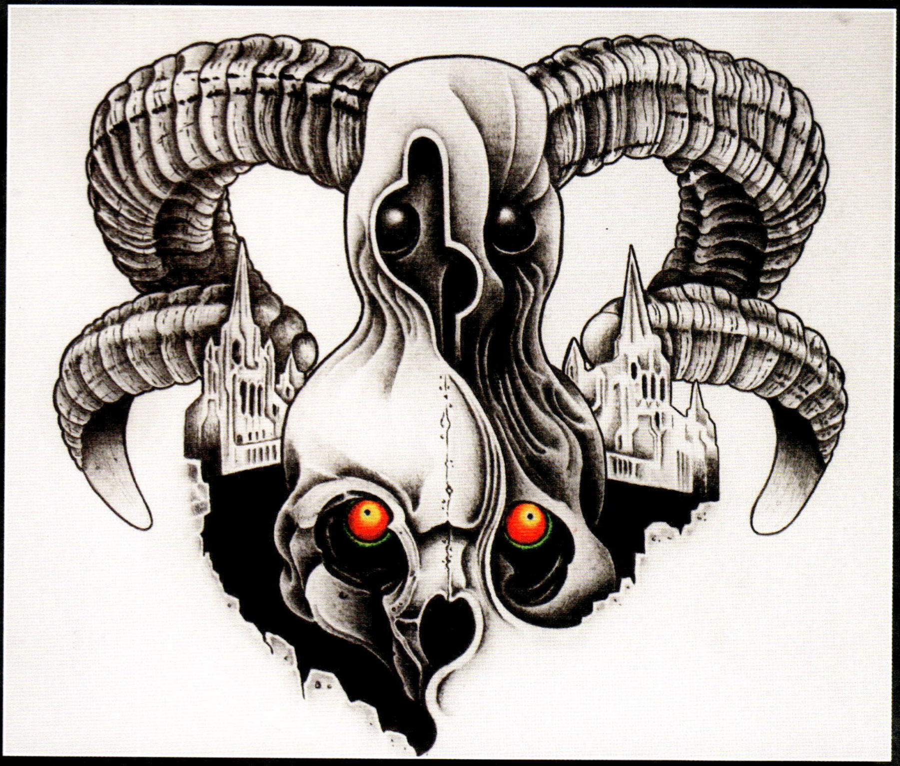
Chris Winterson

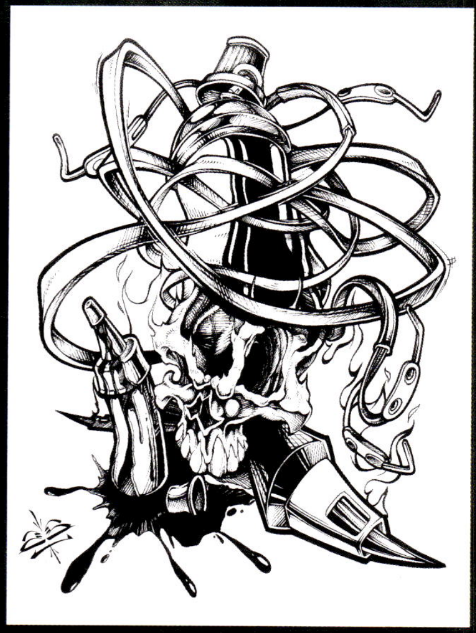
Richard Munoz

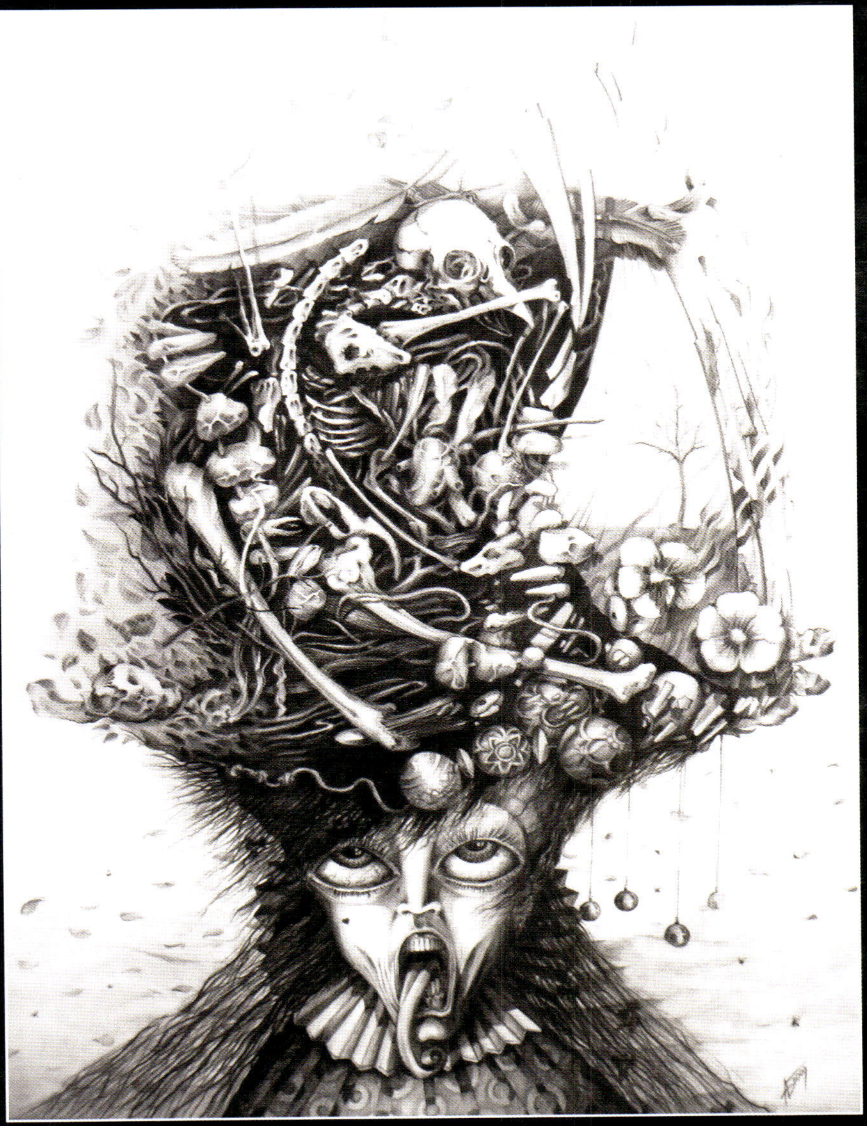

Andrey Barkov

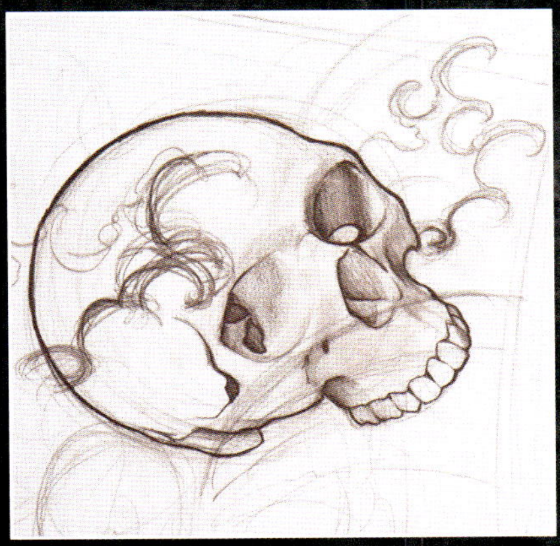

Craig Driscoll

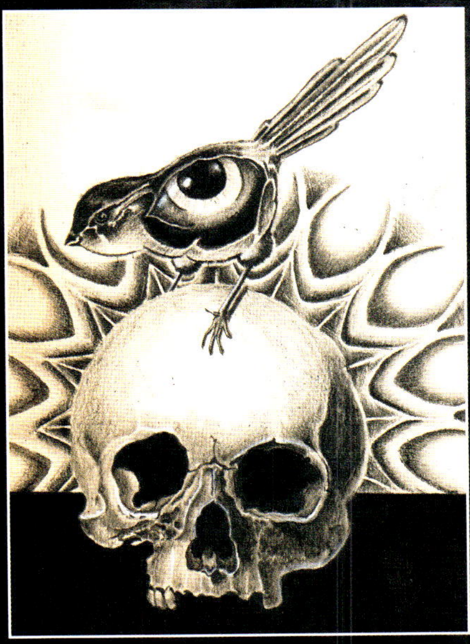

Chris Winterson

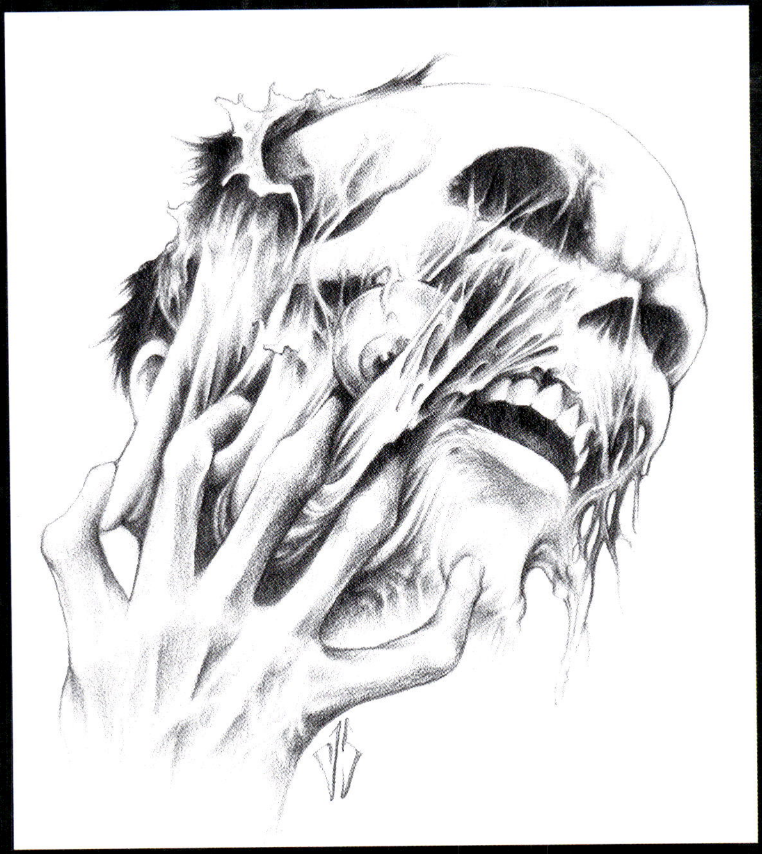

Jesse Santos

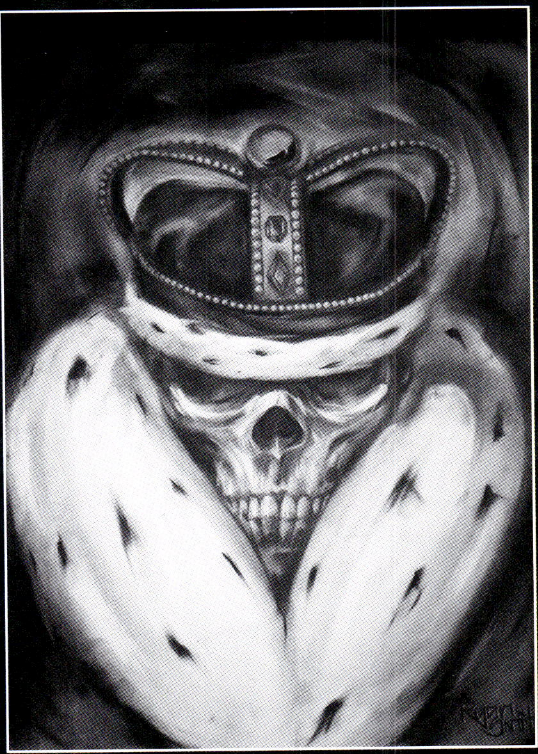

Ryan Smith

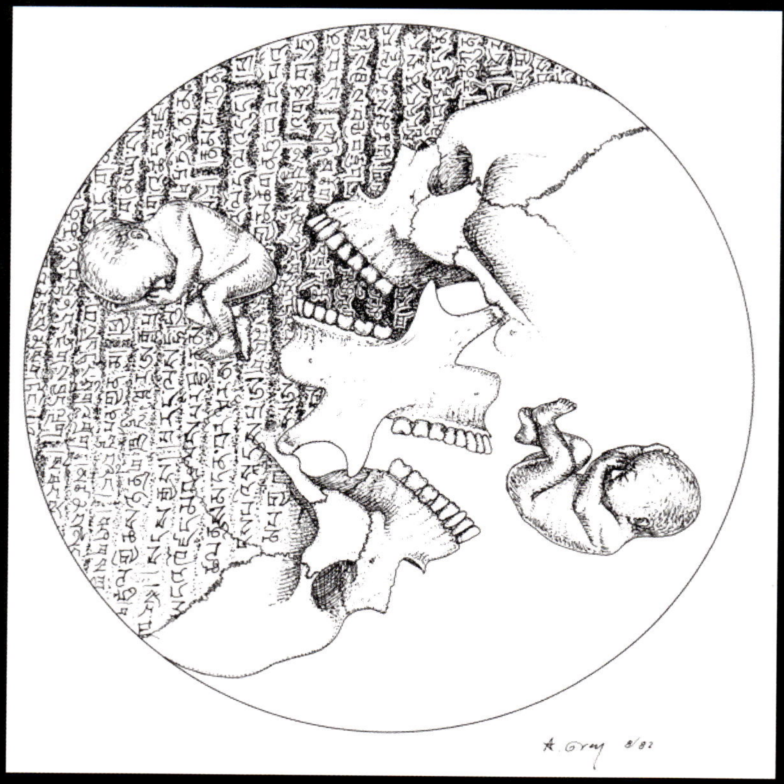

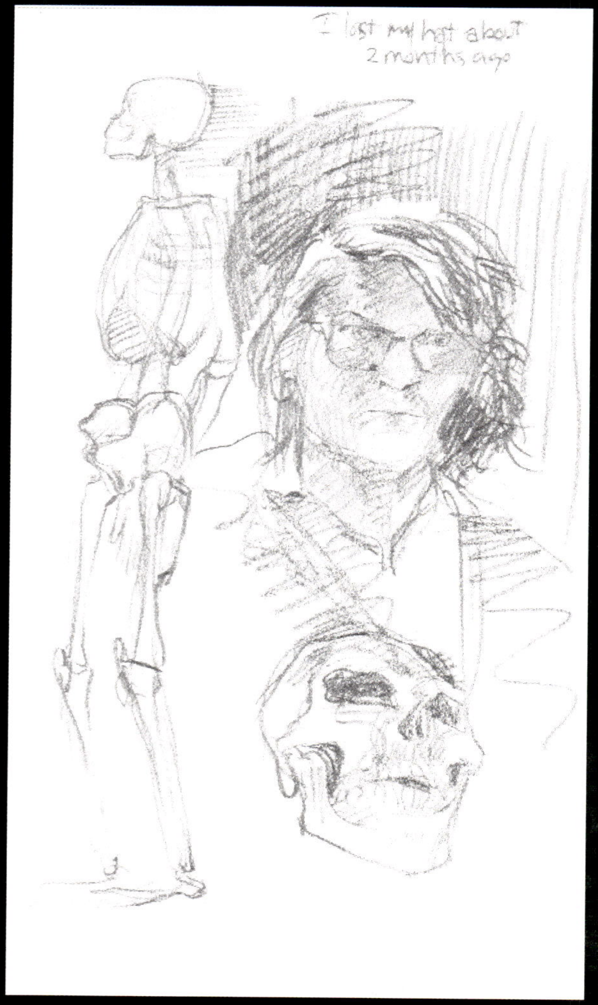

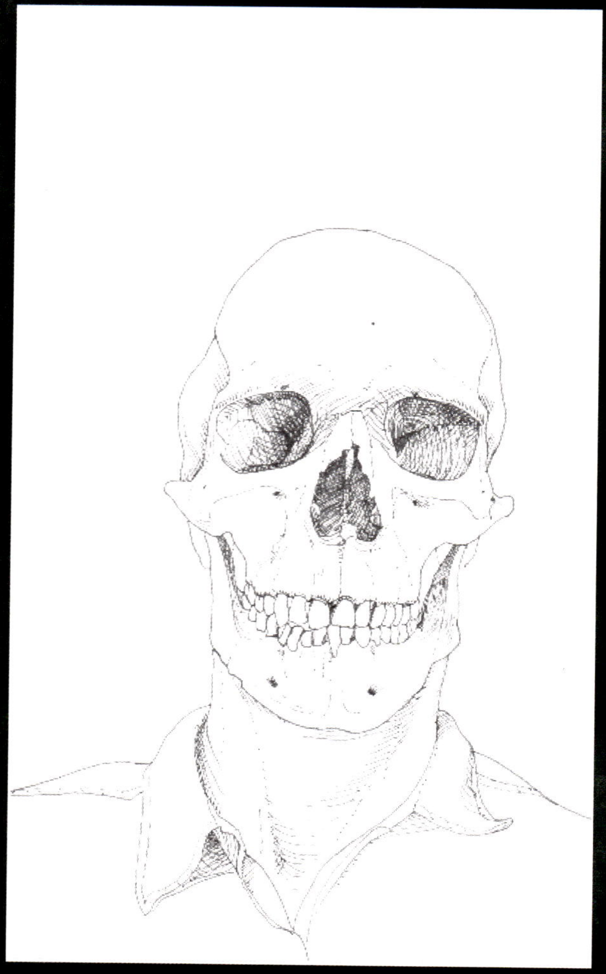

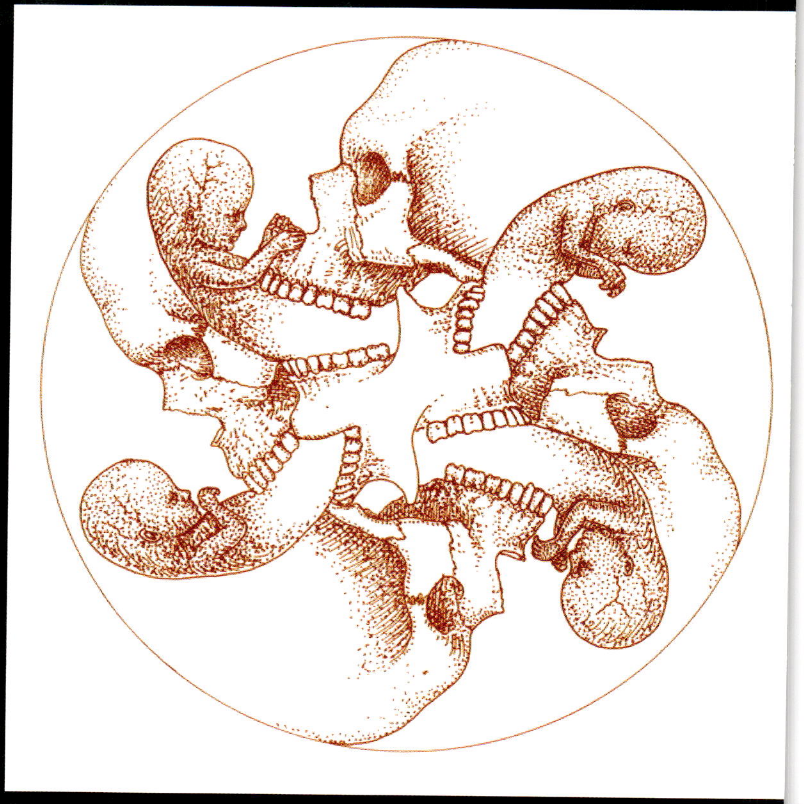

Alex Grey

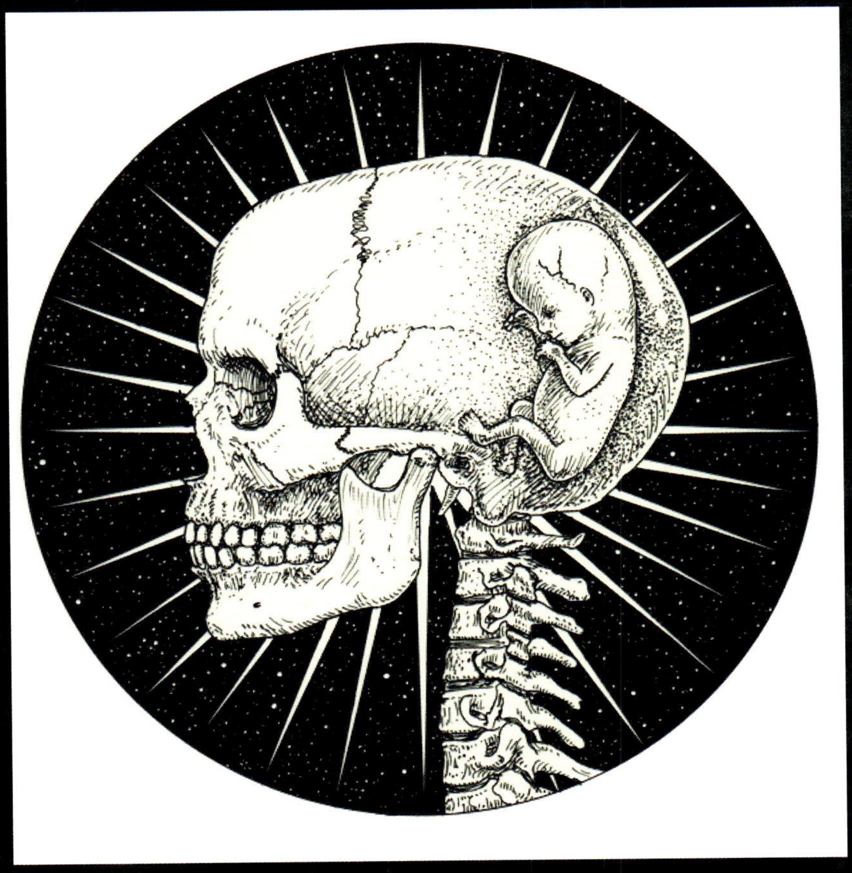

Alex Grey

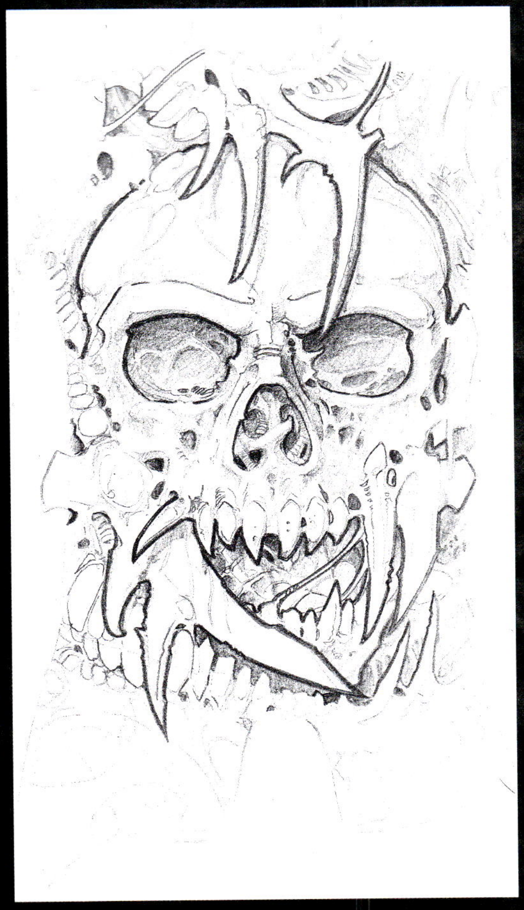

Marco Valezquez

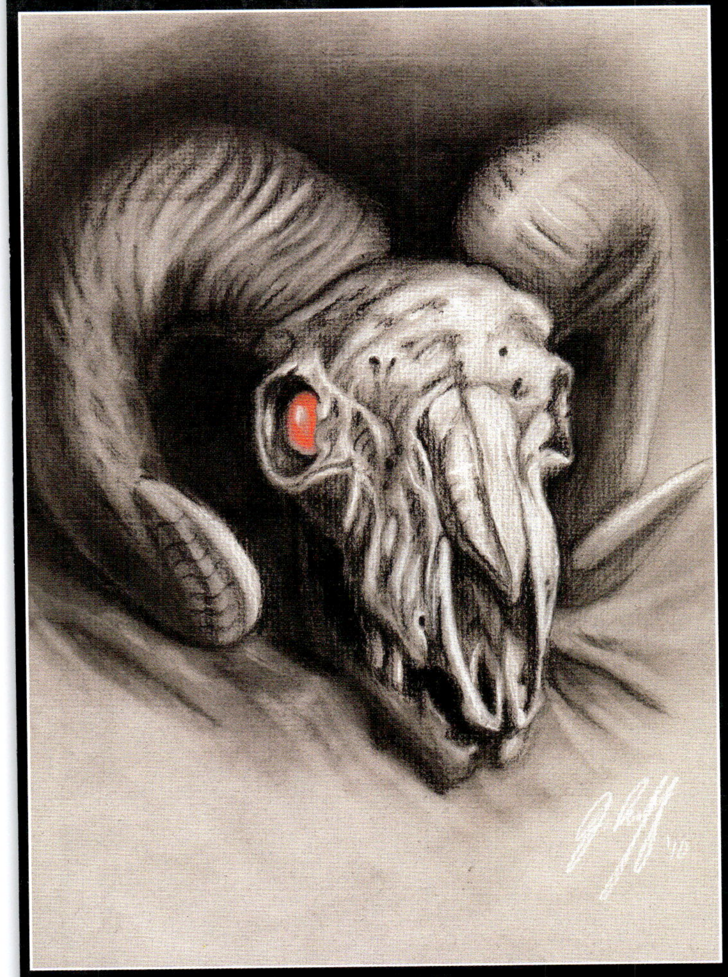

Josh Ruff

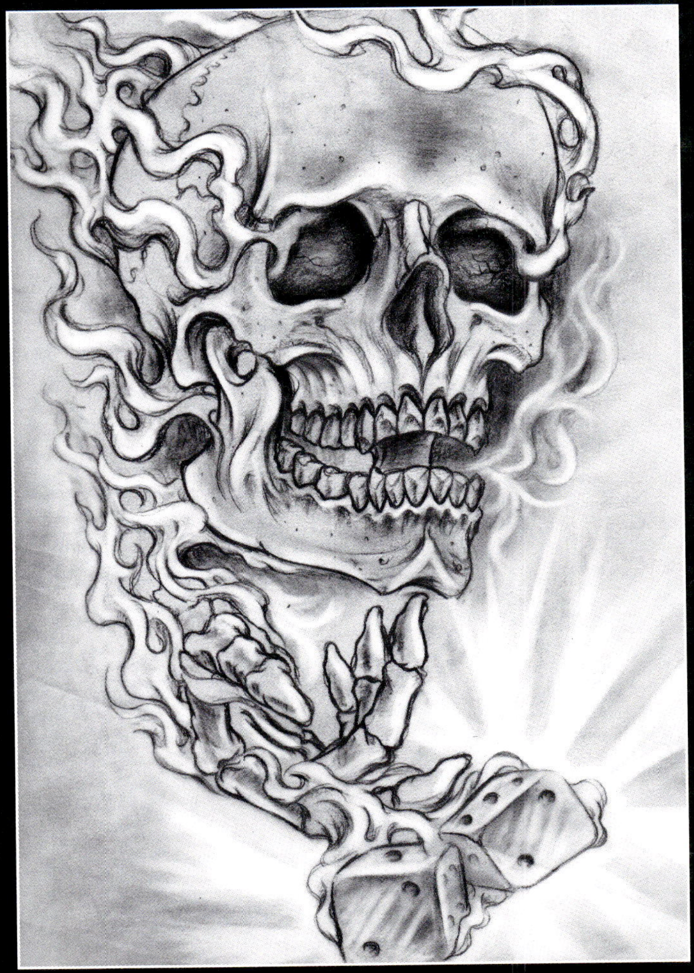

Edwin Marin

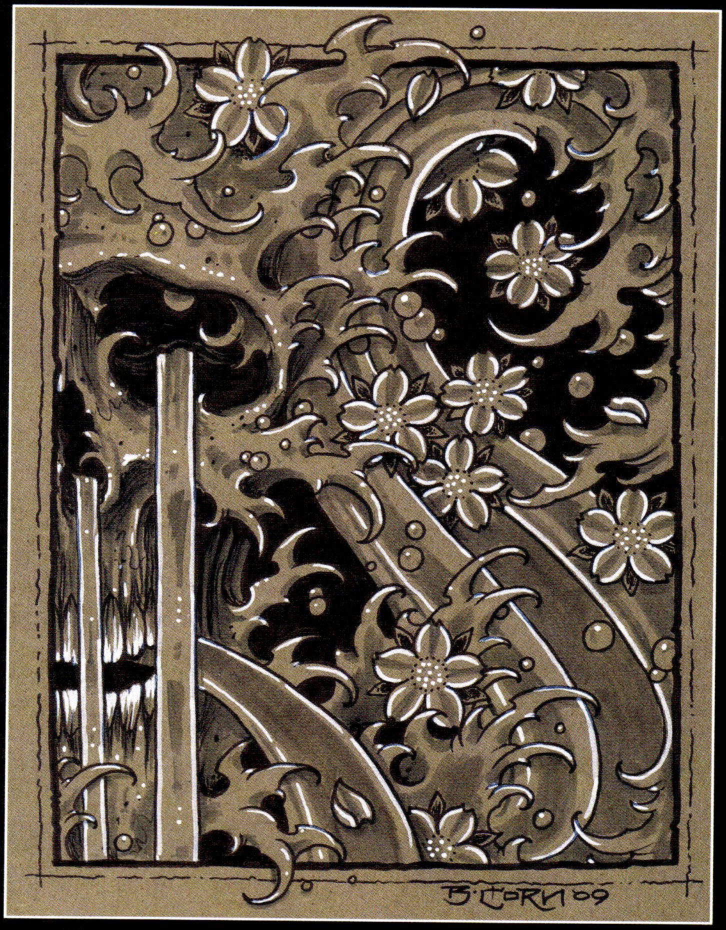

Ben Corn

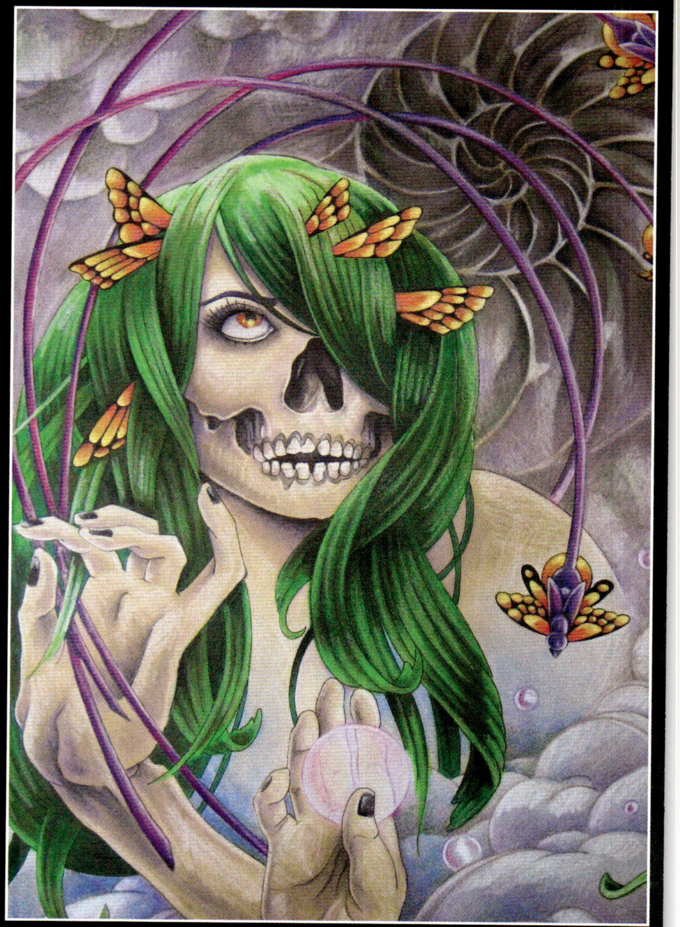

Jessica Rocha

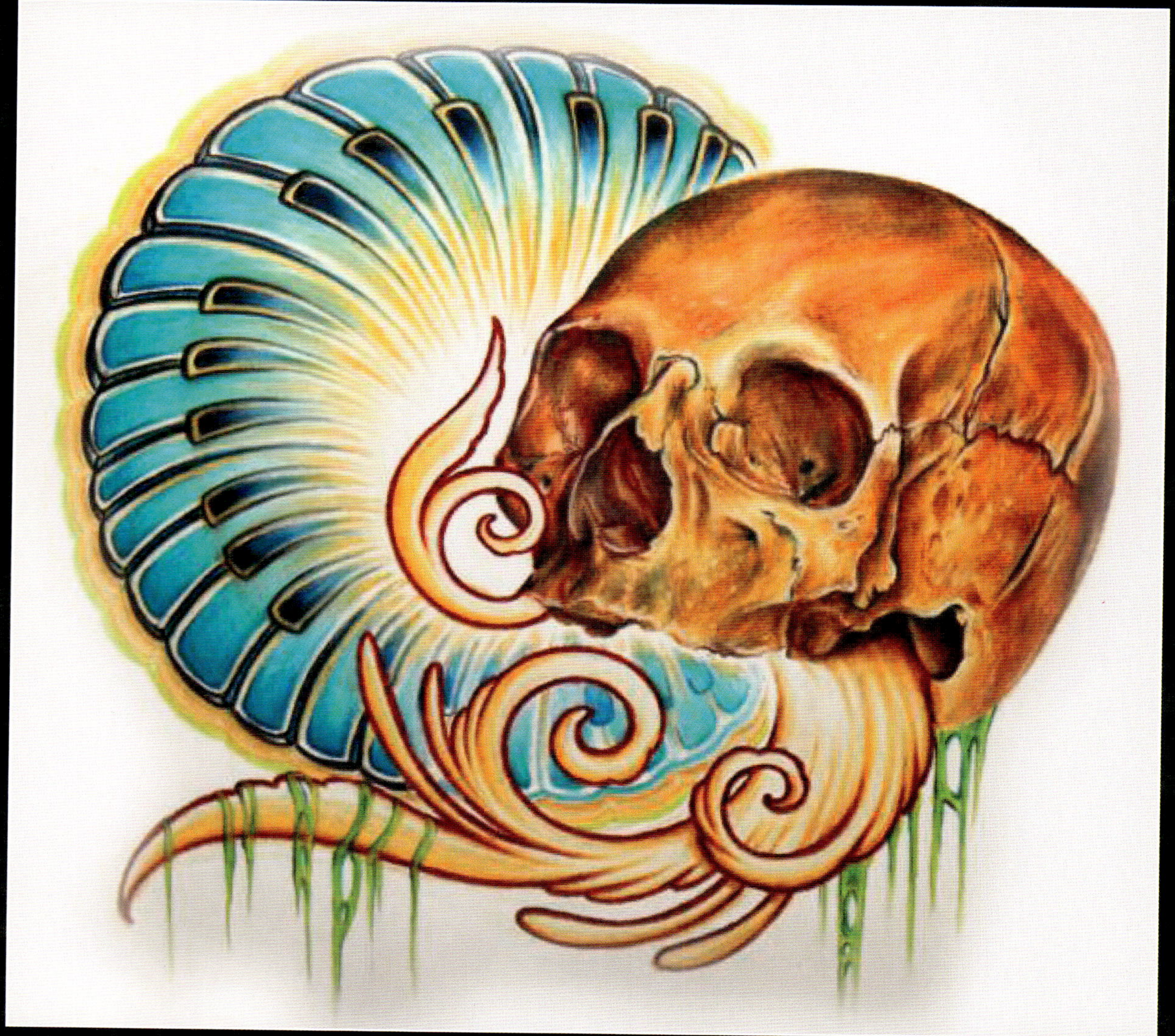

London

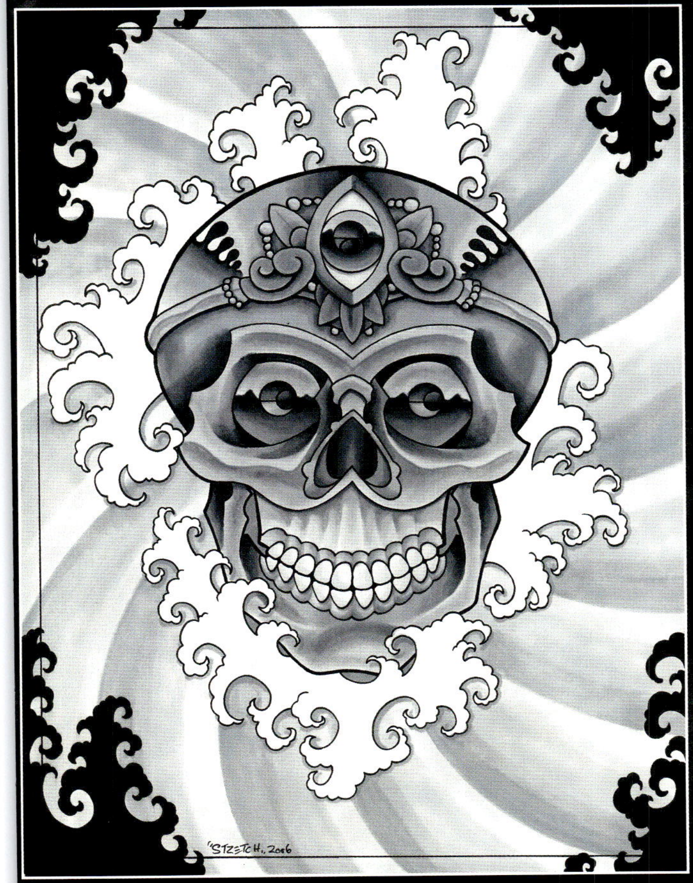

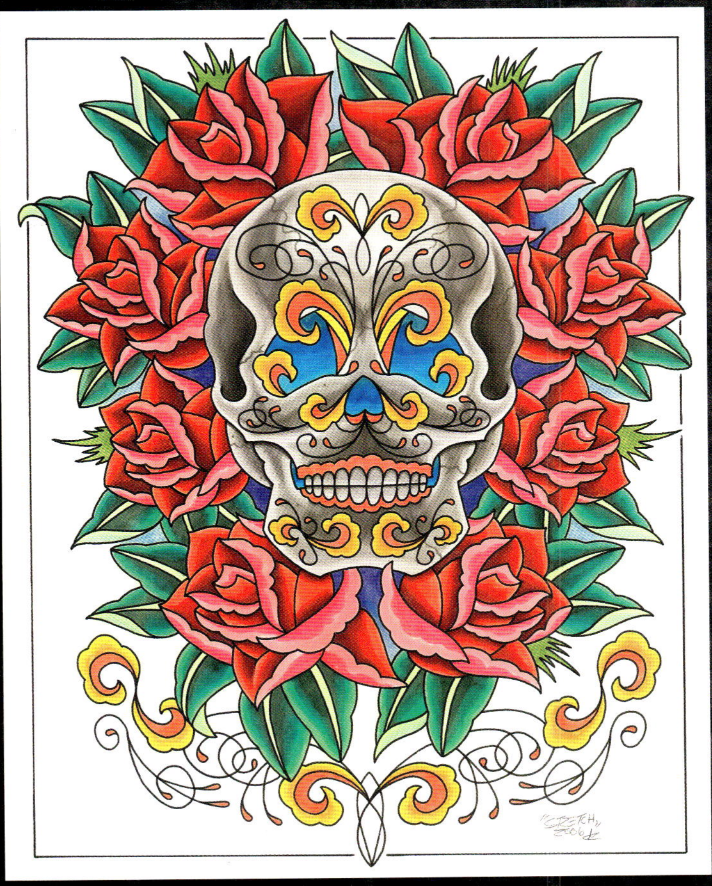

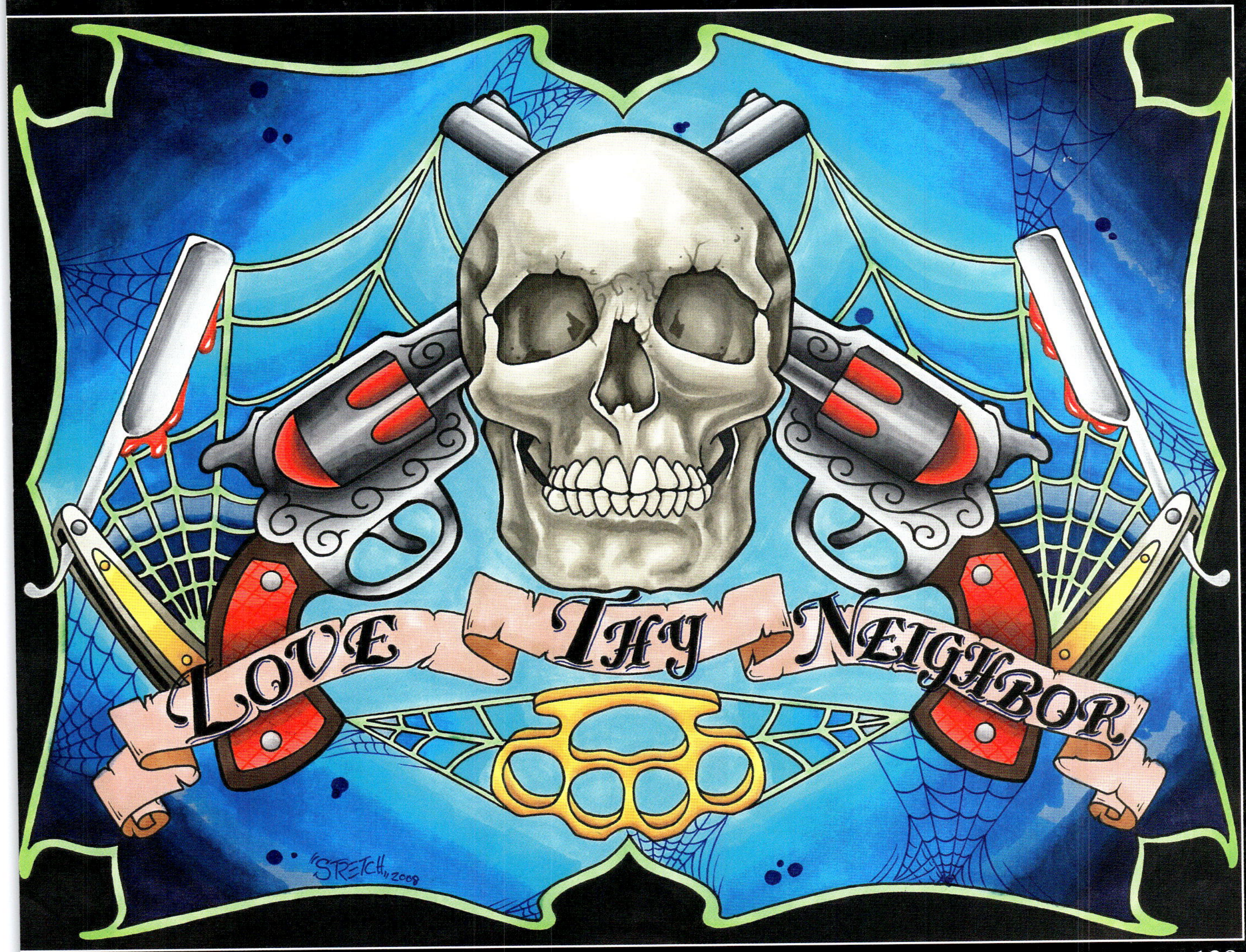
LOVE THY NEIGHBOR

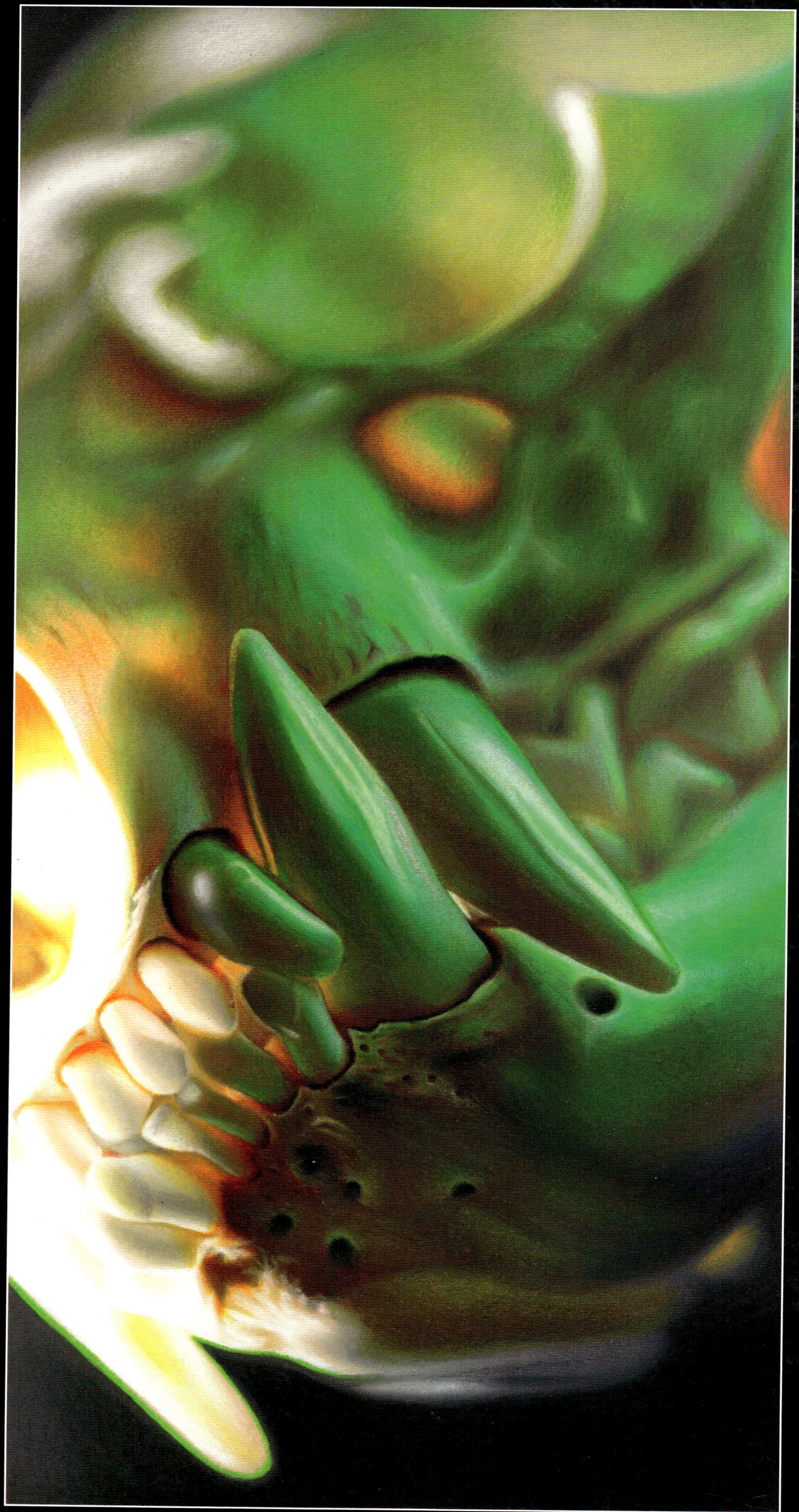

Jeff Ensminger

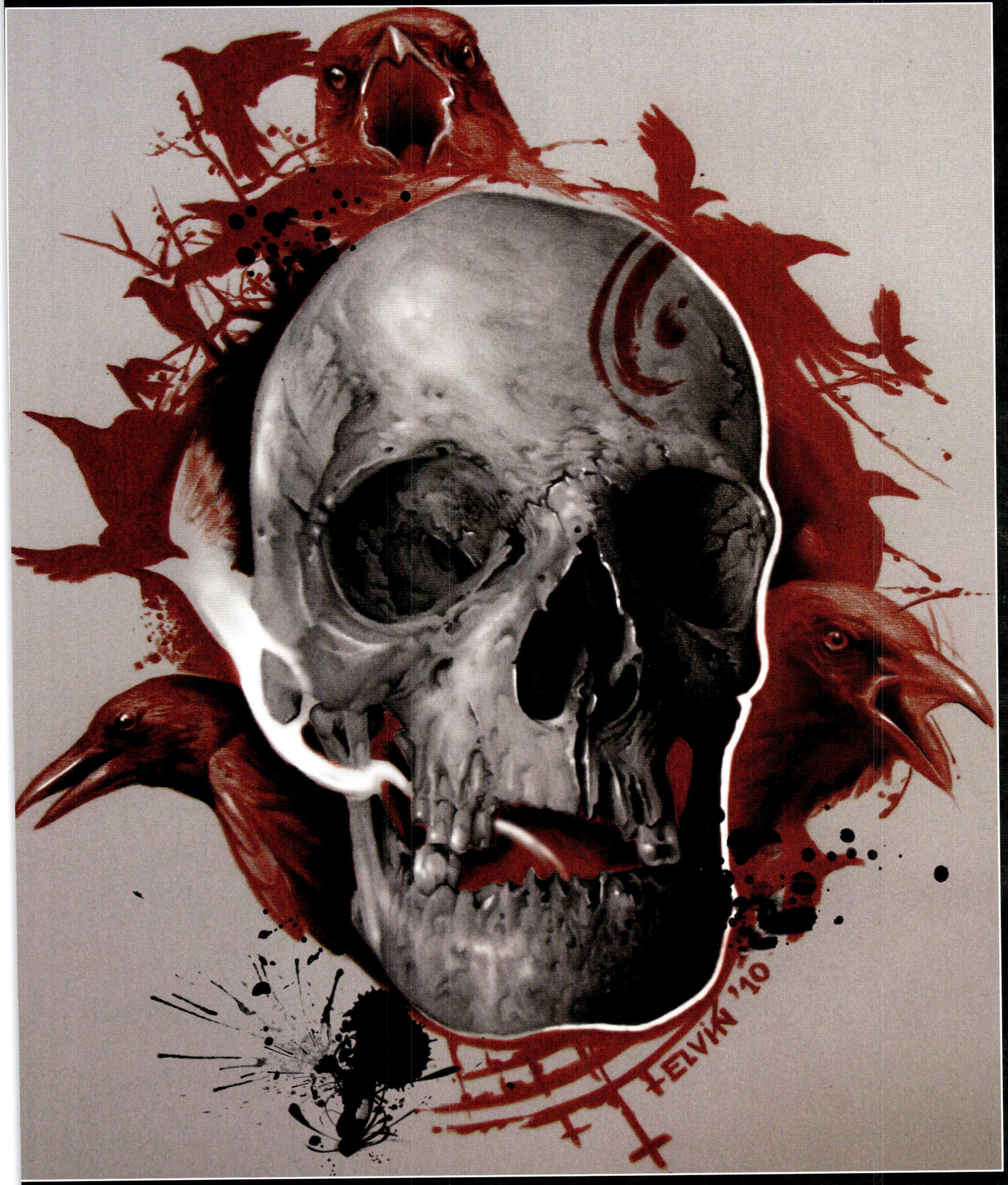

Elvin Yong

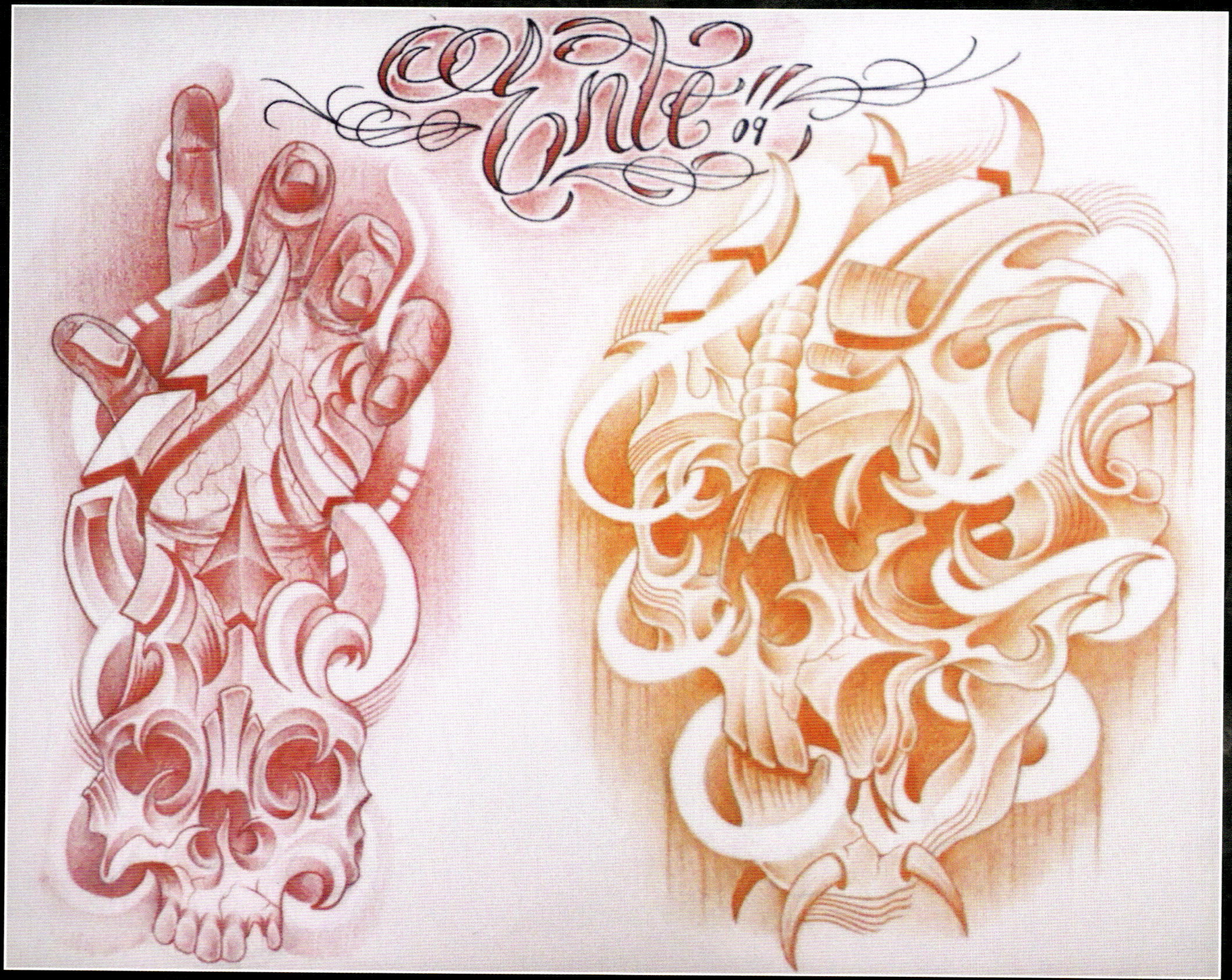

Juan Gollaz

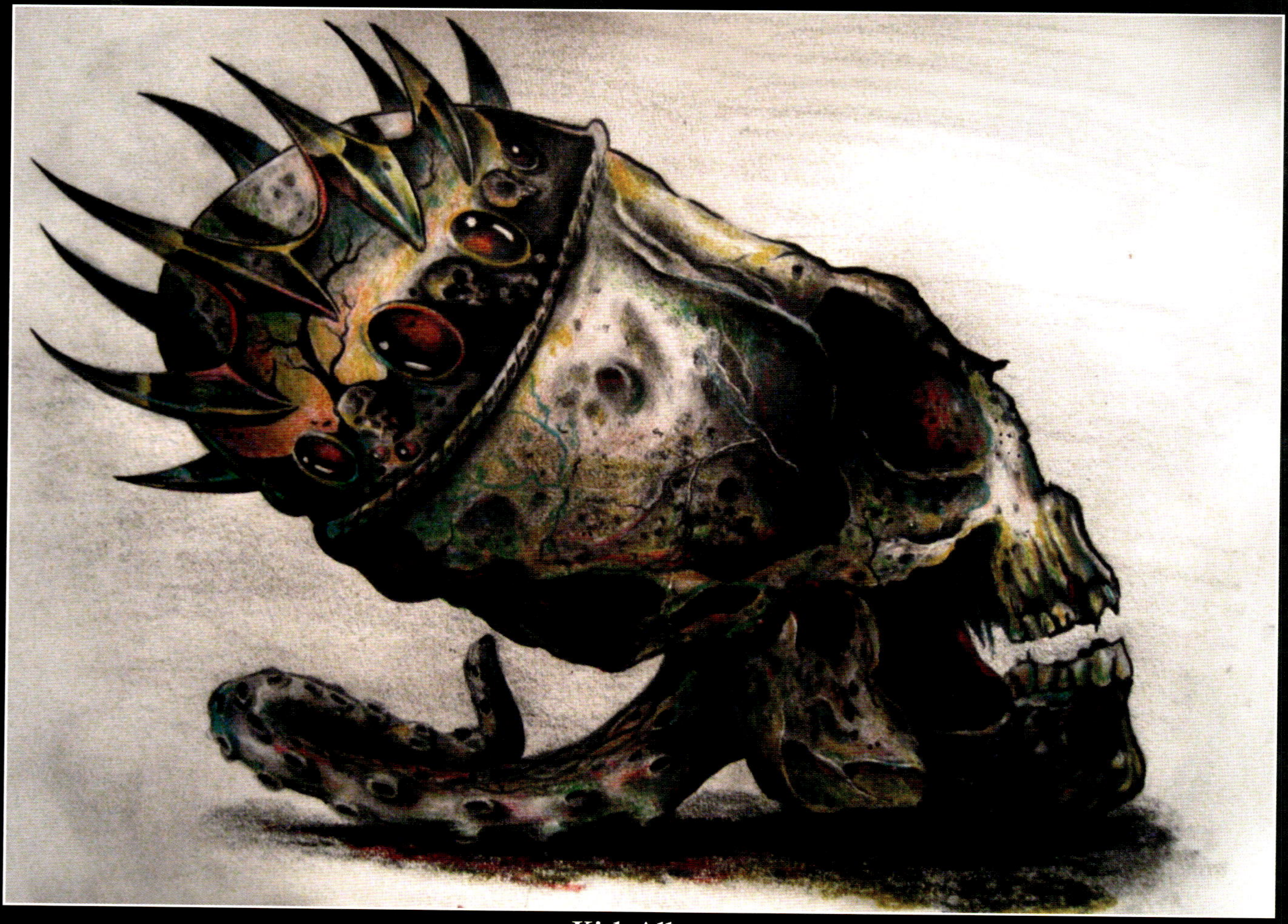

Kirk Alley

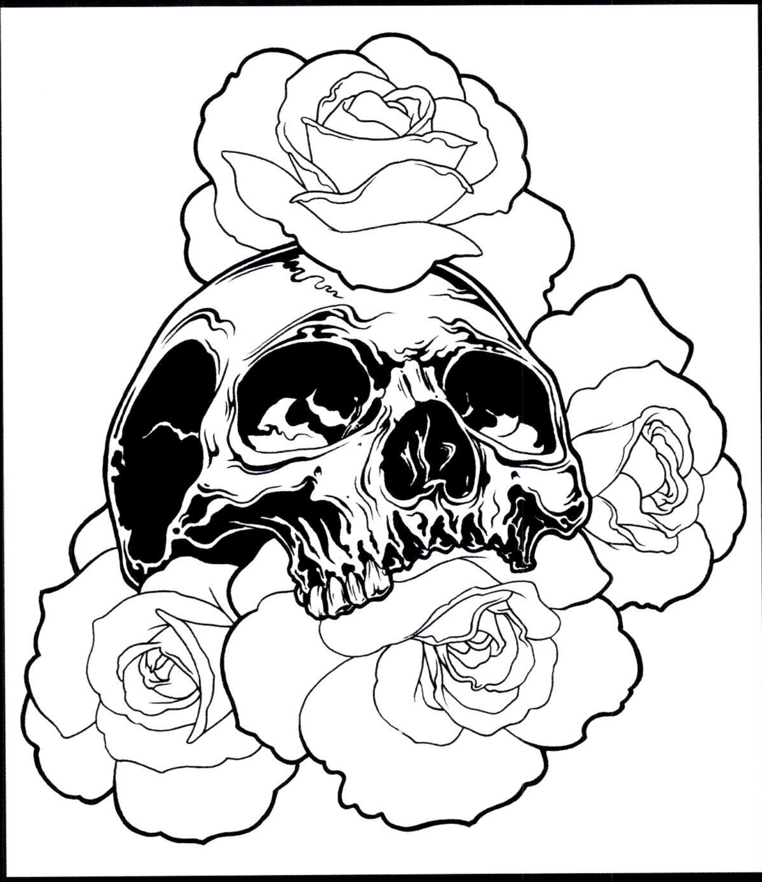

Josh Duffy

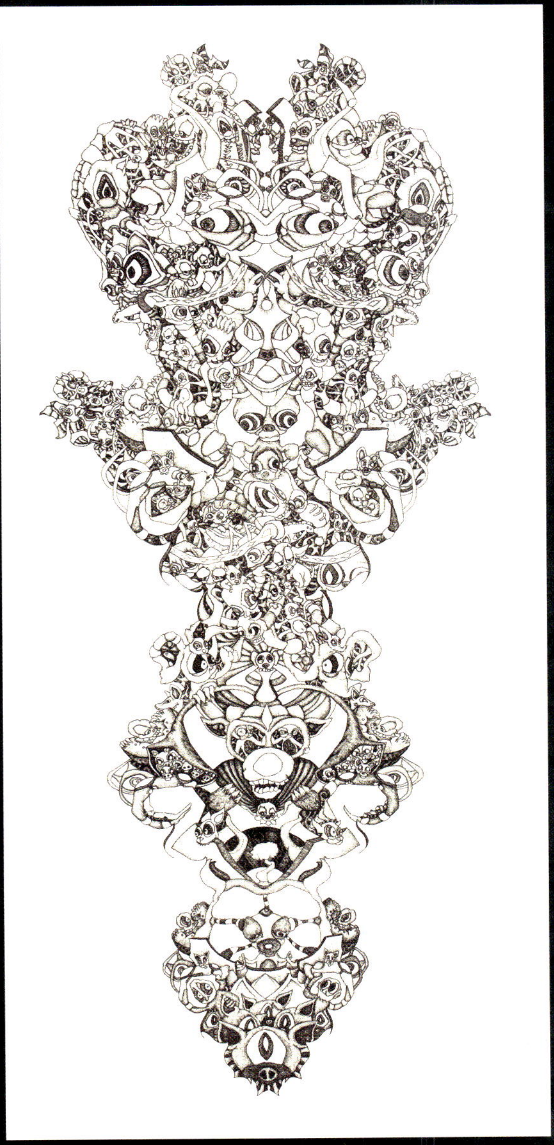

Michael Sexauer

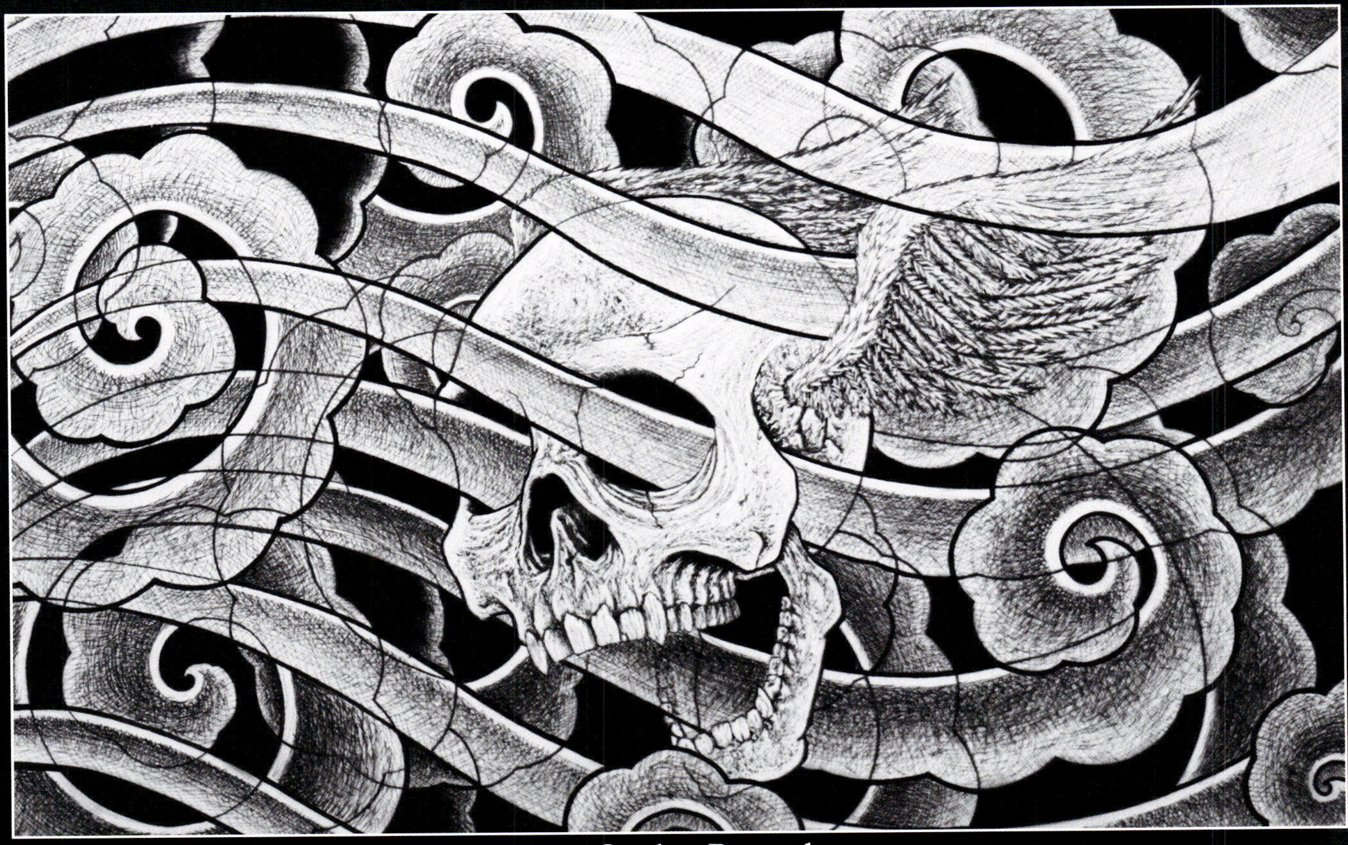

Stephen Barnard

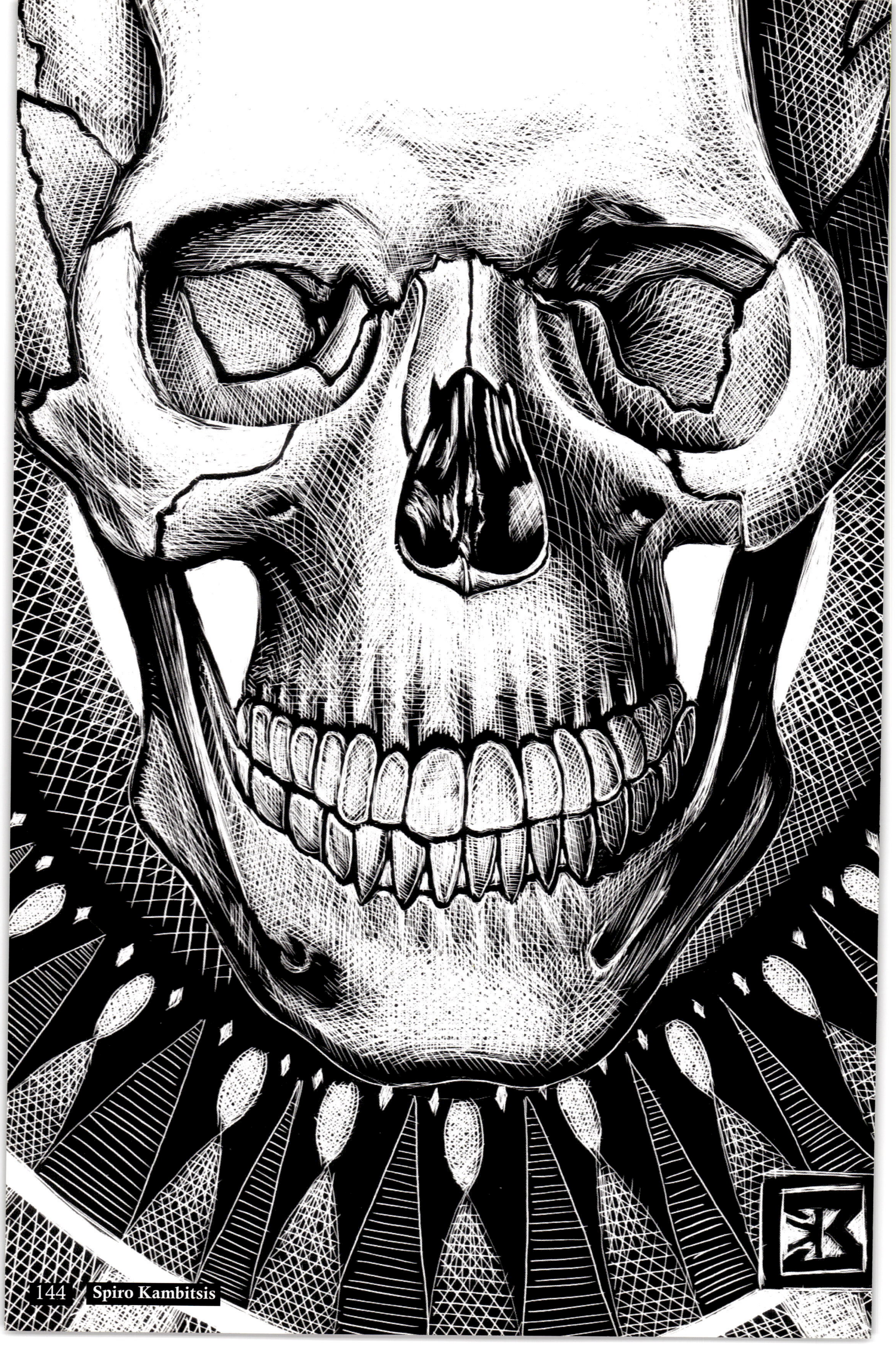

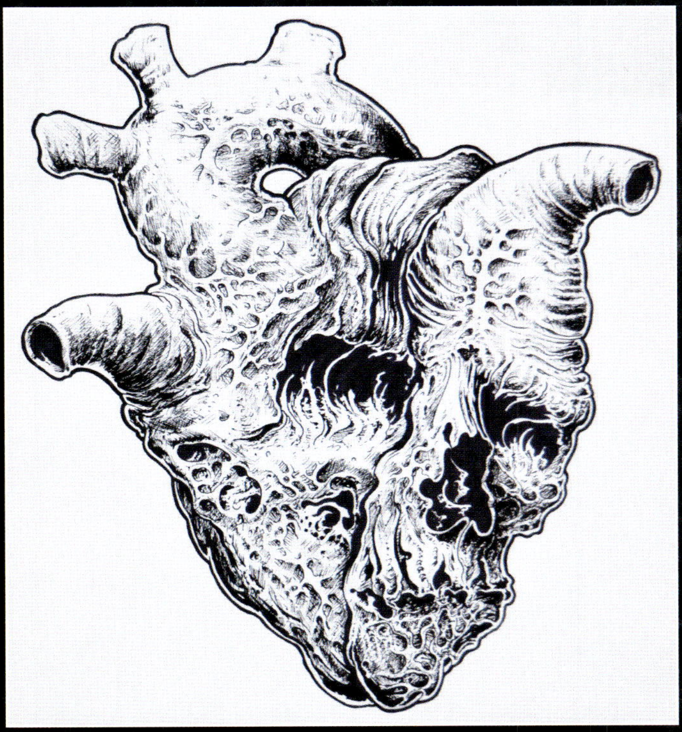

Josh Duffy

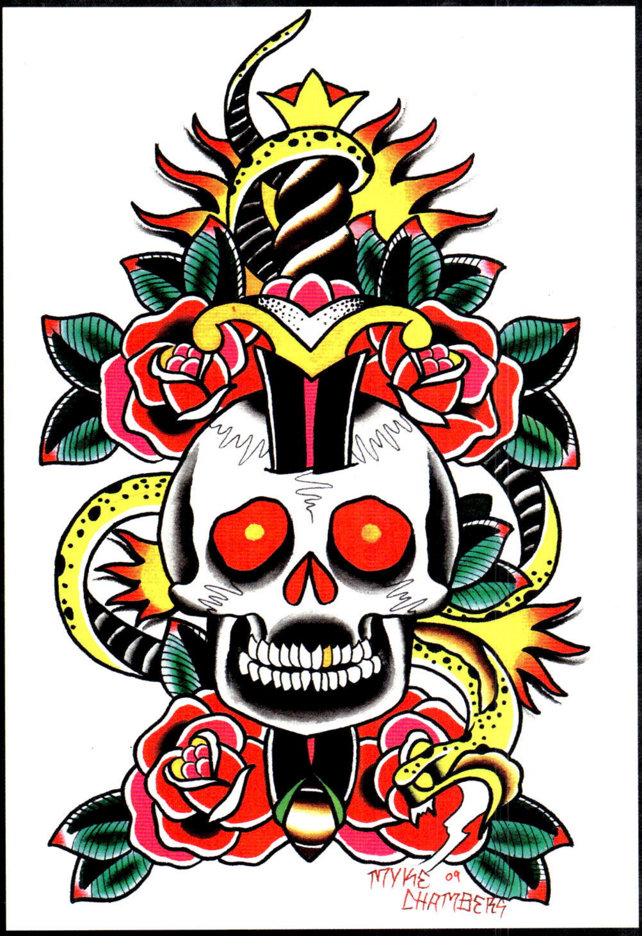

Myke Chambers

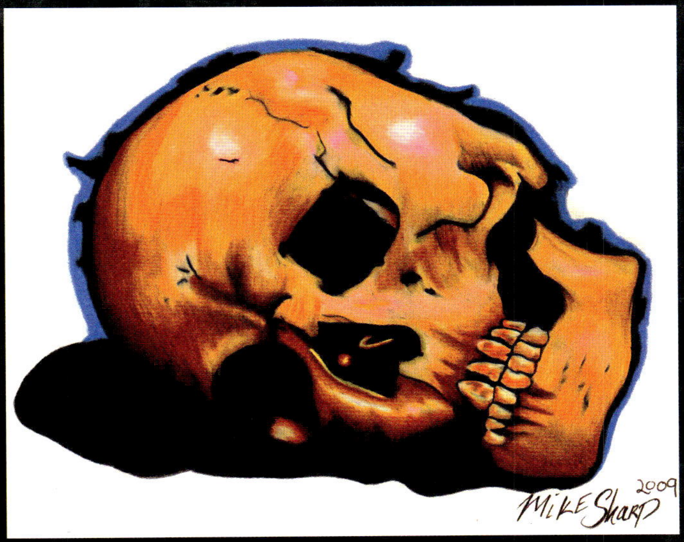

Mike Sharp

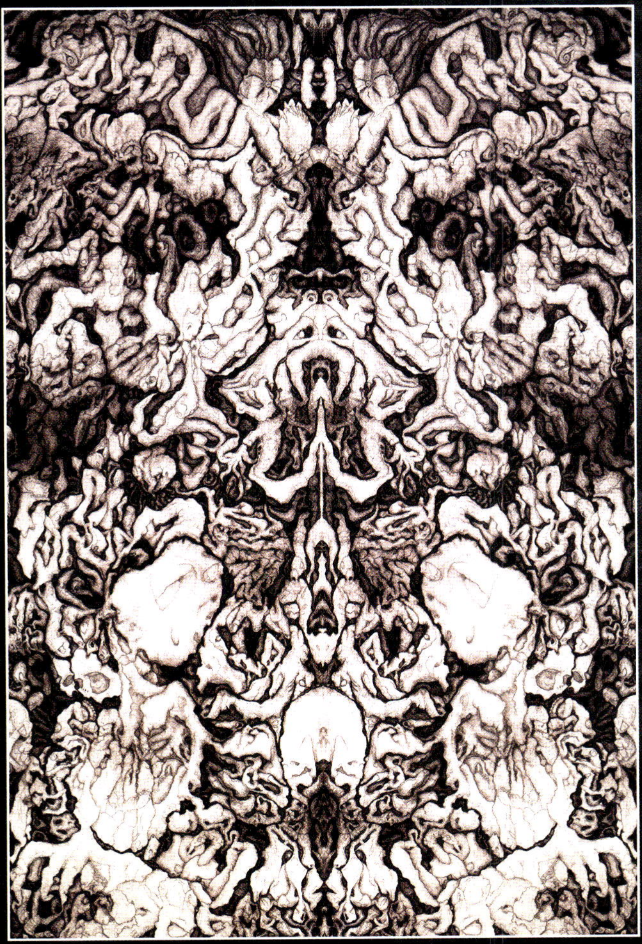

Michael Sexauer

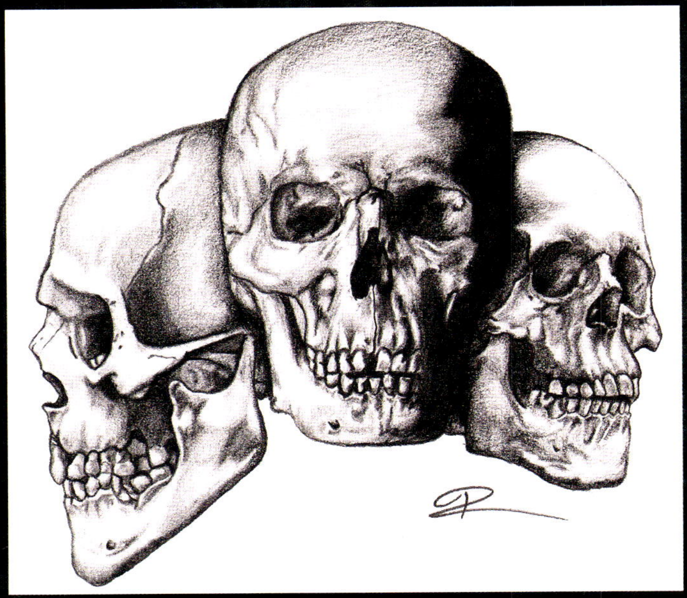

Ricky Rivera

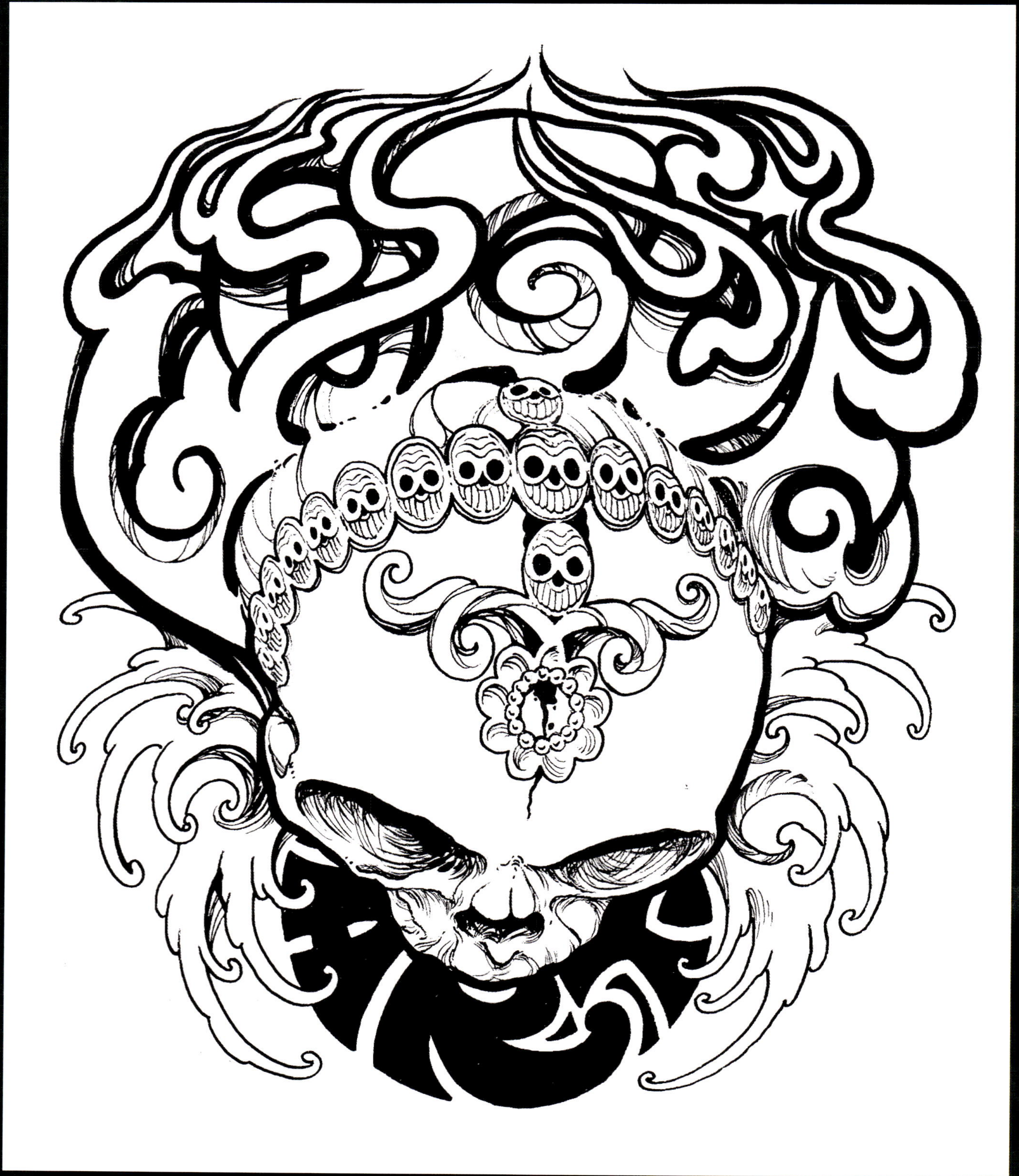

Jeff Gogué

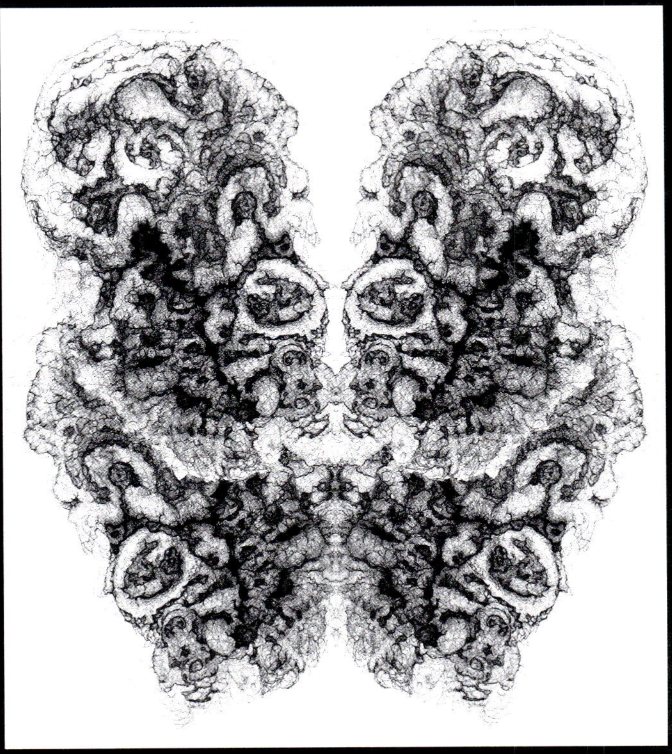

Michael Sexauer

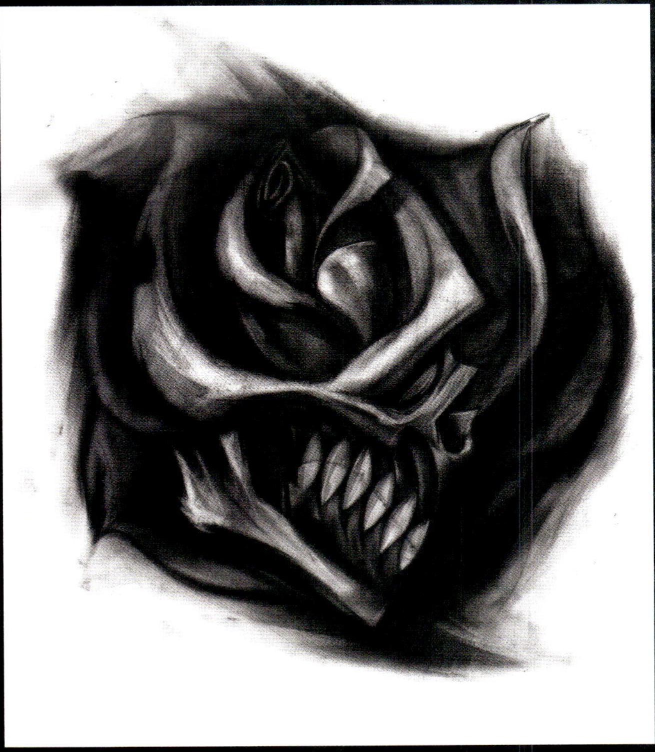

Ryan Smith

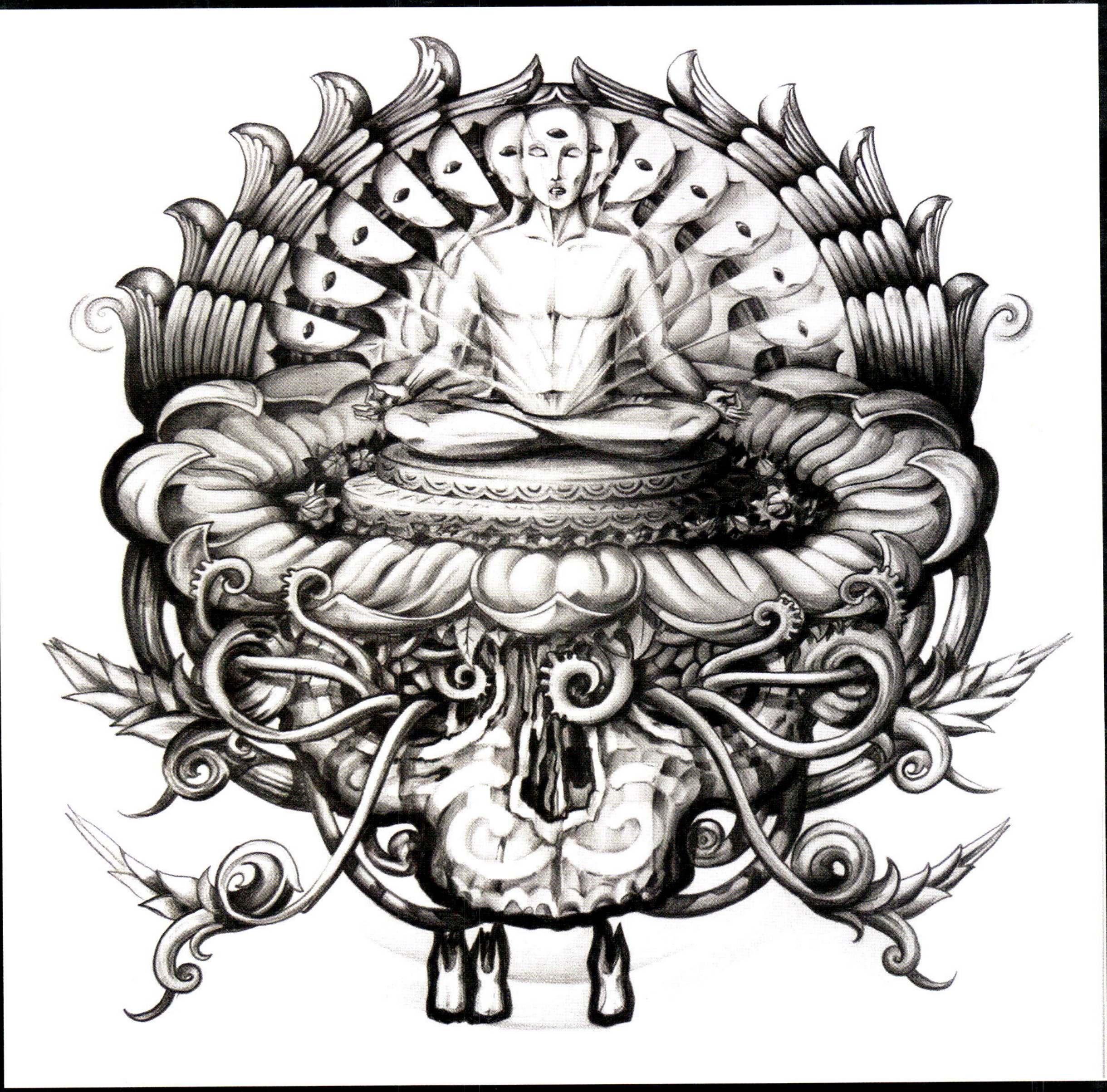

Andrey Barkov

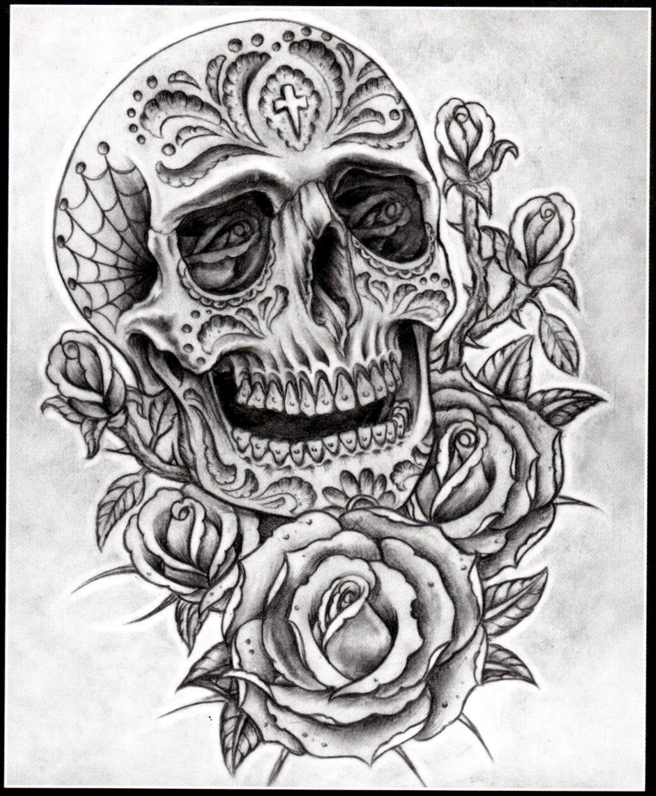

Edwin Marin

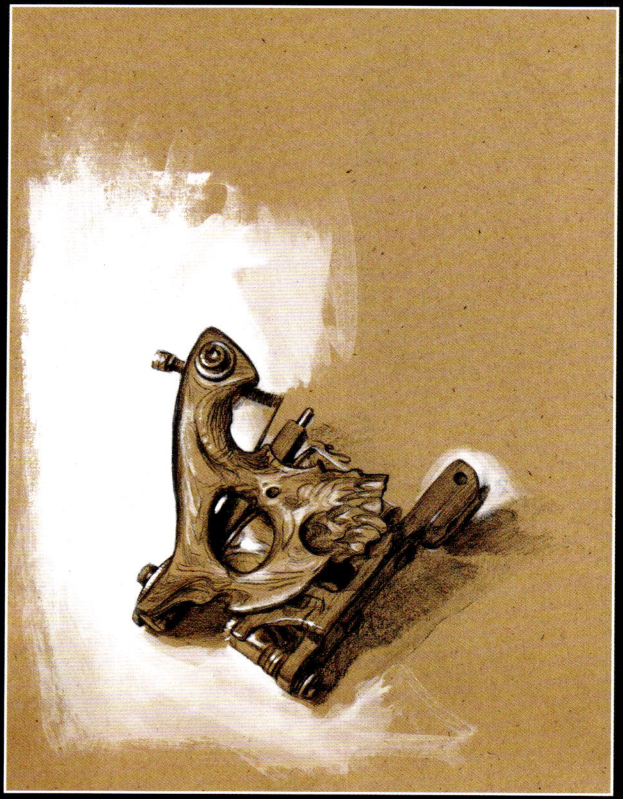

Shawn Barber

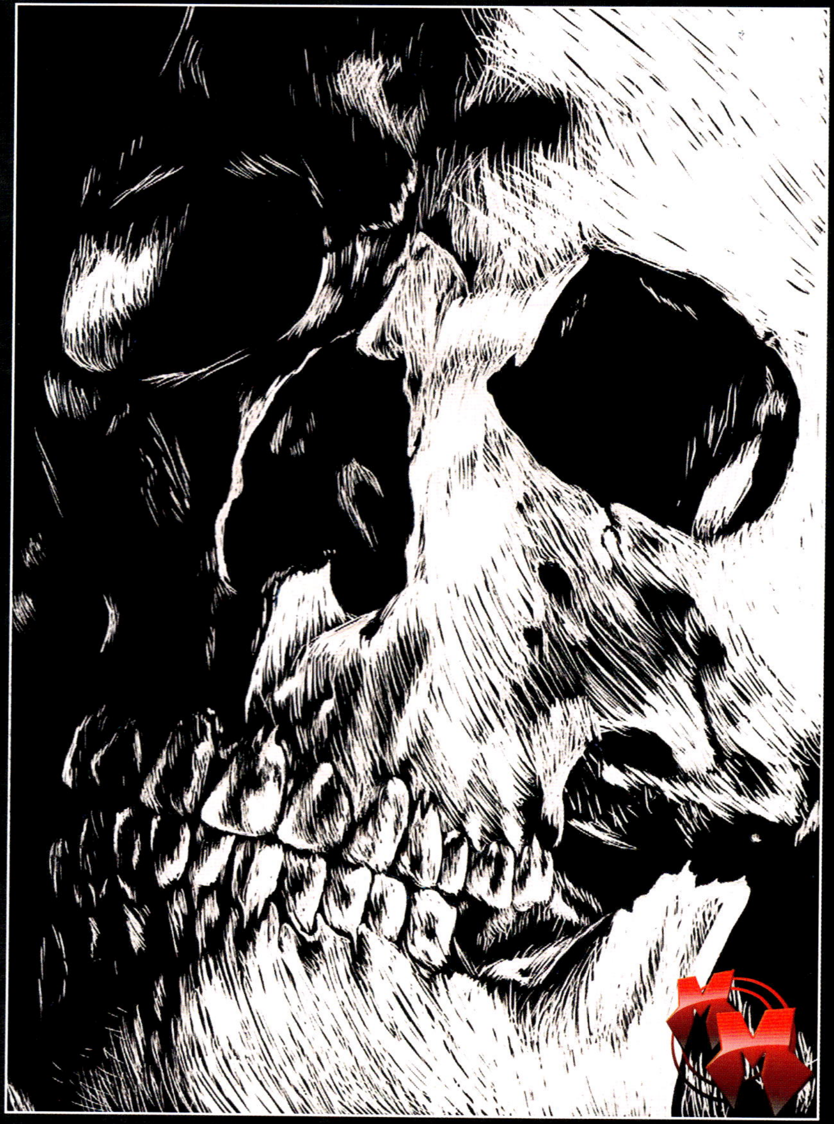

Mattie Macabre

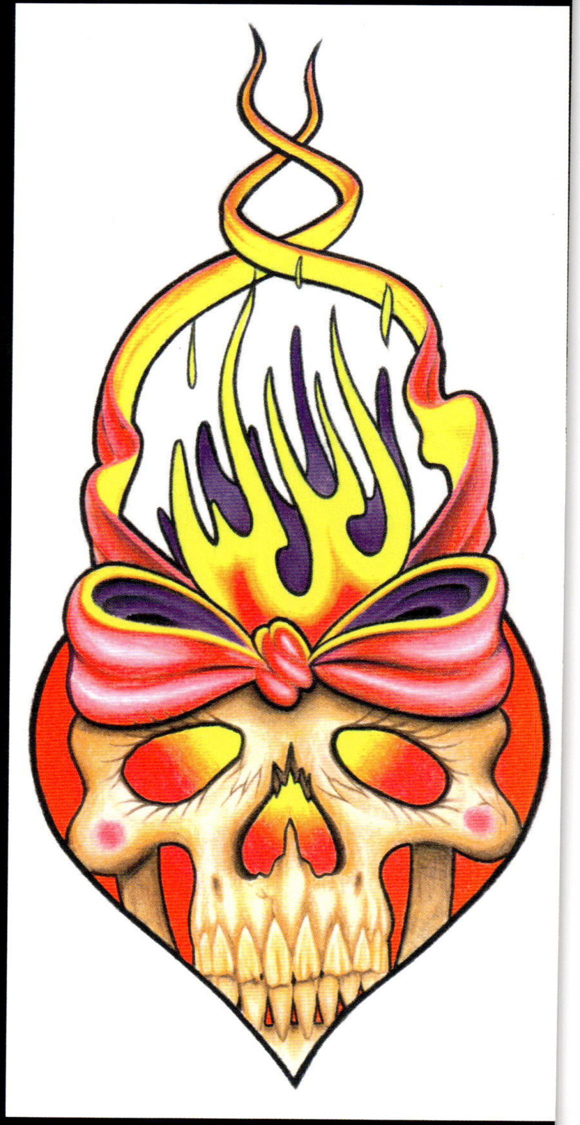

Spiro Kambitsis

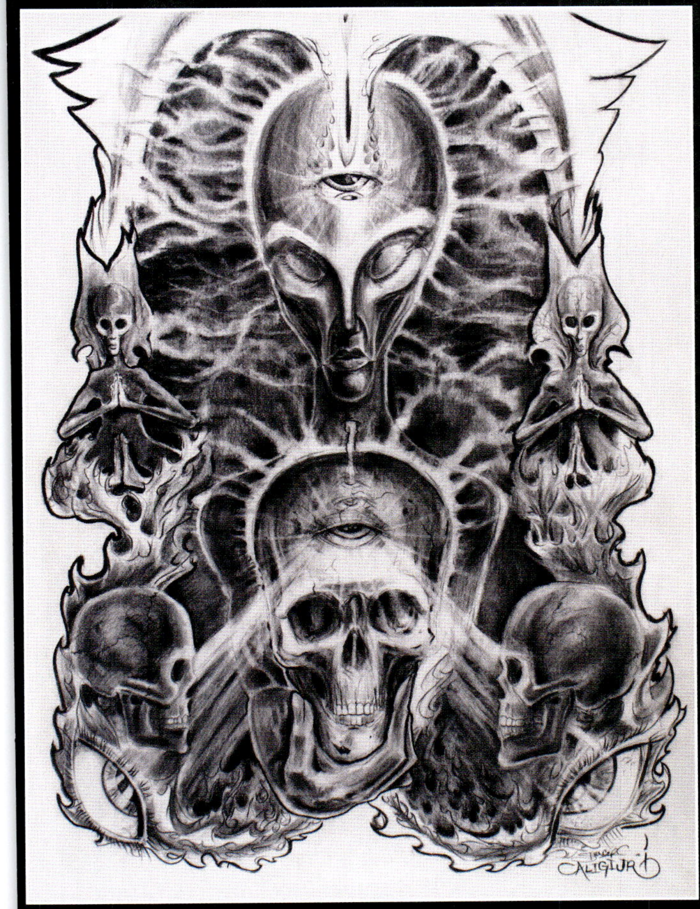

Spencer Caligiuri

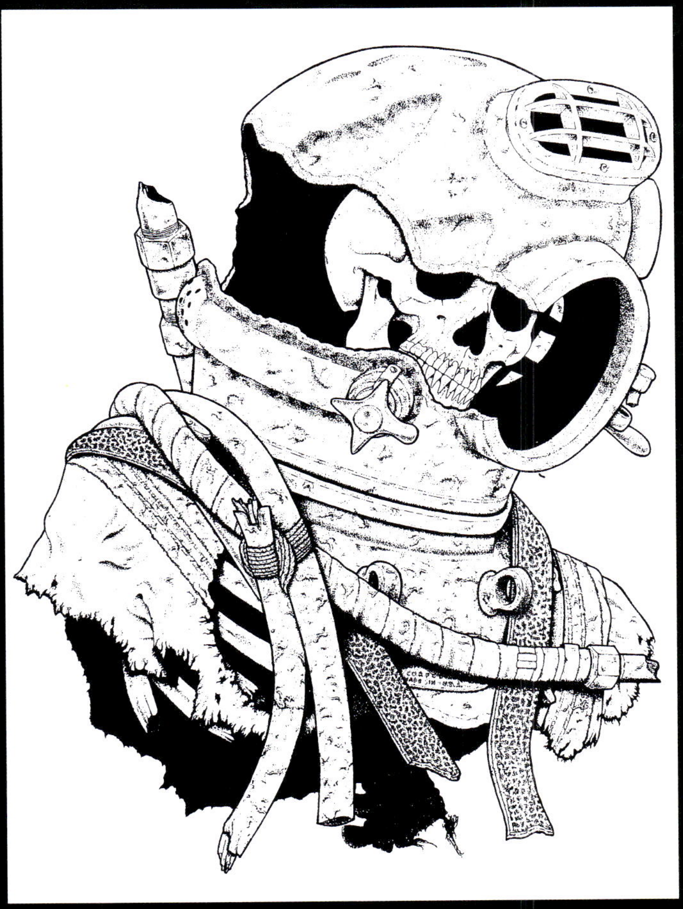

Micheal A. Simonetti

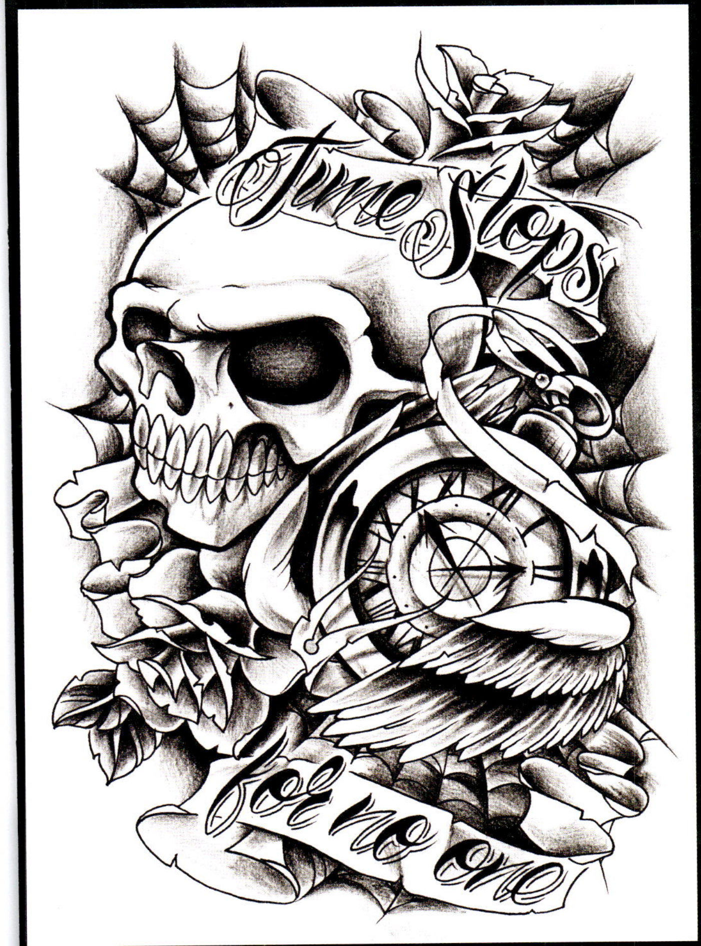

Willem Janssen

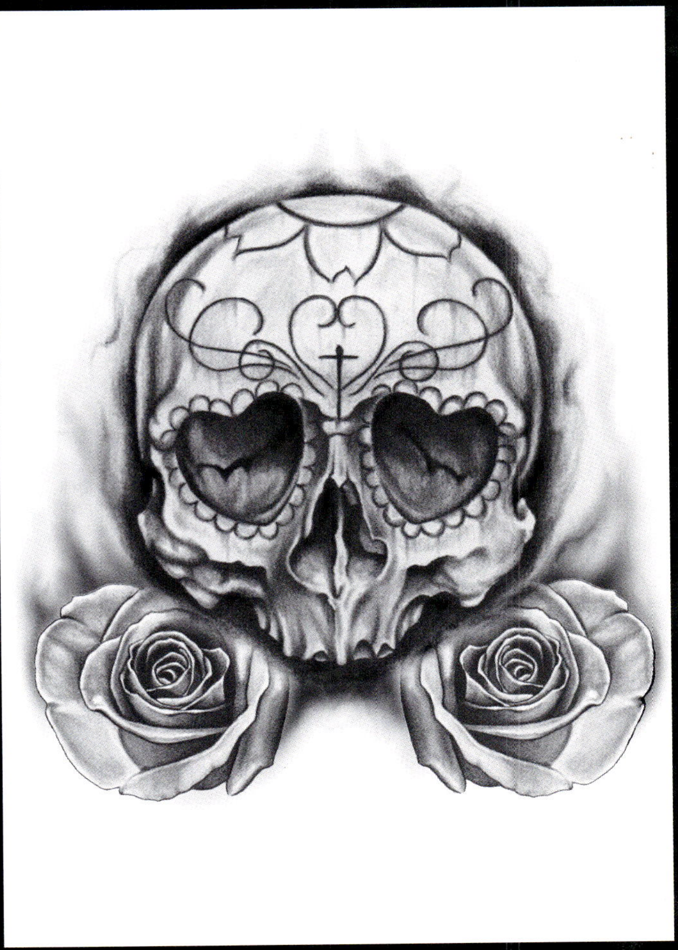

Mogie

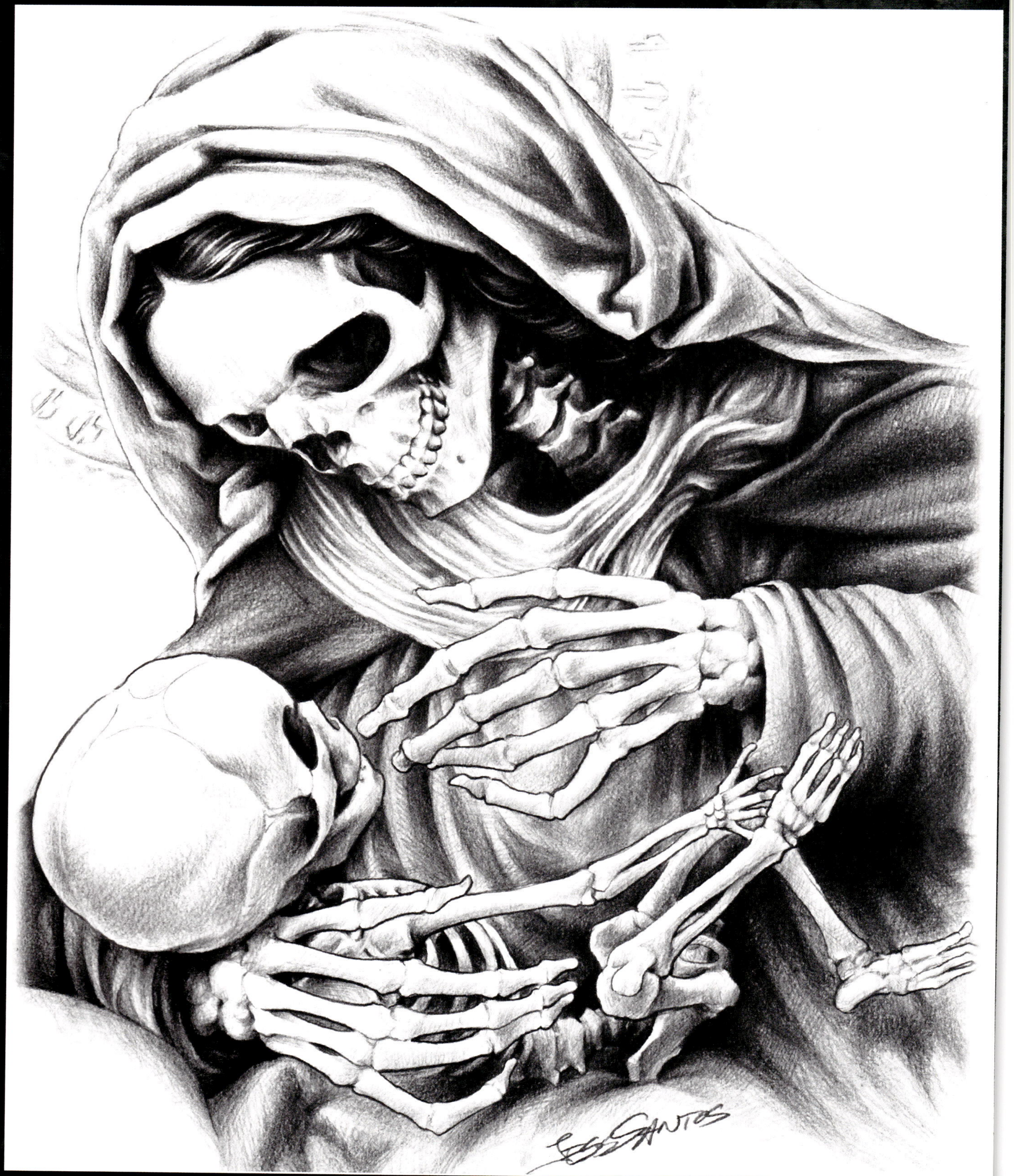

Jesse Santos

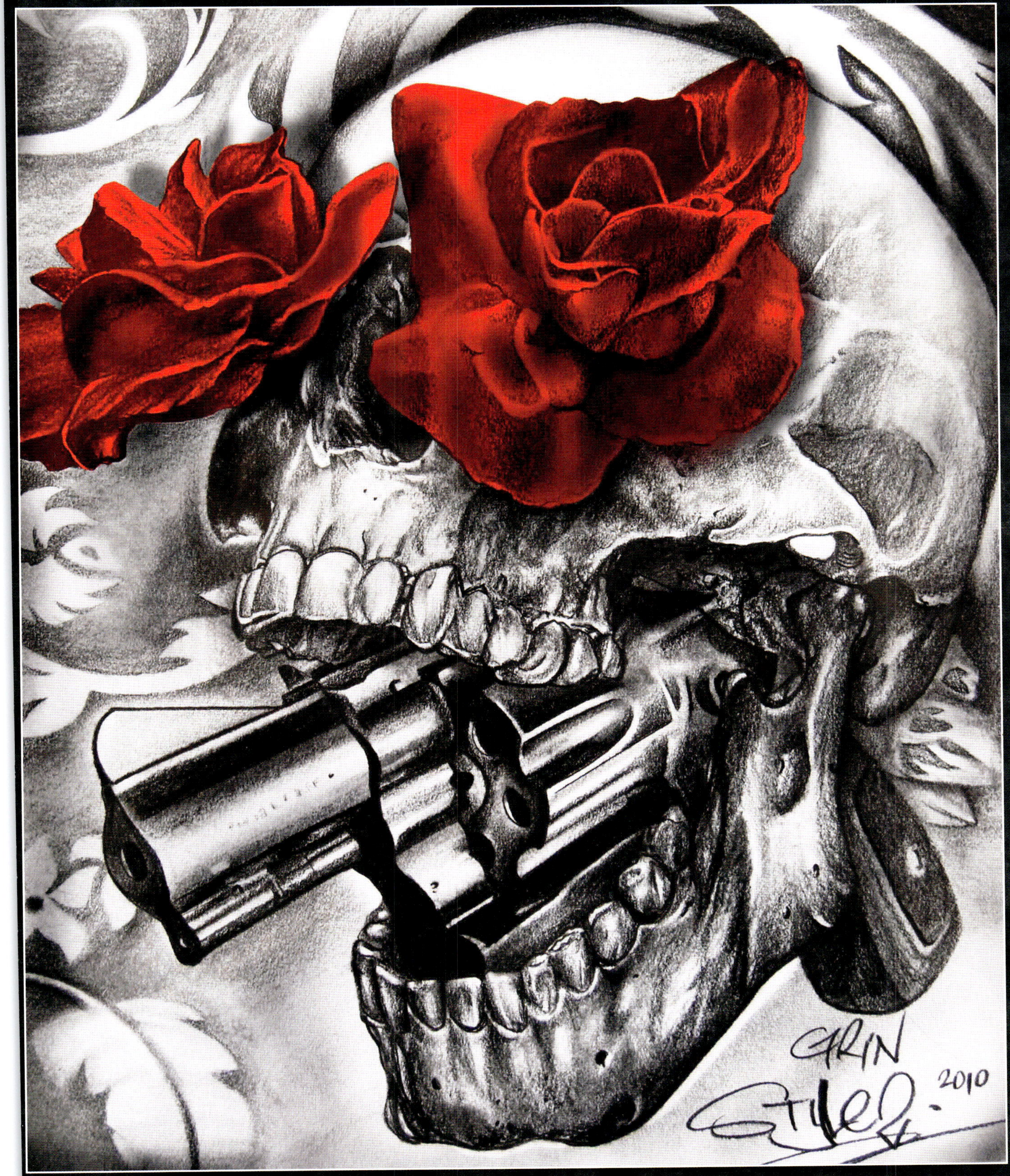

Grin Stylez

·MIXED MEDIA· ·DIGITAL ART· ·GRAFFITI·

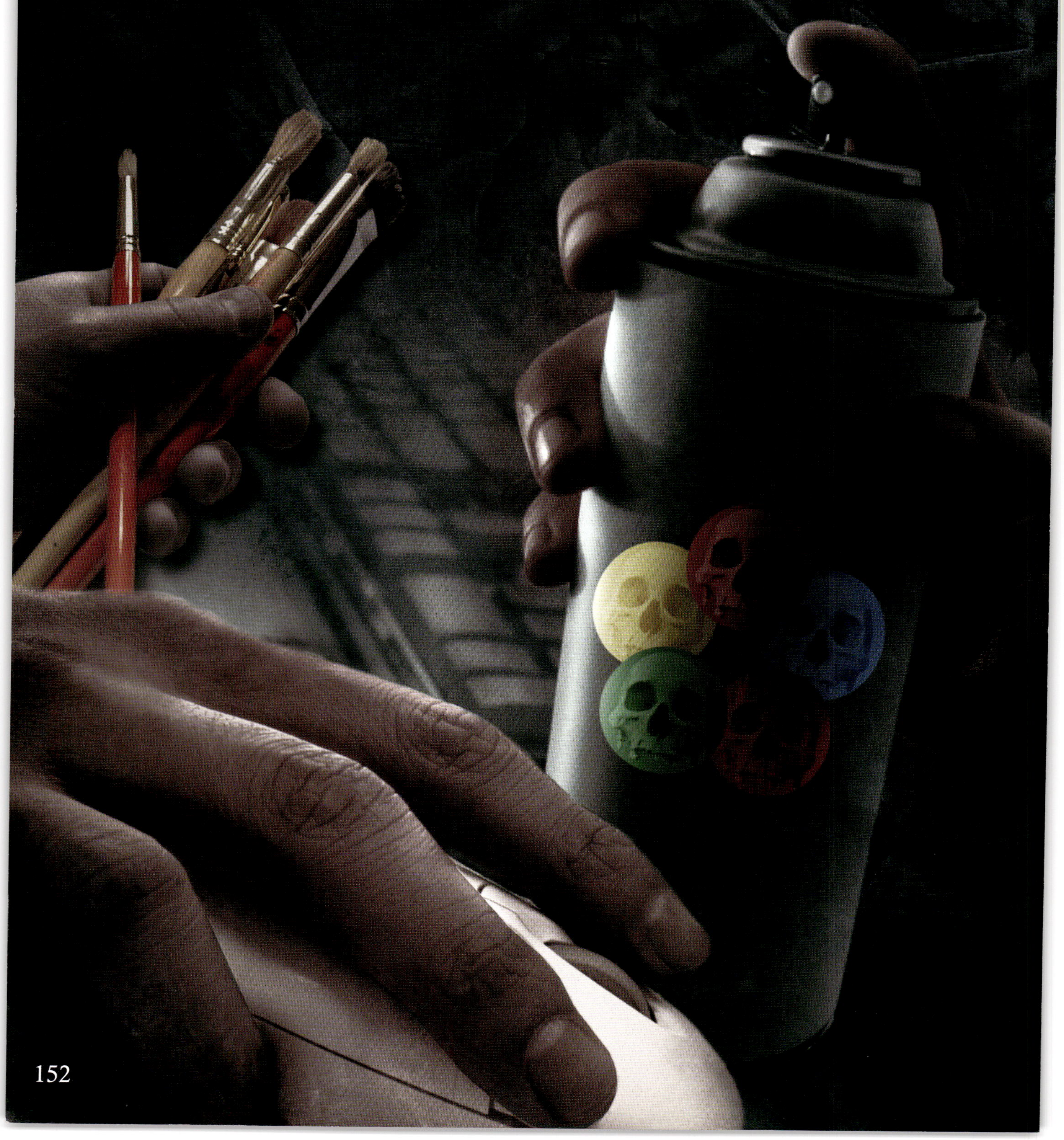

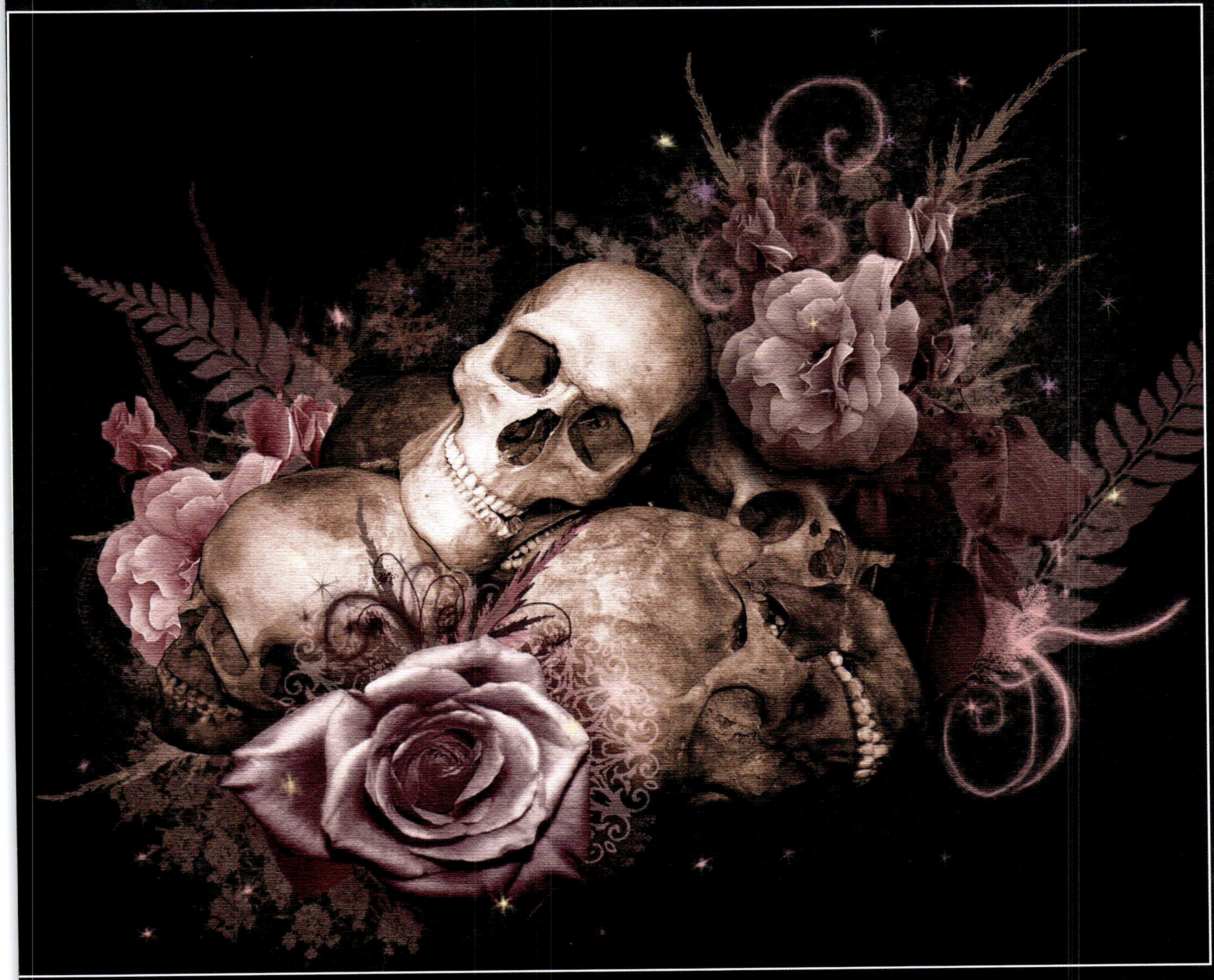

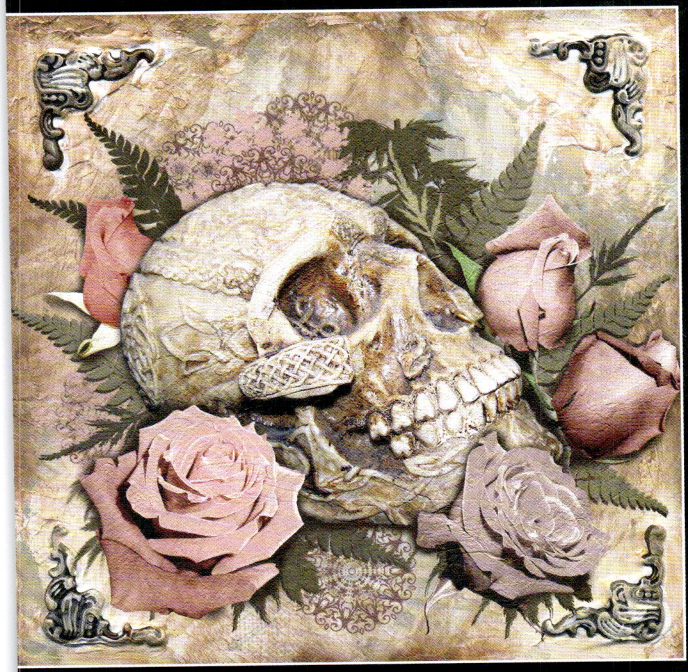

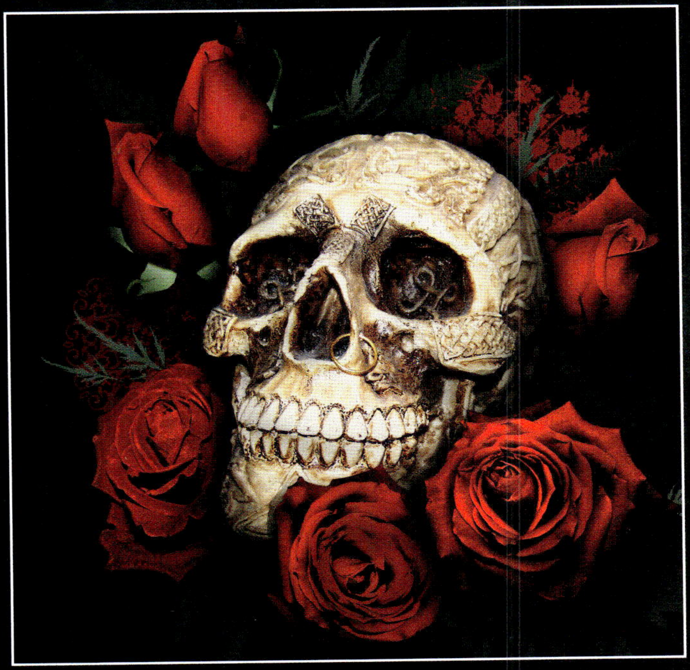

Jodi Fry

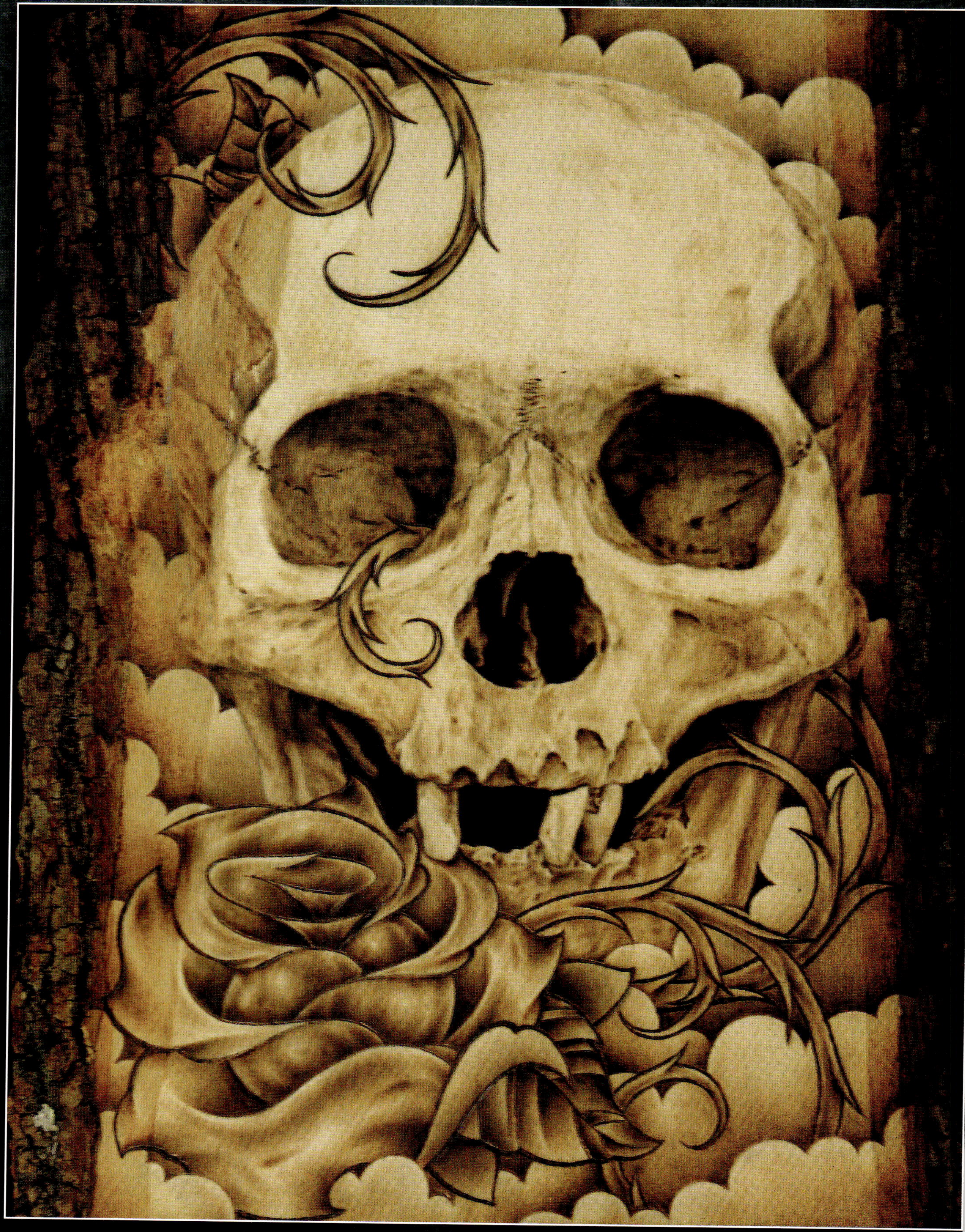

Scott Kalina

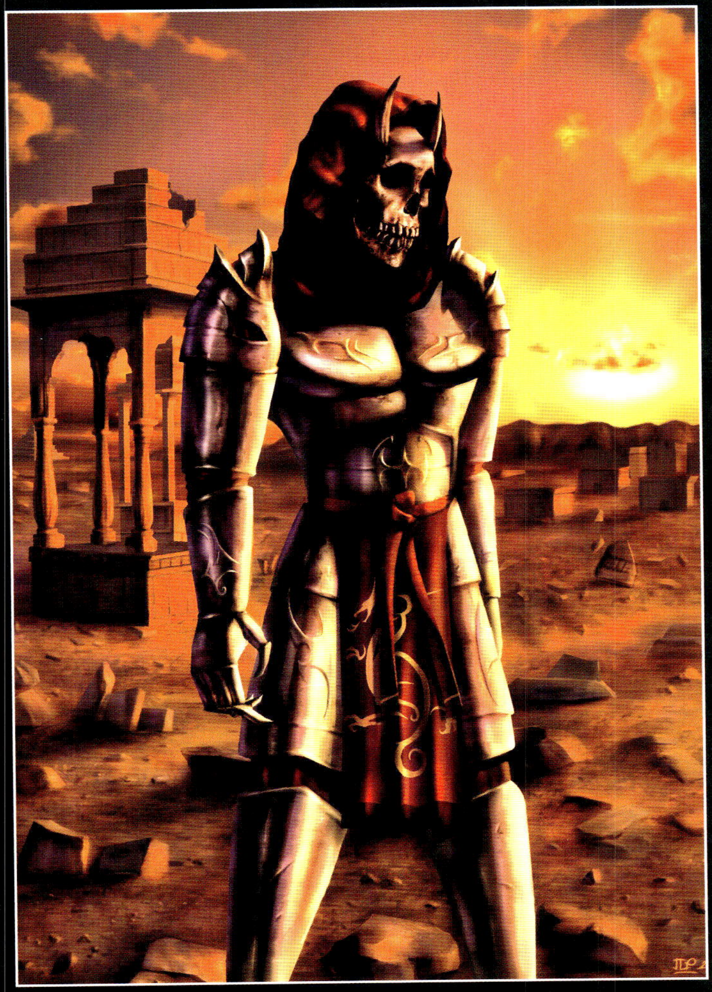
Jamie Lee Parker

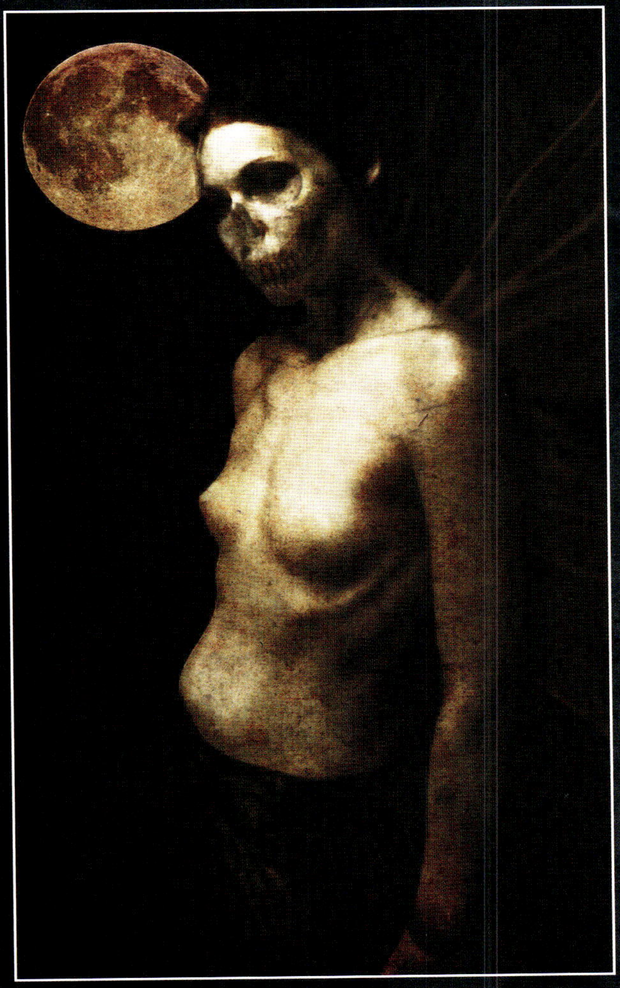
Josh Winton

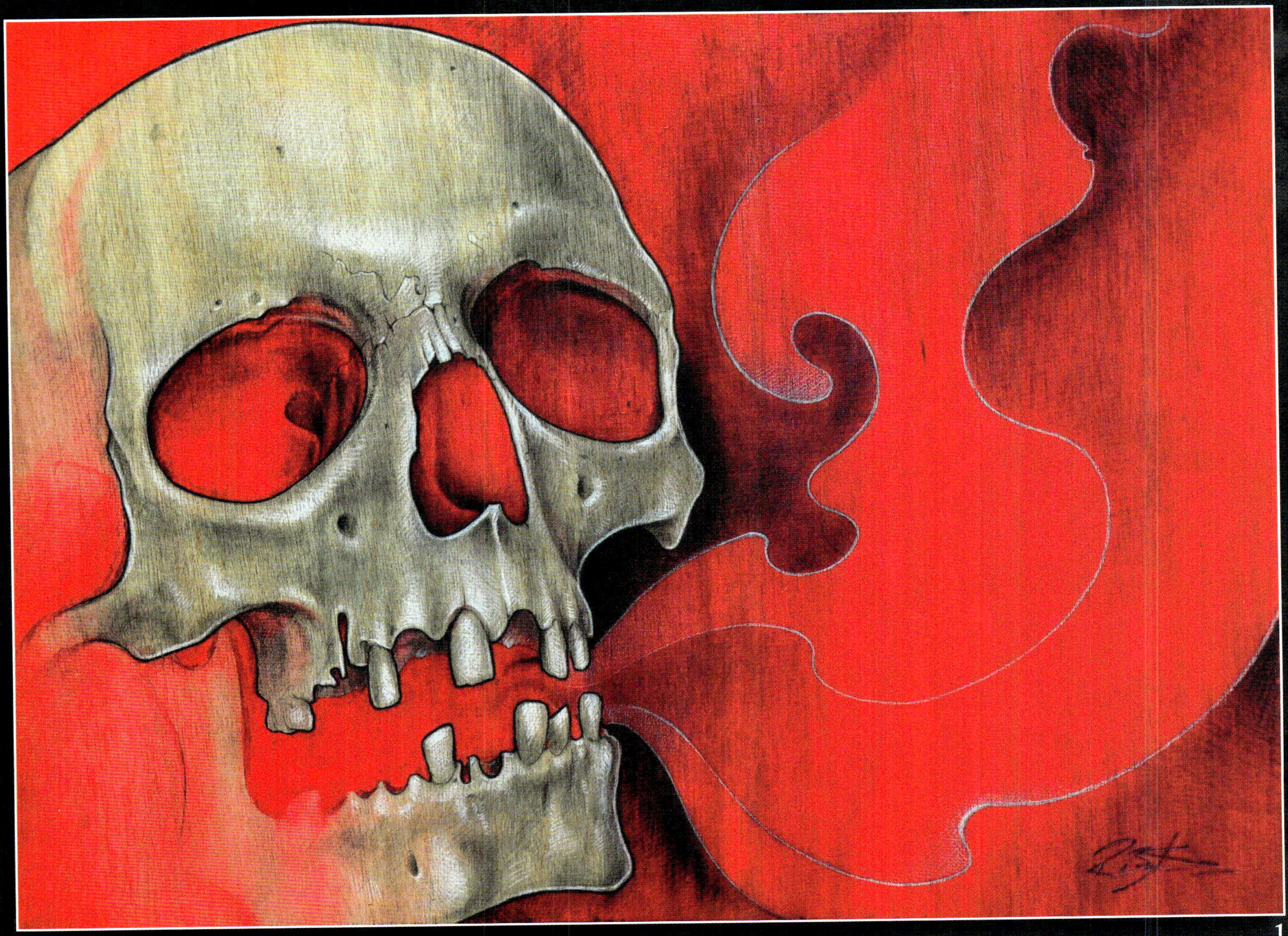
Bez

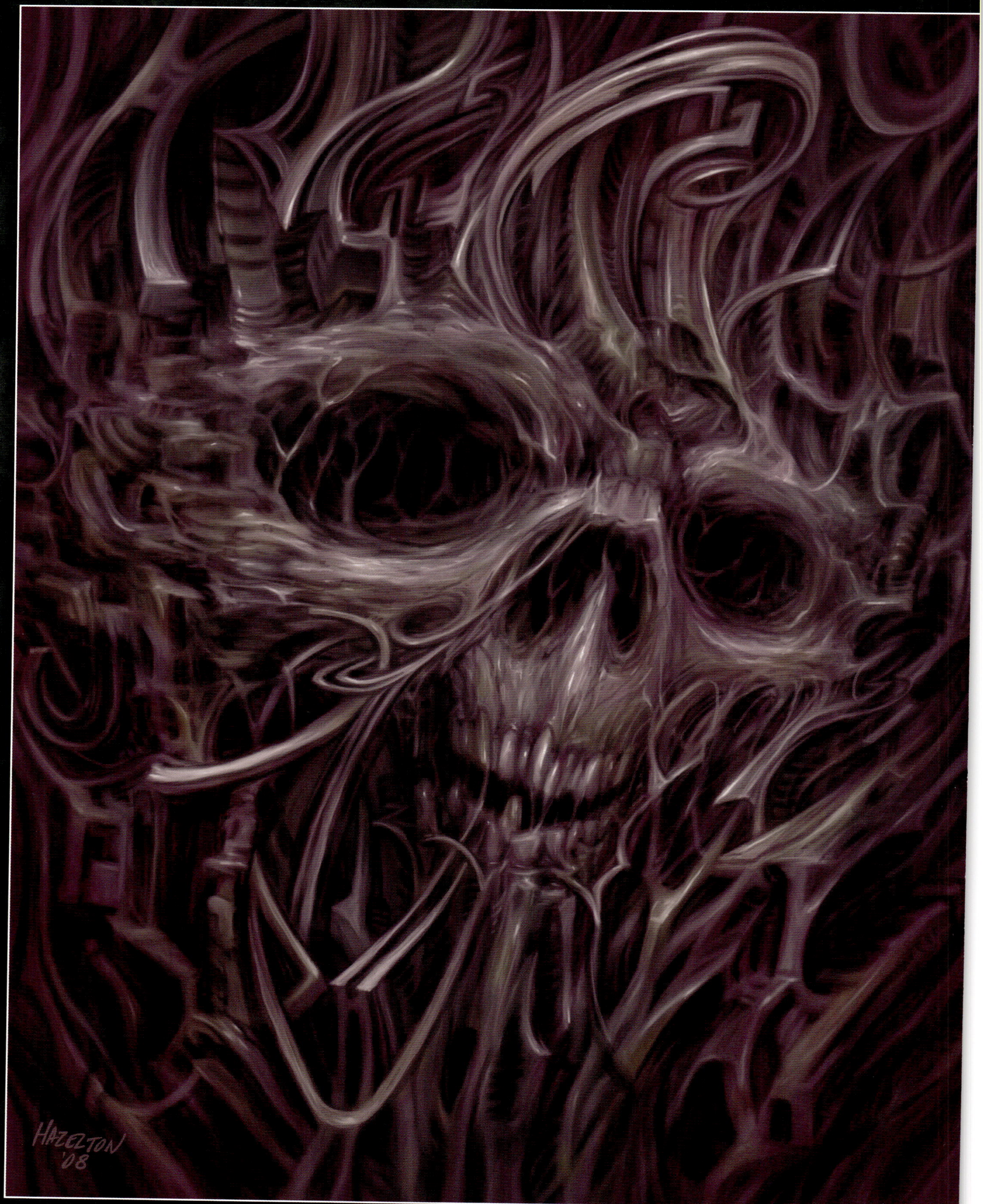

Dan Hazelton

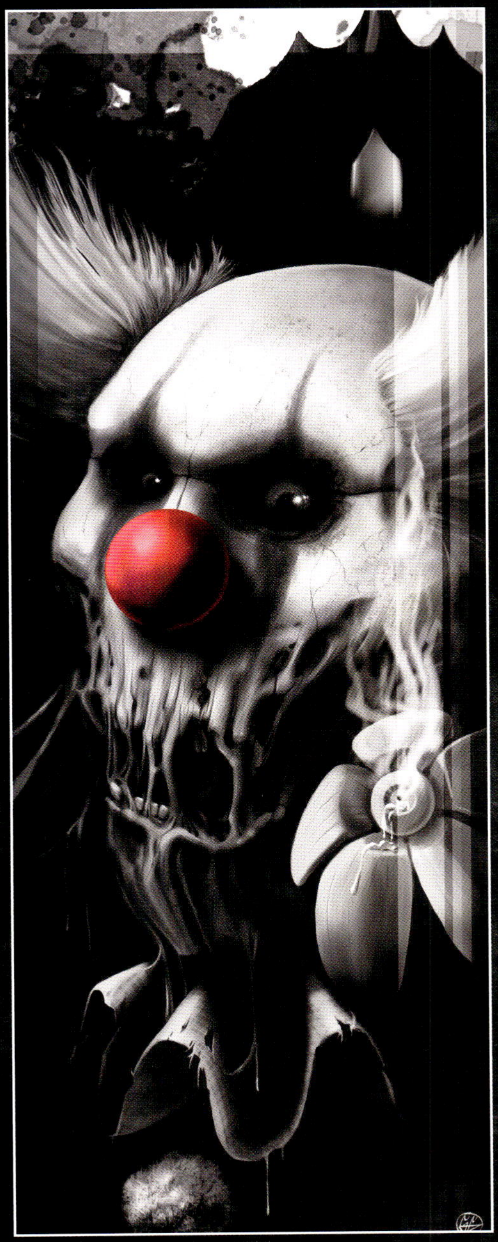

Sven Gnida

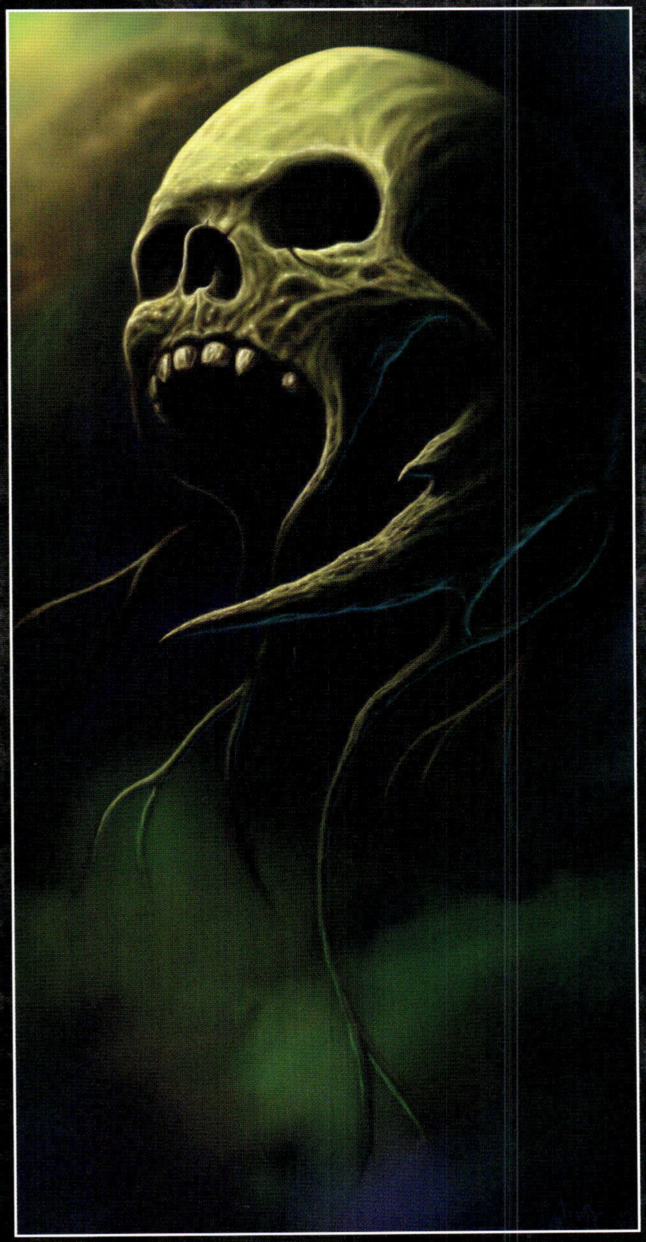

Josh Ruff

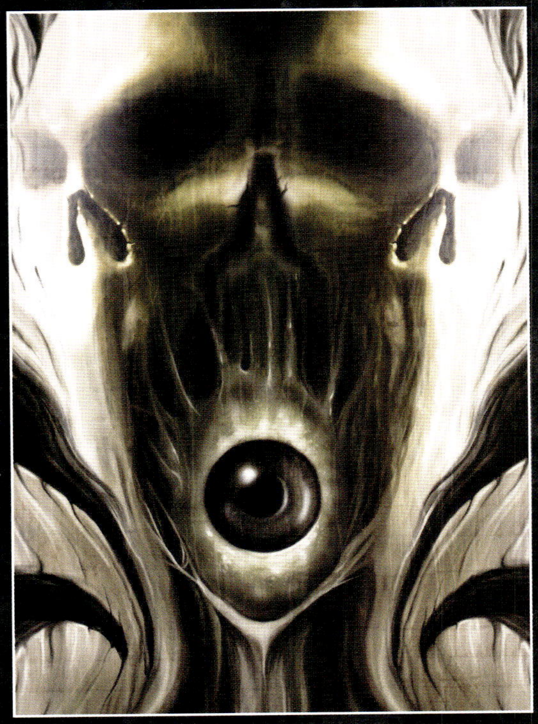

Sven Gnida

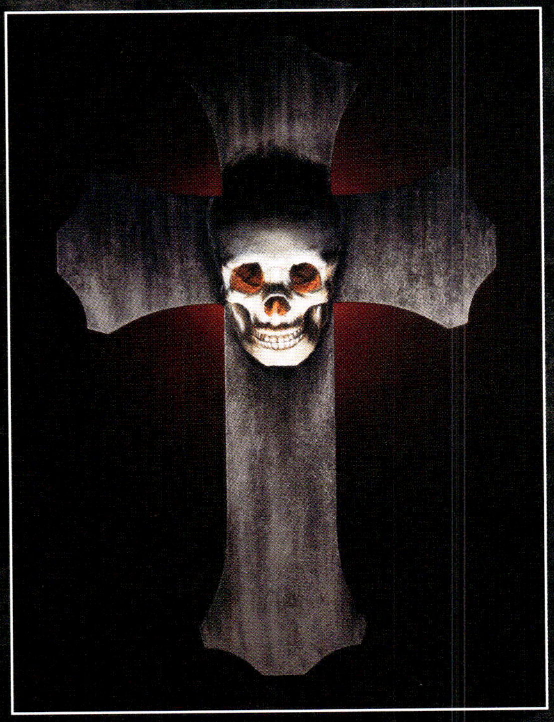

Will Thompson

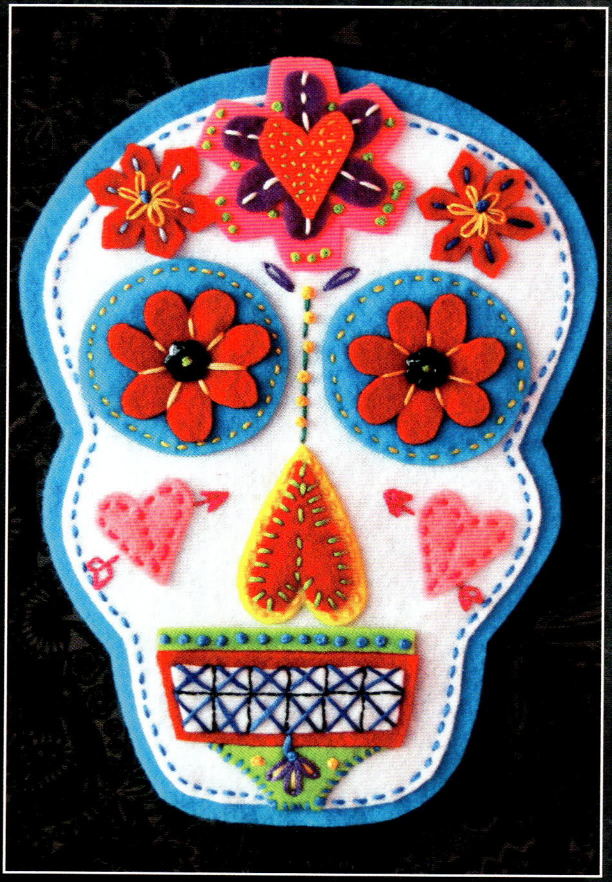
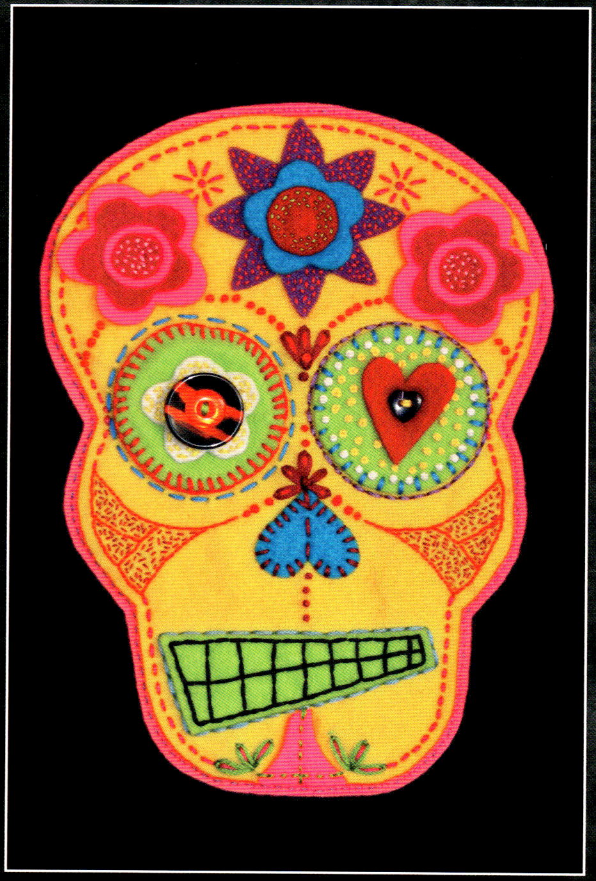
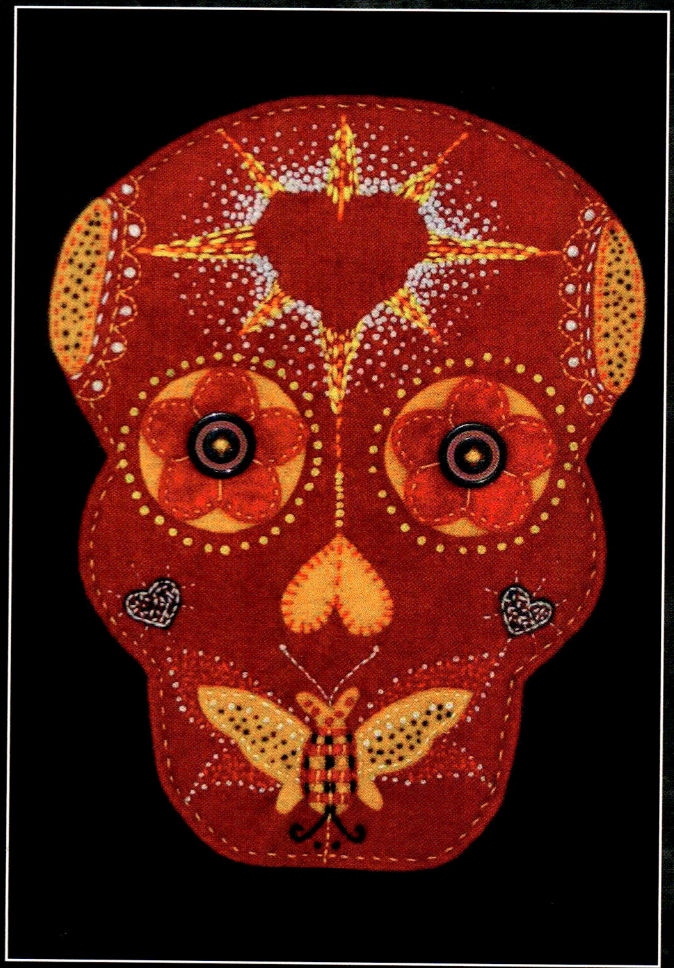
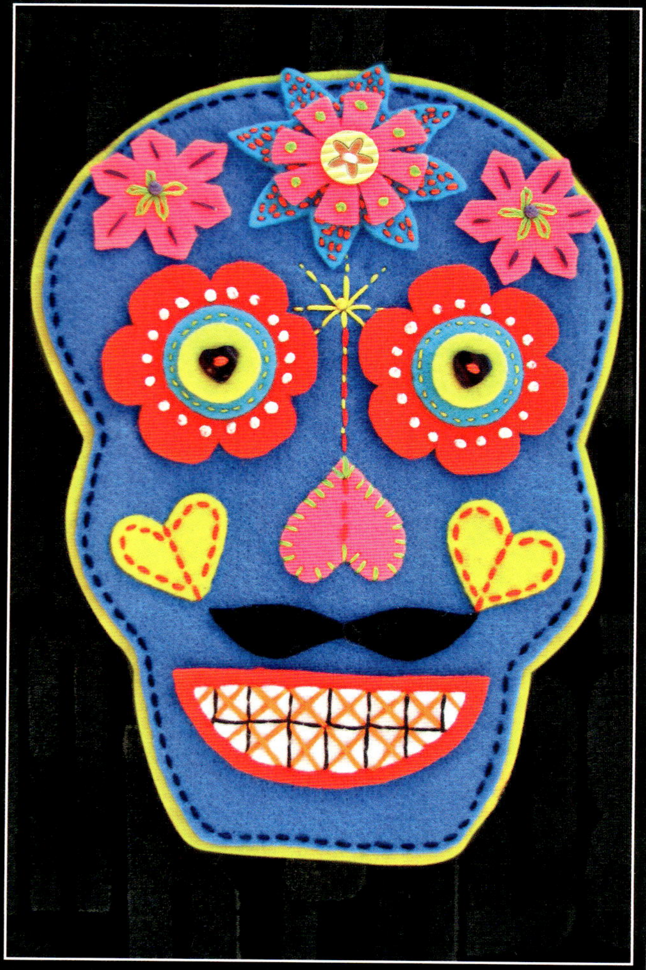

Nicola

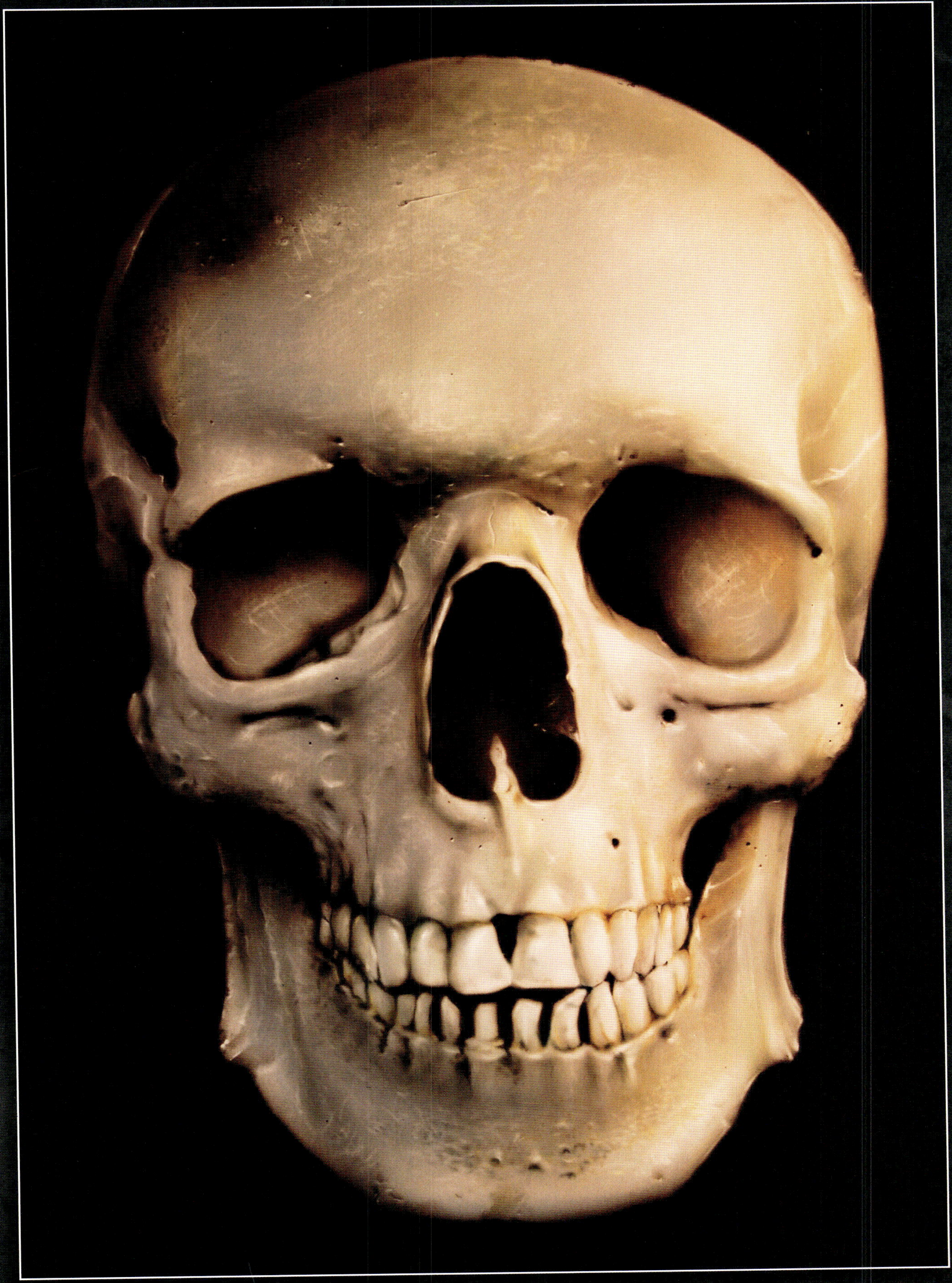

Alex De Pase

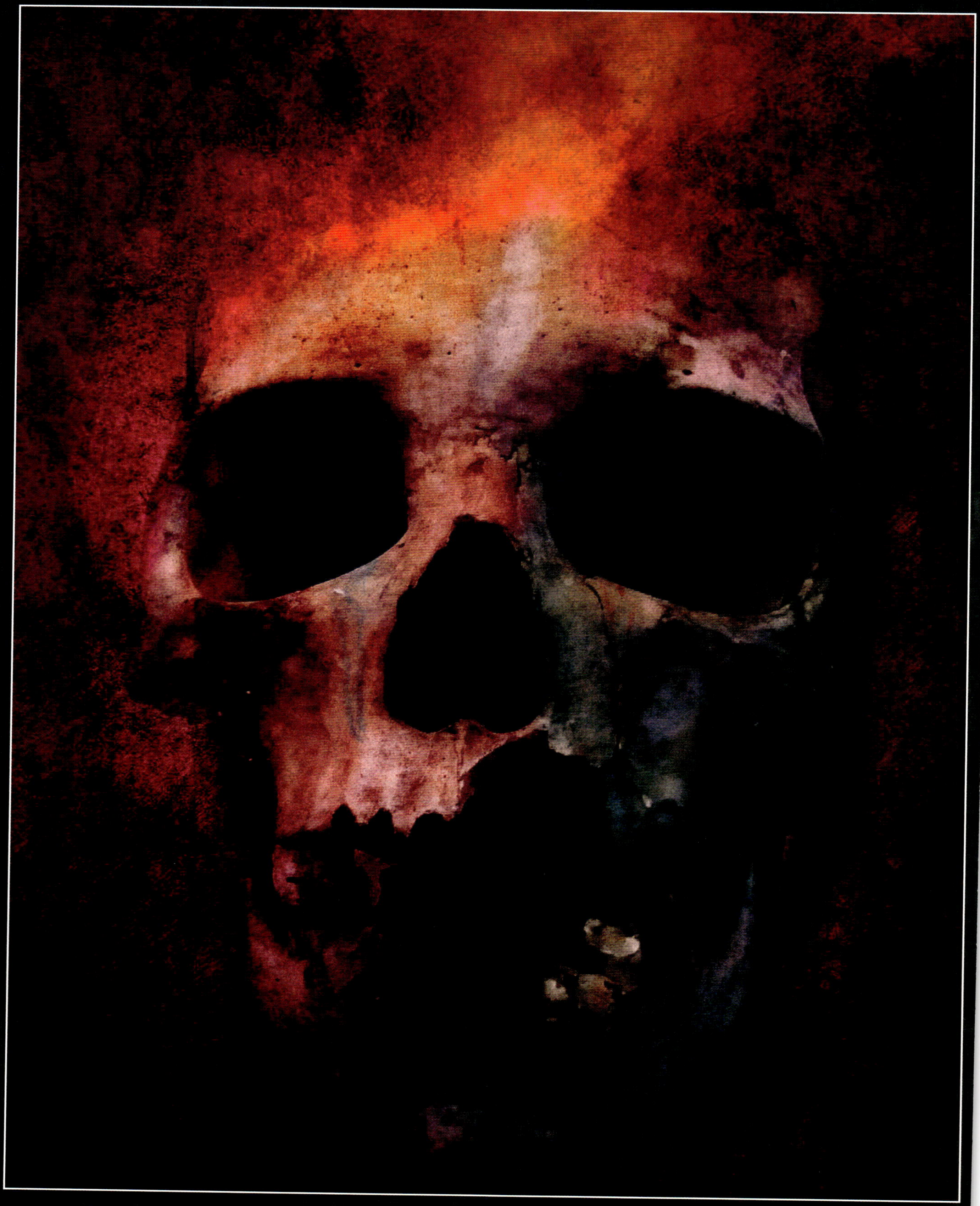

Gena "Exile In Blonde" Creazzo

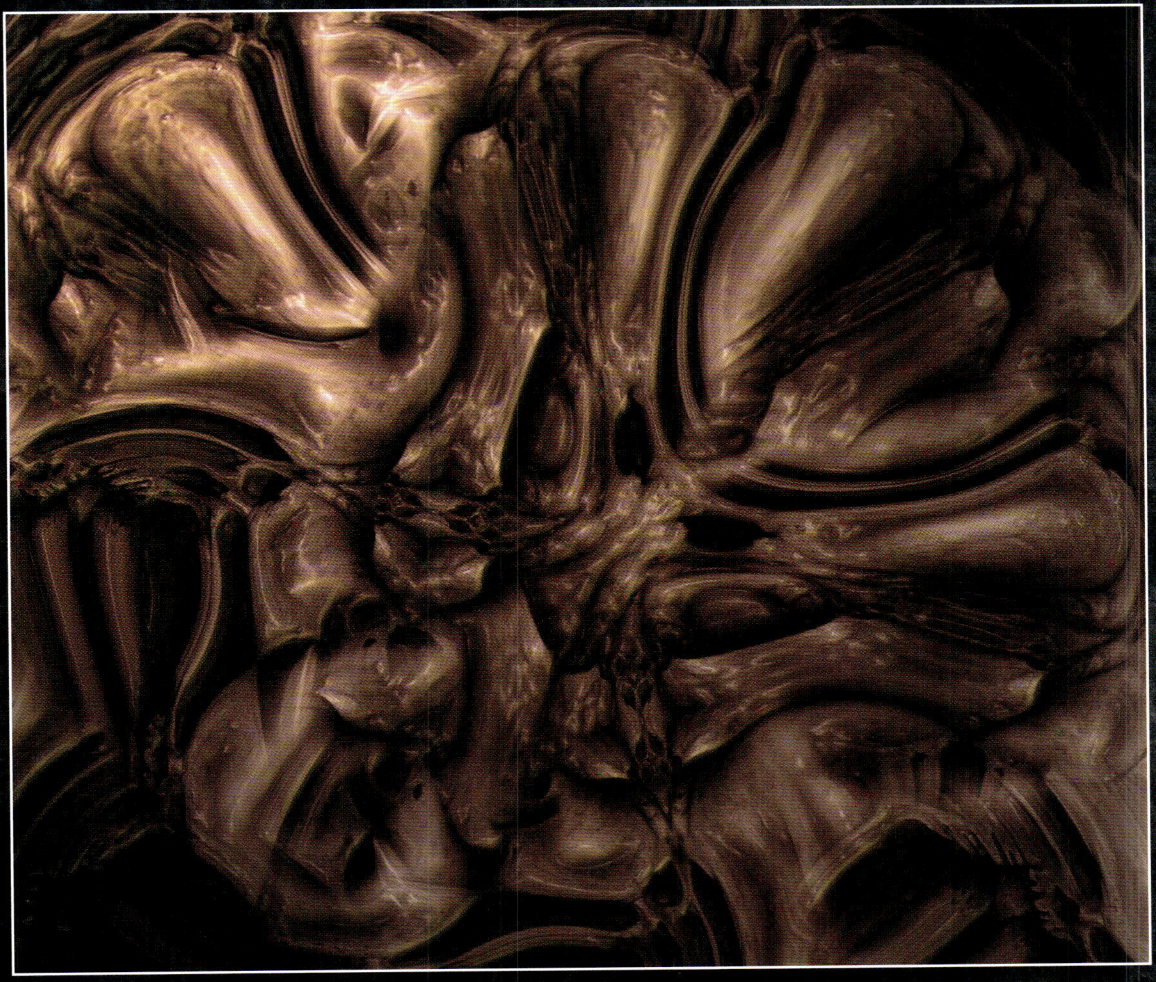

Benjamin Moss

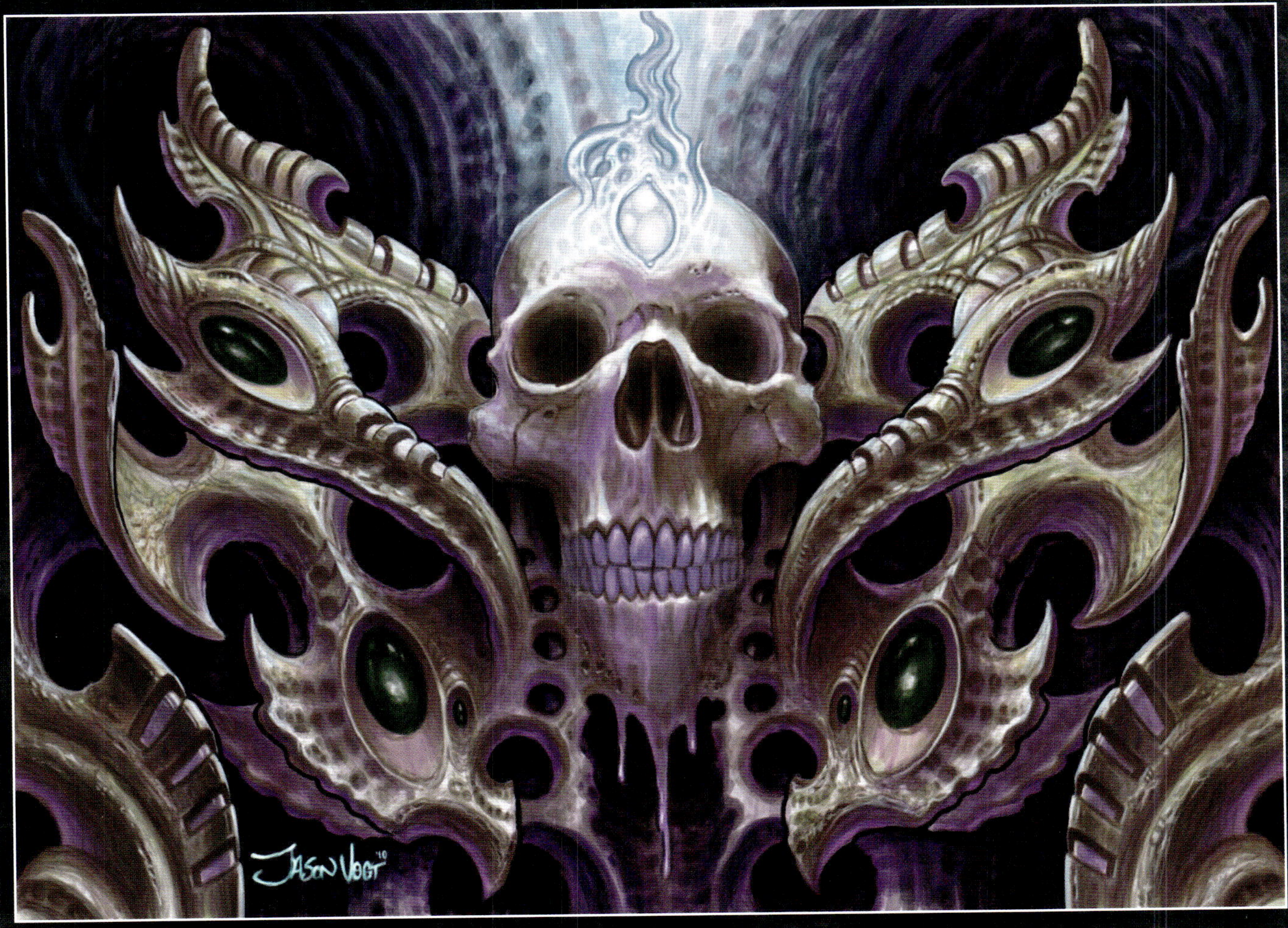

Jason Vogt

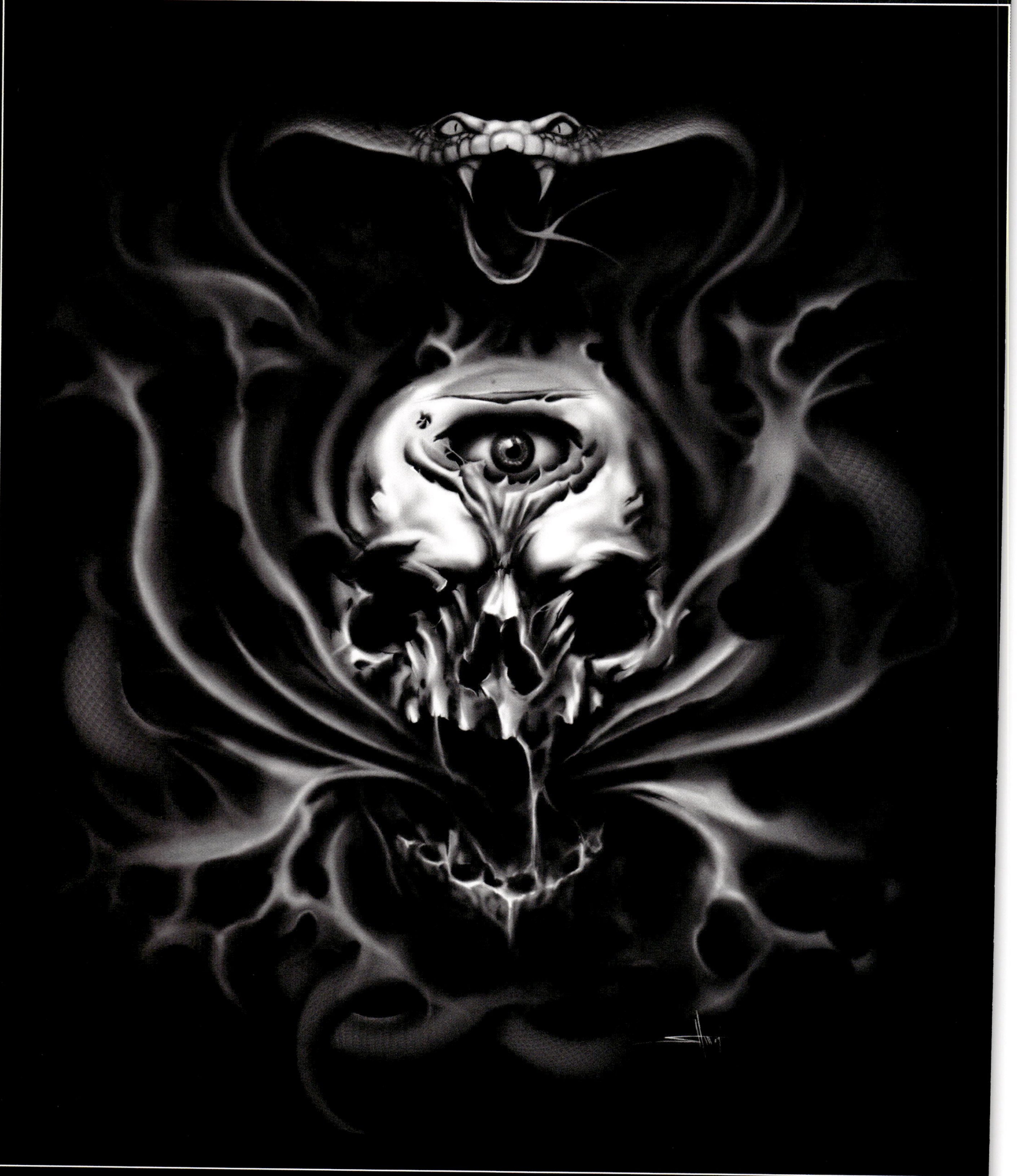

Simon Hayag

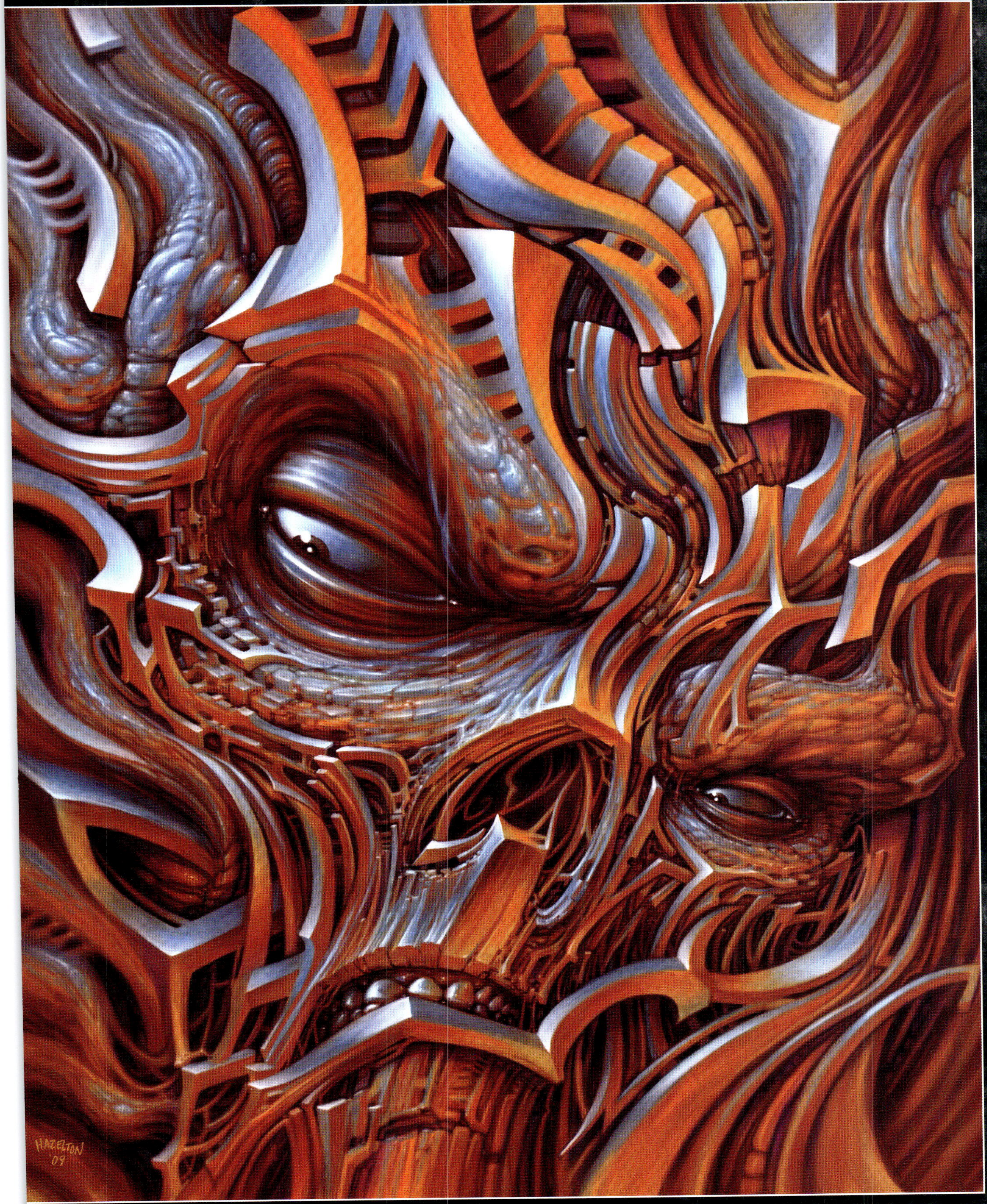

Dan Hazelton

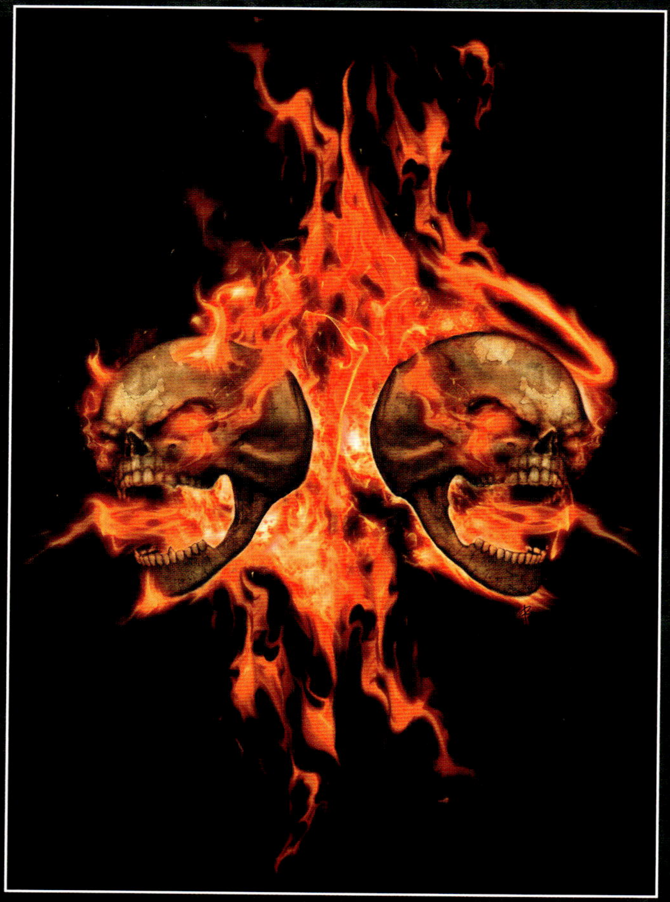

Roger Pister

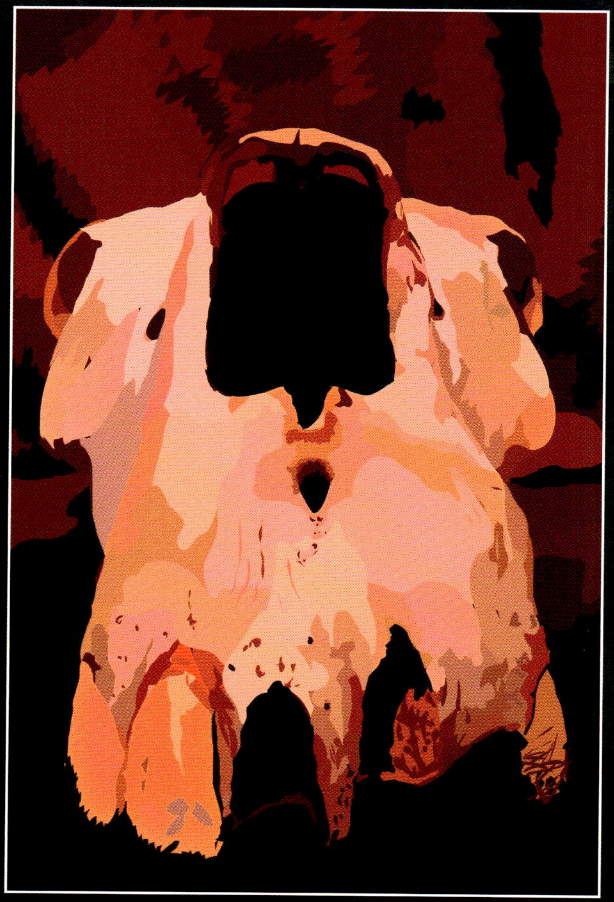

Keith Cross

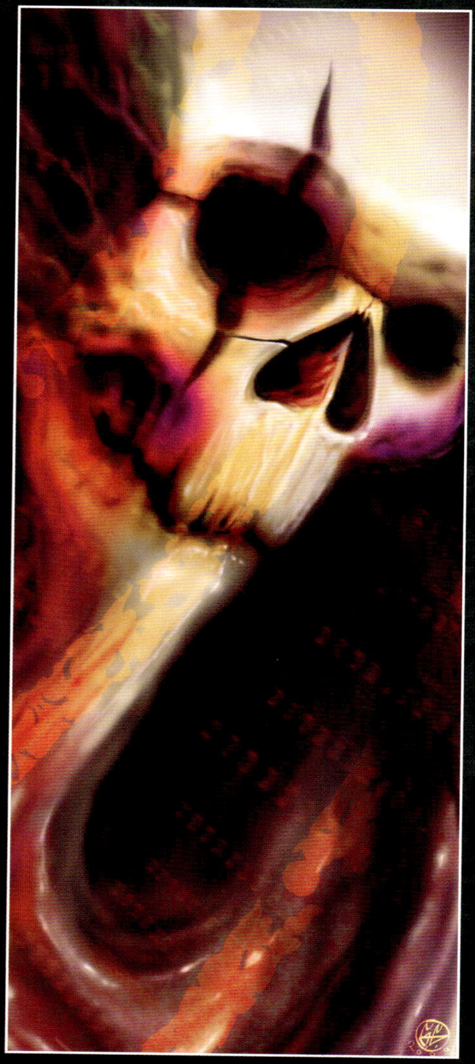

Sven Gnida

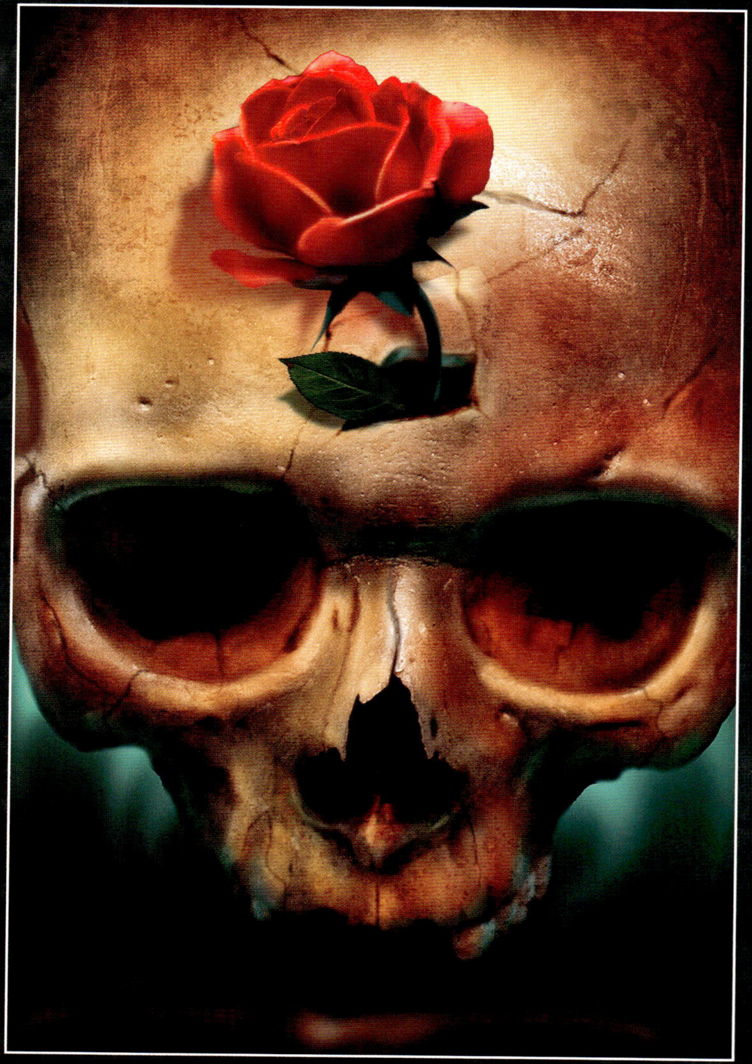

Jamie Lee Parker

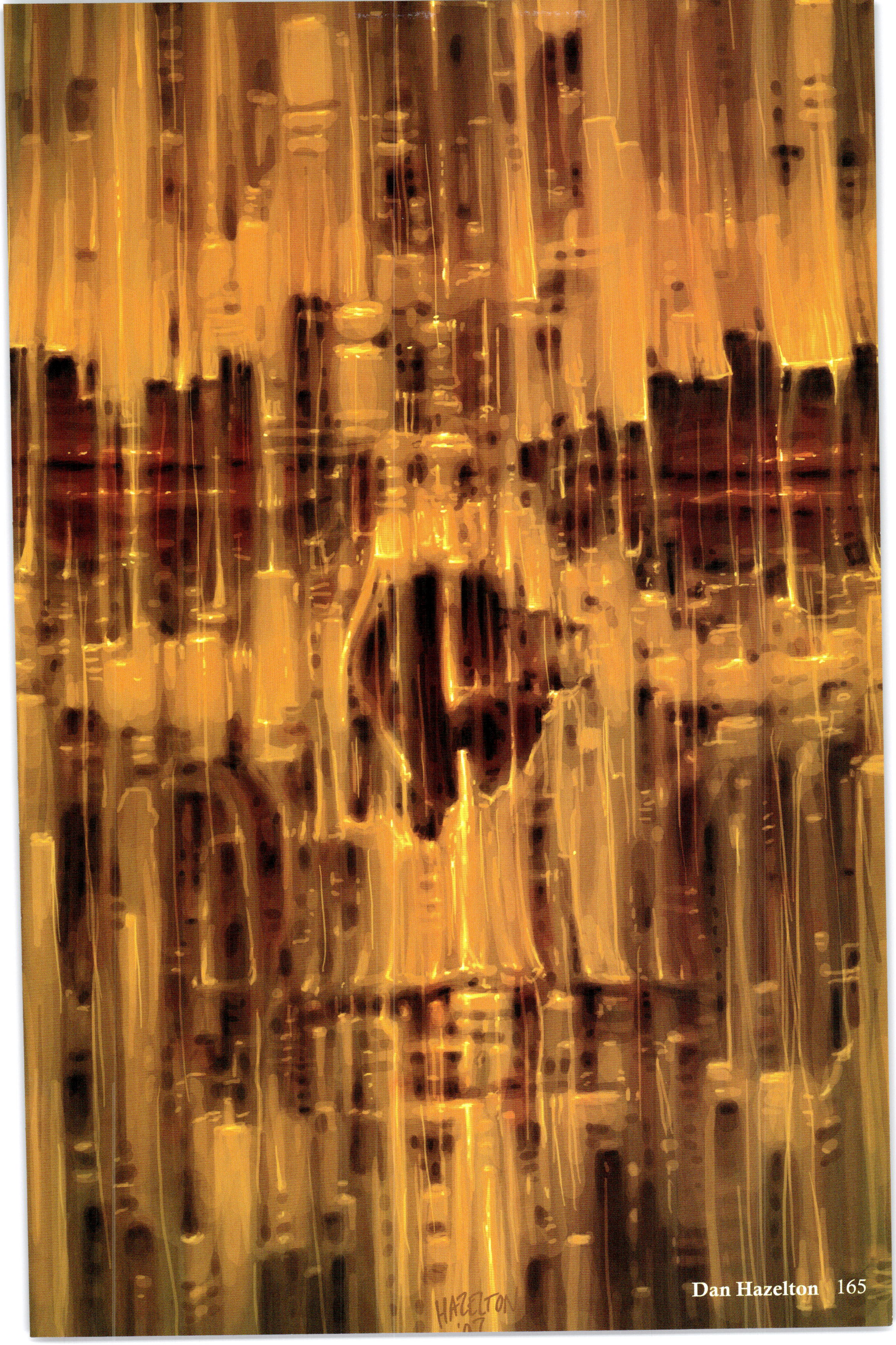
HAZELTON '07

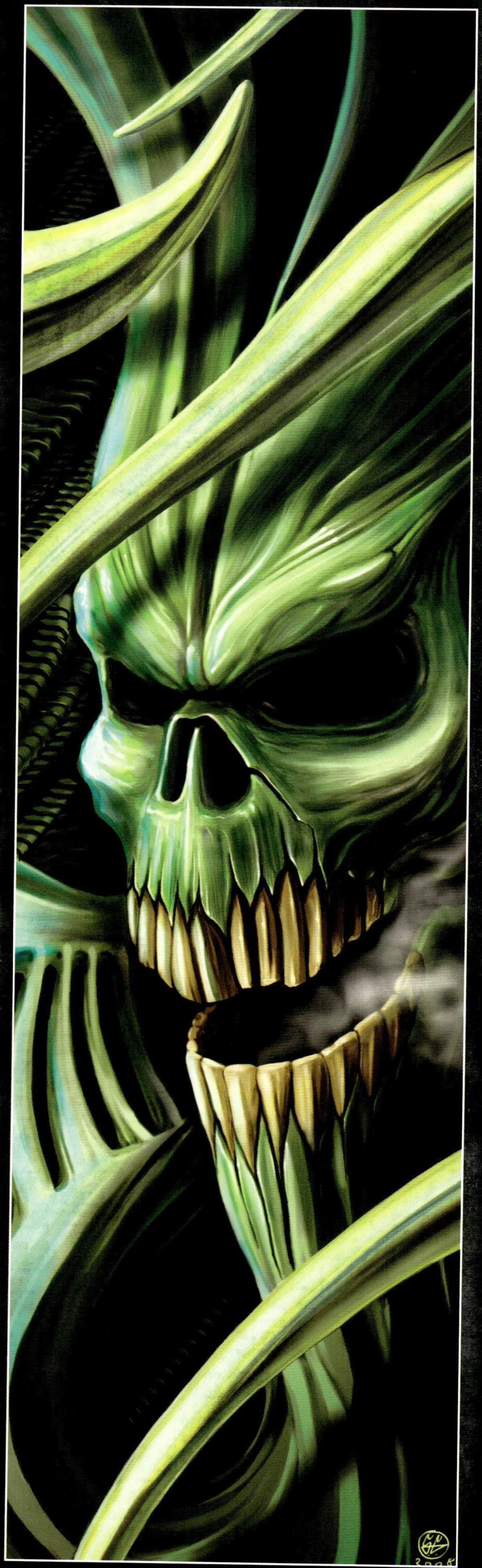

Sven Gnida

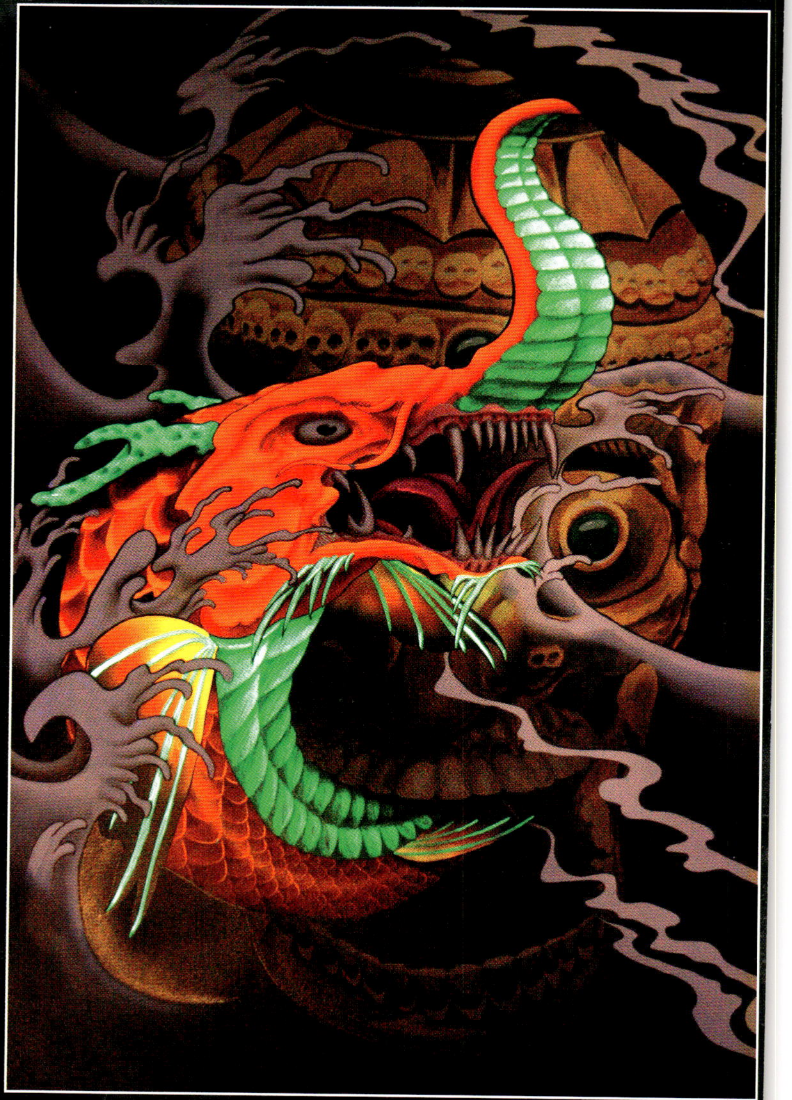

Vince Wishart

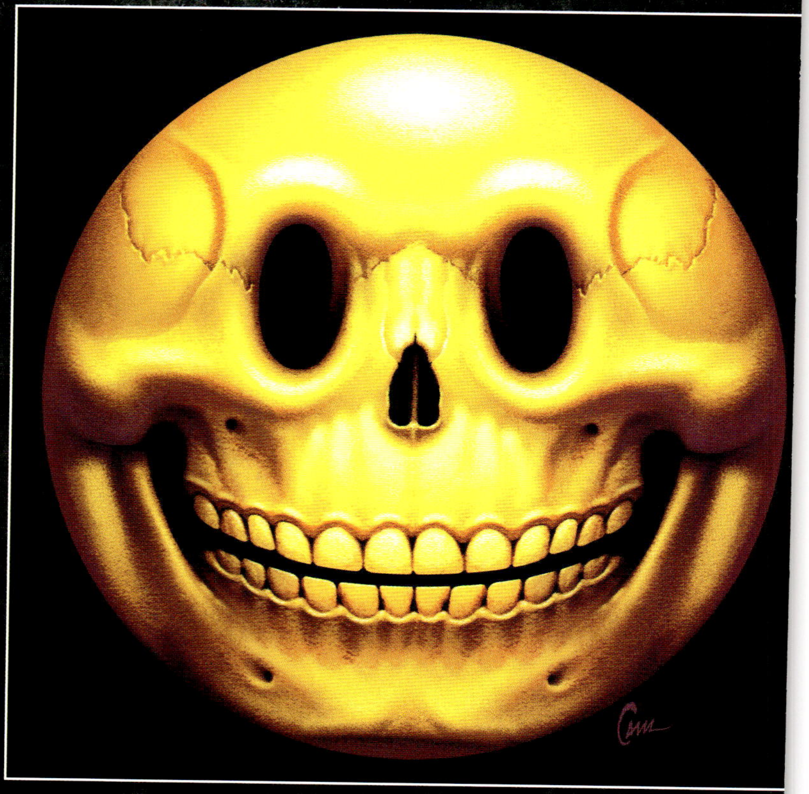

Cam de Leon

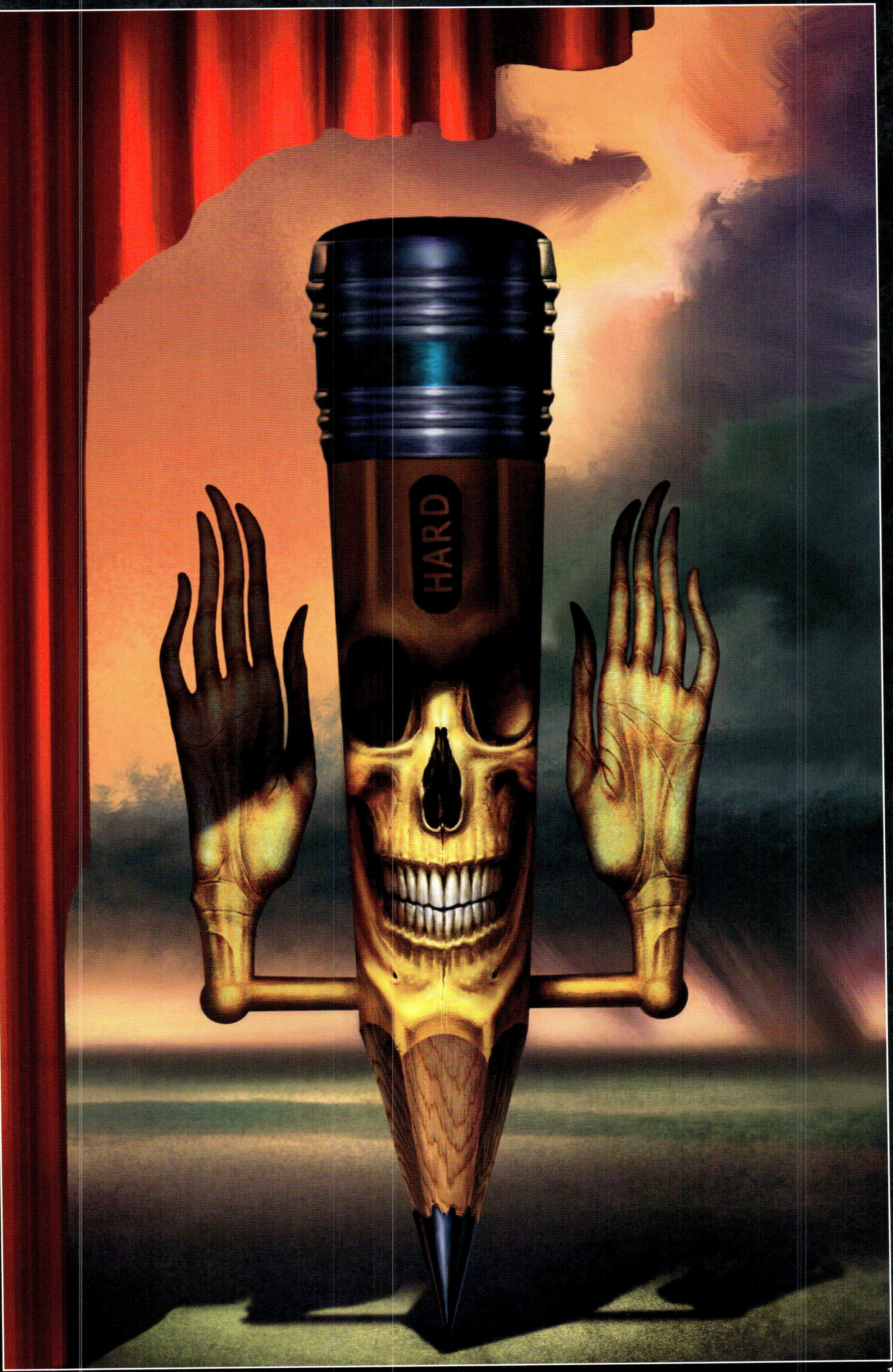

Cam De Leon

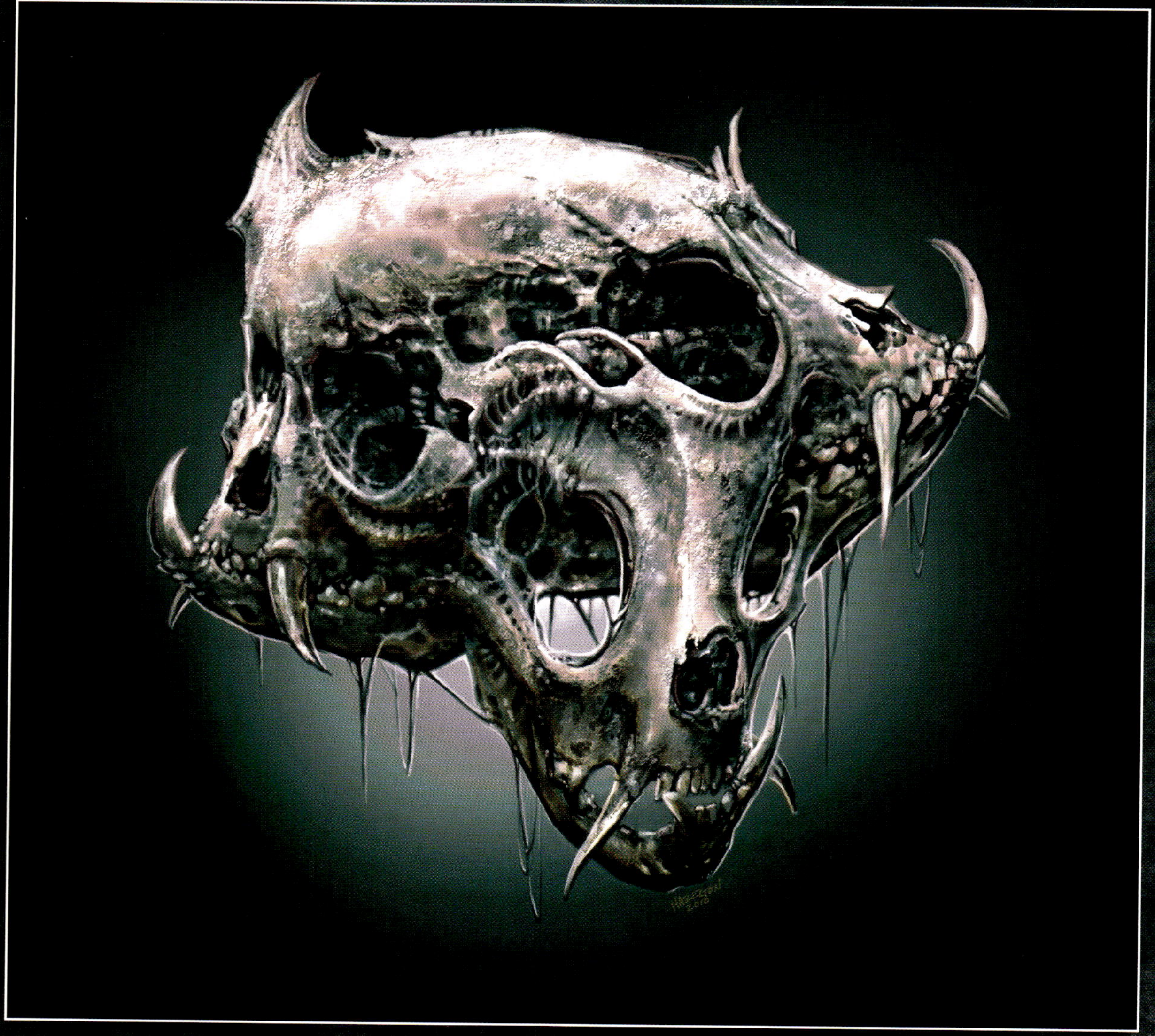

Dan Hazelton

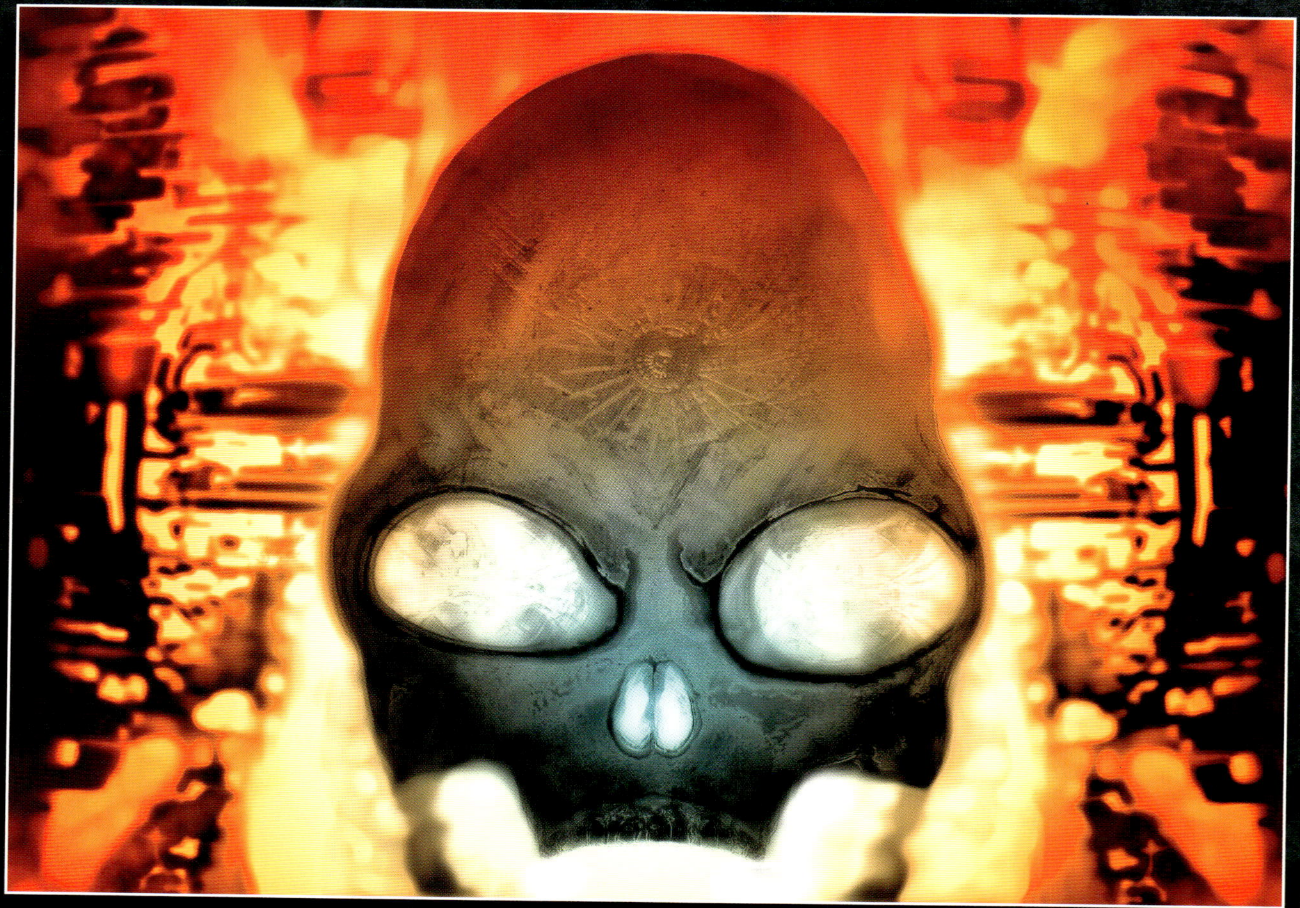

Chris Roi

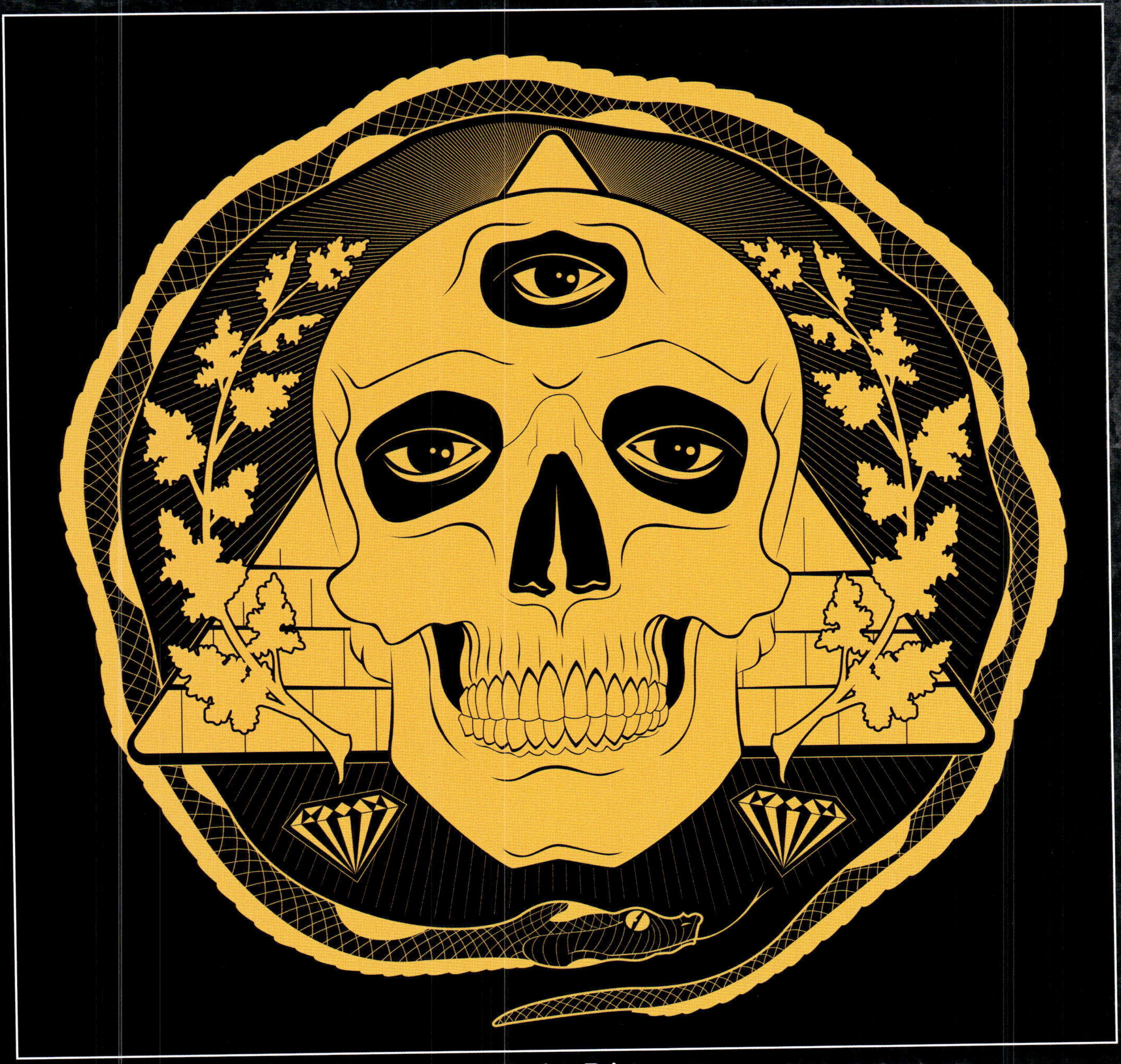

Reuben Briggs

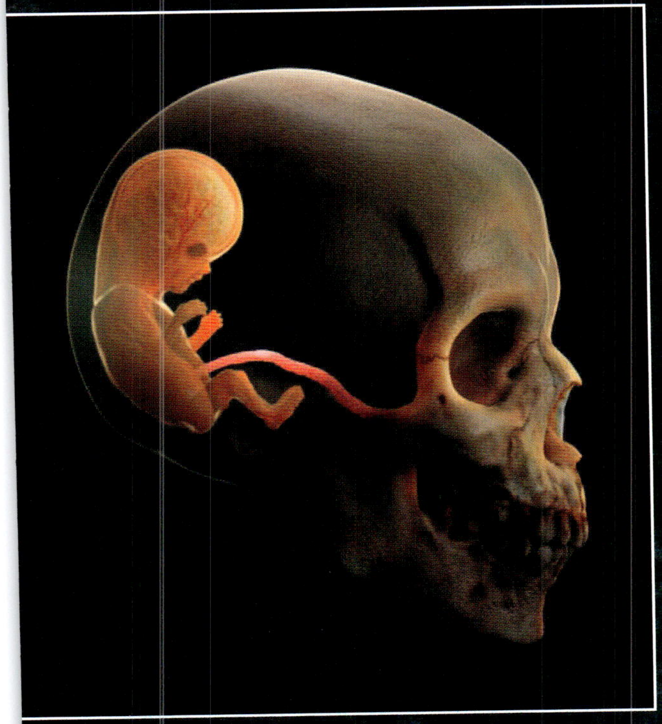

Alex Grey

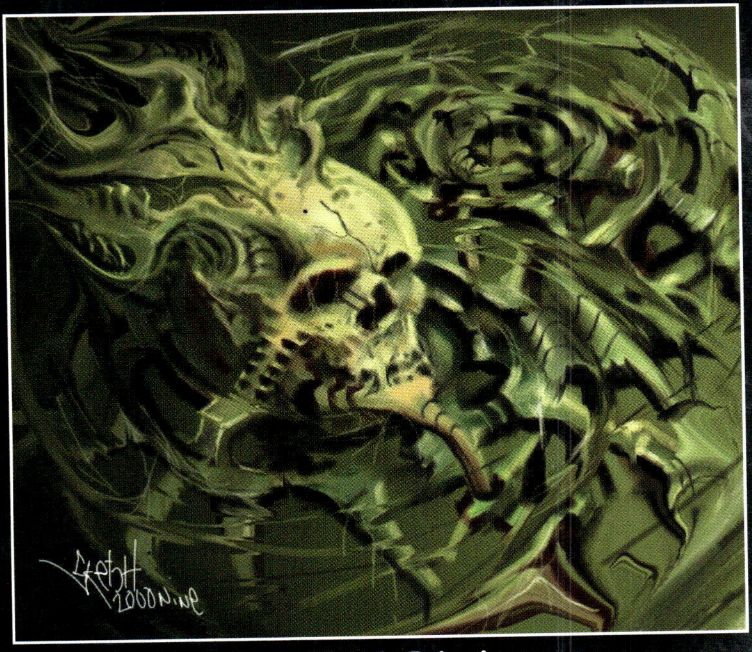

Sketch Grinnis

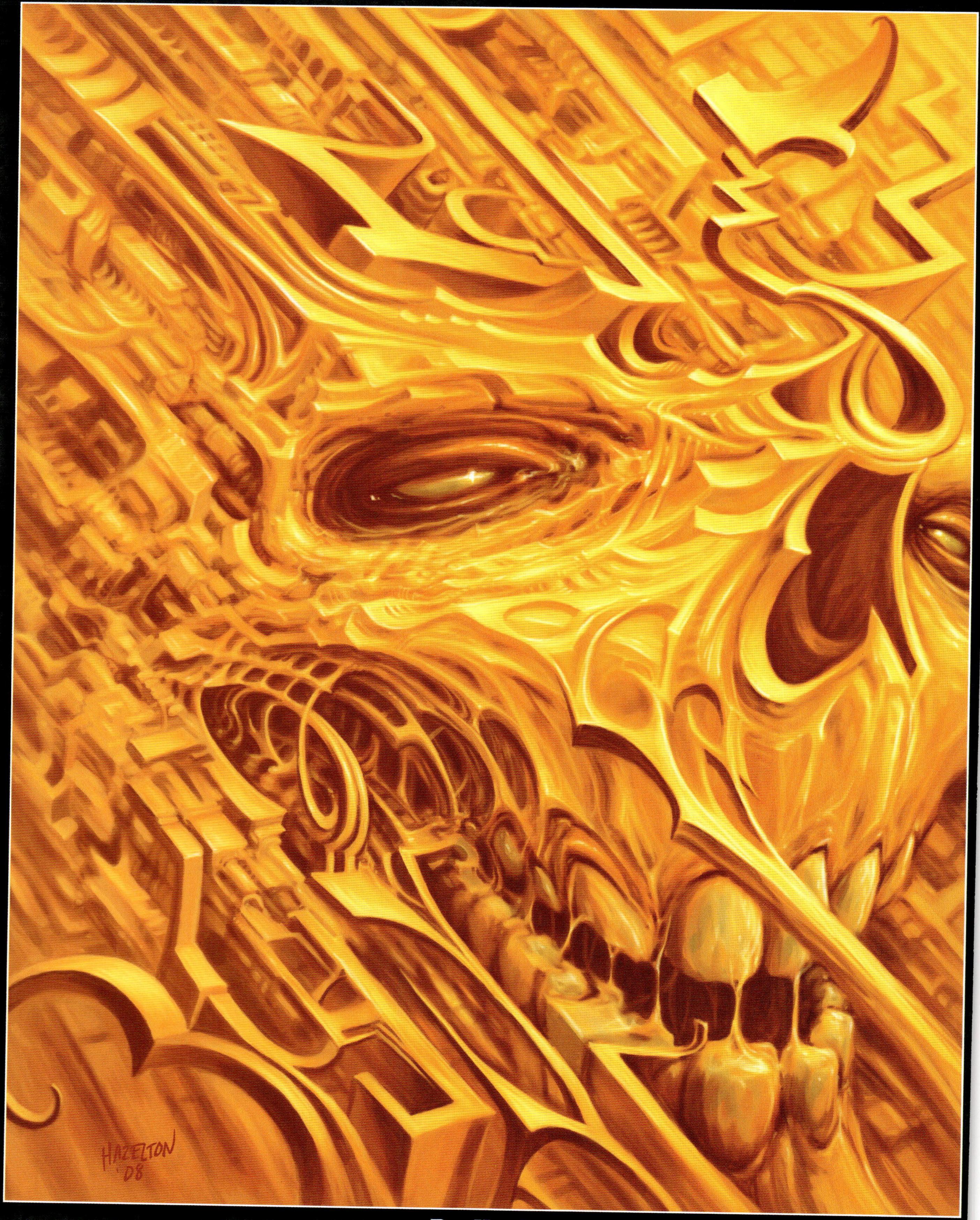

Dan Hazelton

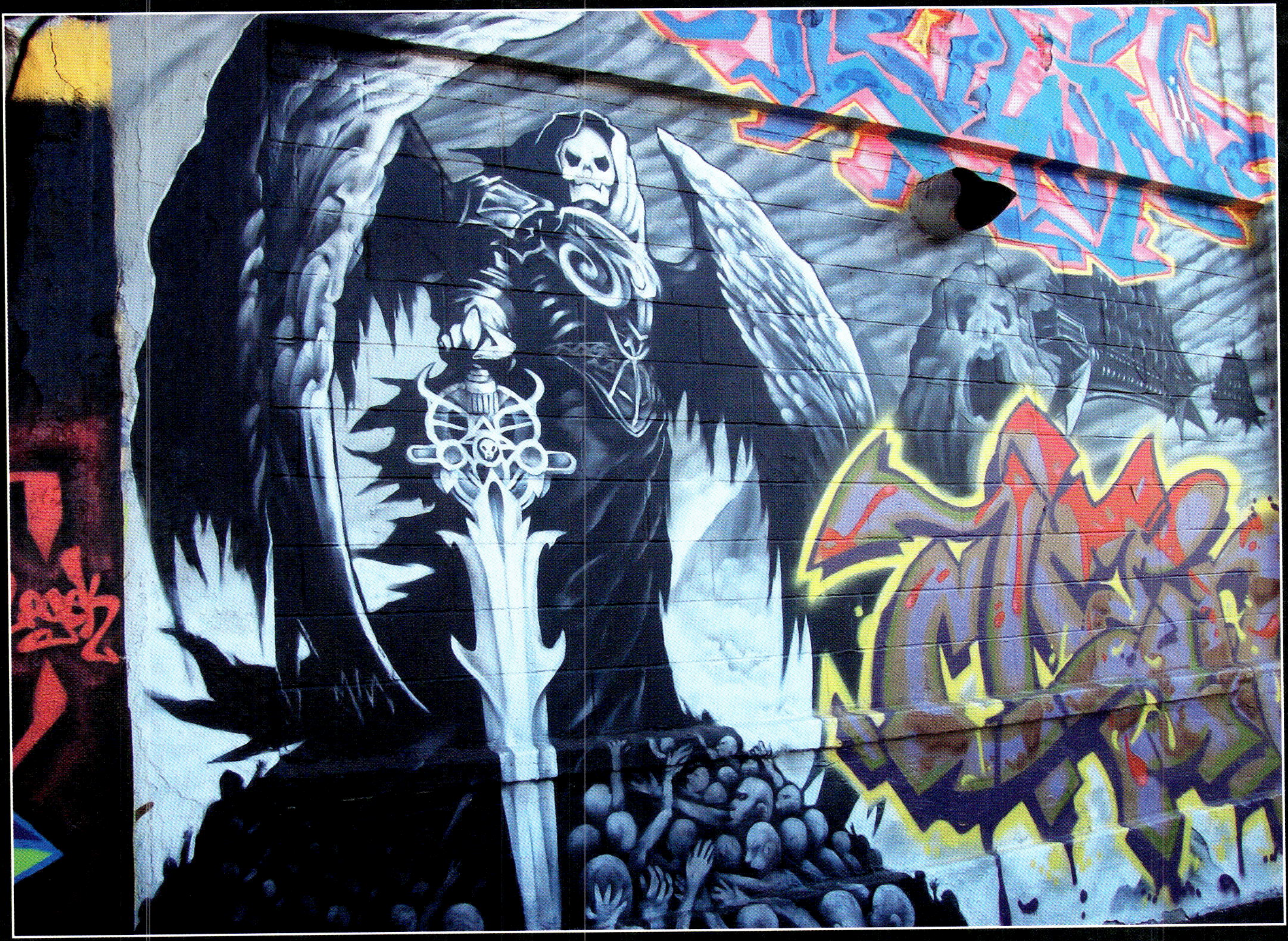

Bron Theron

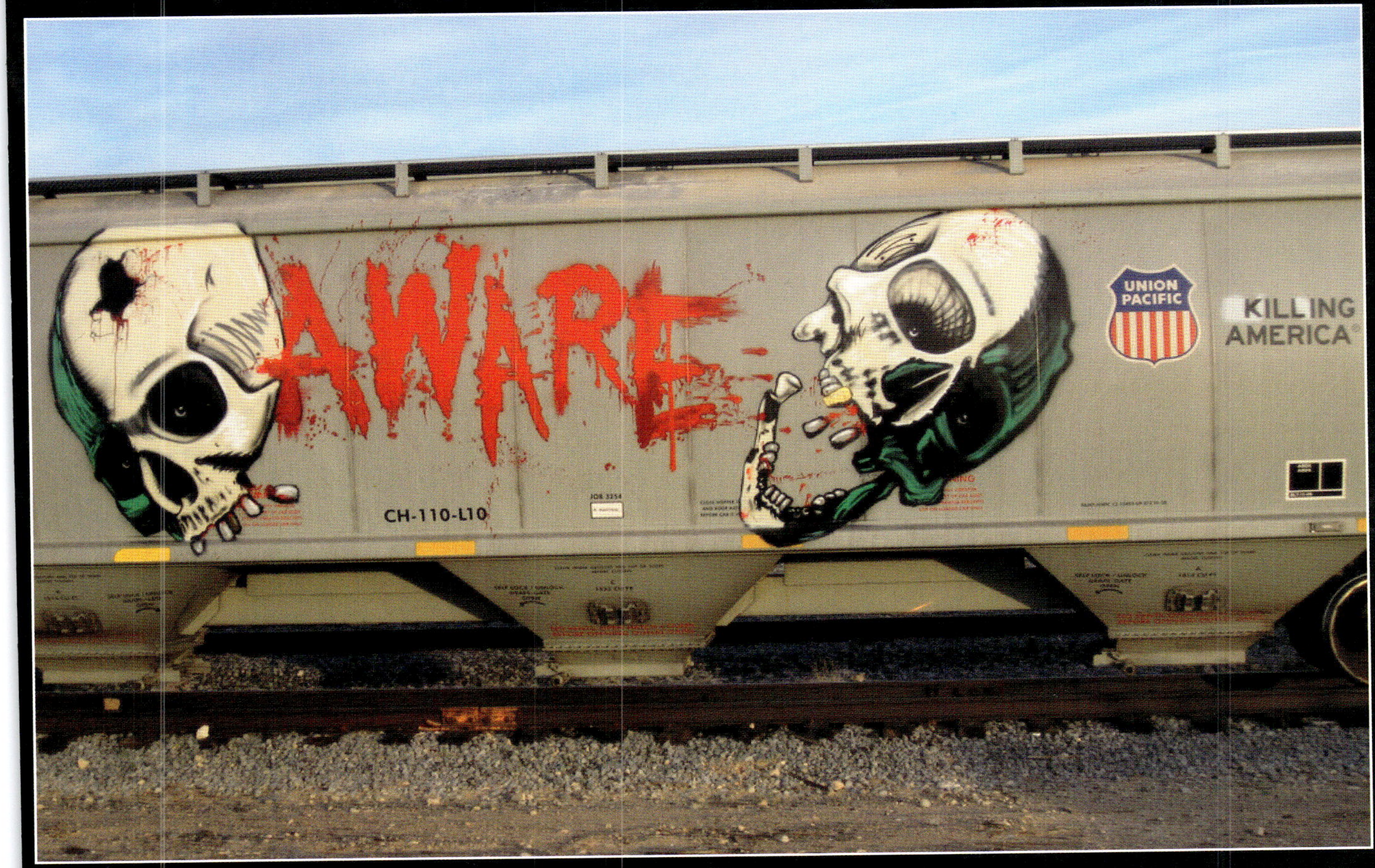

Bron Theron

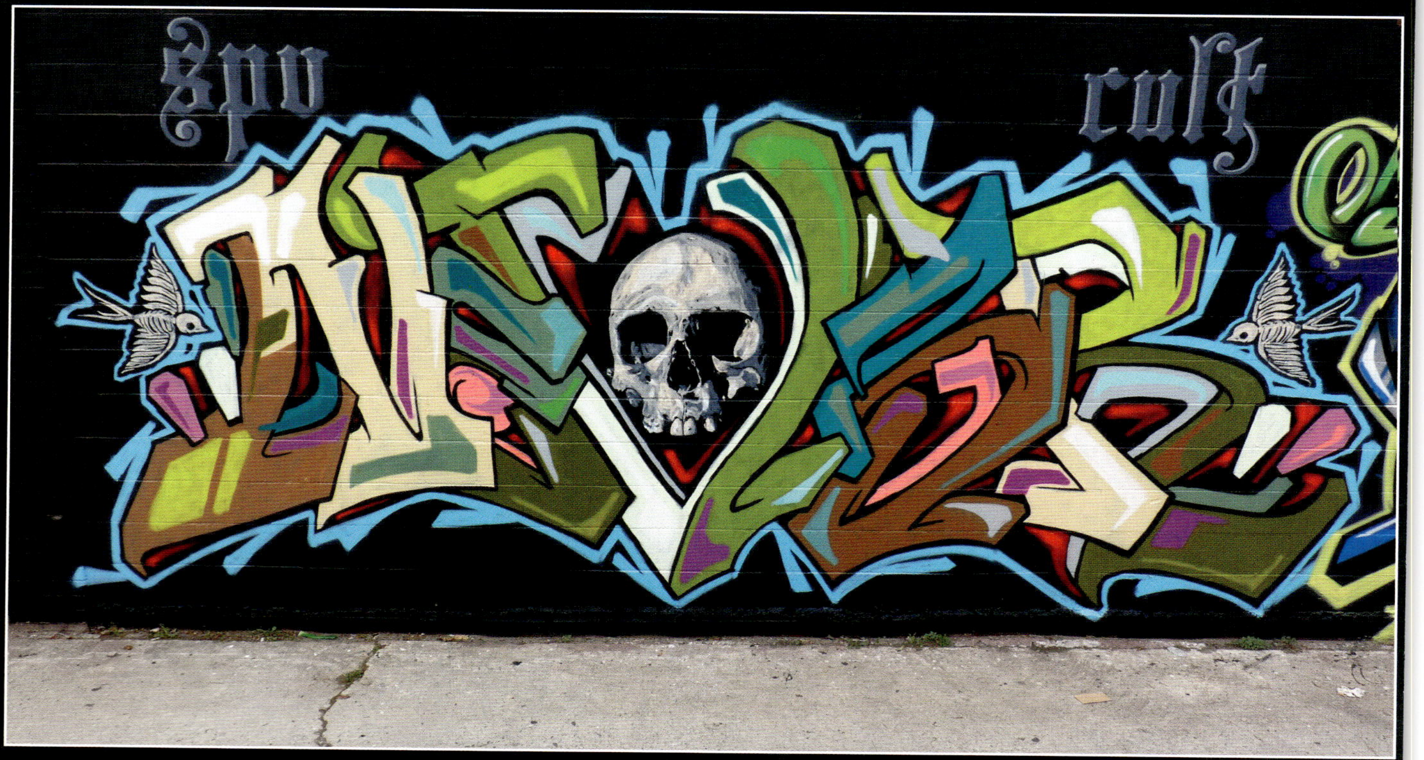

Jonas Never

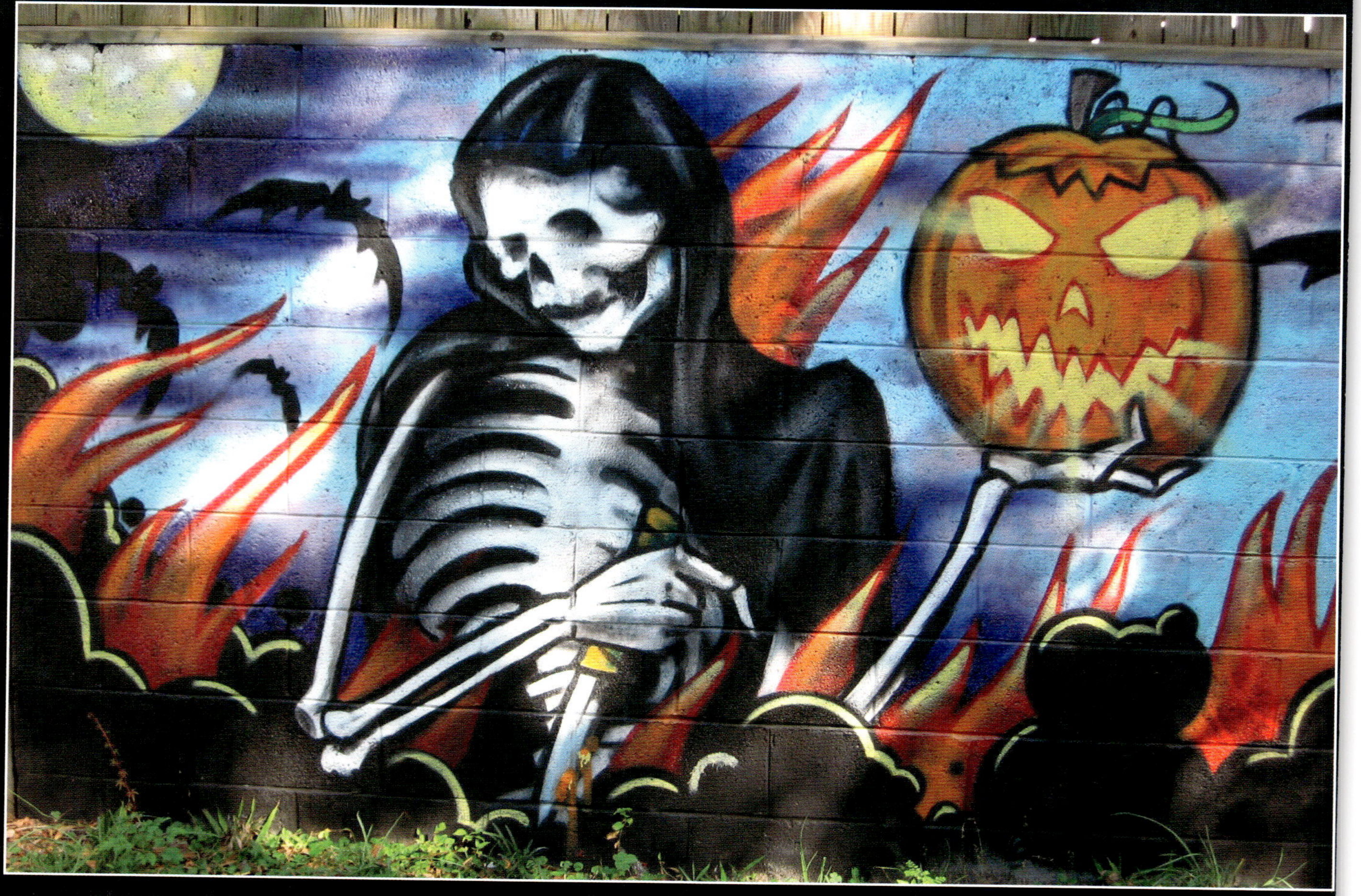

Richard Mcgee

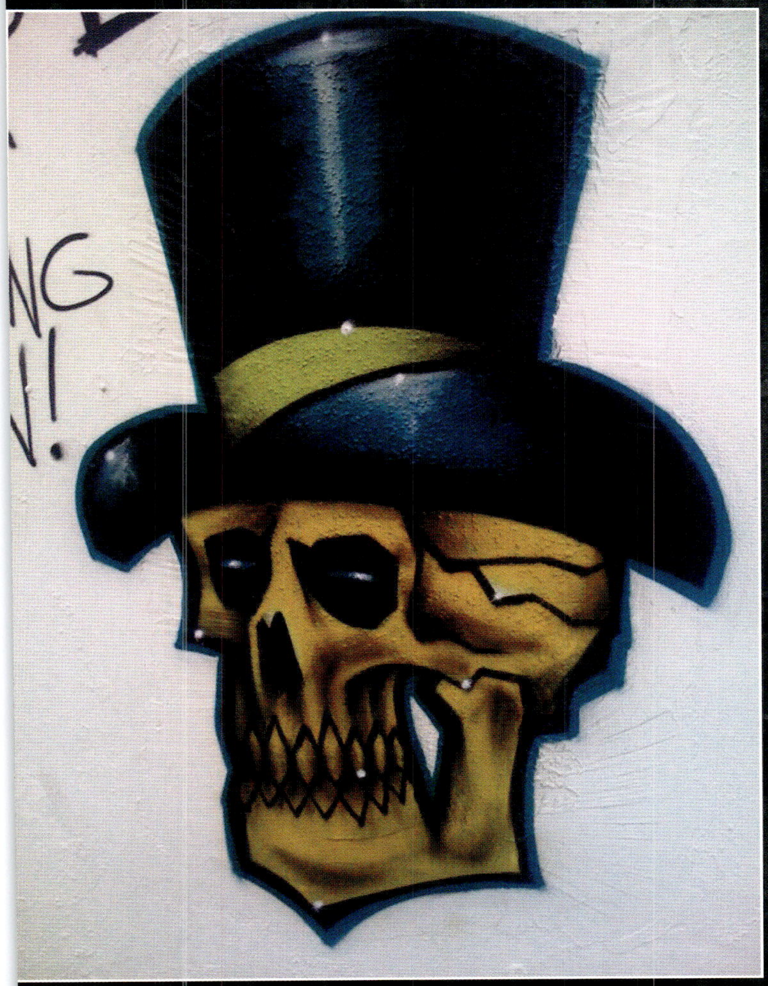

MFN8

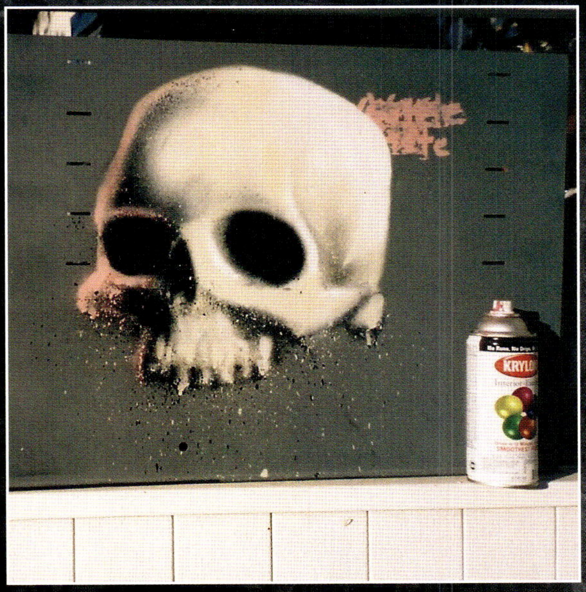

Bron Theron

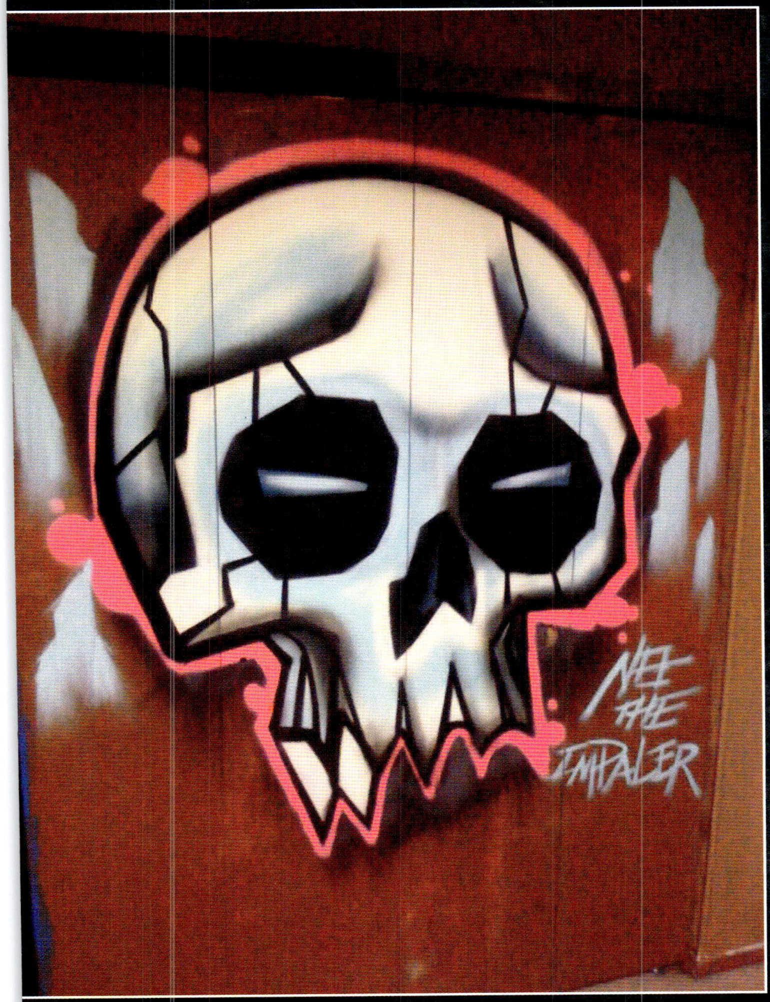

MFN8

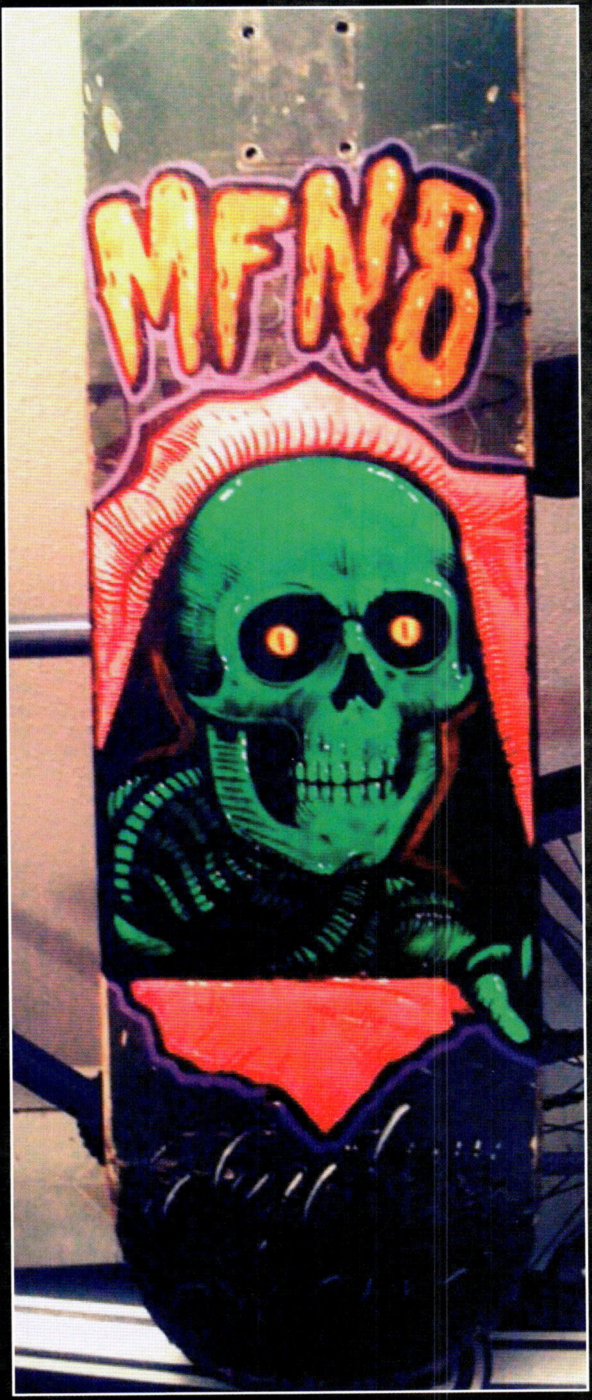

MFN8

PHOTOGRAPHY

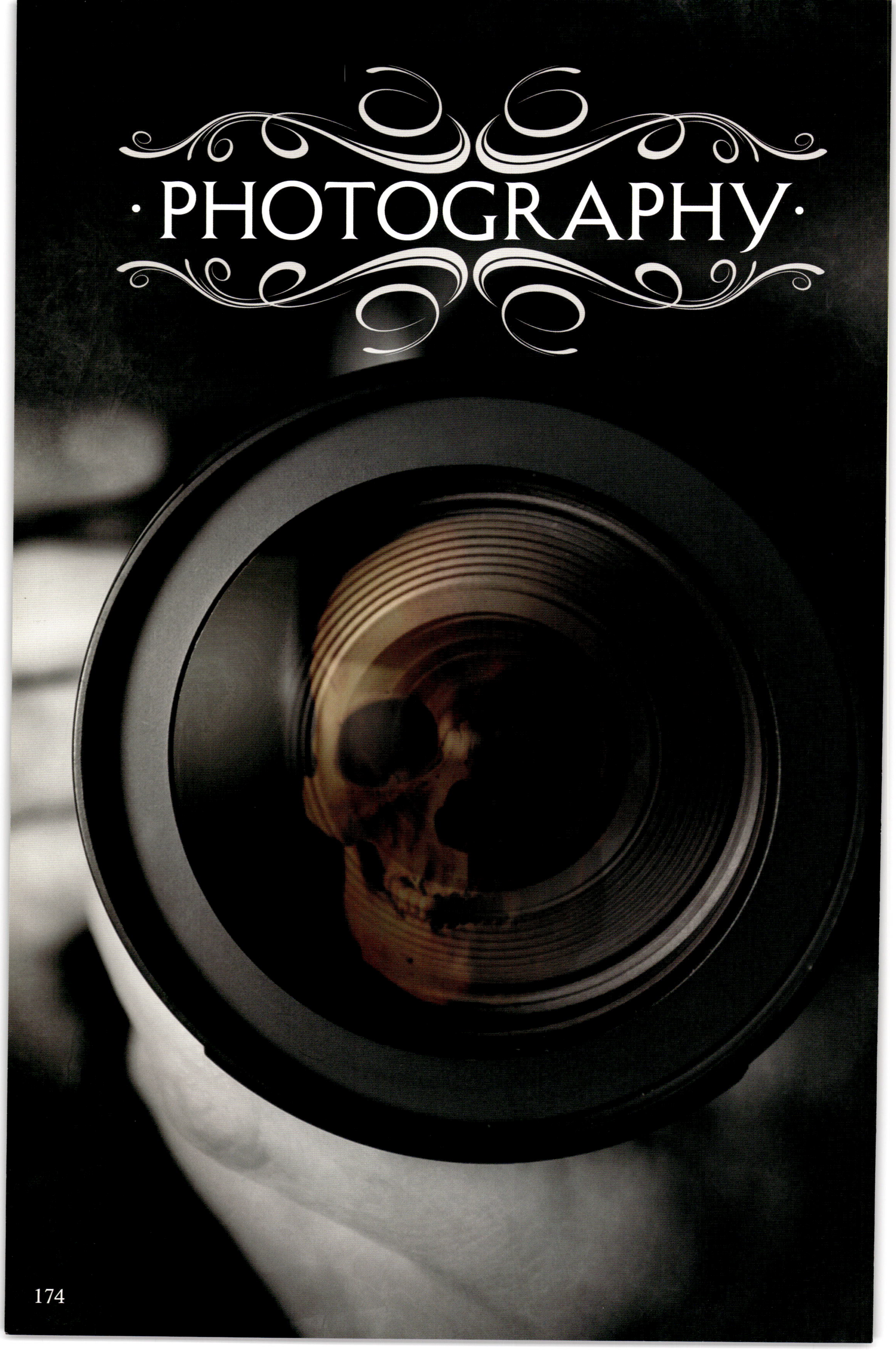

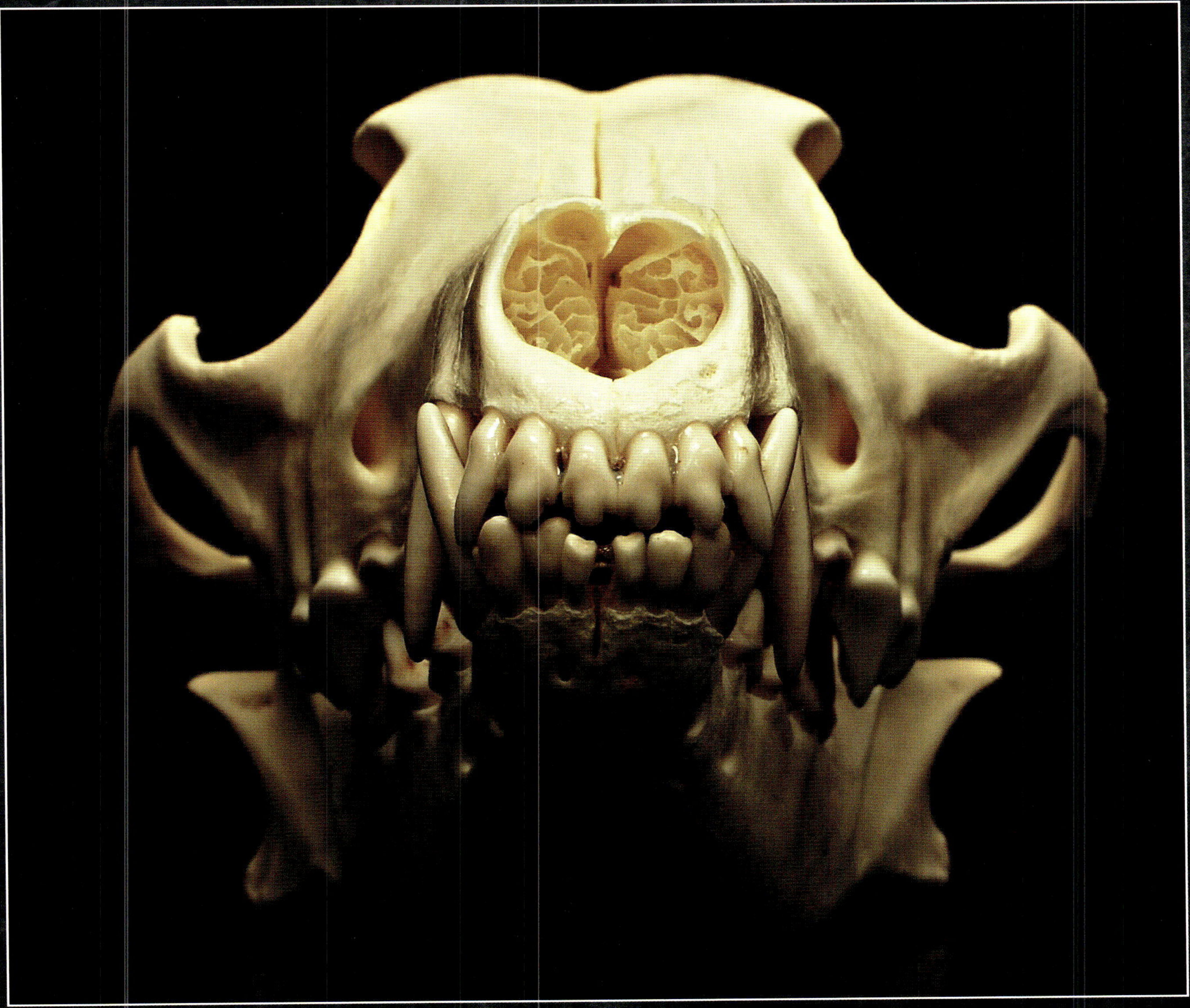

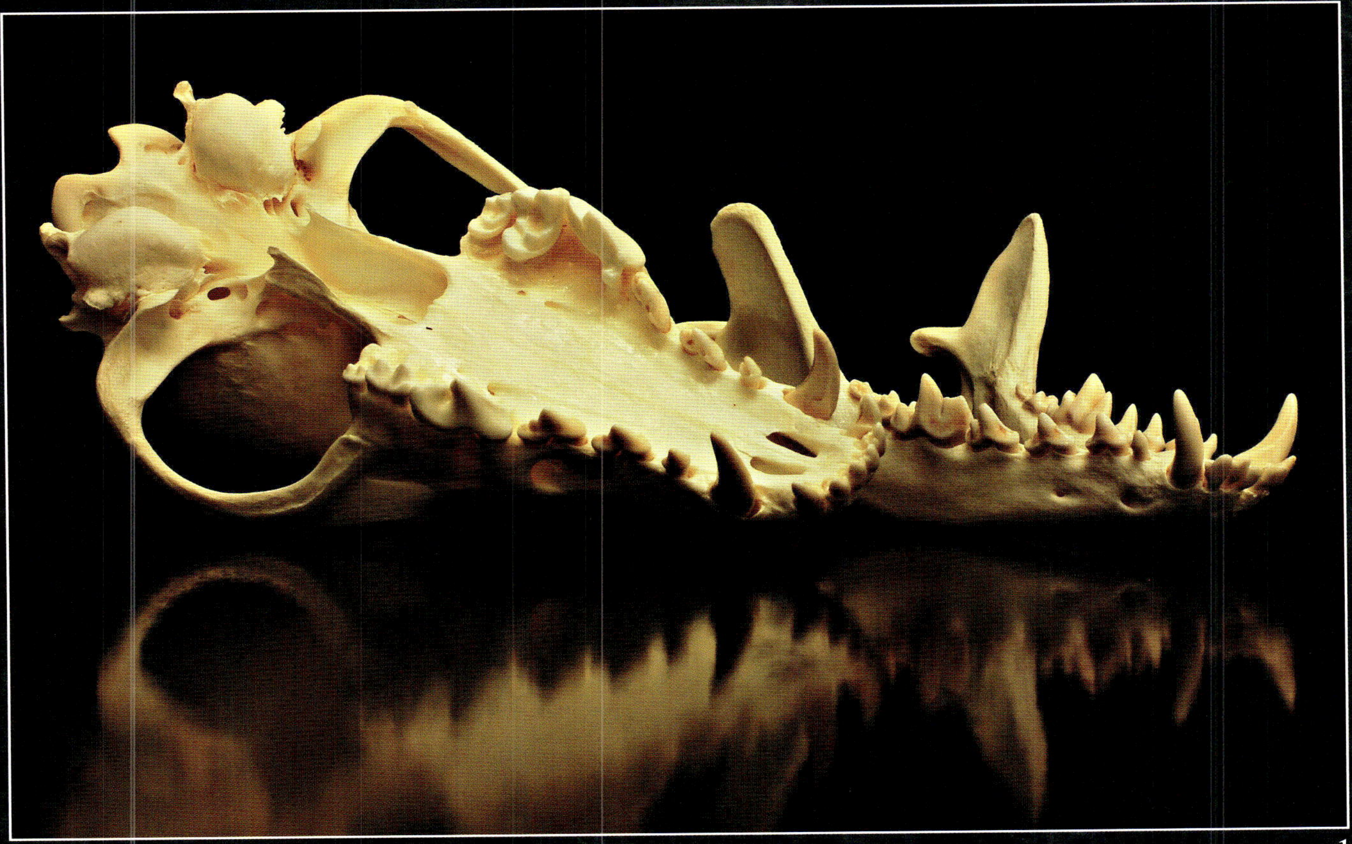

Russel Haig

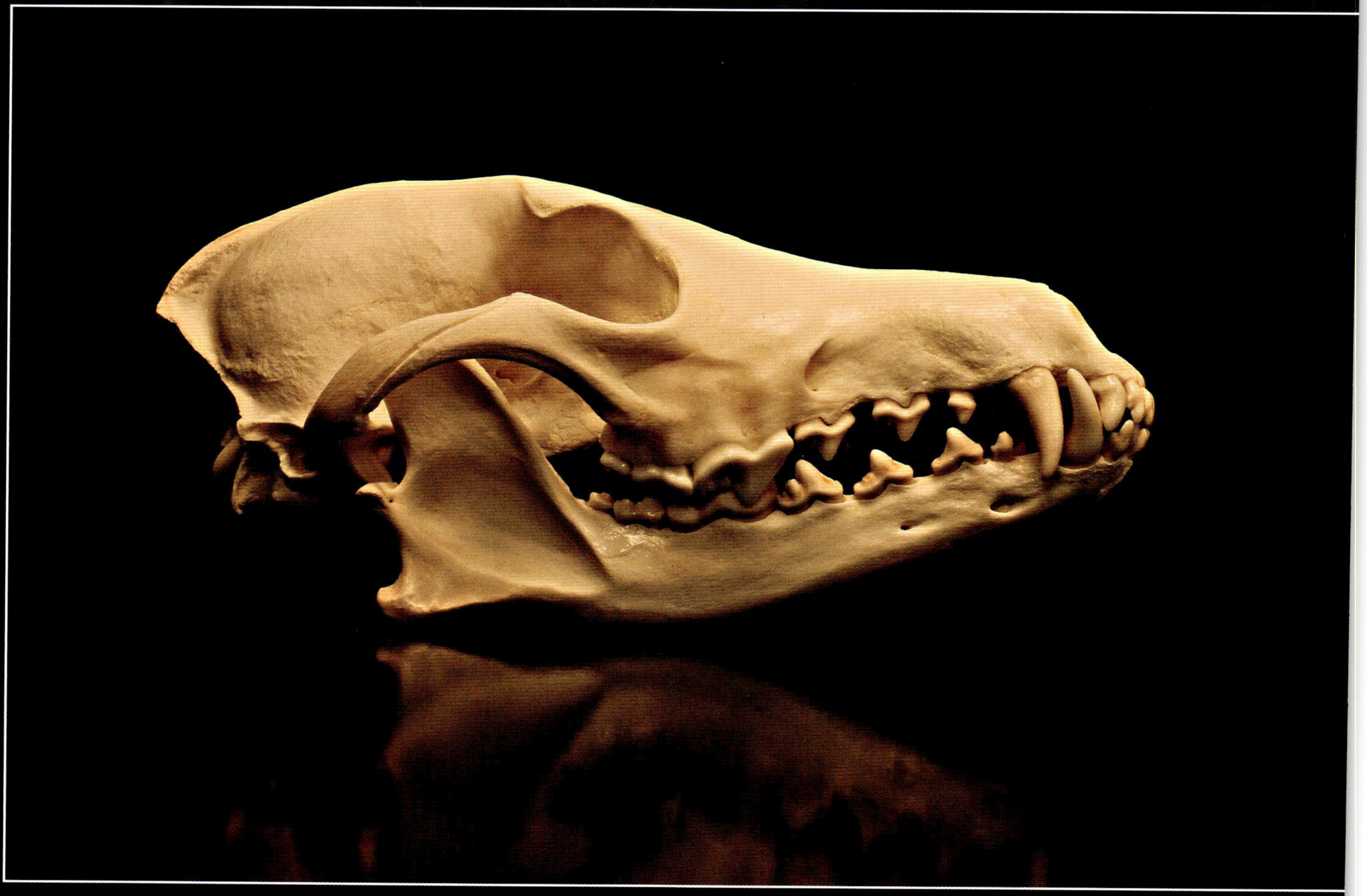

Russel Haig

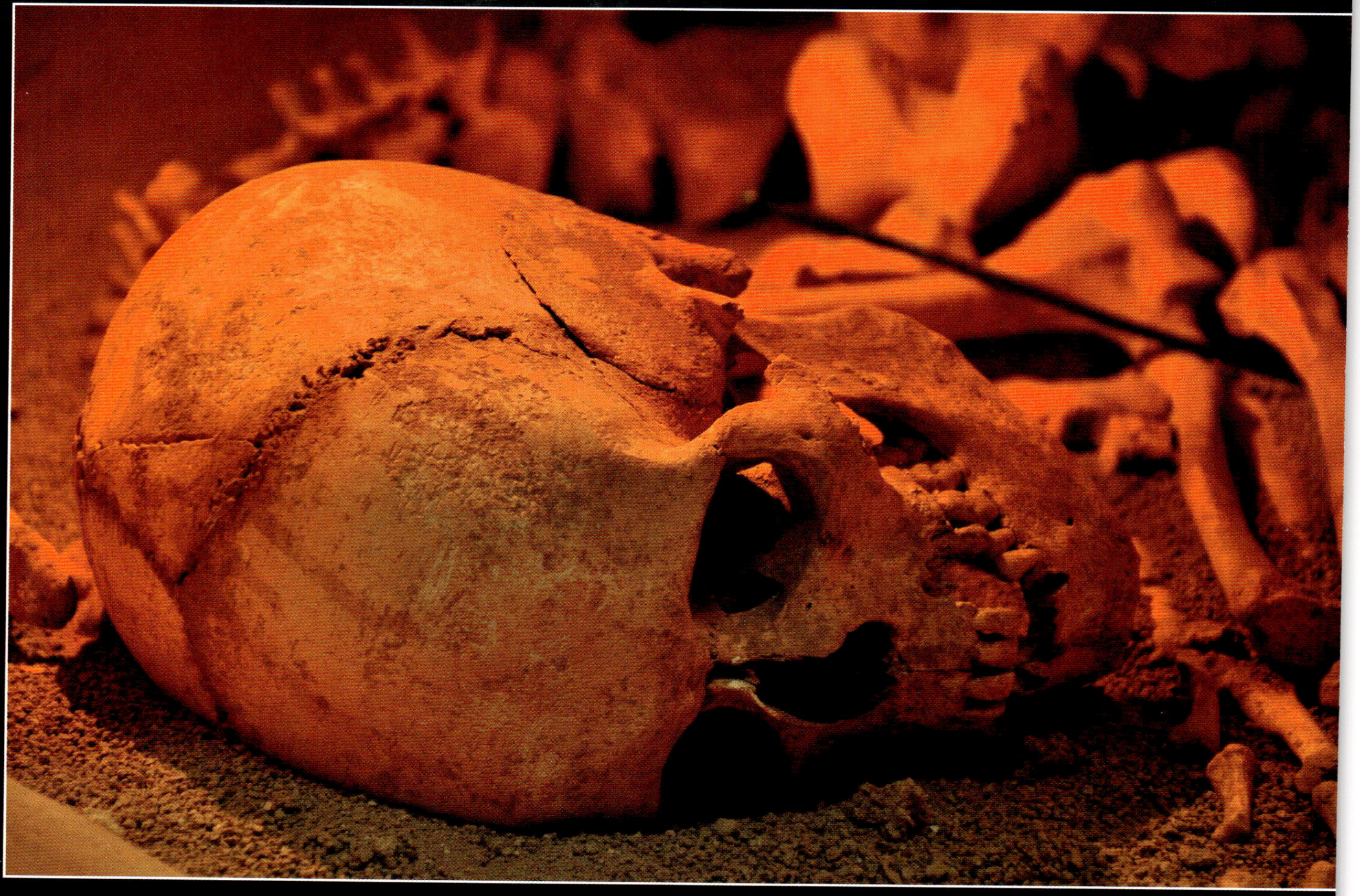

Patrice Perez

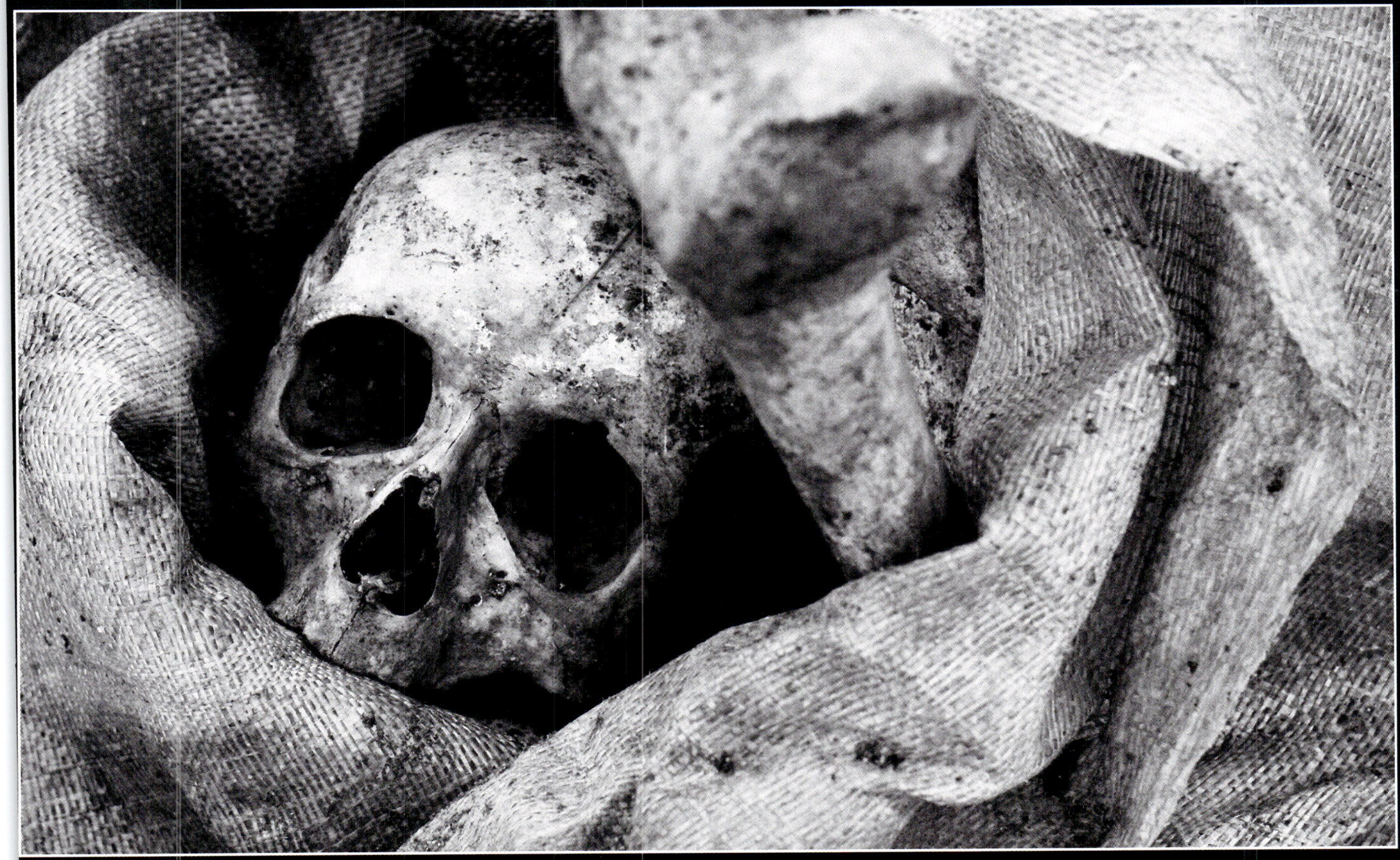

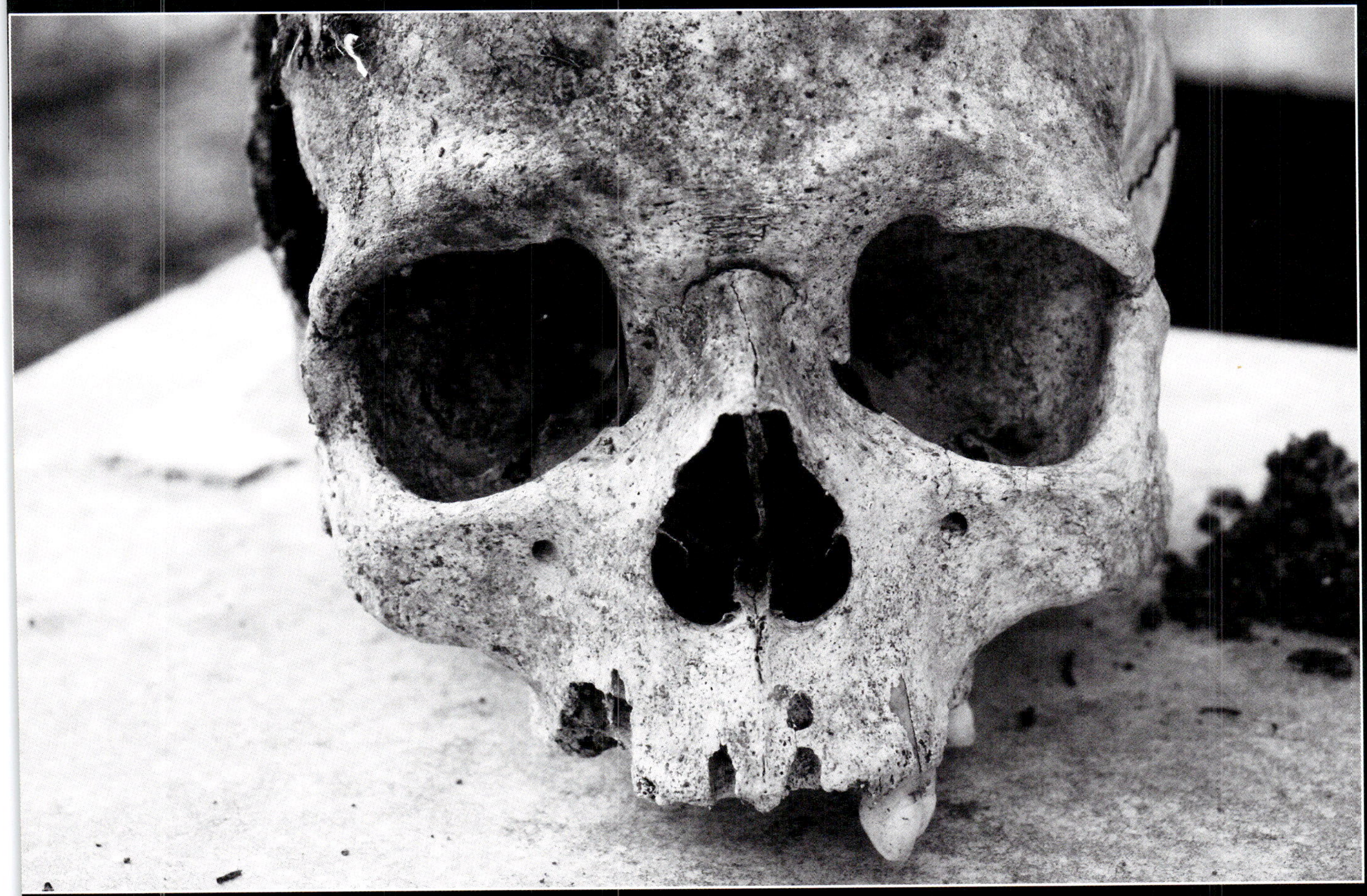

Julius Motal

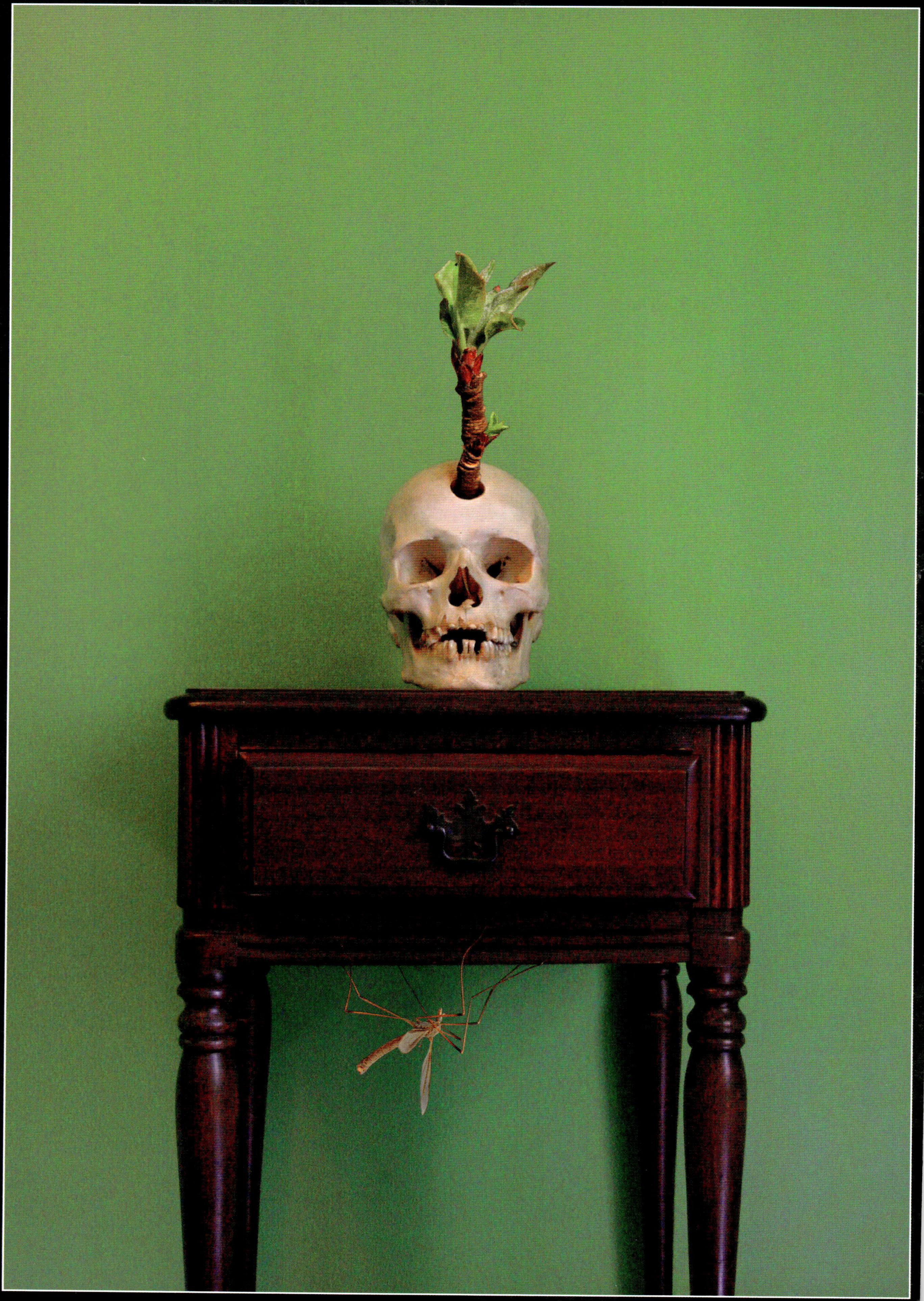

Nick Baxter

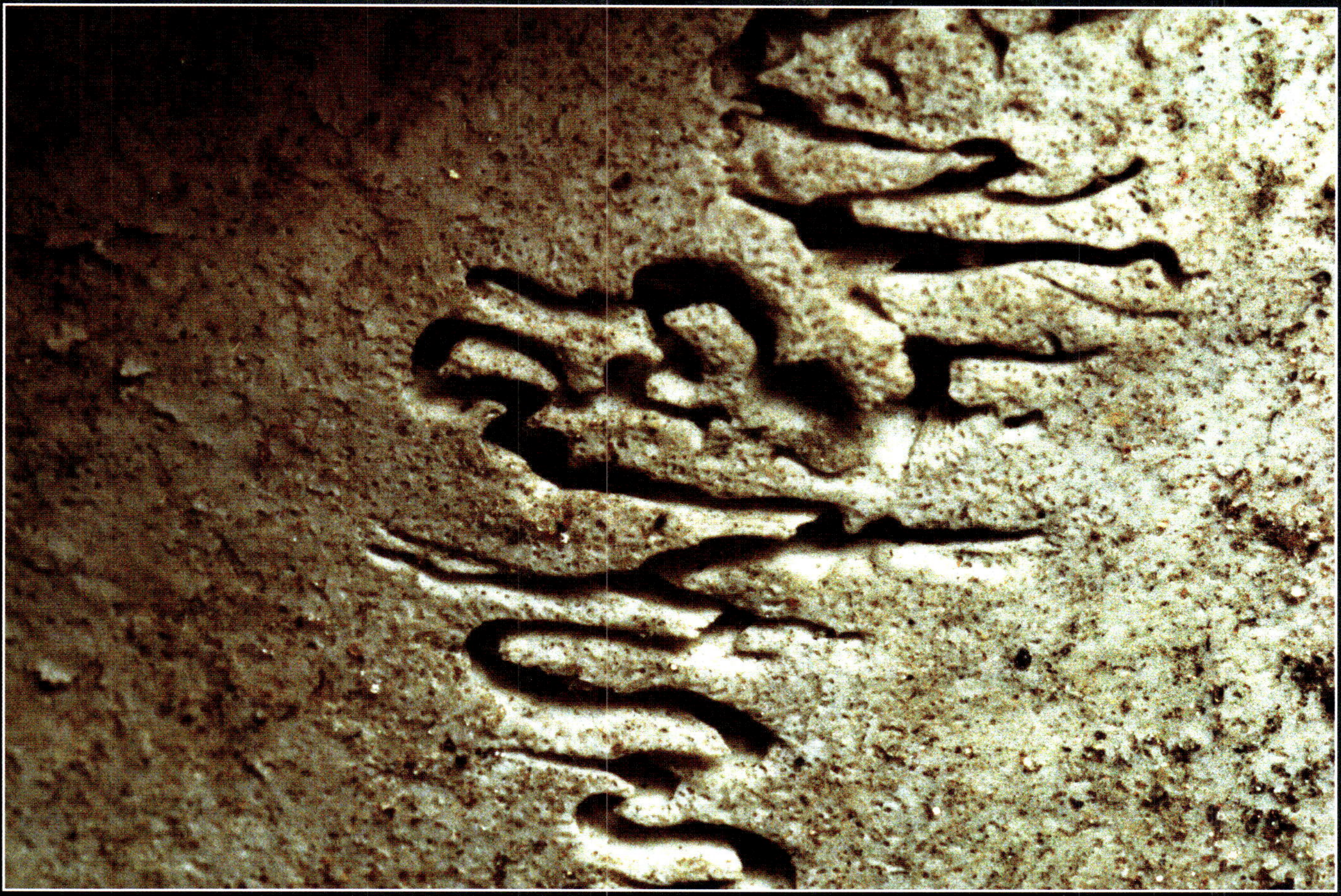

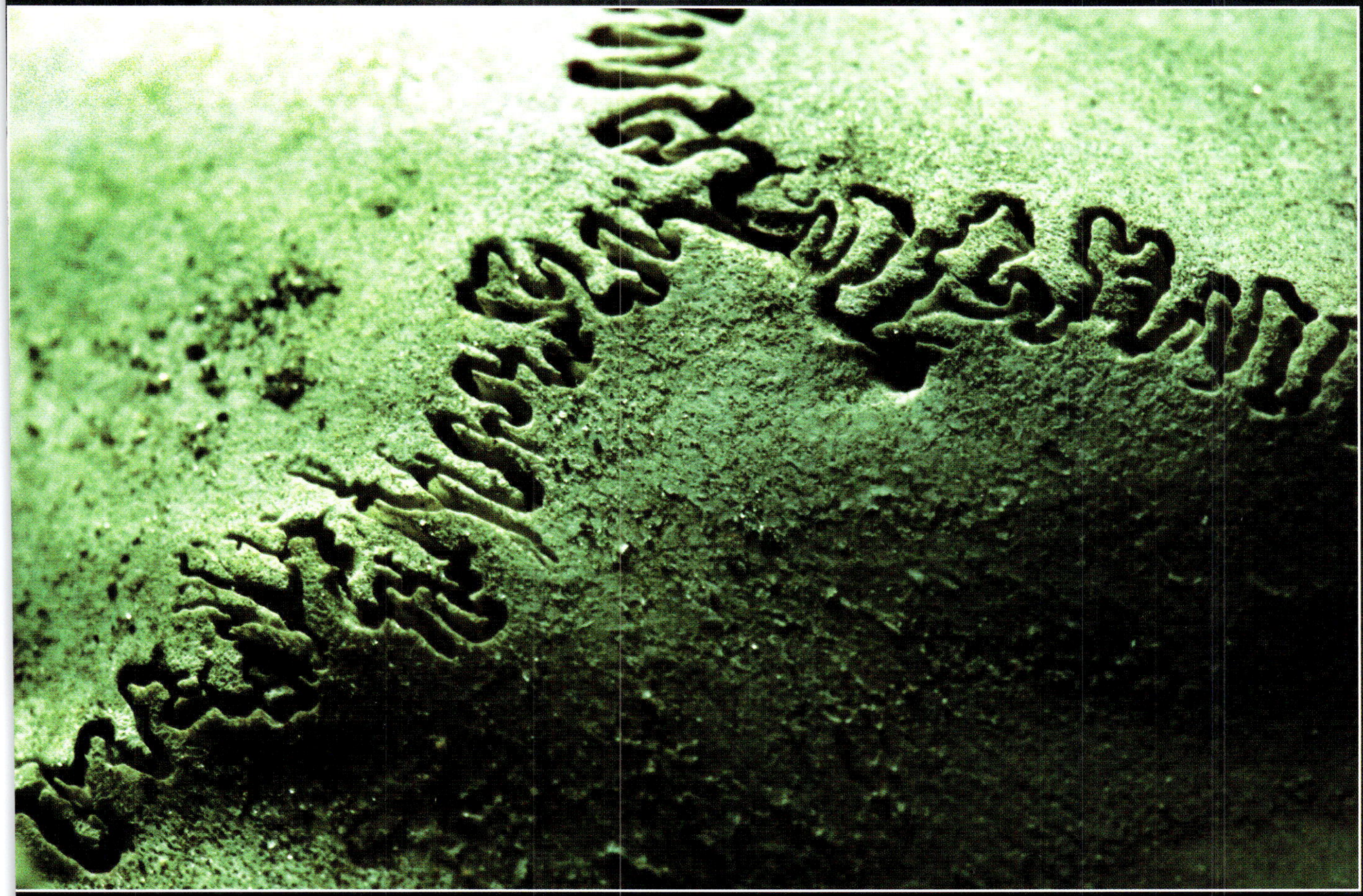

Nick Baxter

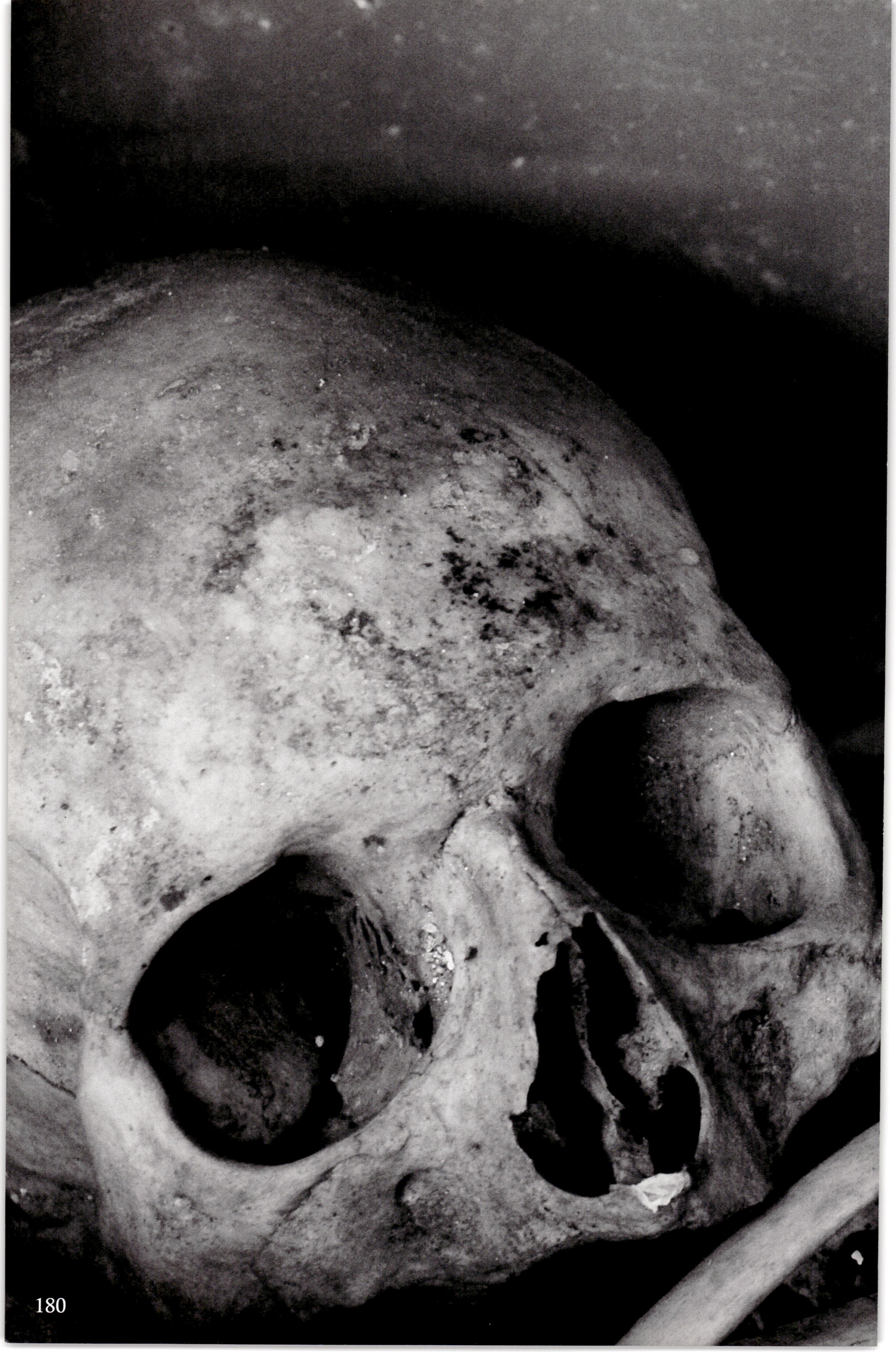

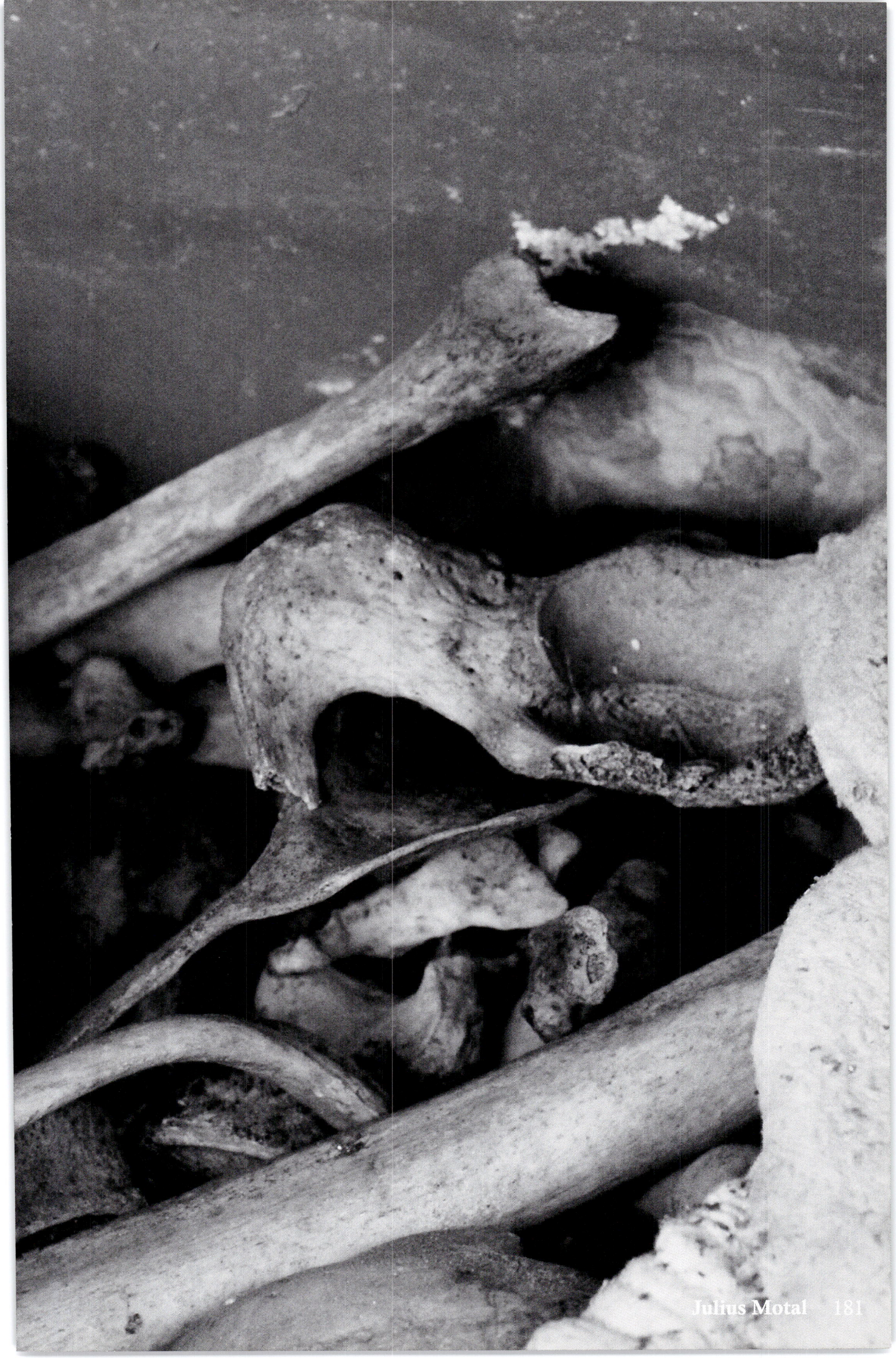

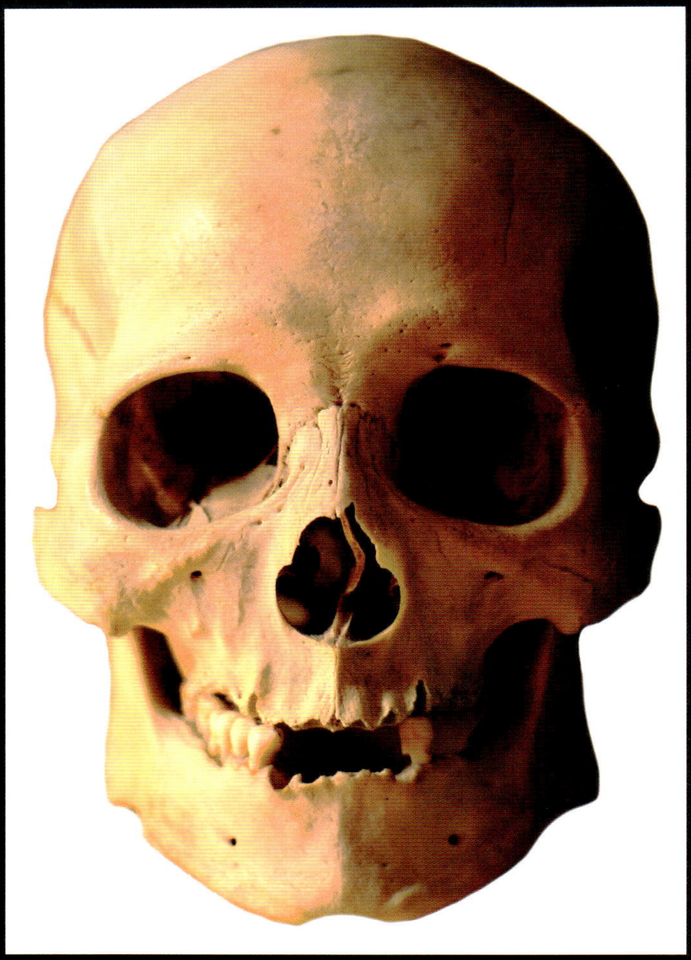
Jeff Ensminger

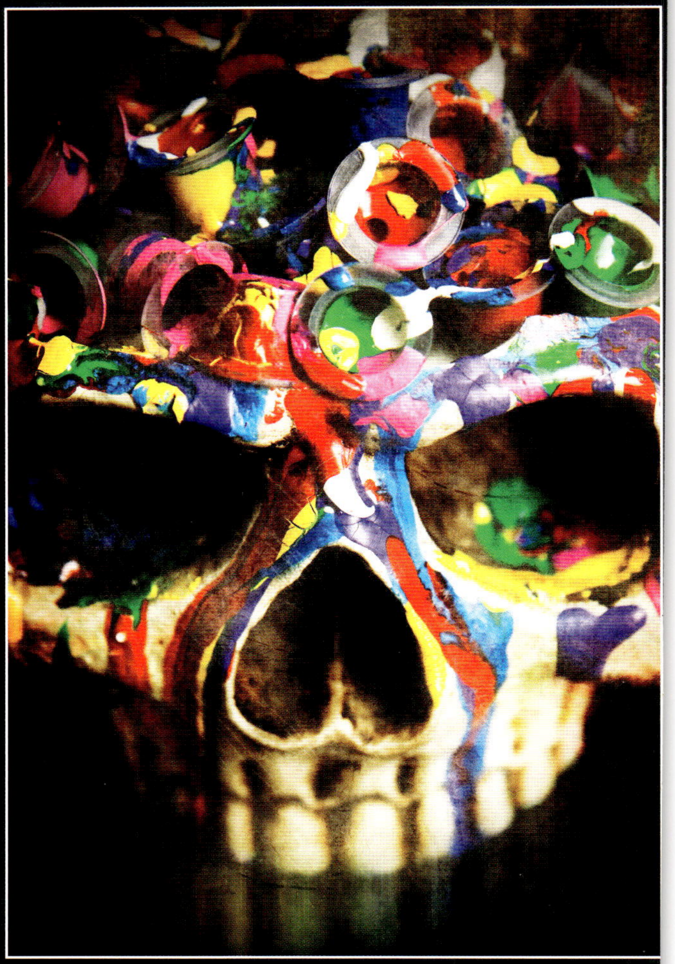
Josh Winton

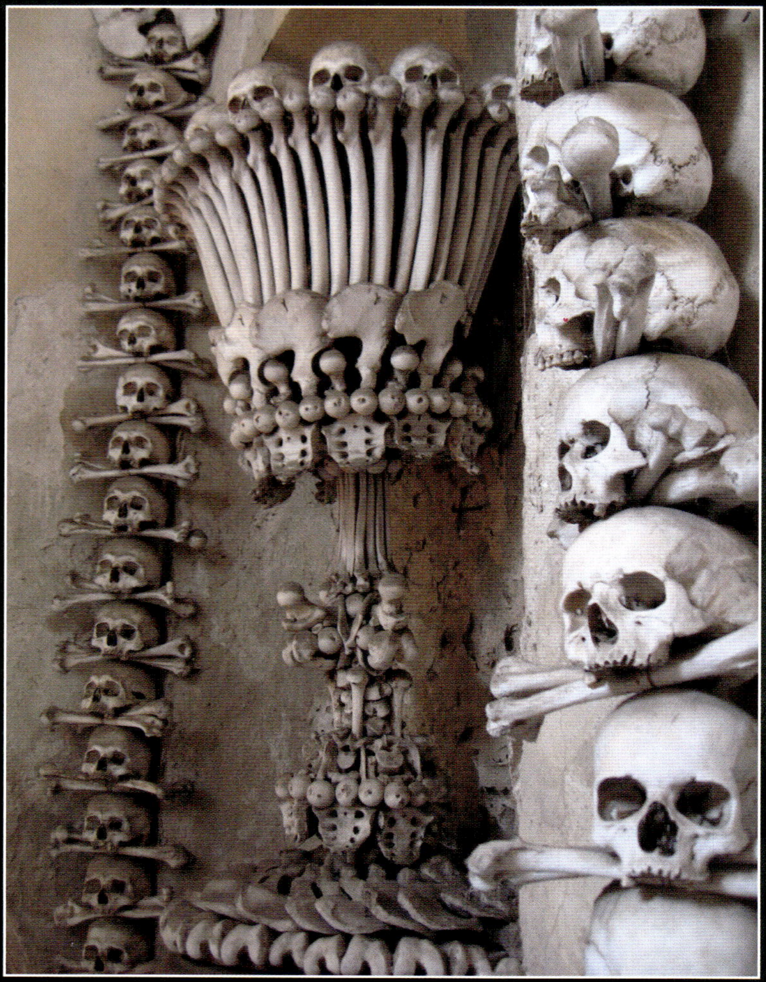
Richard Mcgee

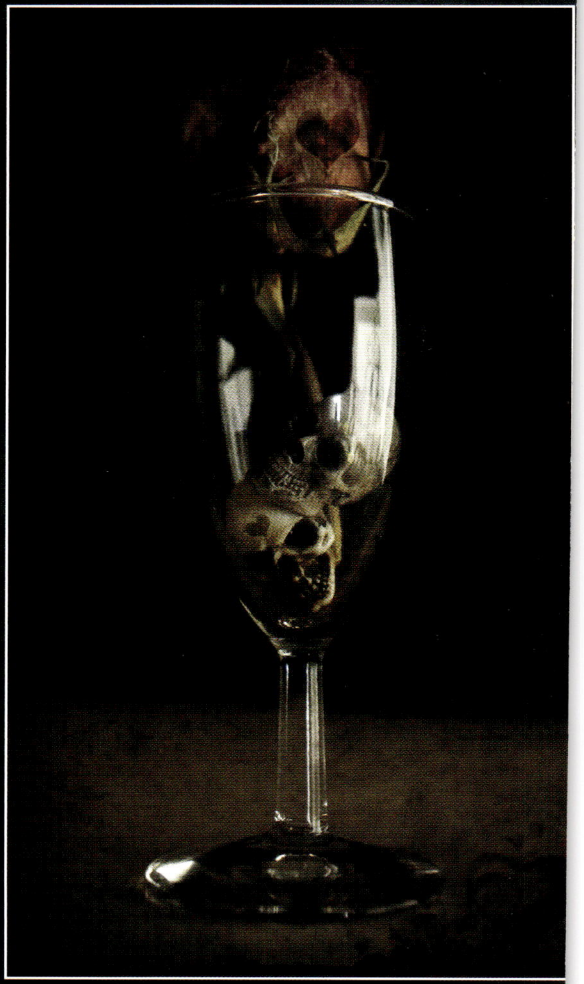
Josh Winton

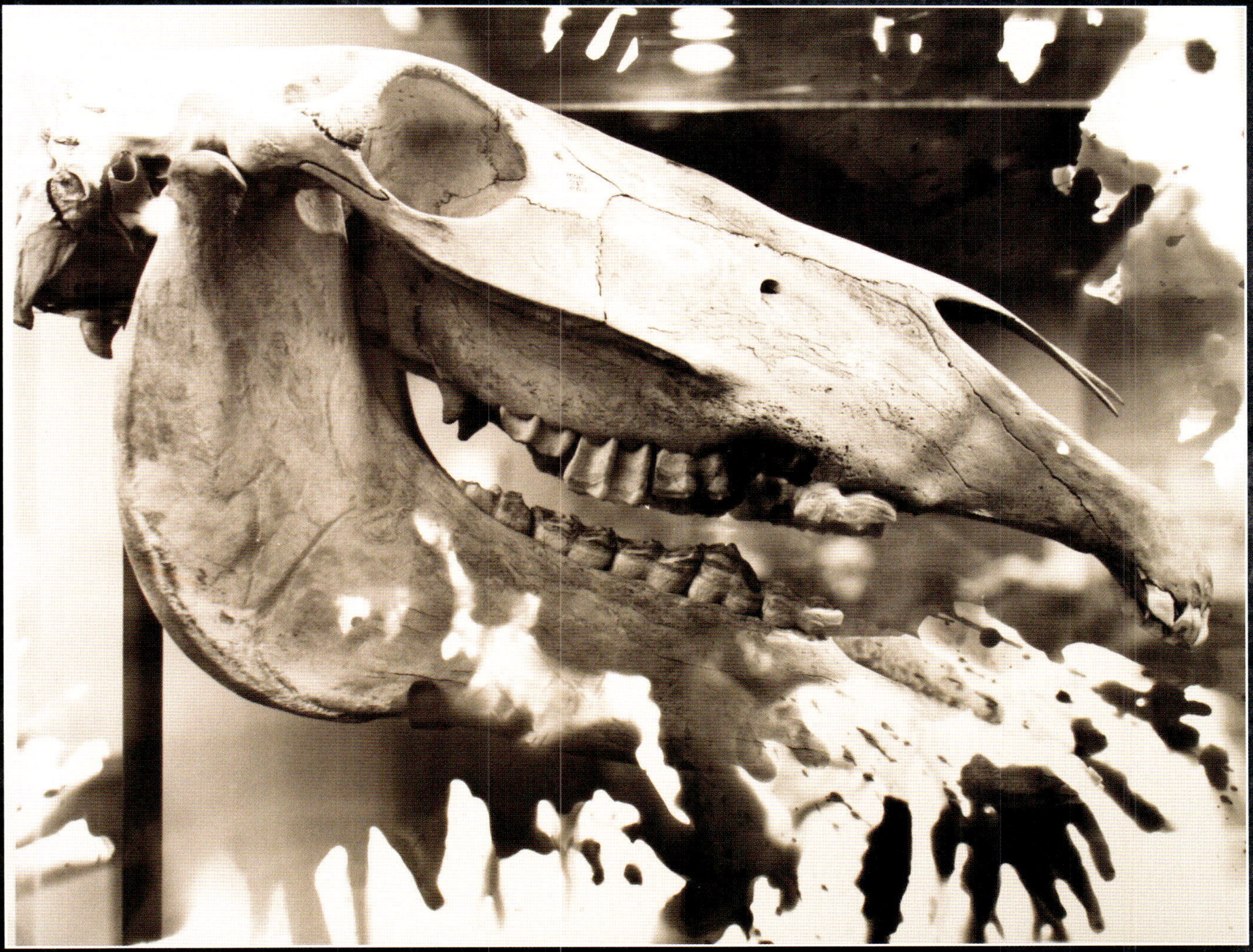

Megan Thompson

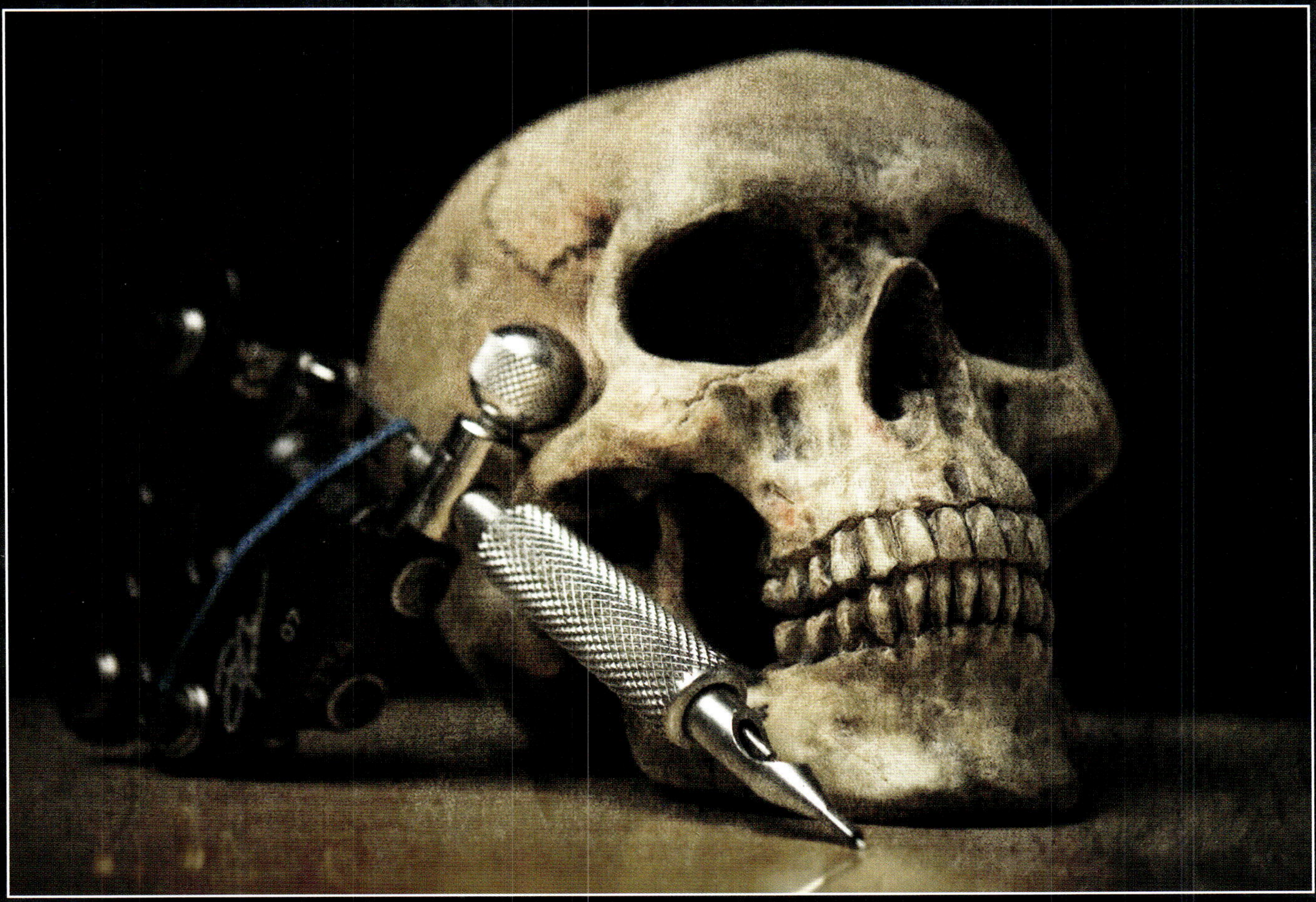

Josh Winton

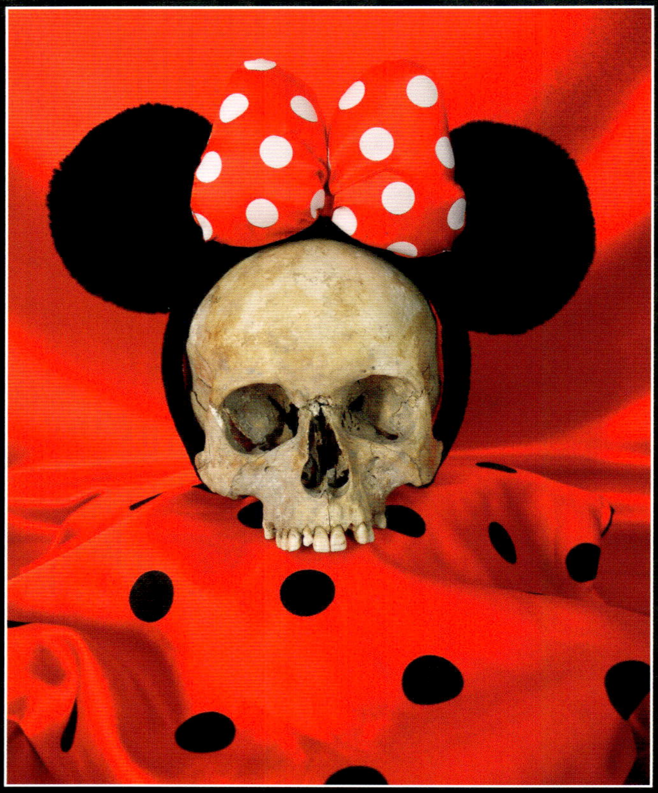

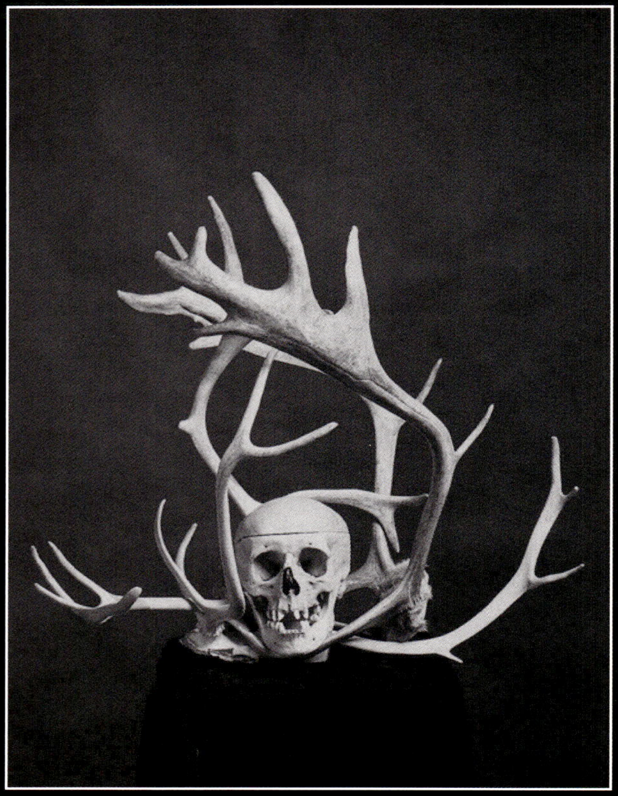

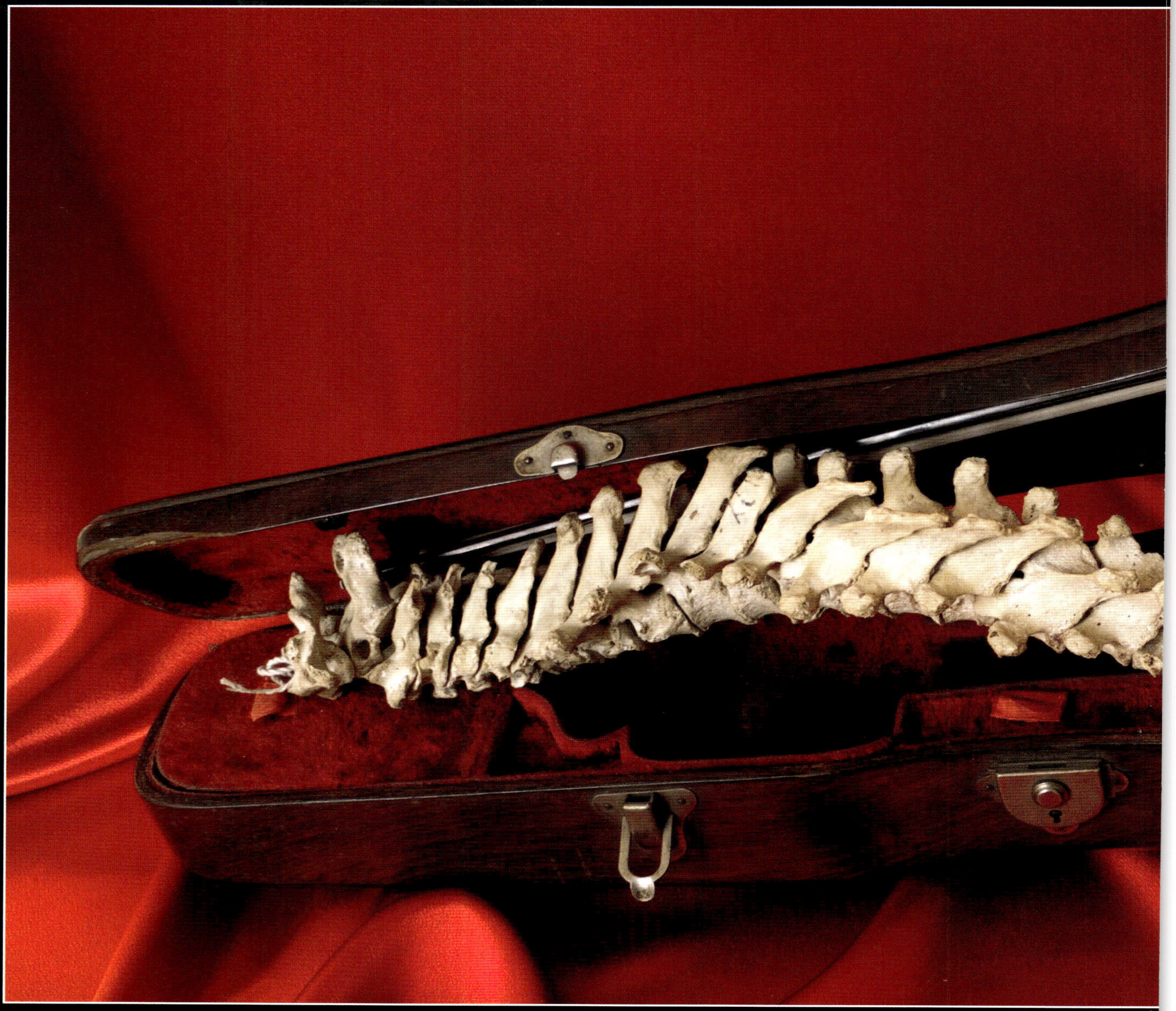

Jean Baptiste Carhaix

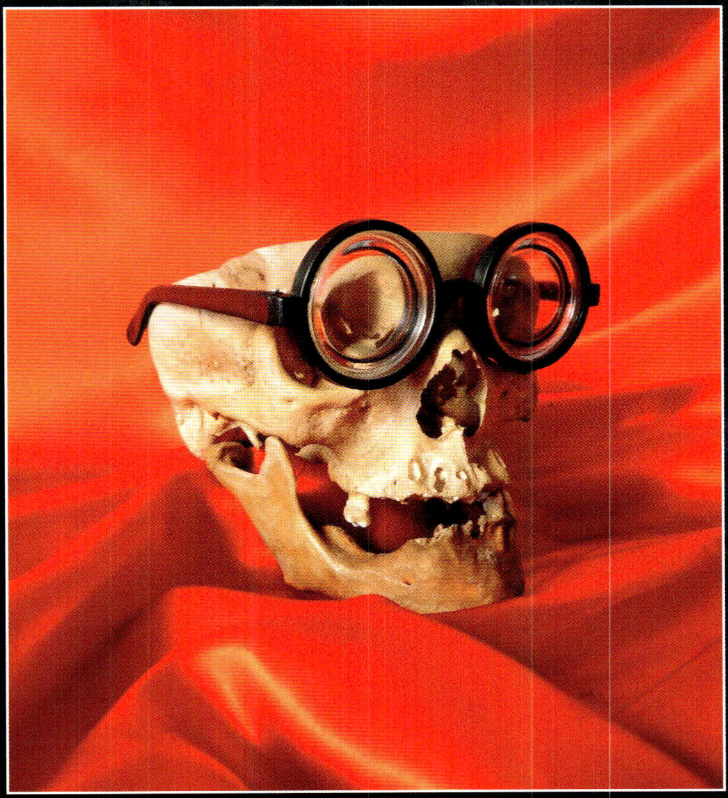

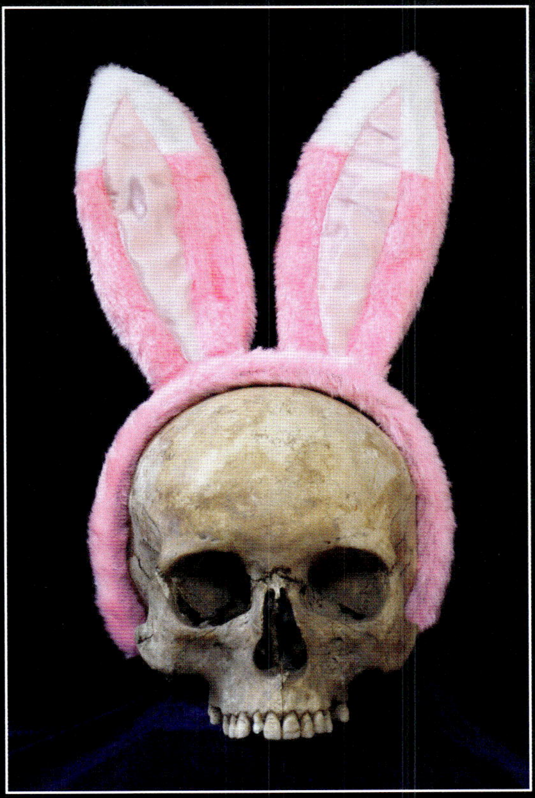

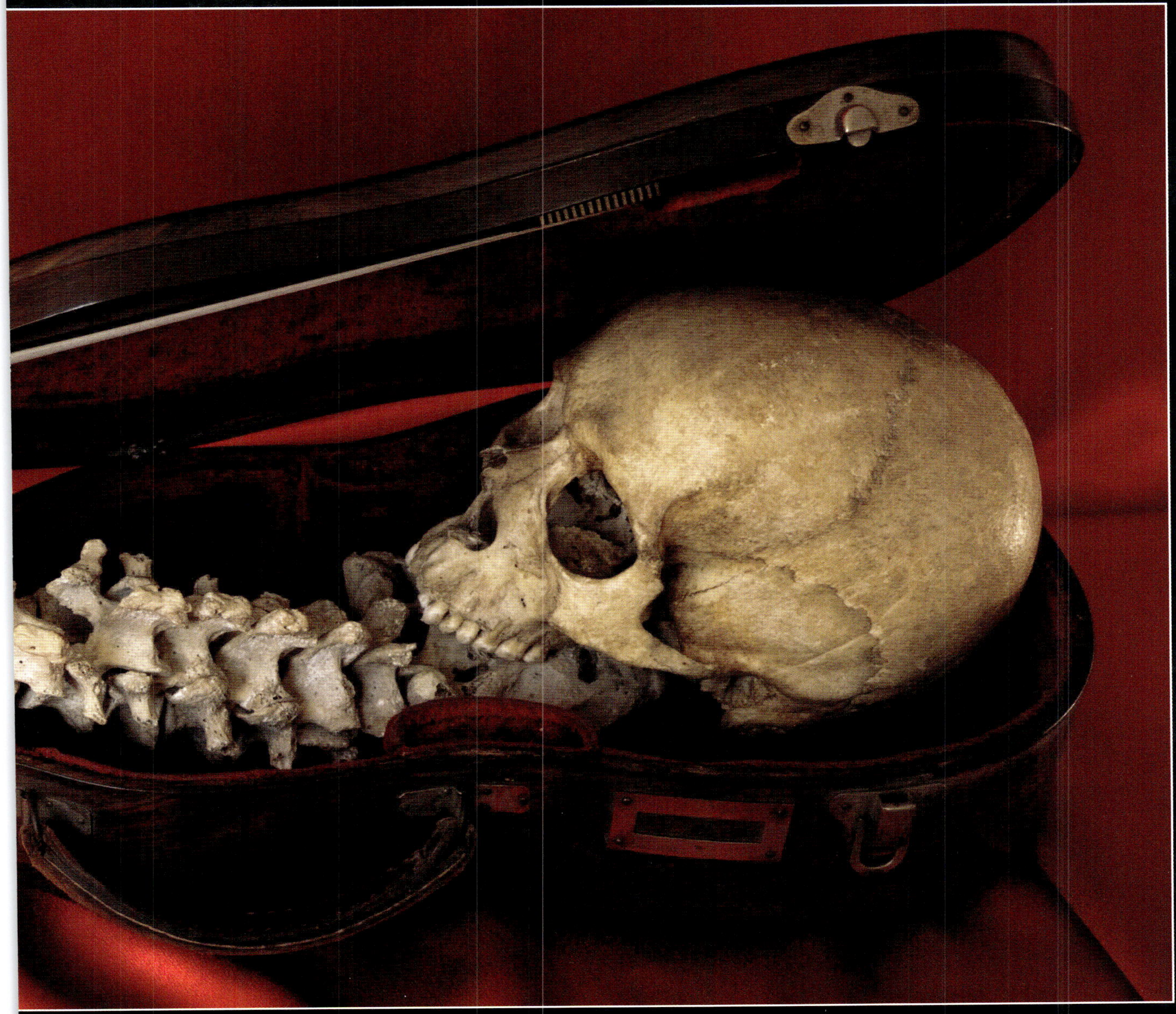

Jean Baptiste Carhaix

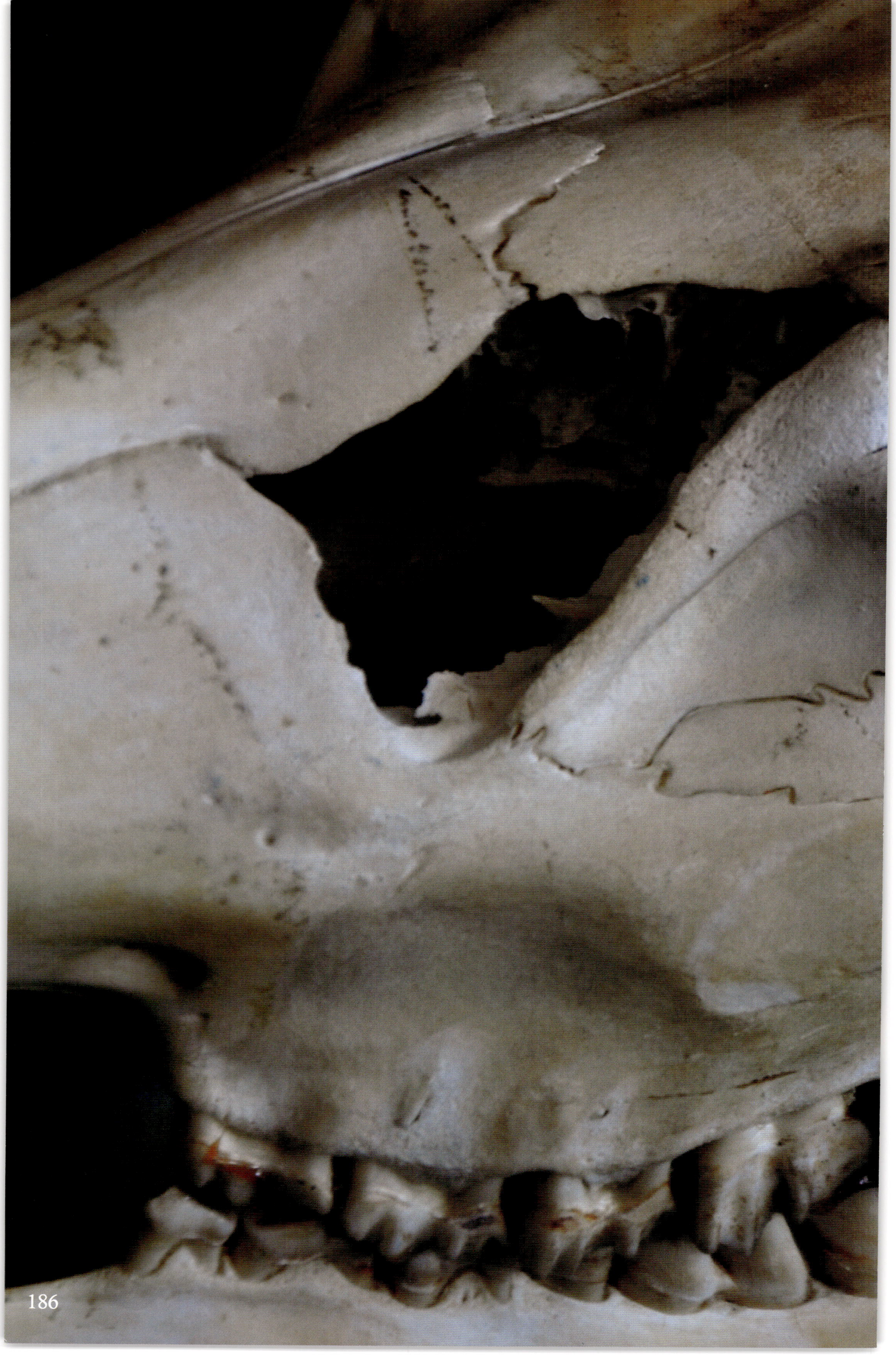

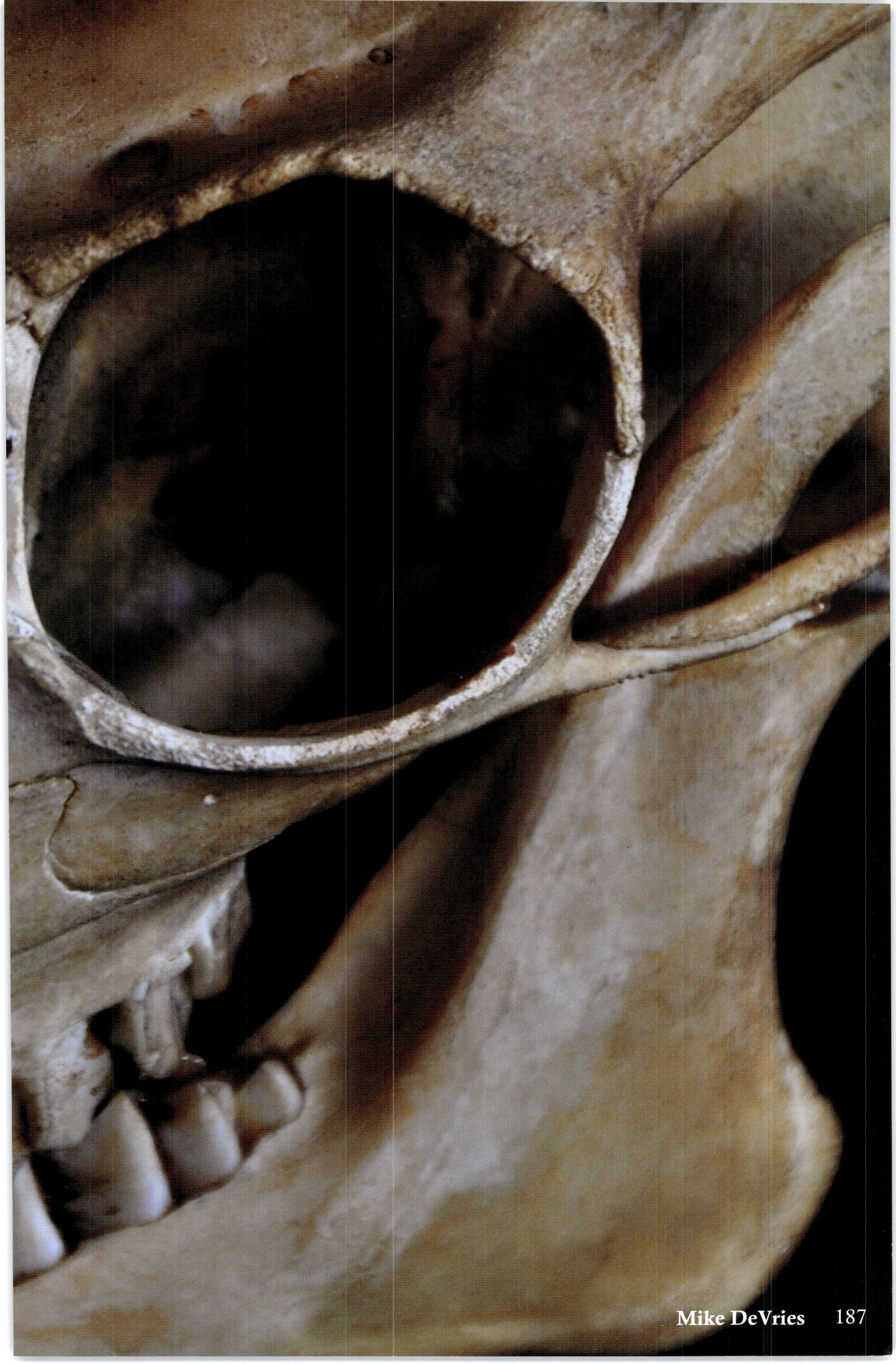

Mike DeVries

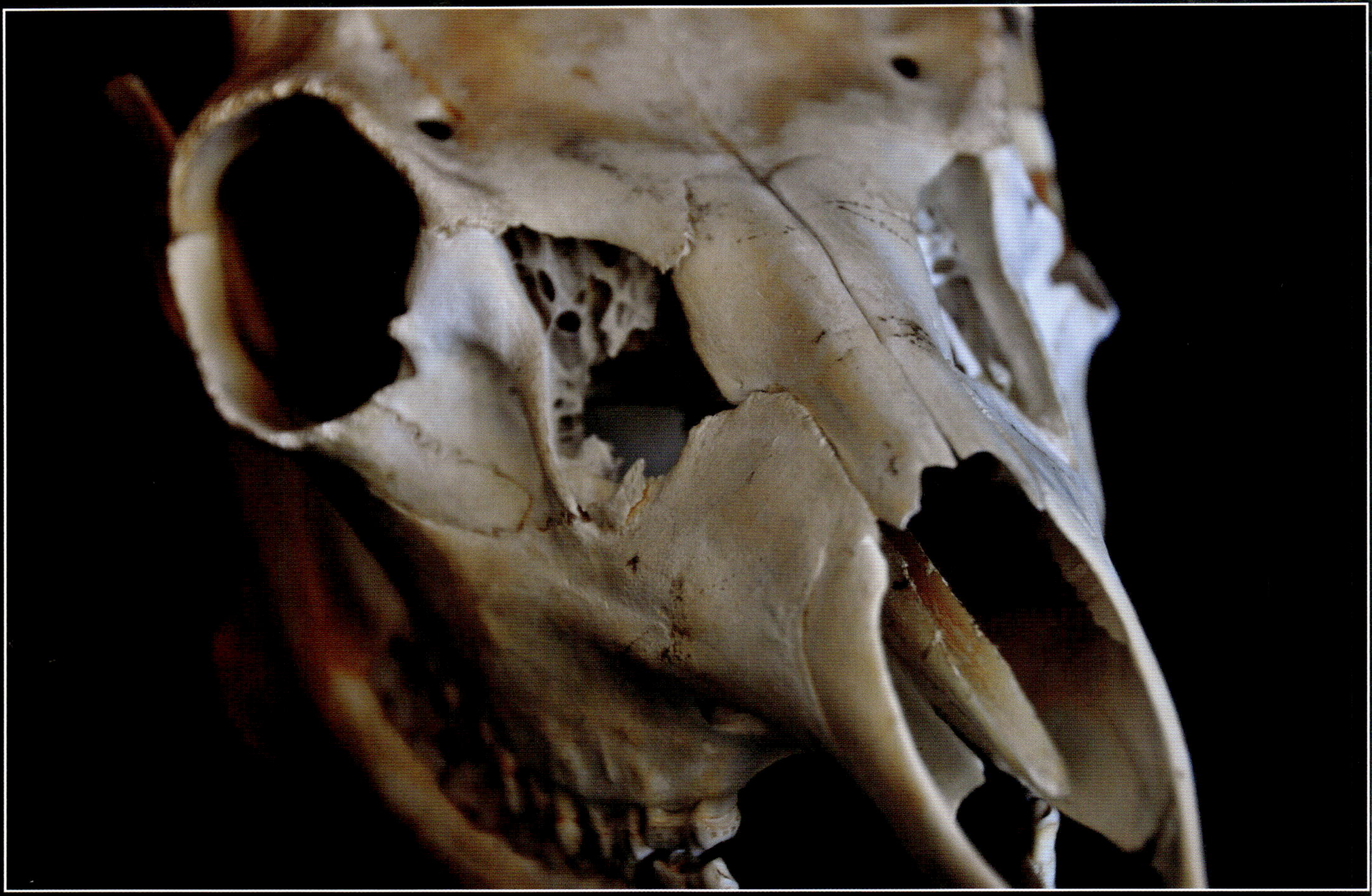

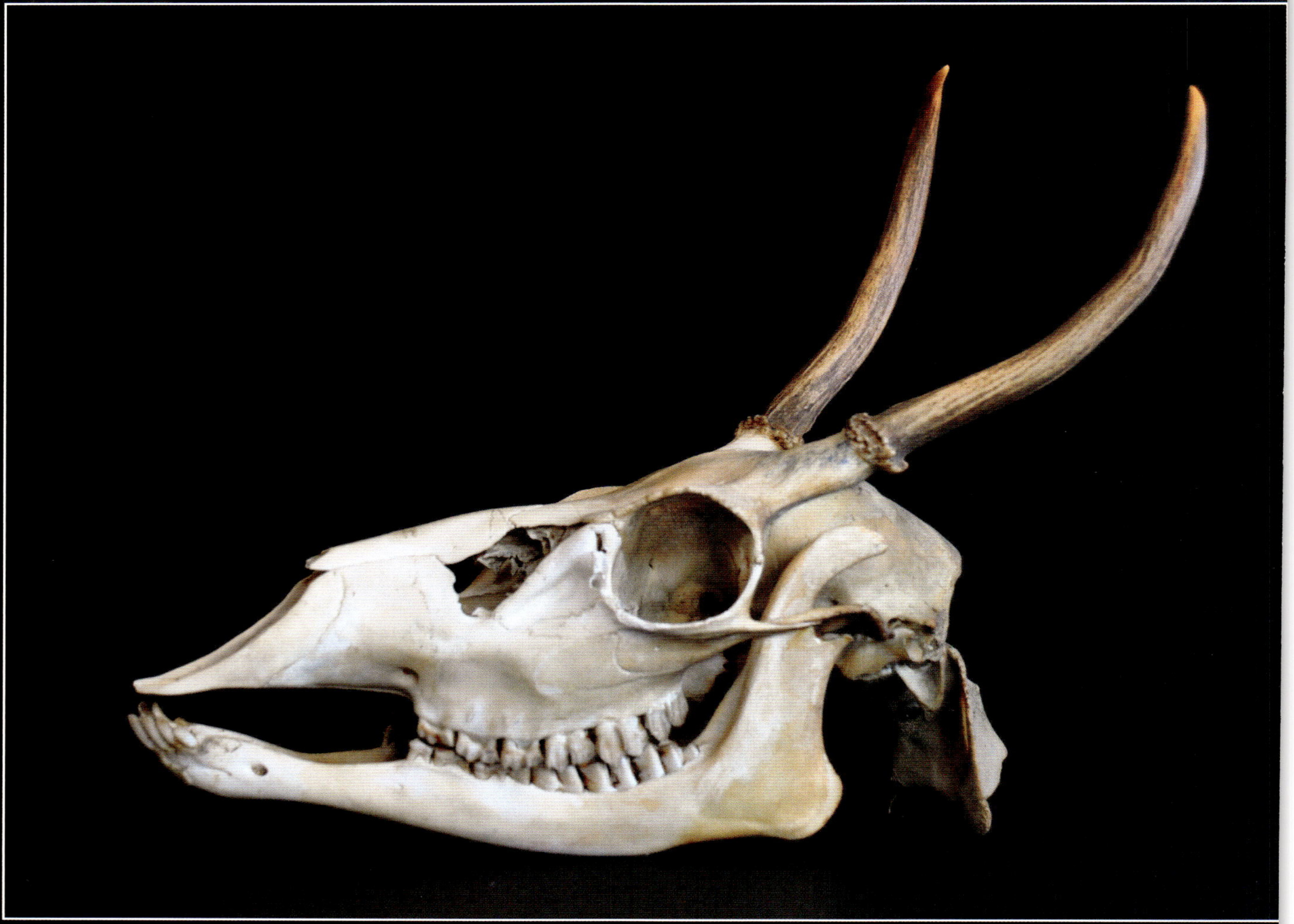

Mike DeVries

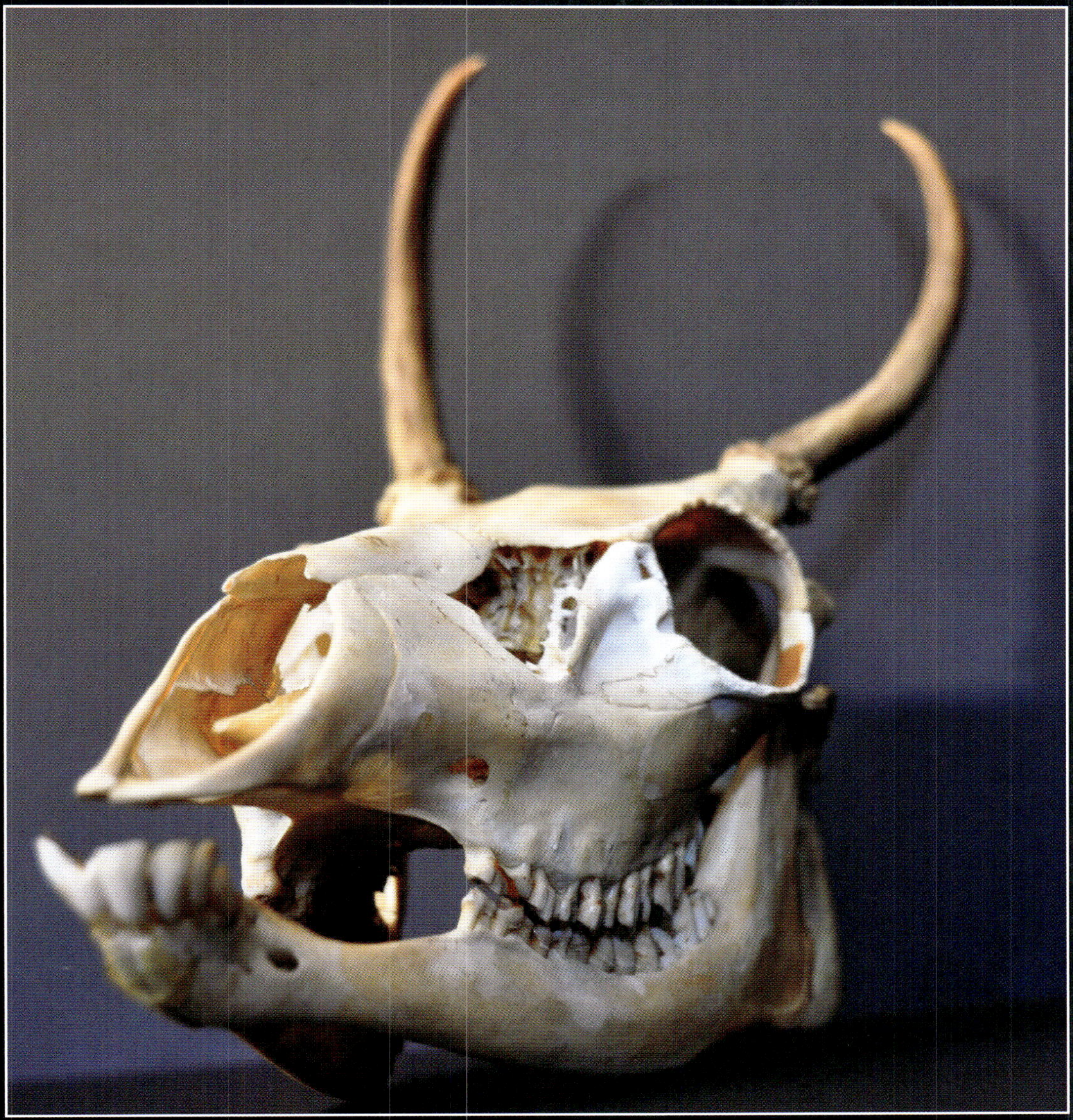

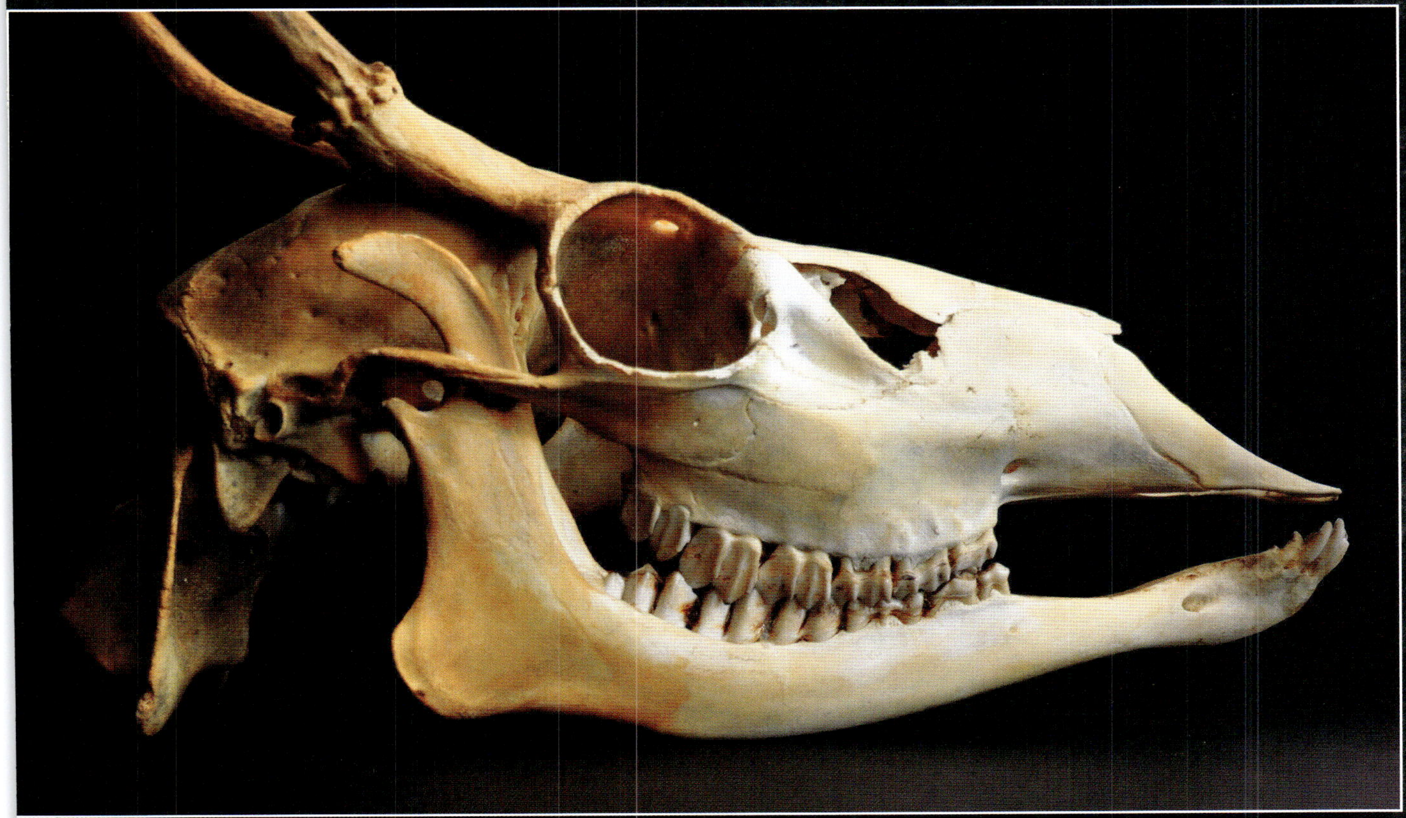

Mike DeVries

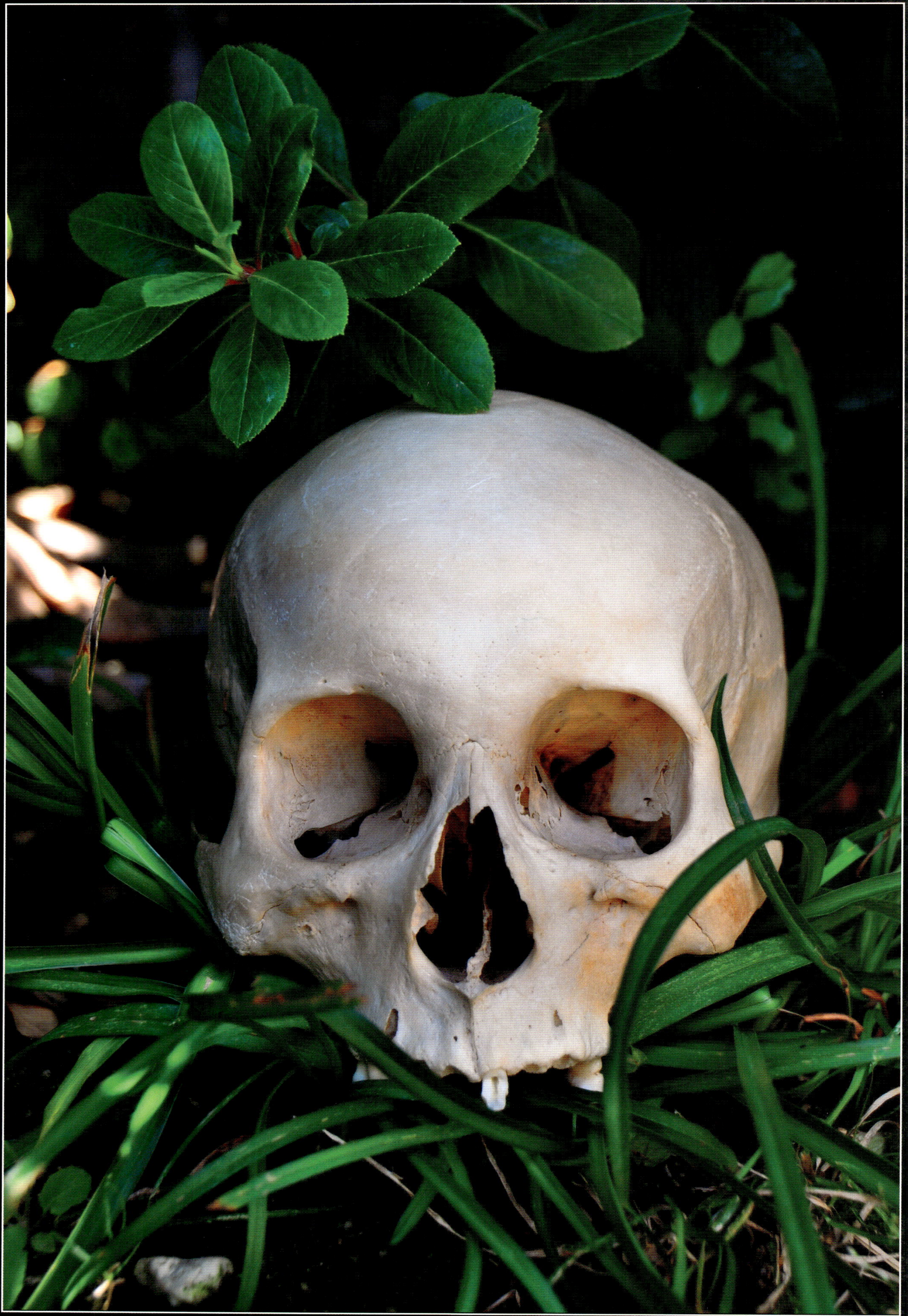

Mike DeVries

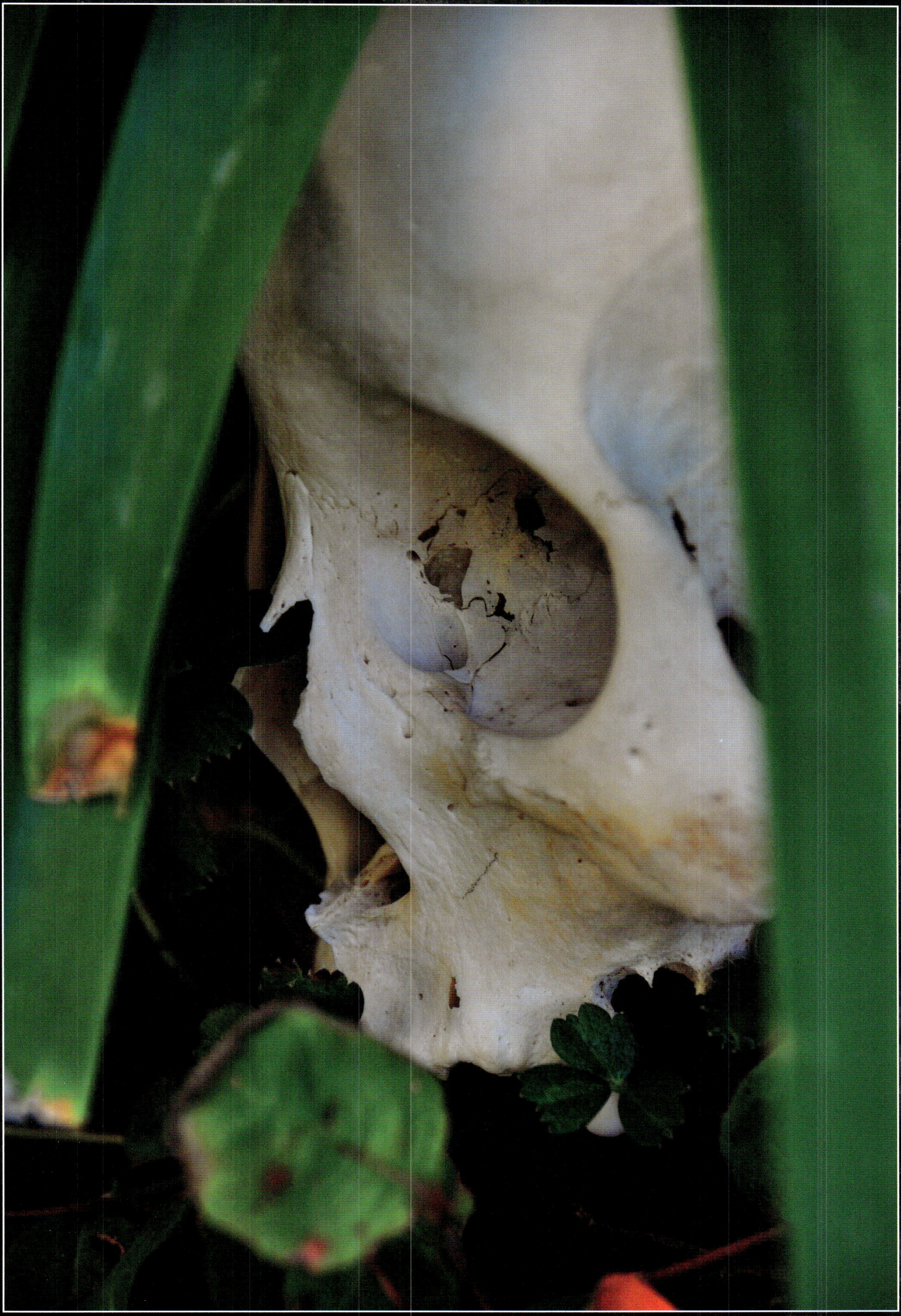

Mike DeVries

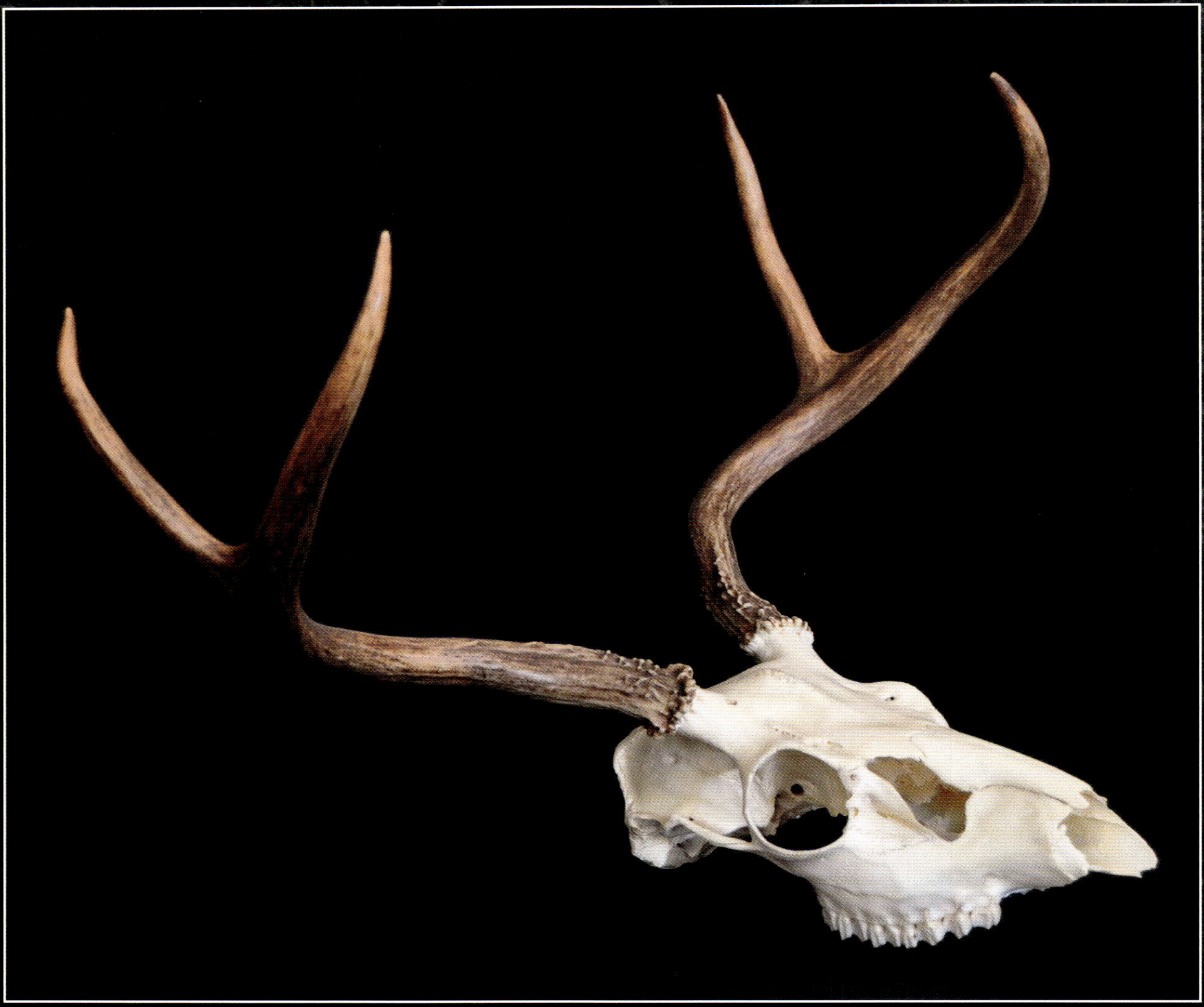

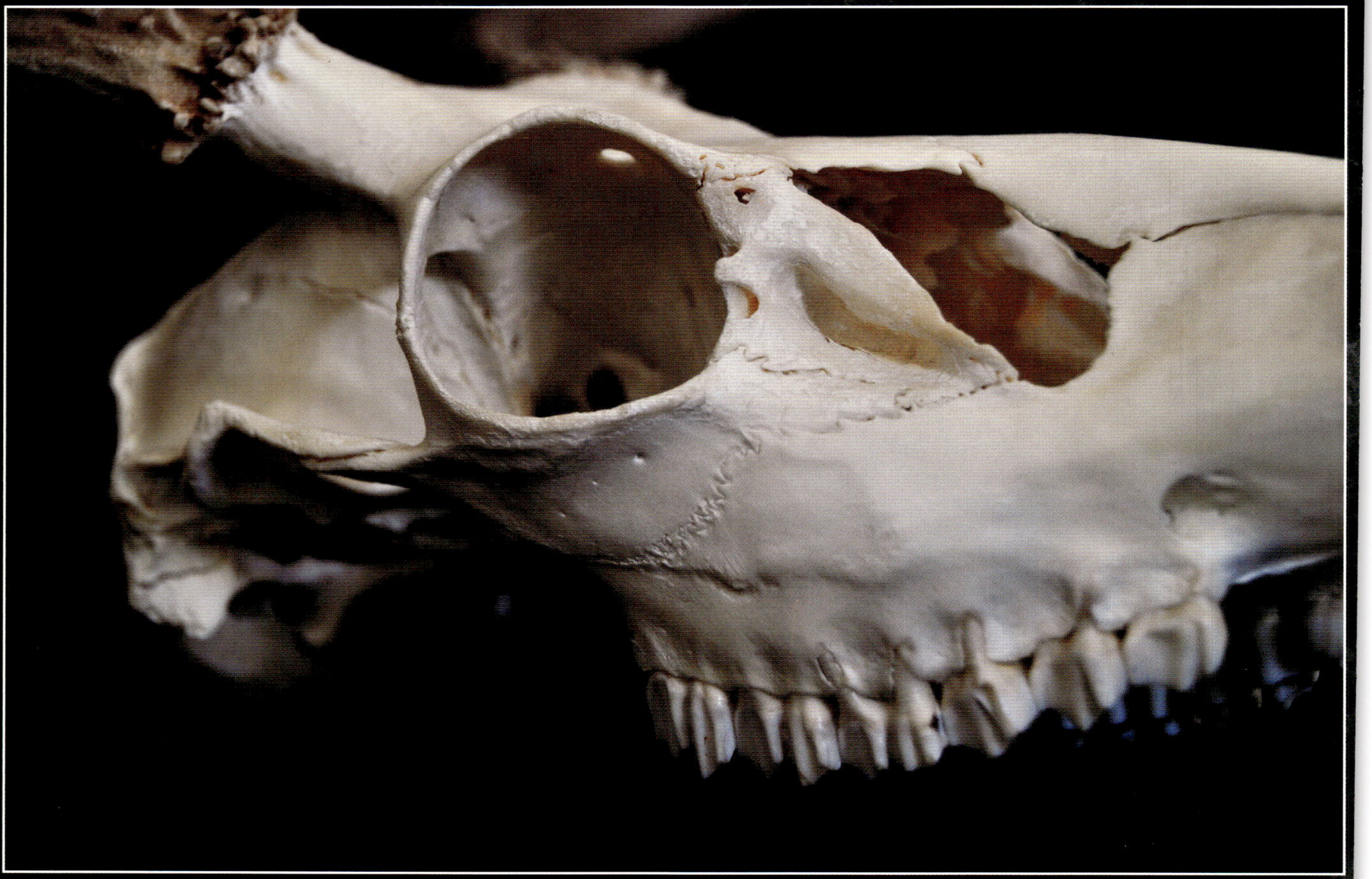

Mike DeVries

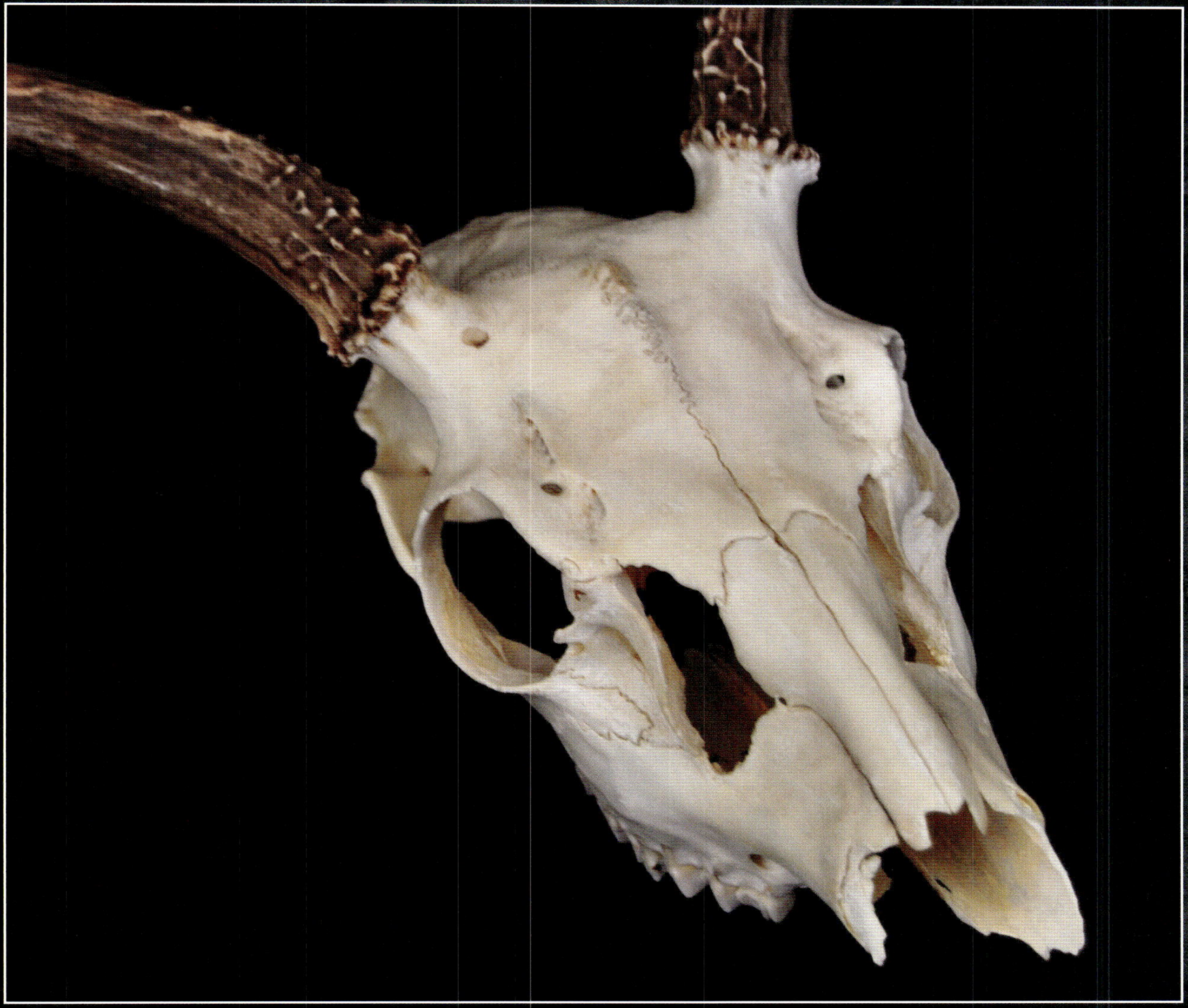

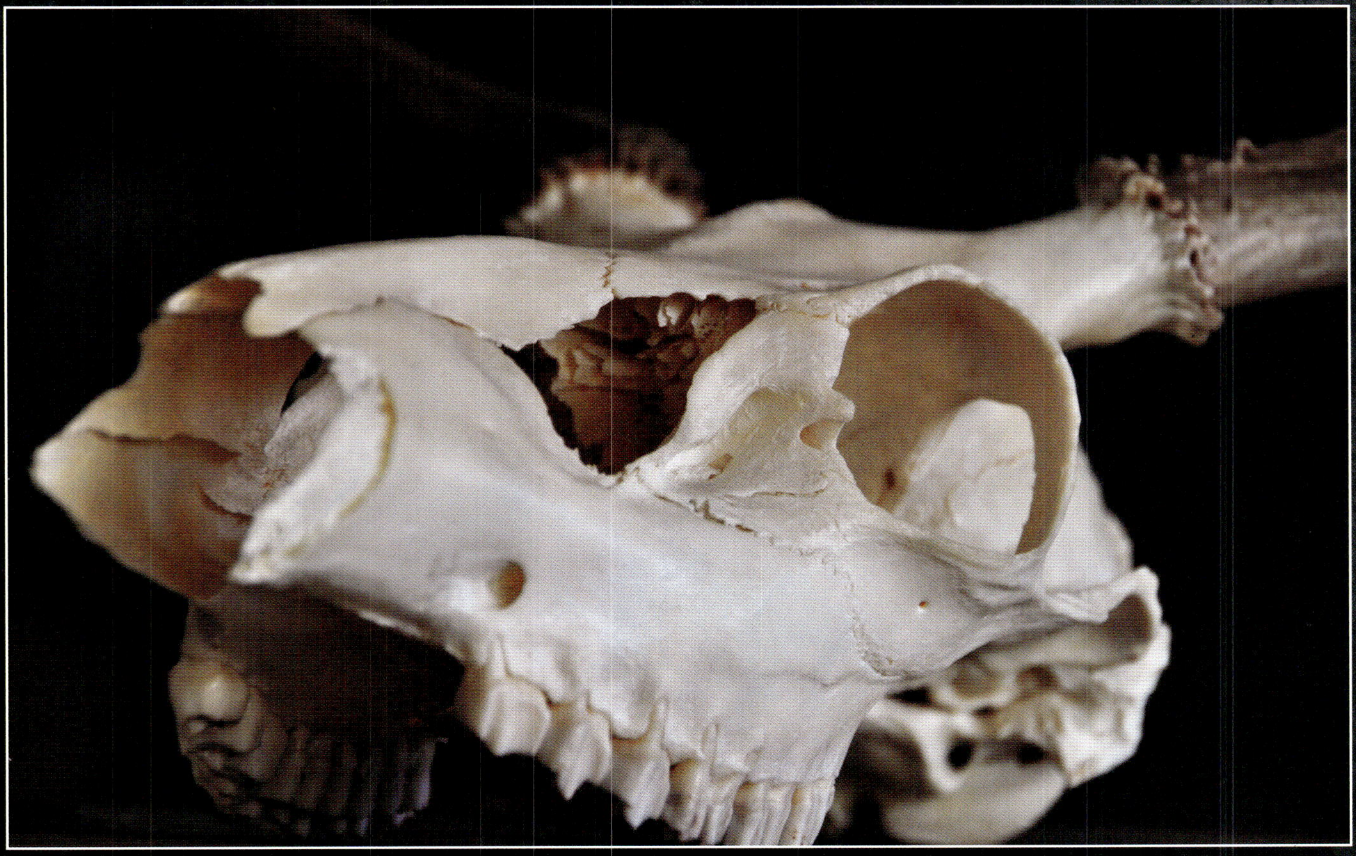

Mike DeVries

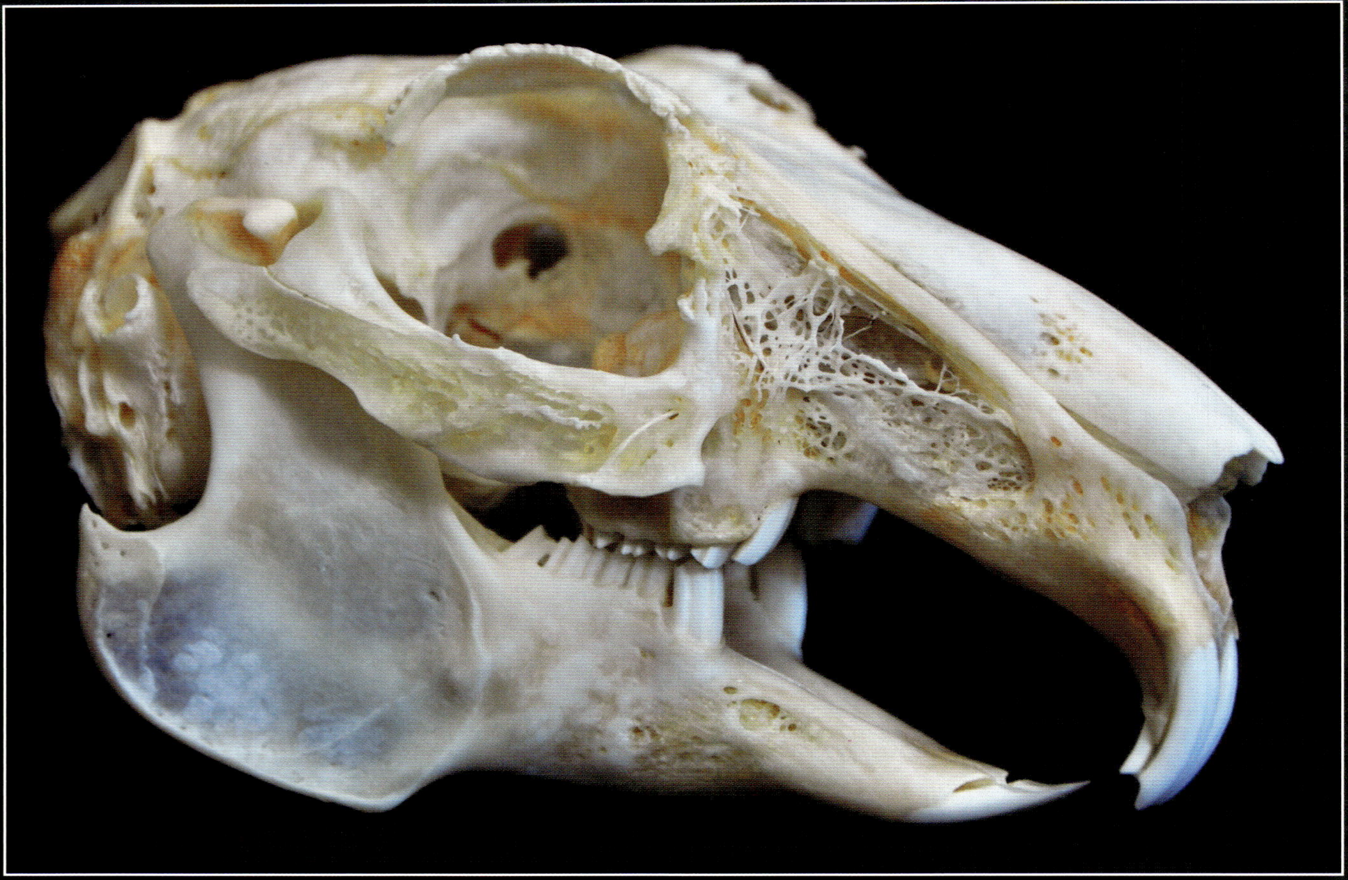

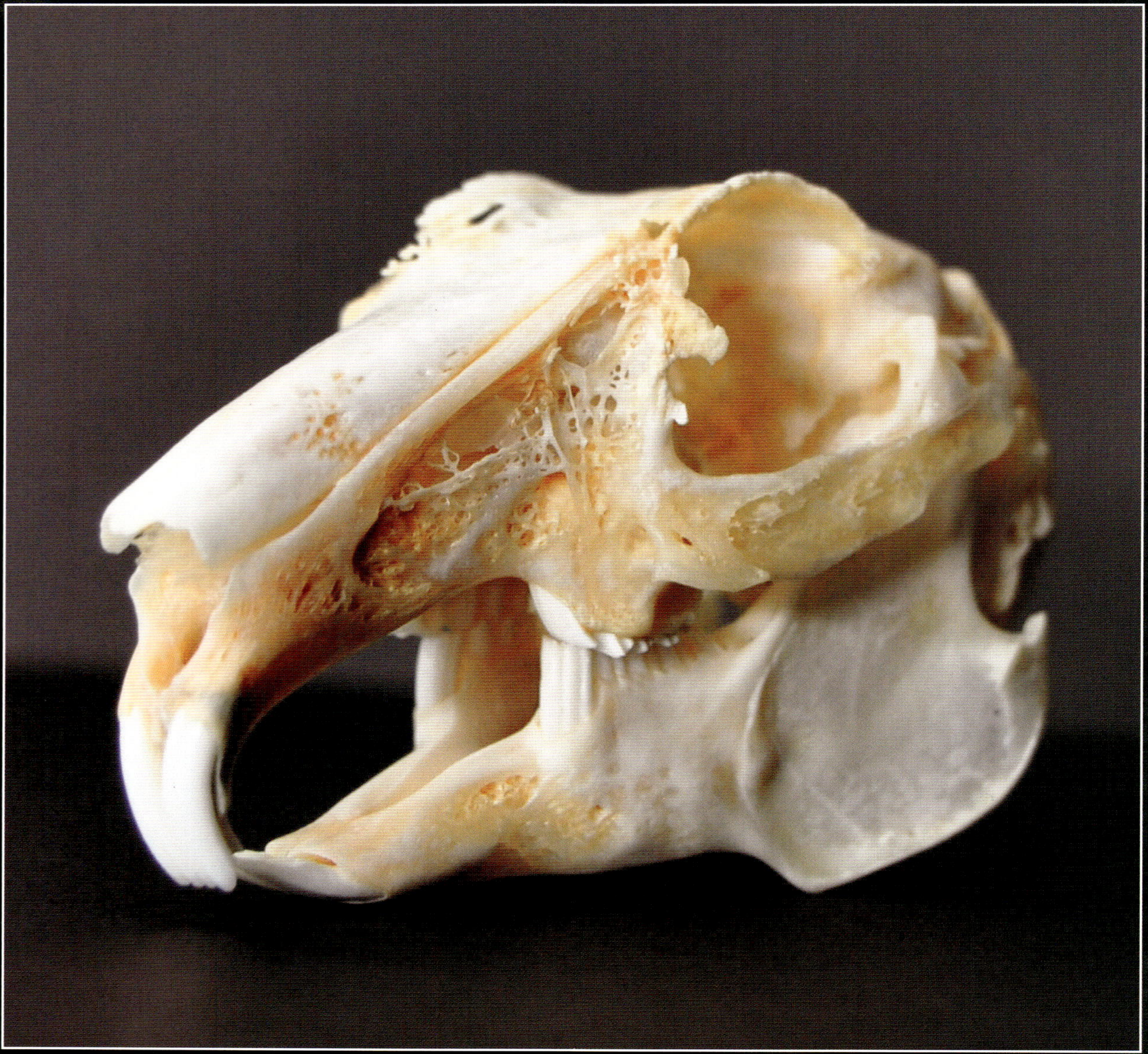

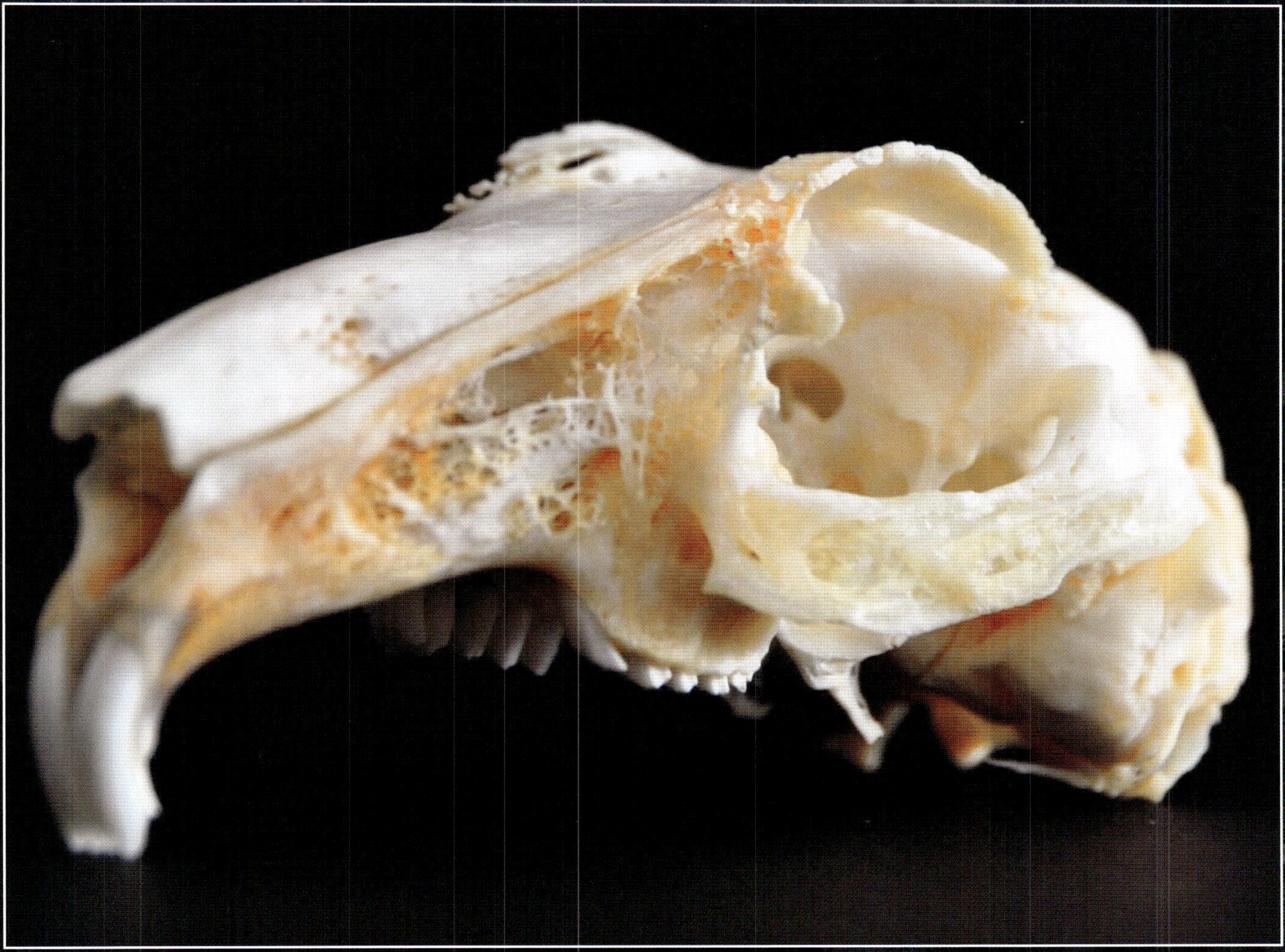

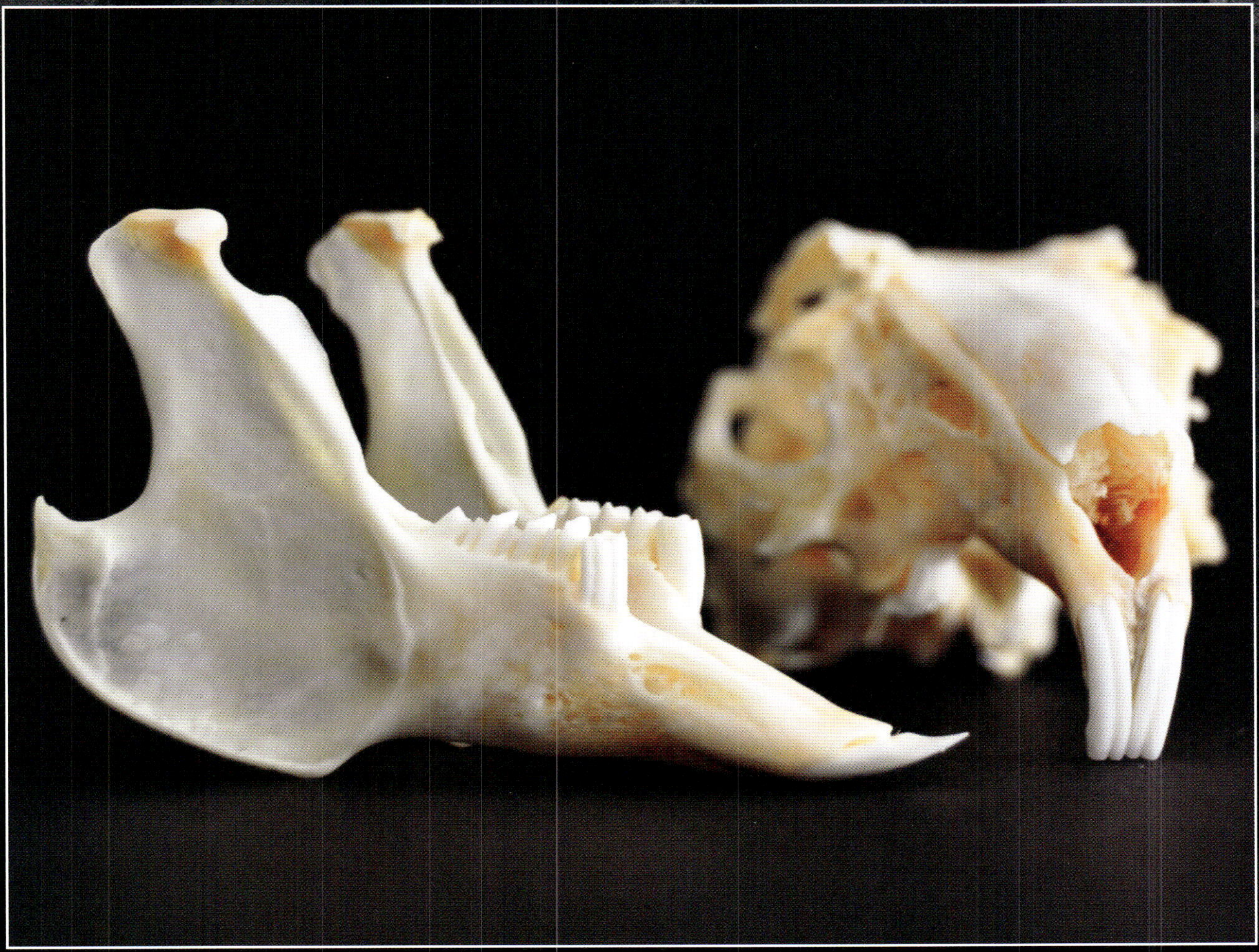

Mike DeVries

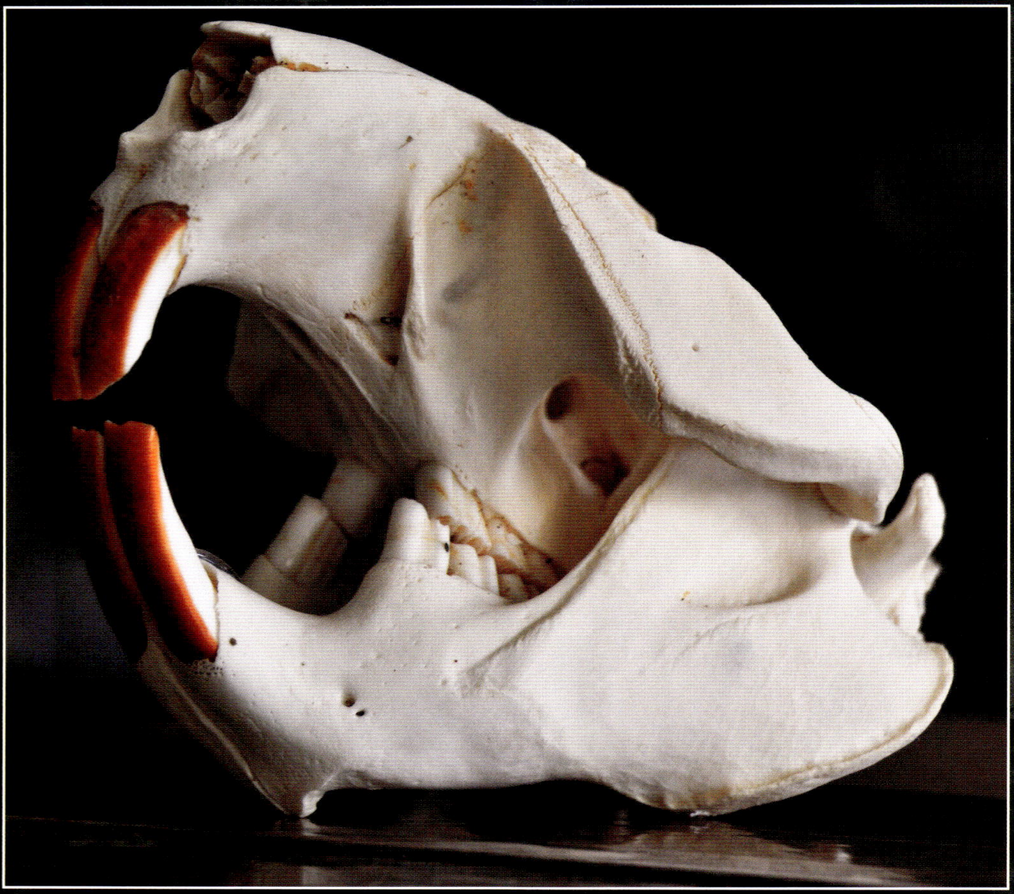

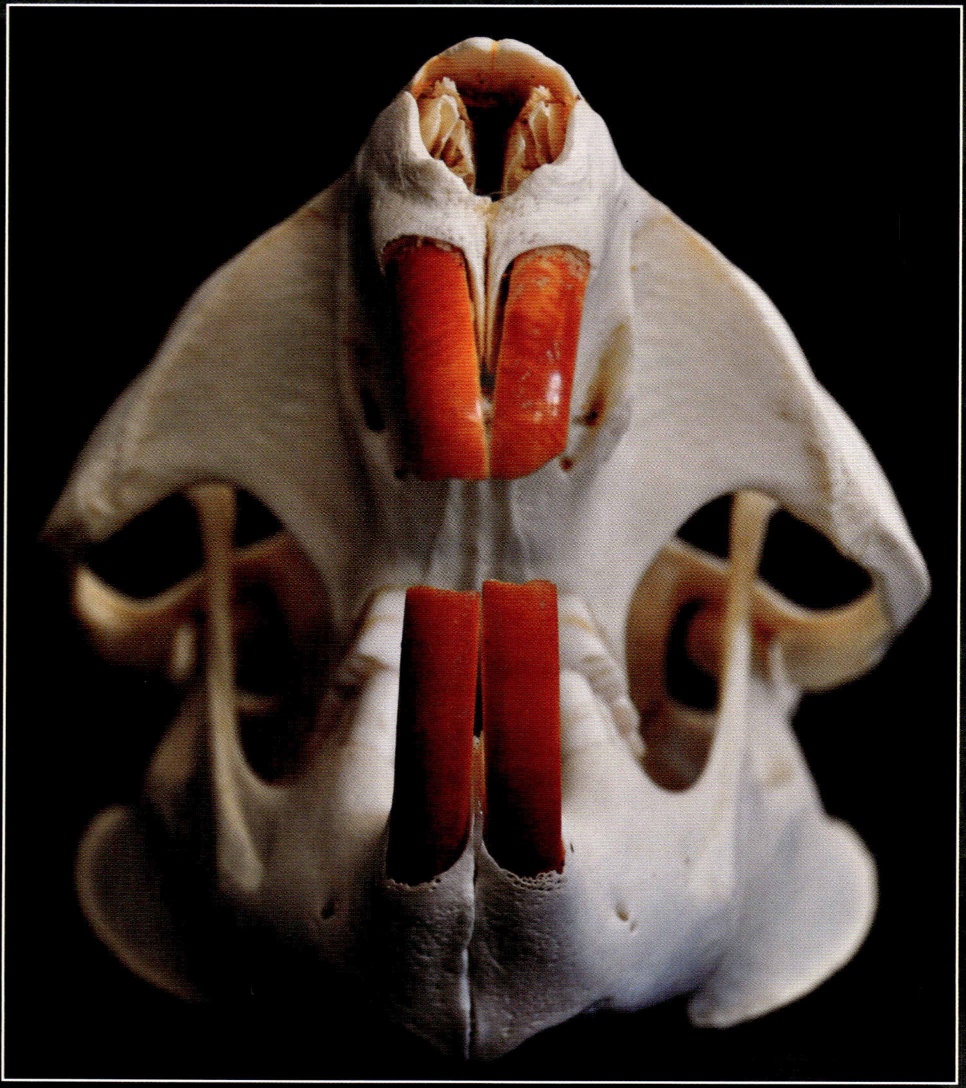

Mike DeVries

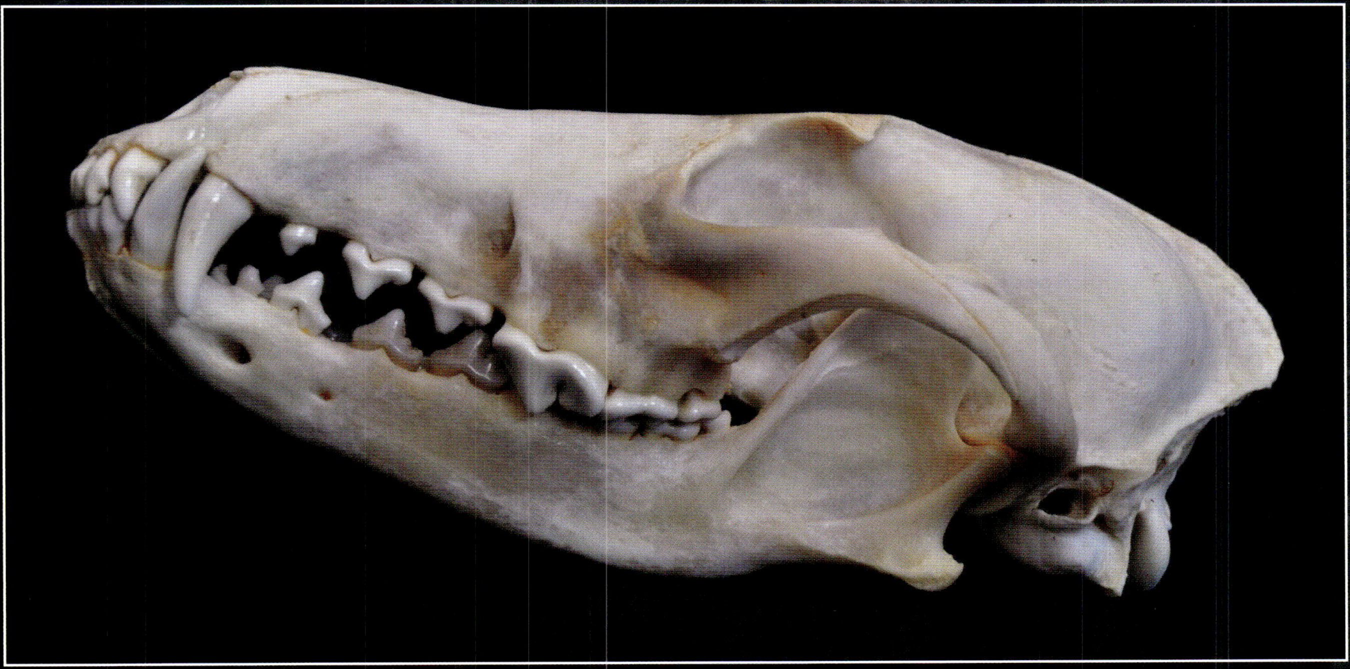

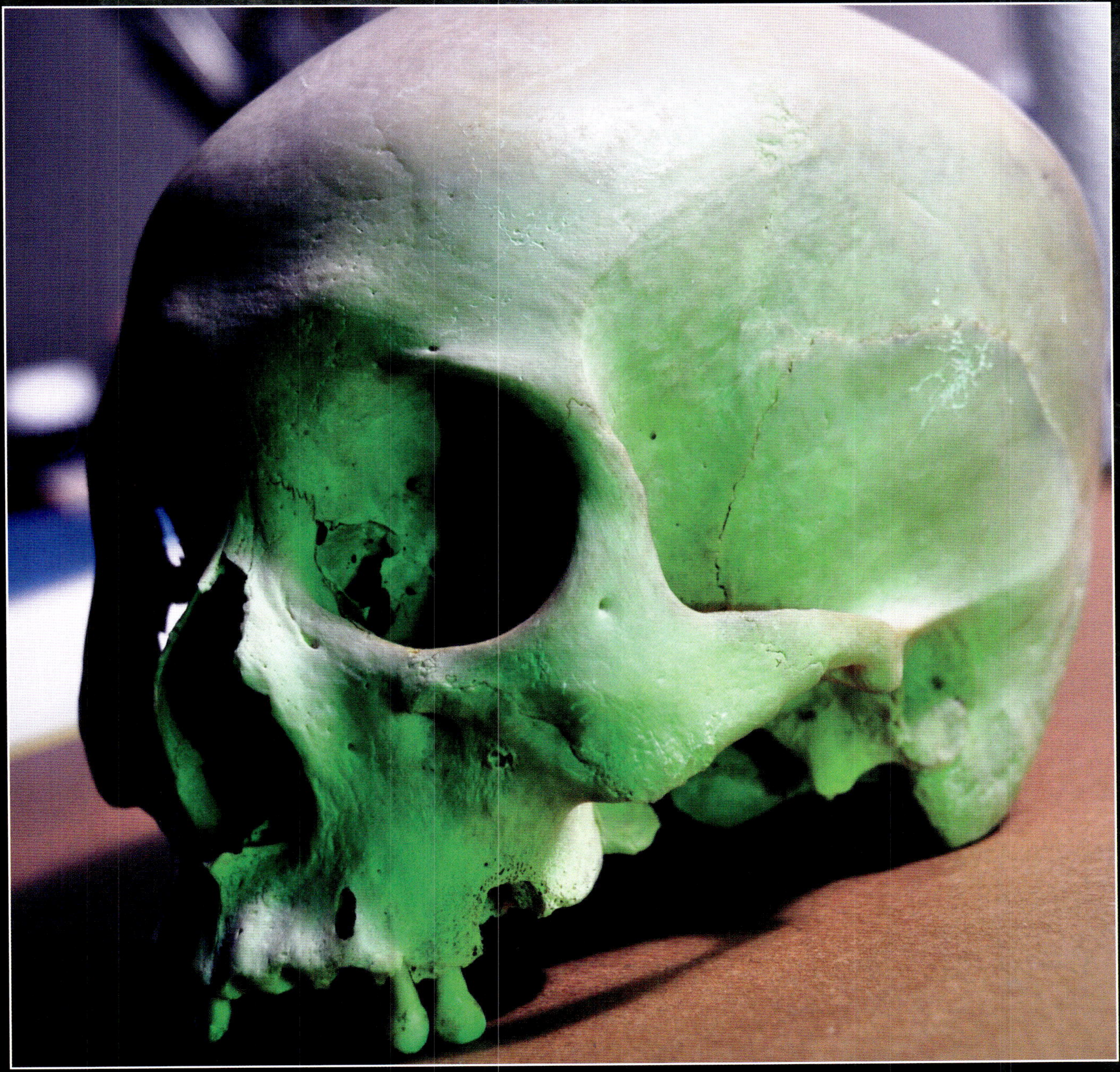

Mike DeVries

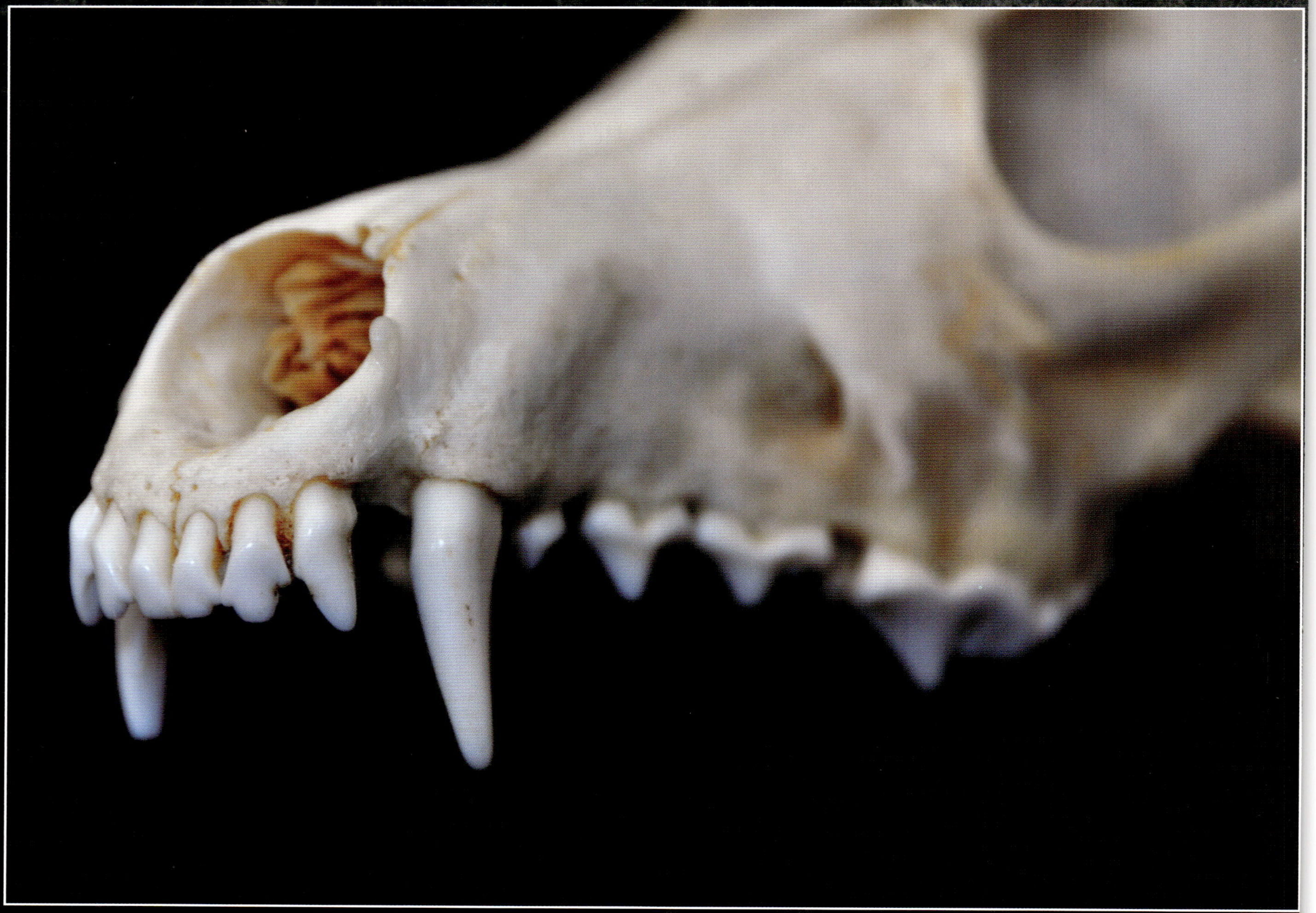

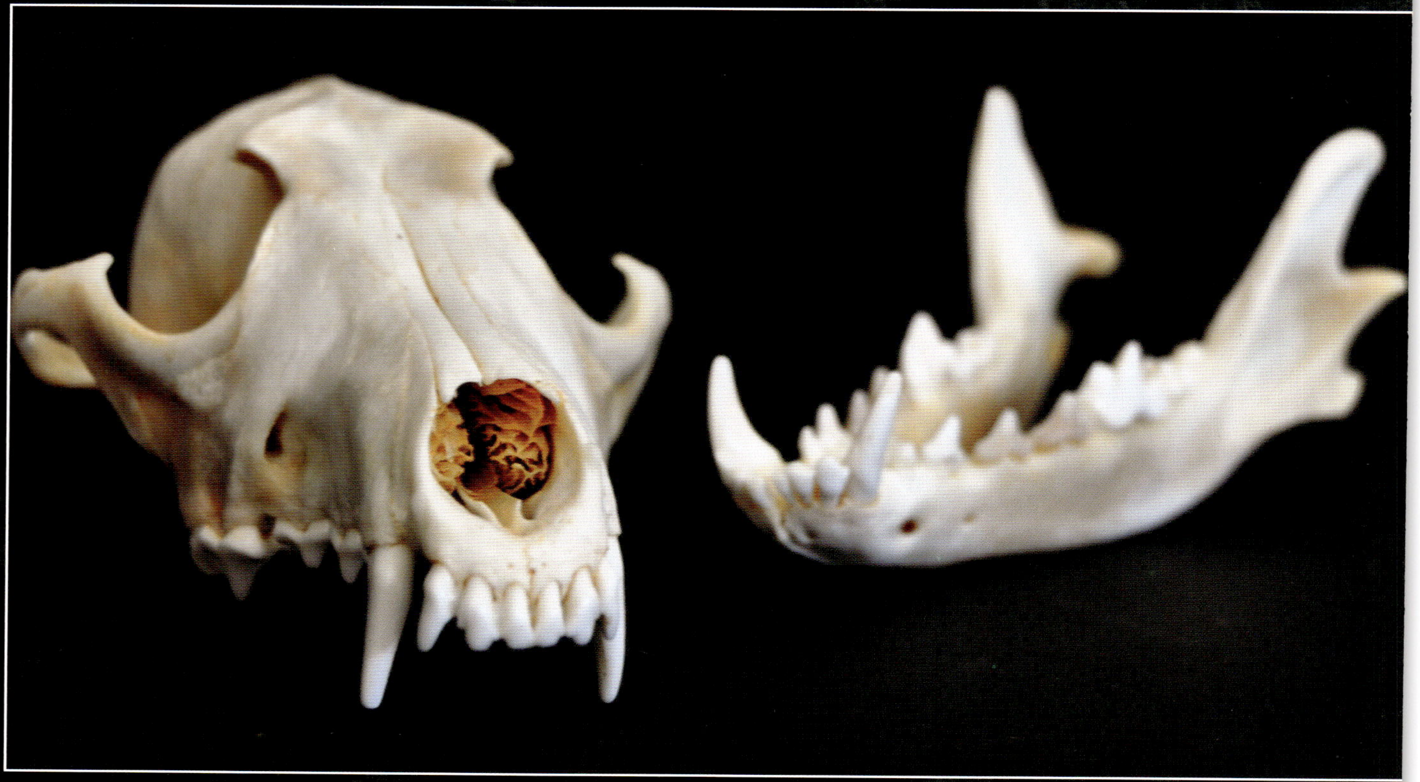

Mike DeVries

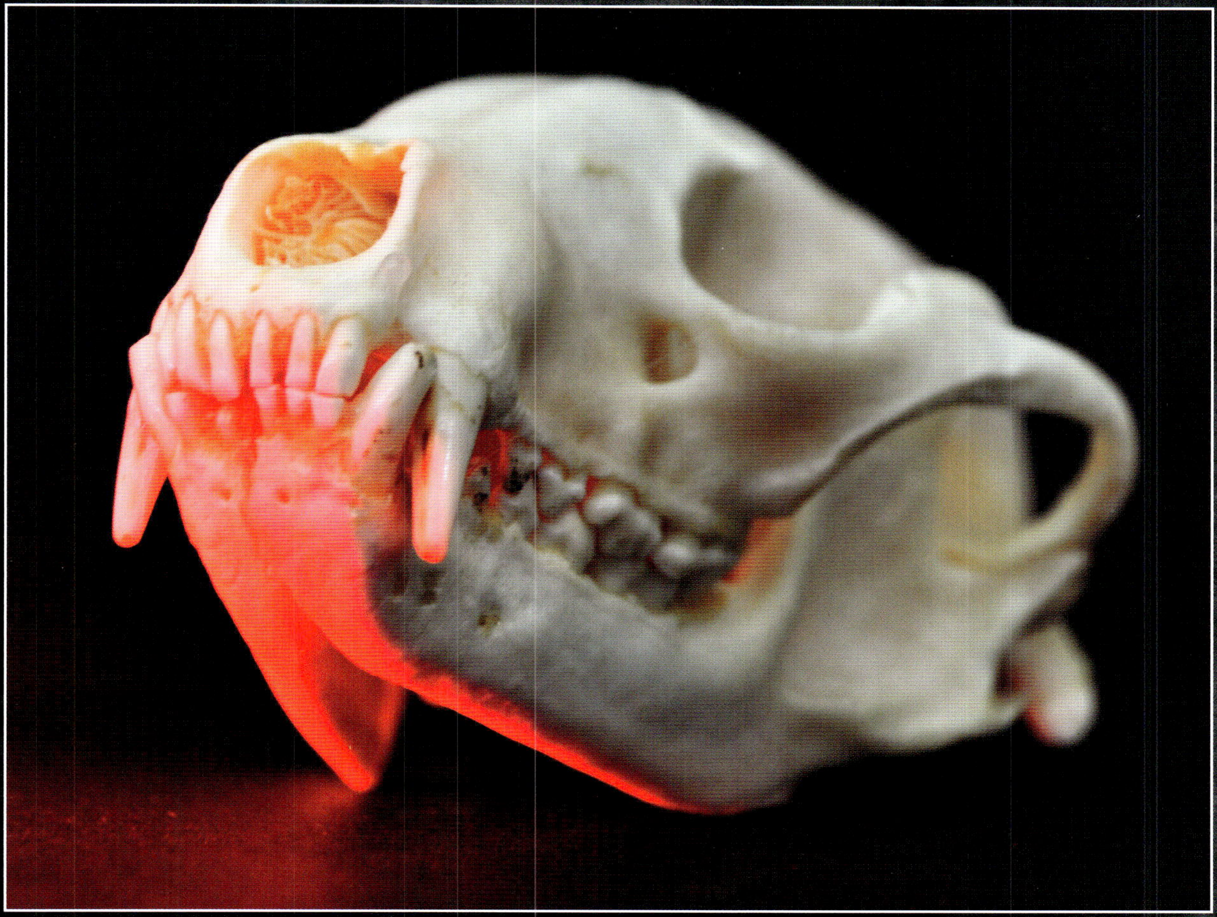

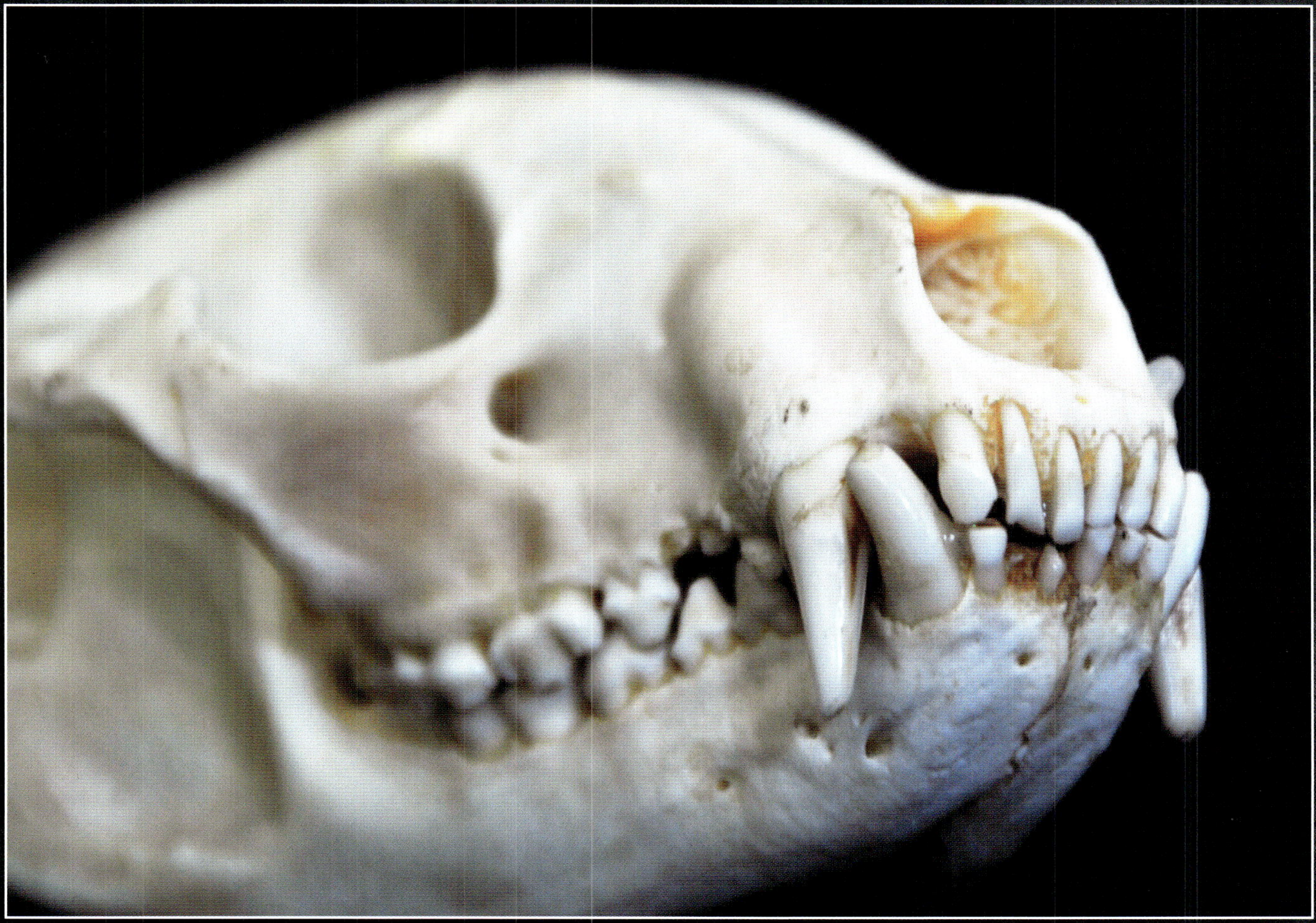

Mike DeVries

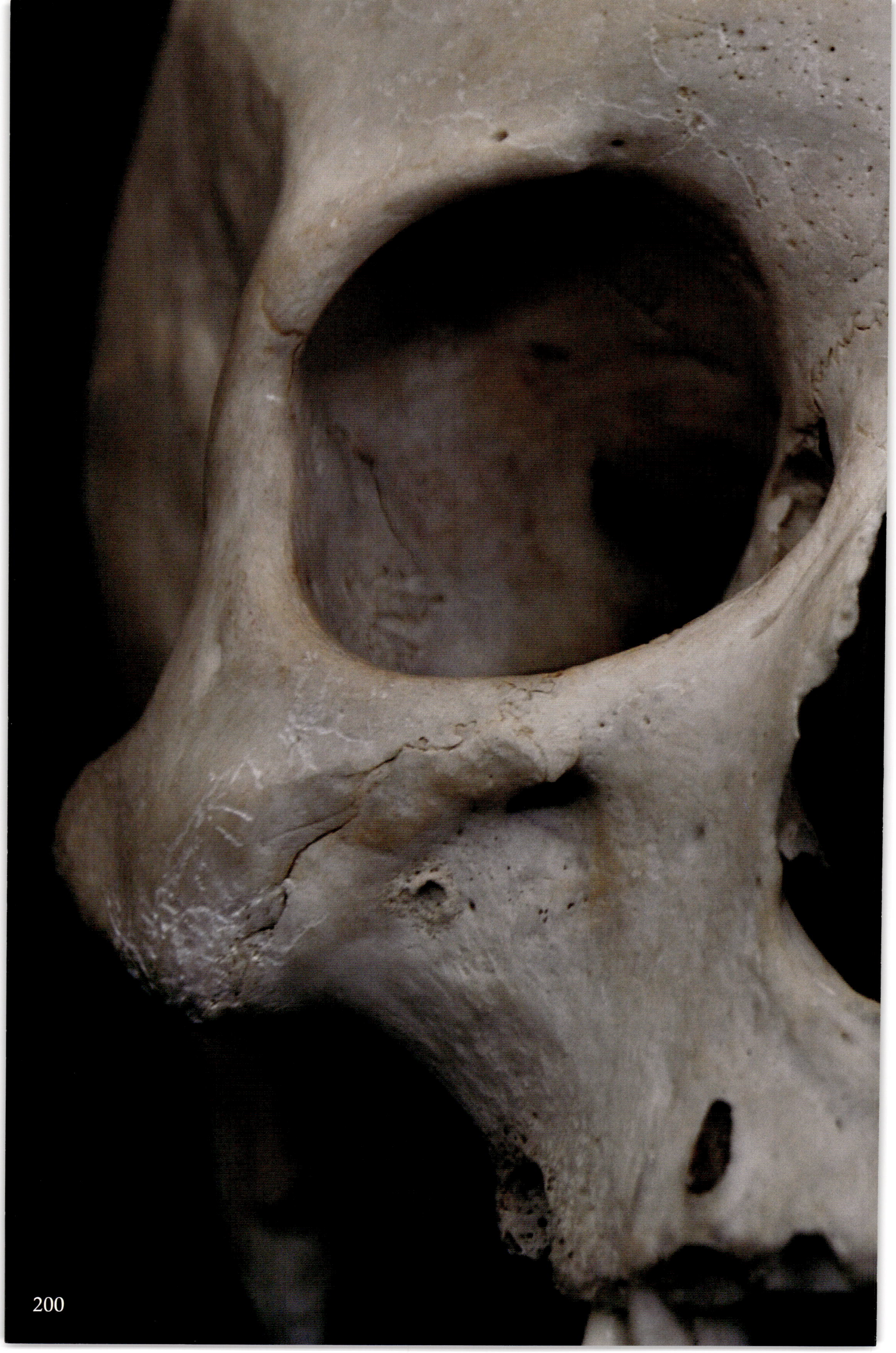

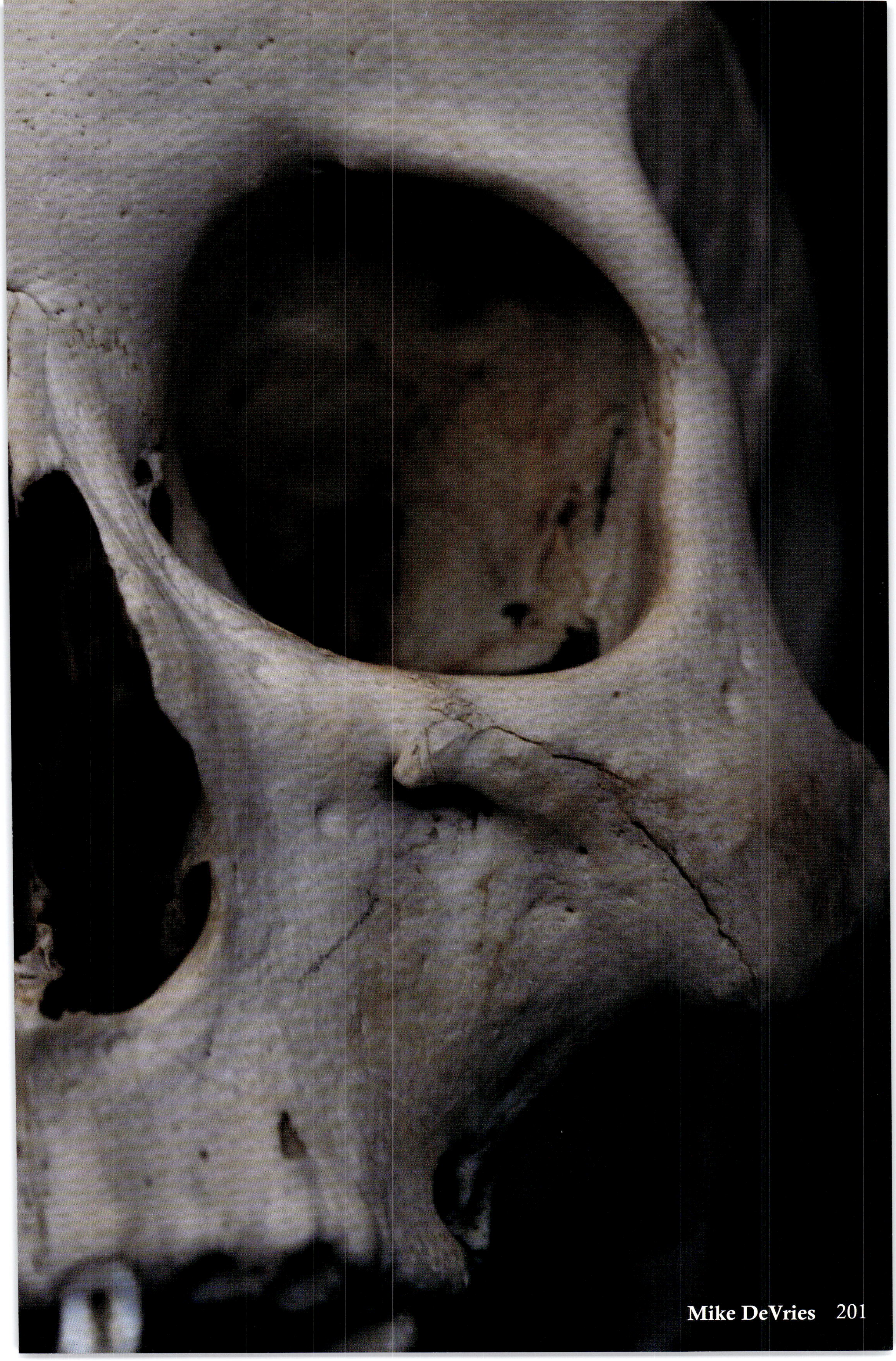
Mike DeVries

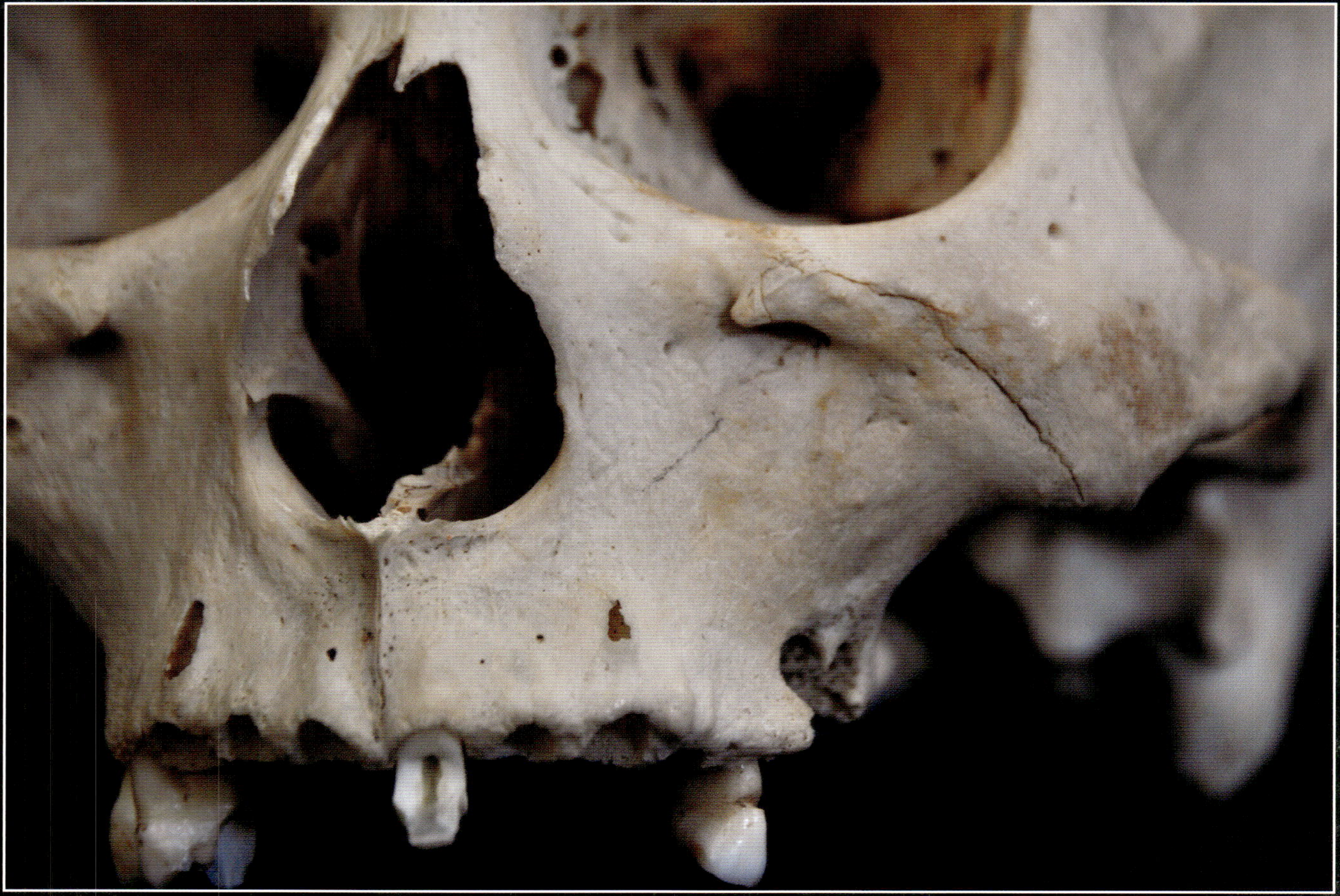

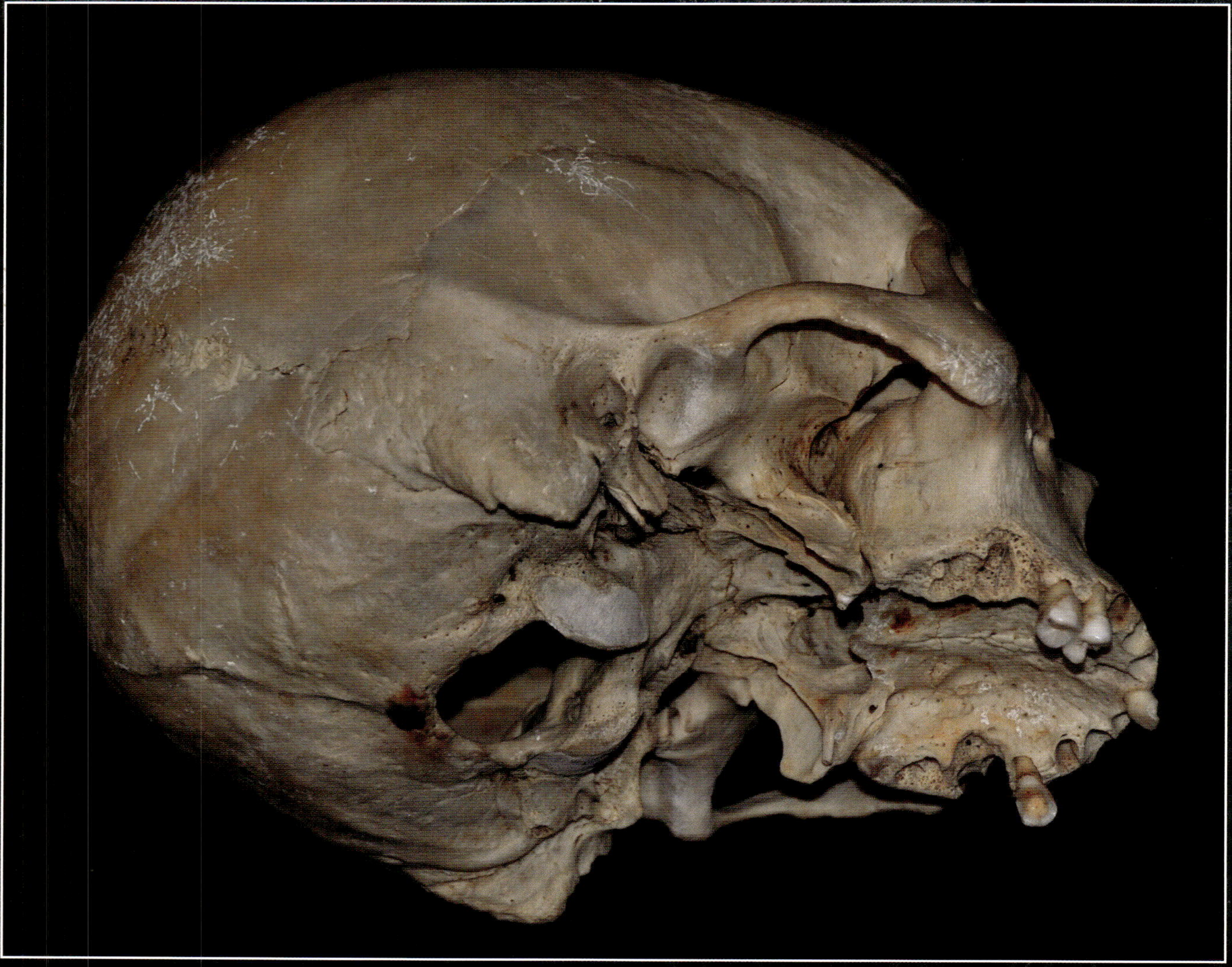

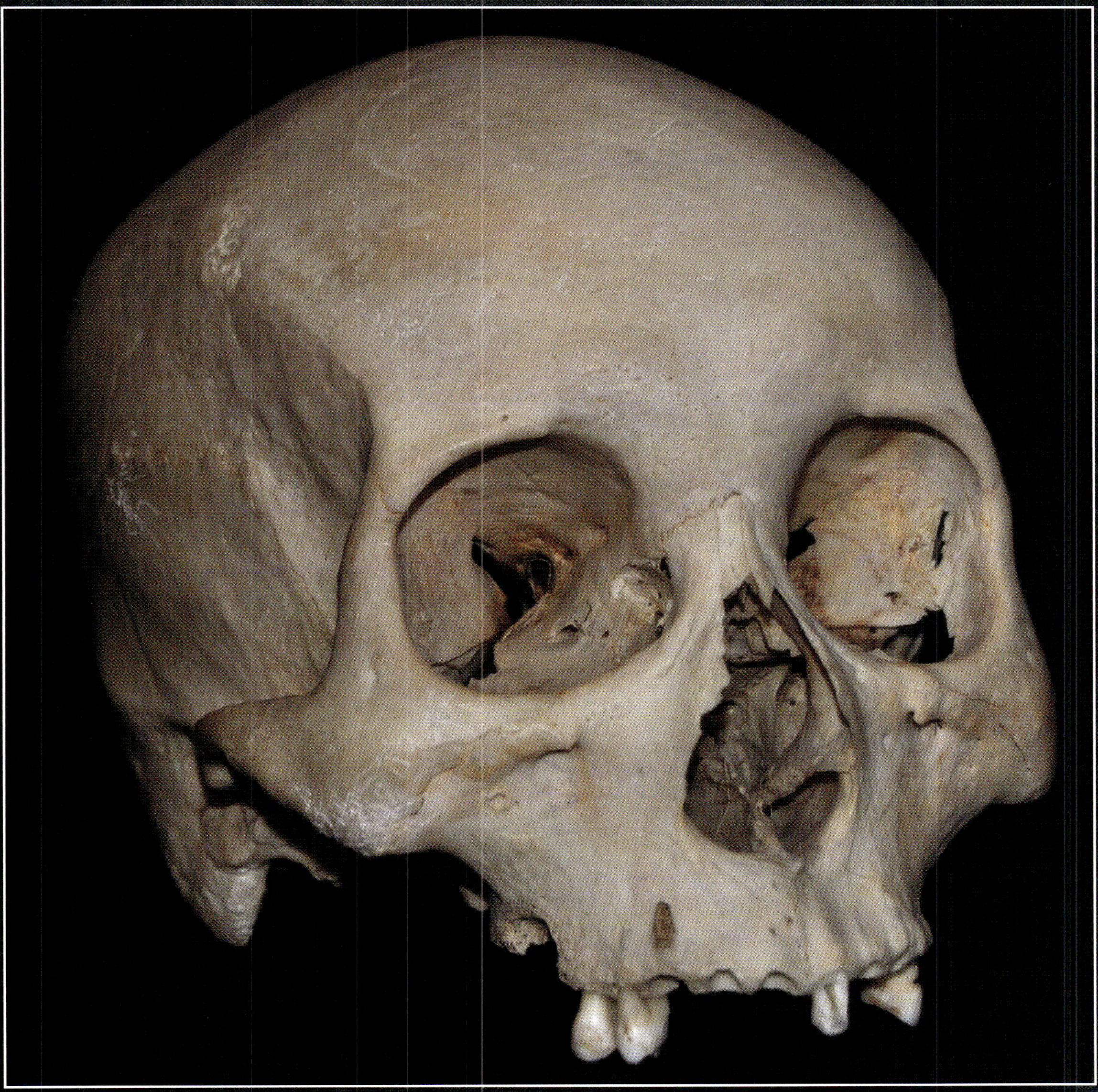

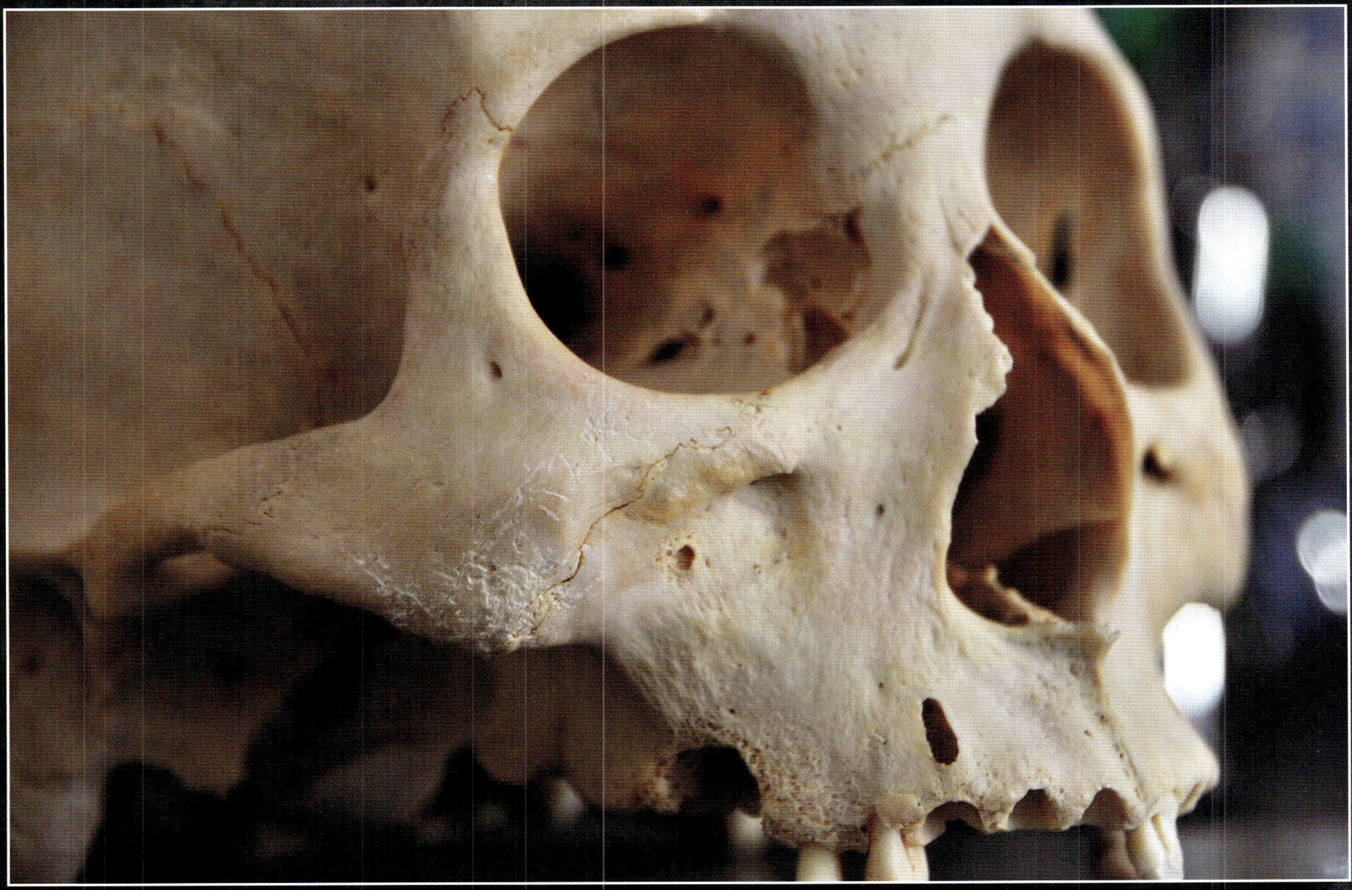

Mike DeVries

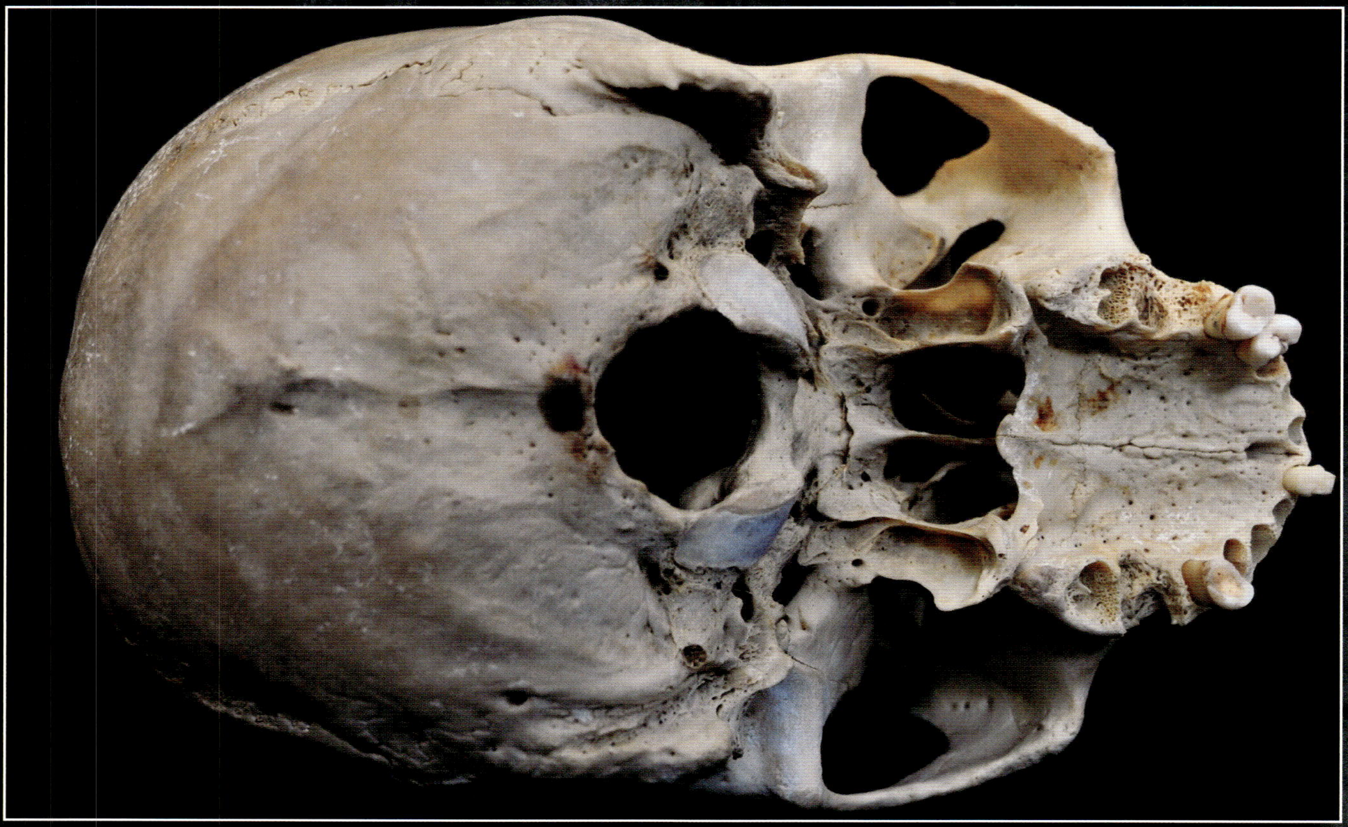

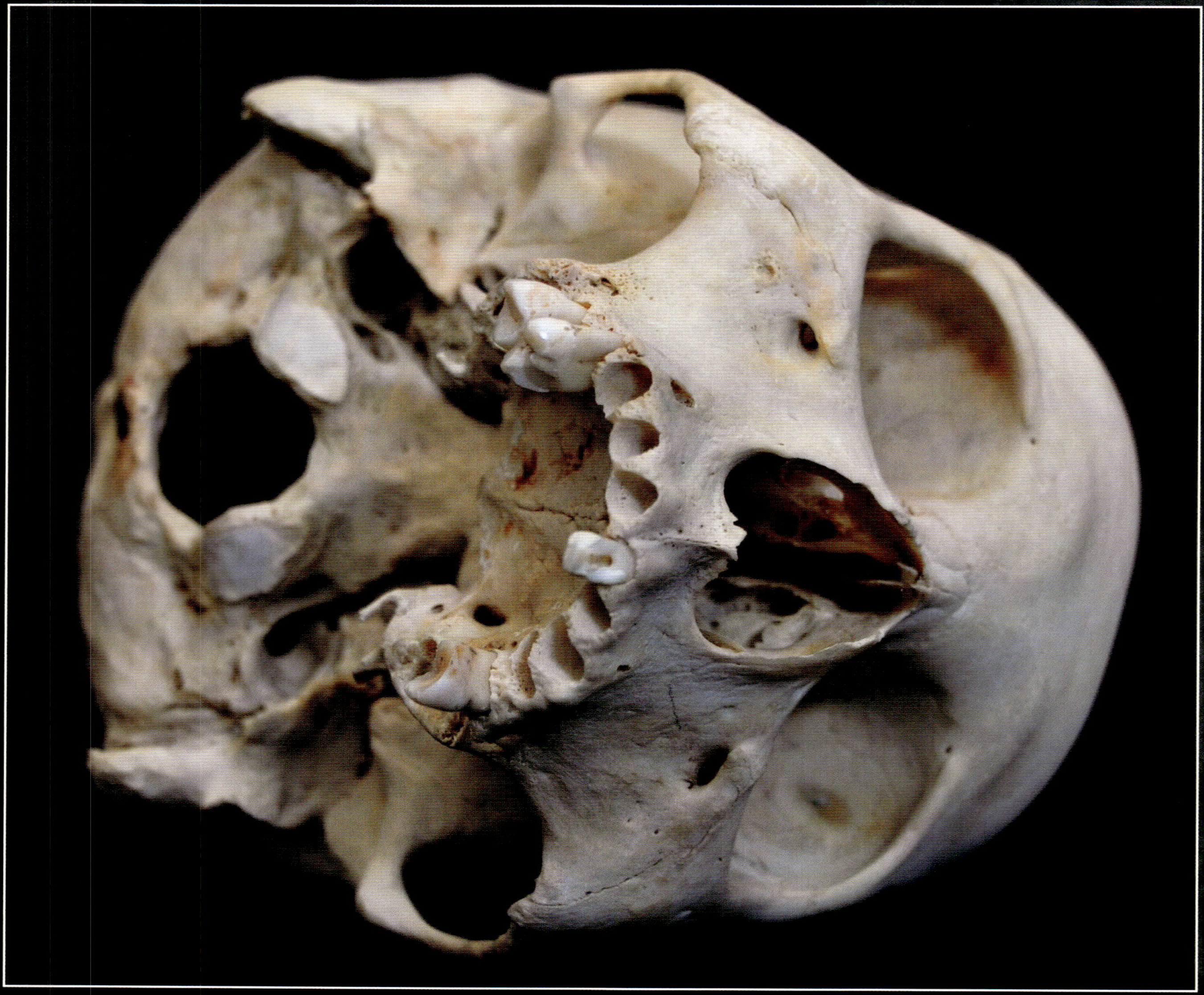

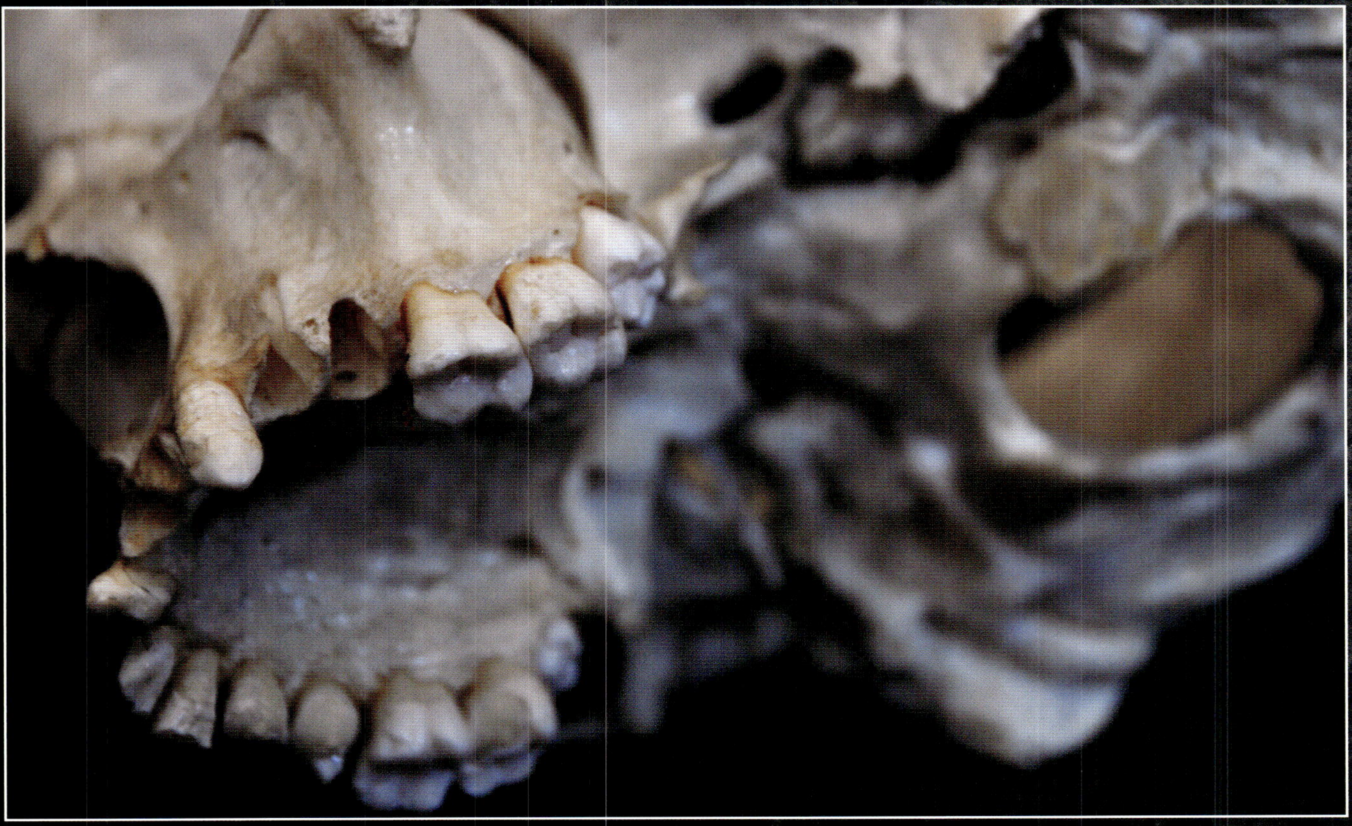

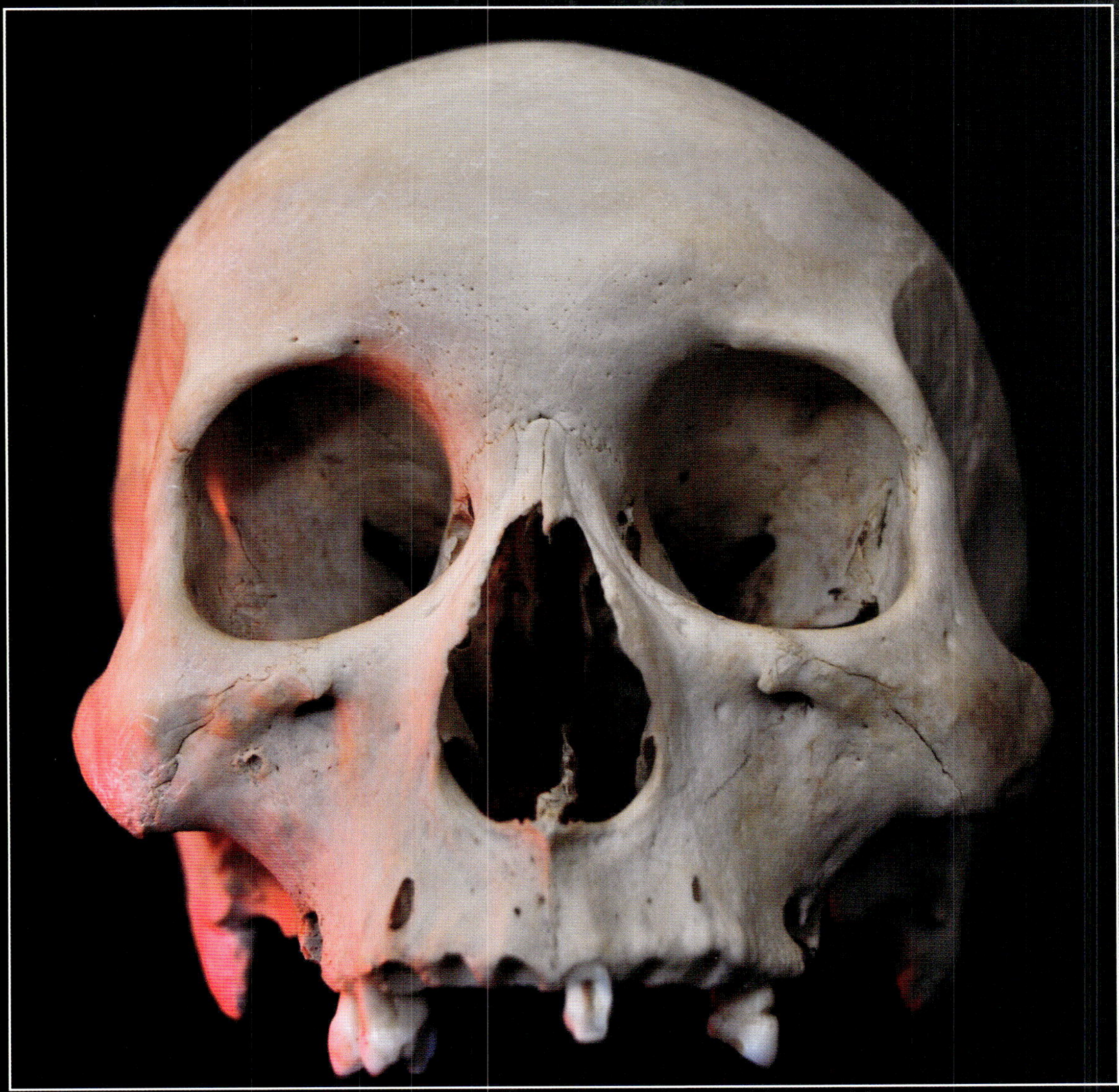

Mike DeVries

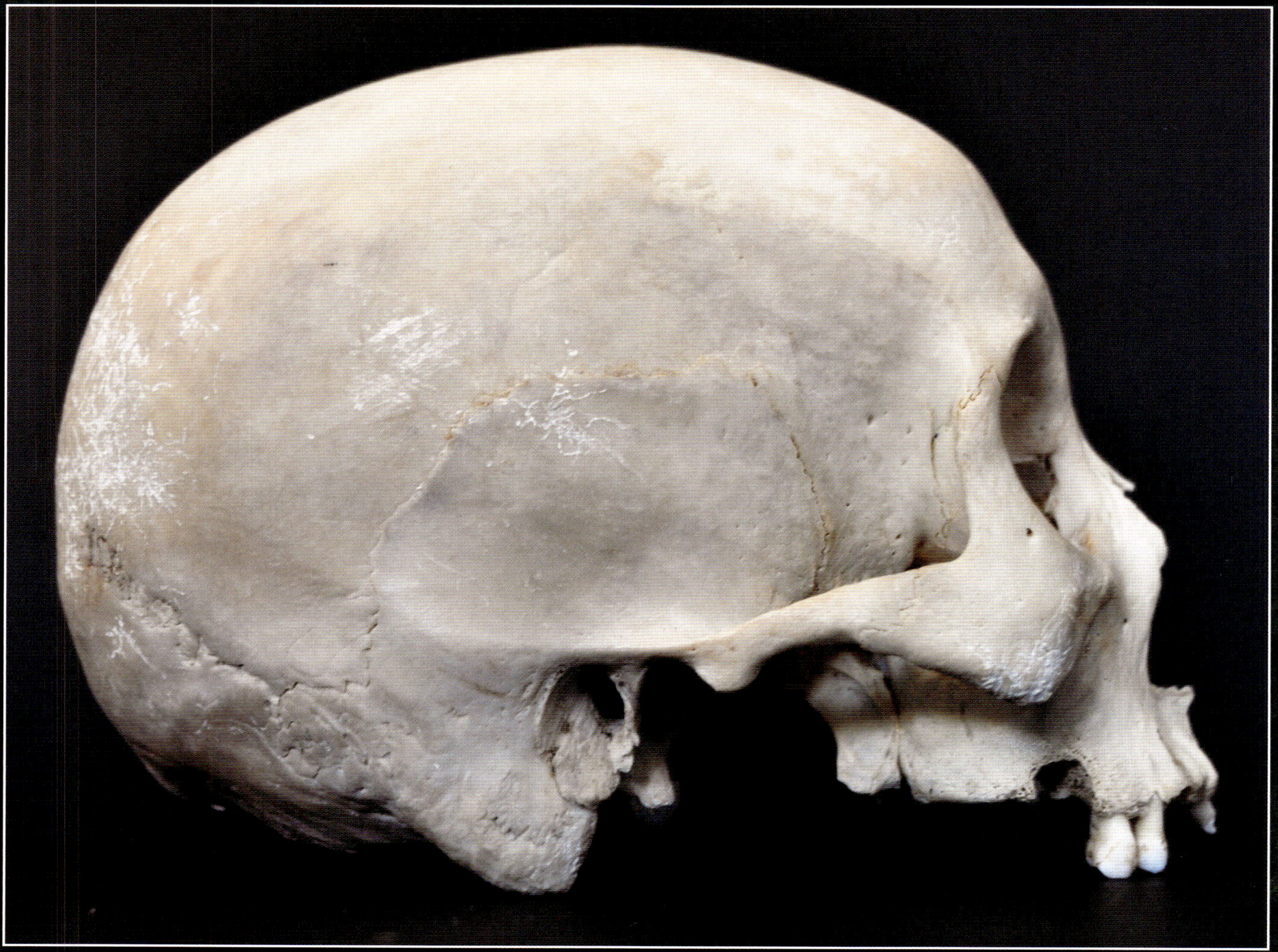

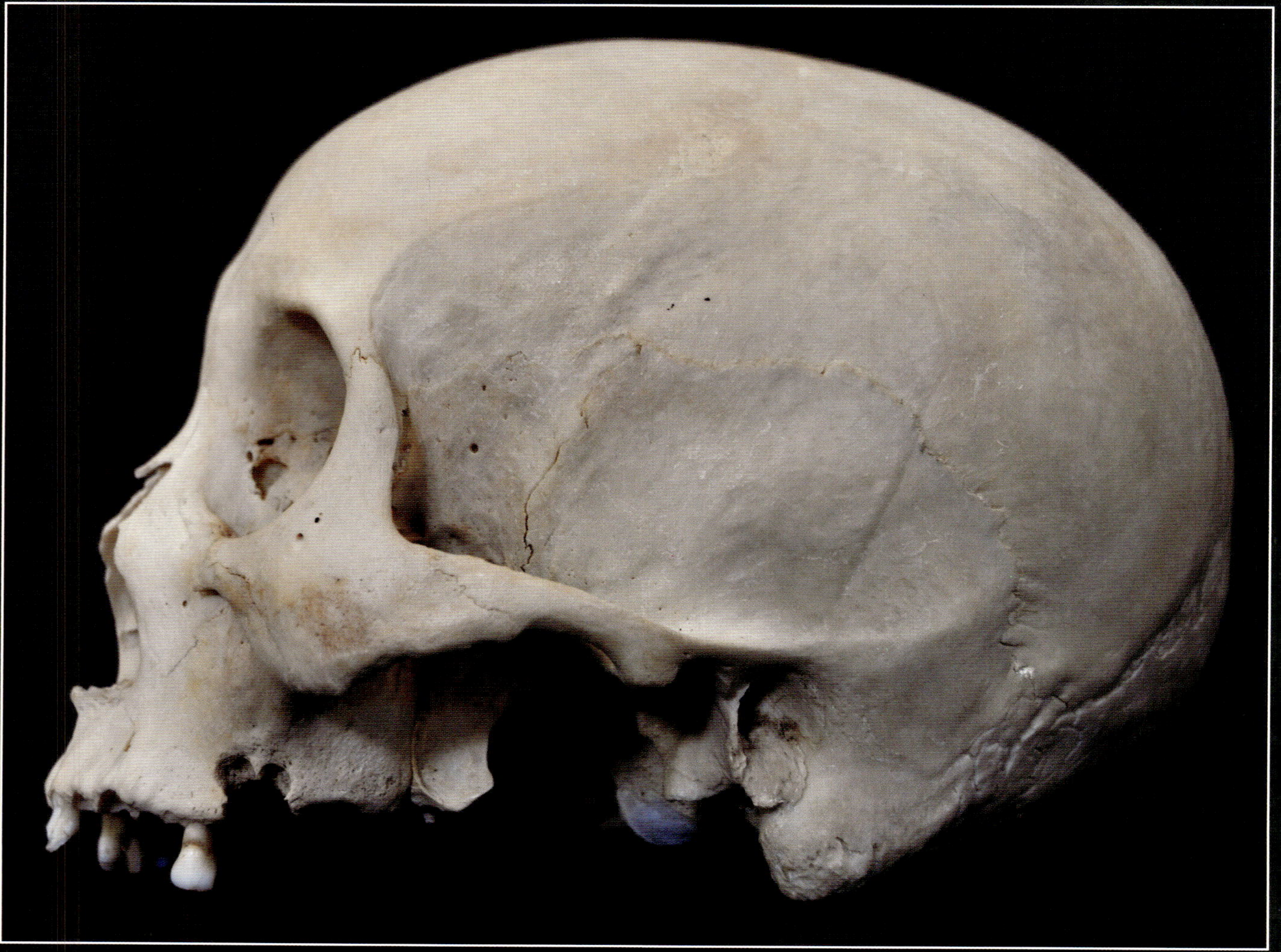

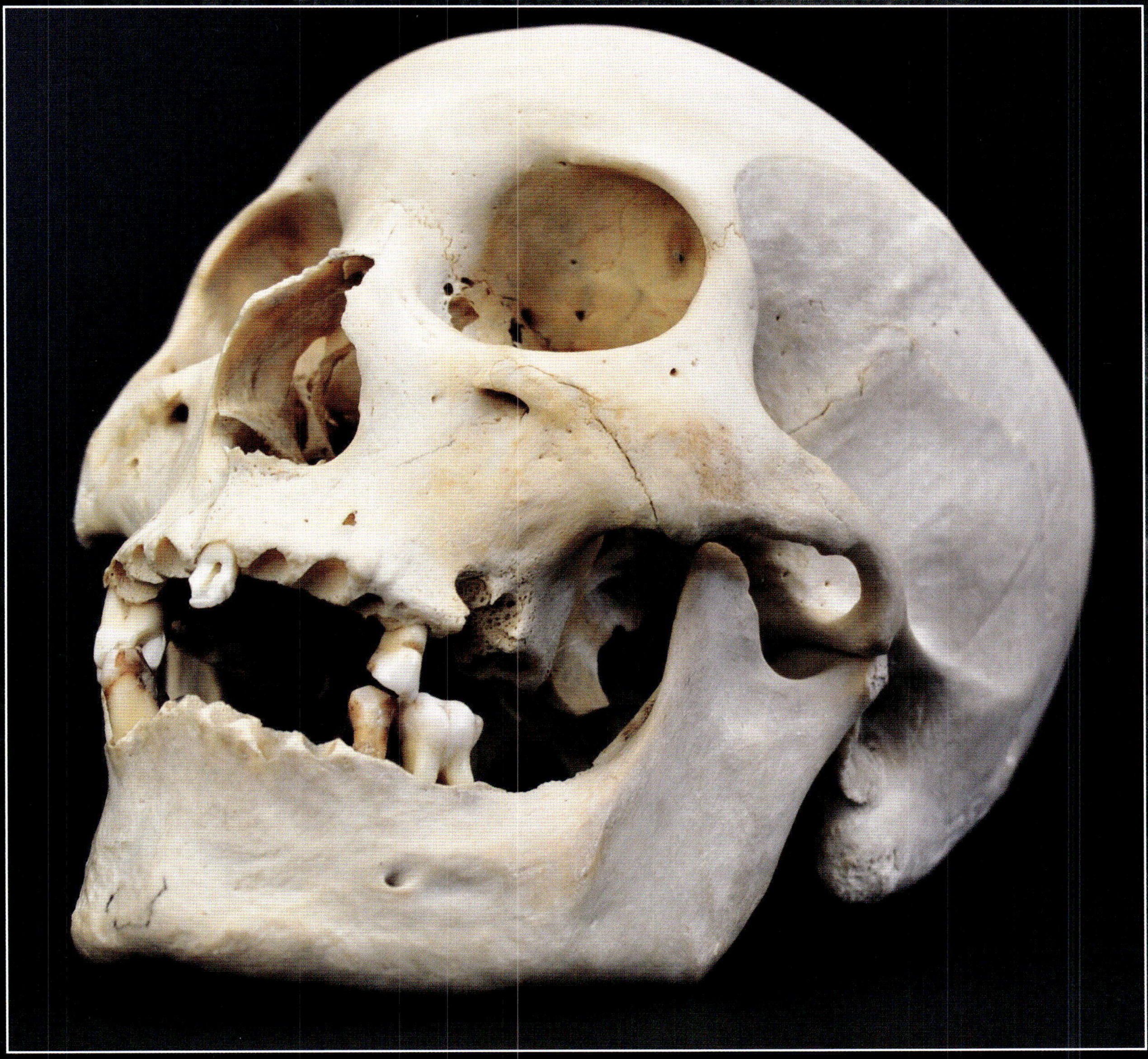

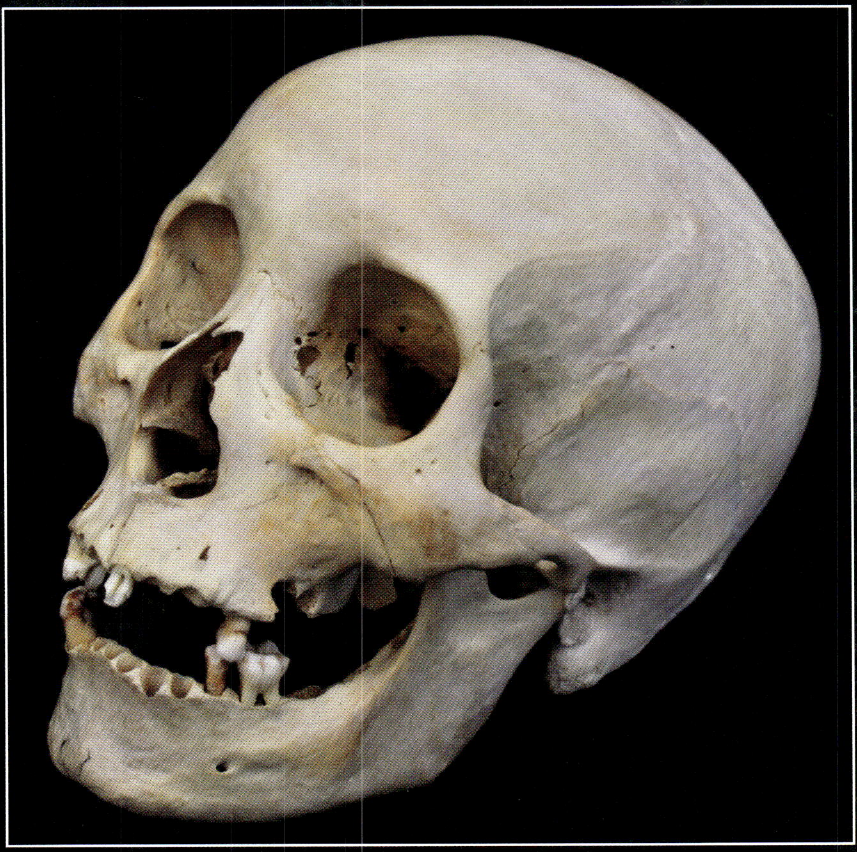

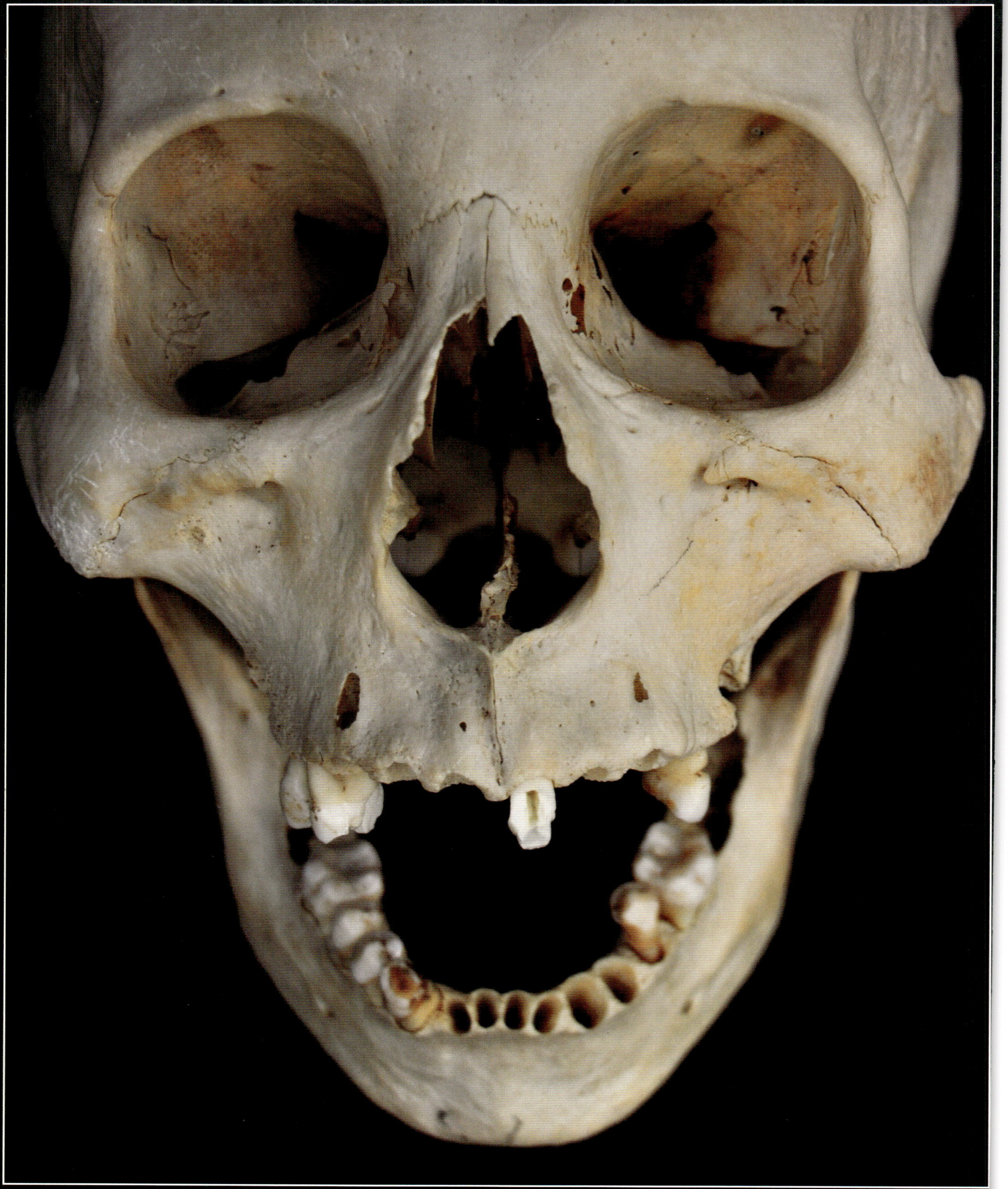

Mike DeVries

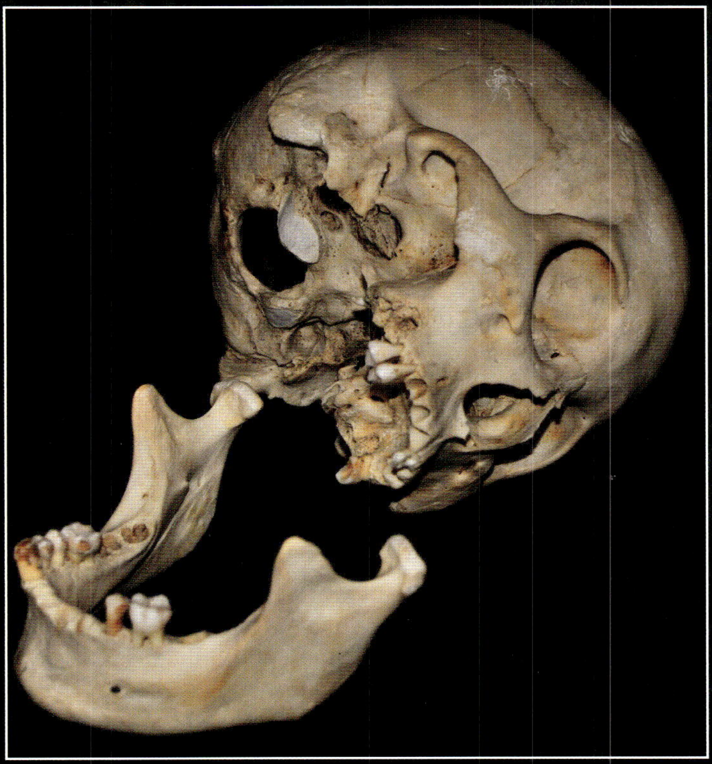

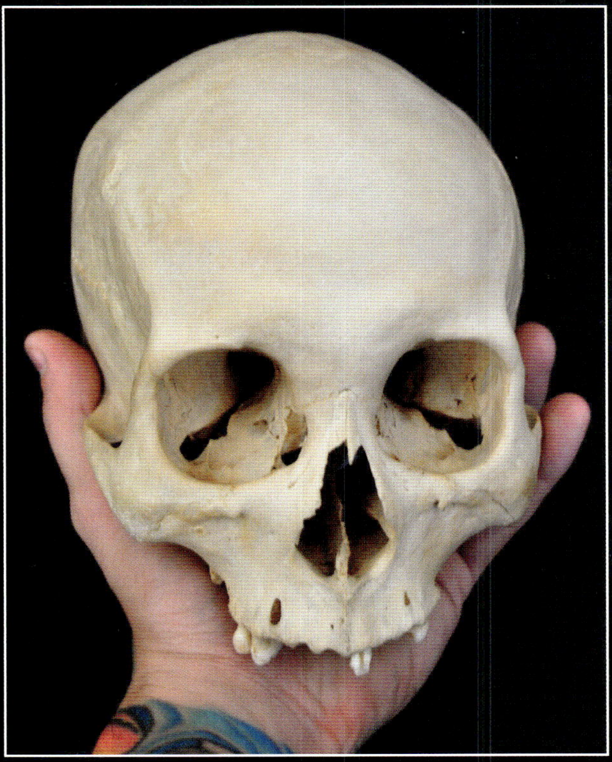

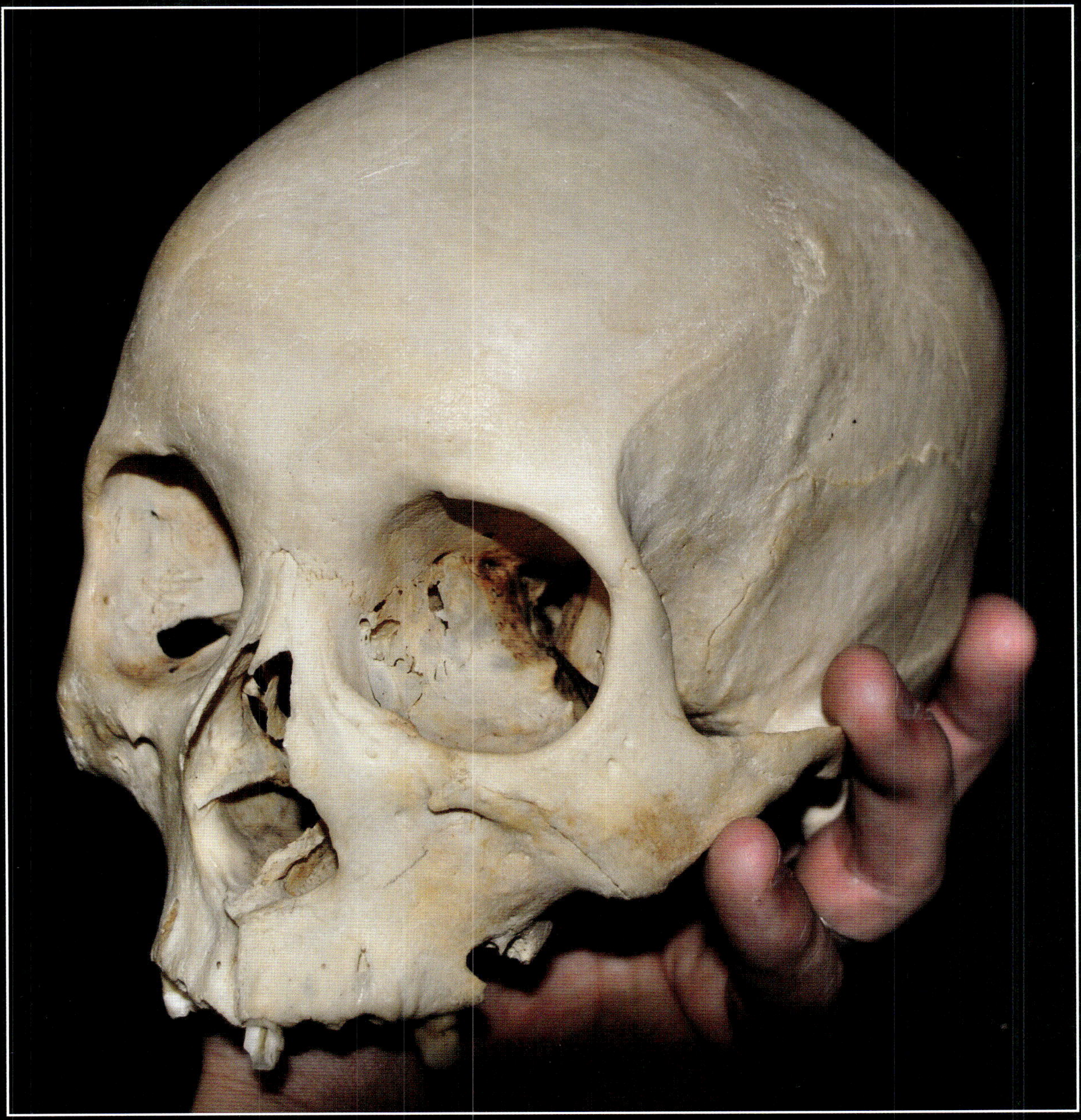

Mike DeVries

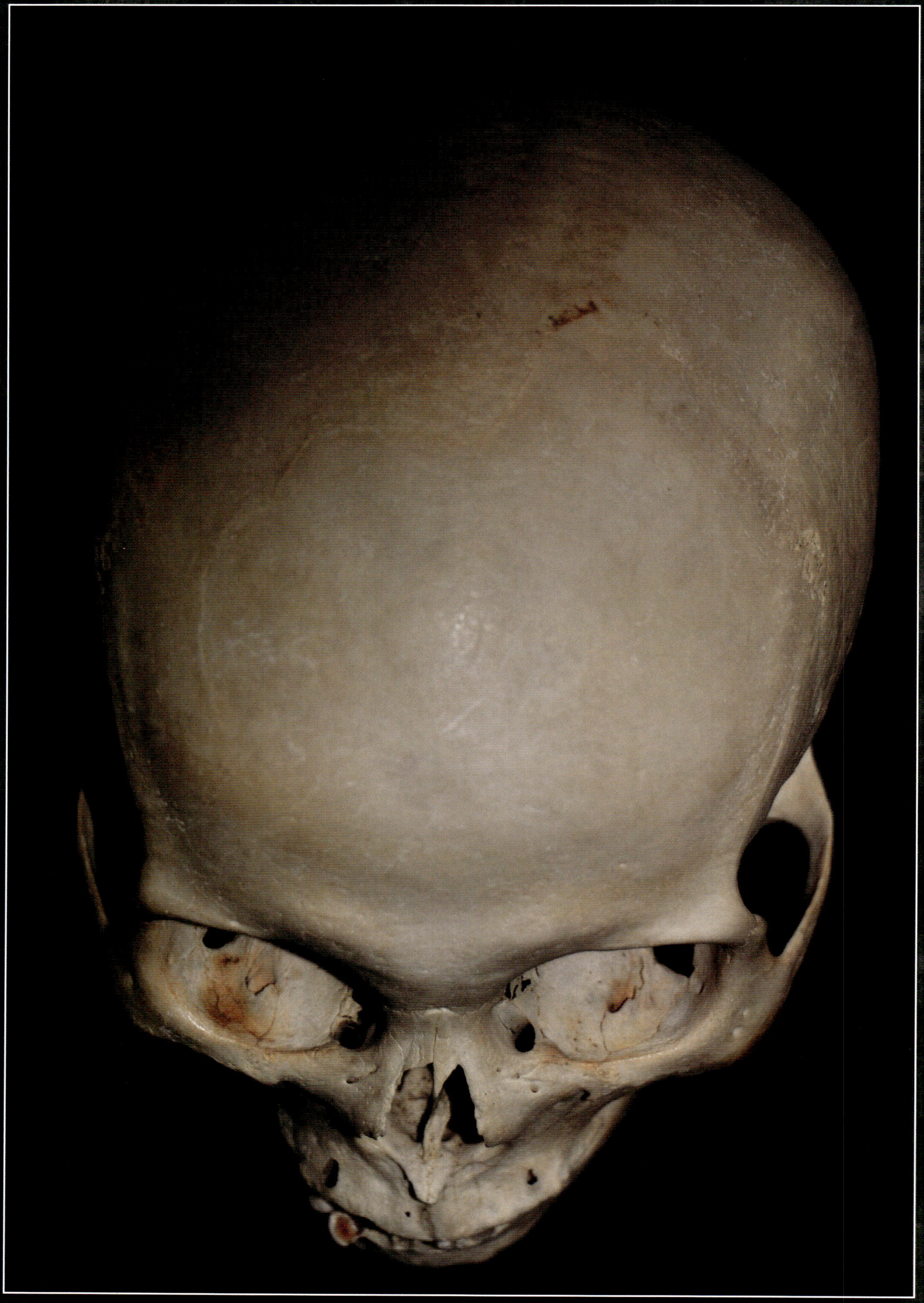

Mike DeVries

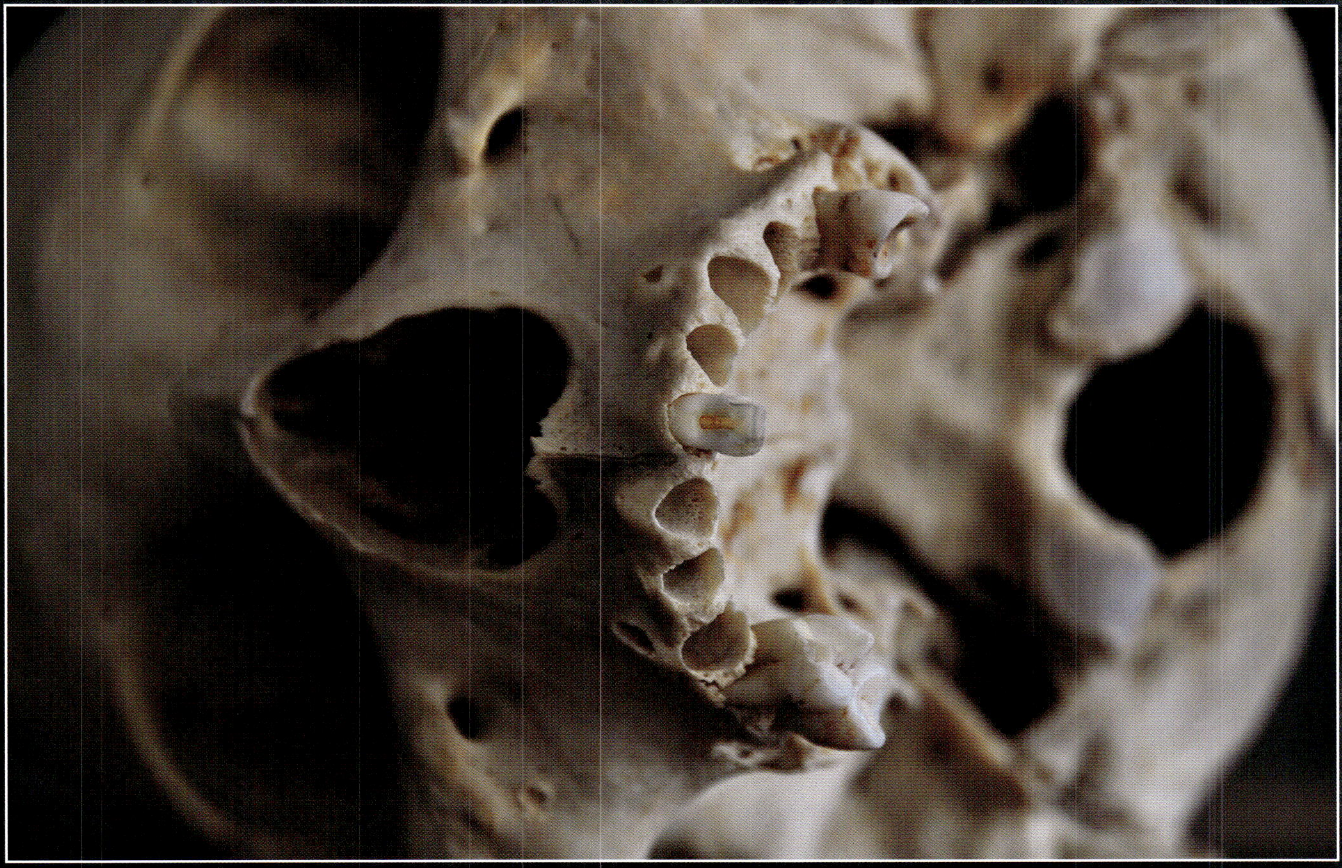

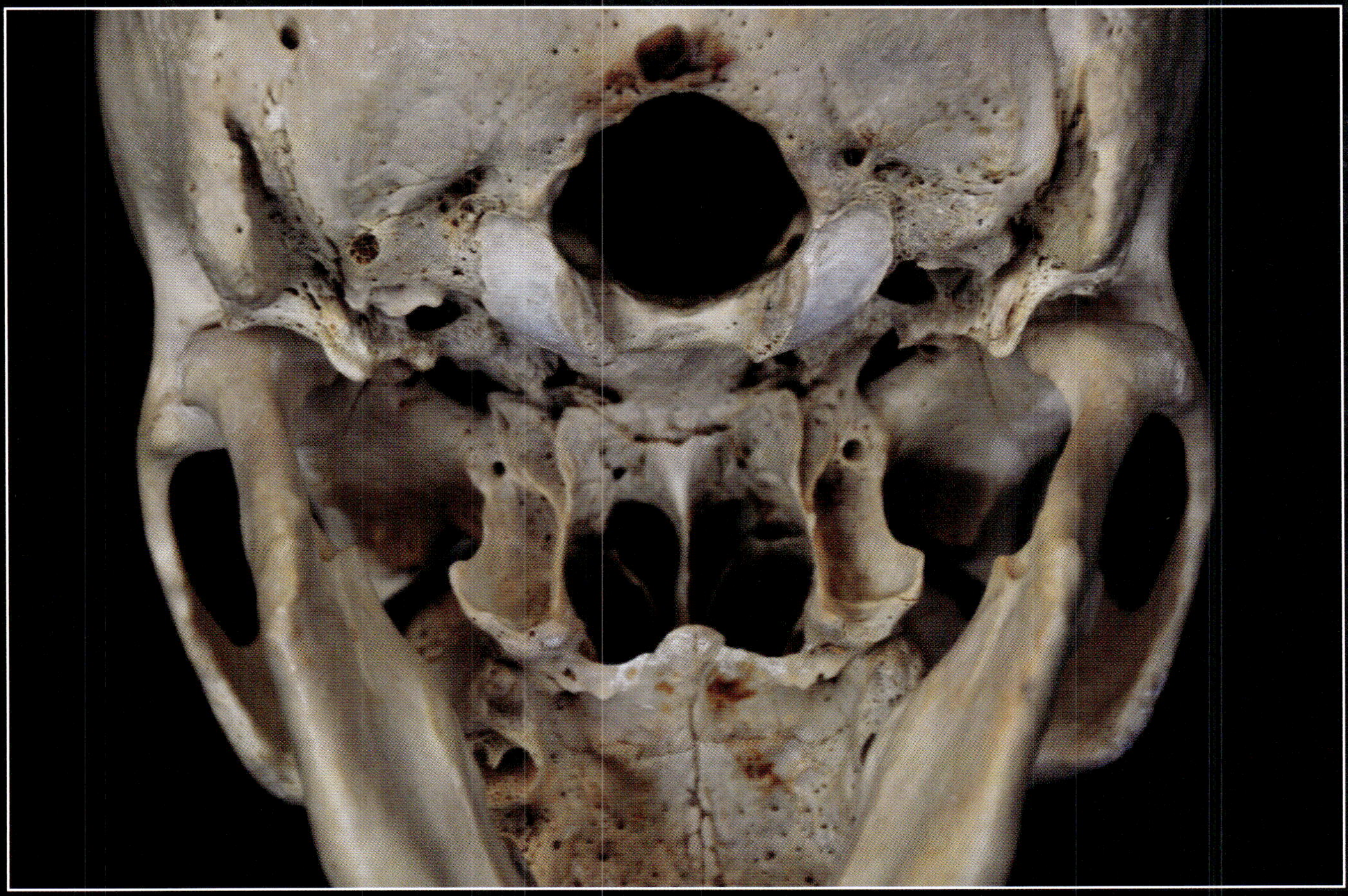

Mike DeVries

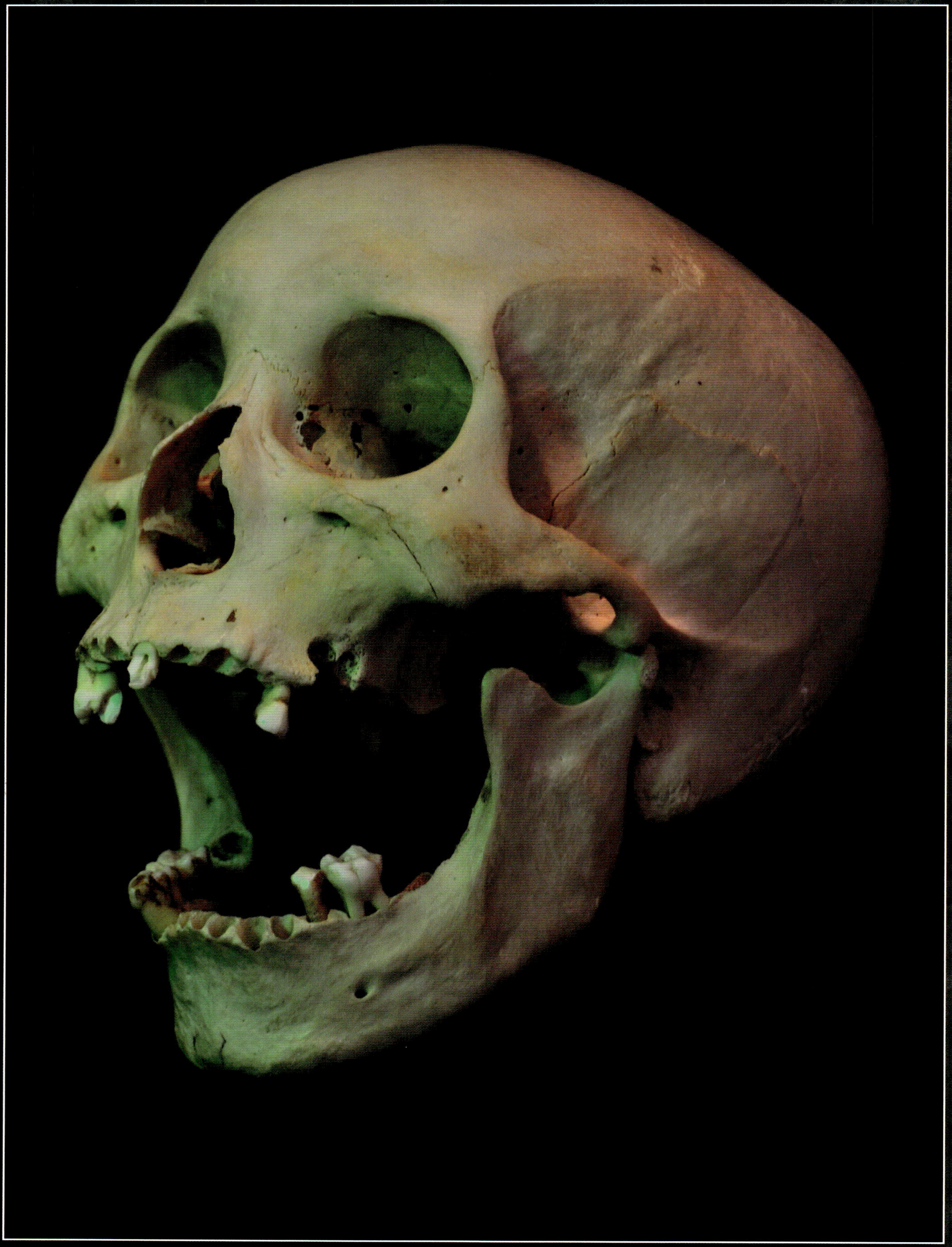

Mike DeVries

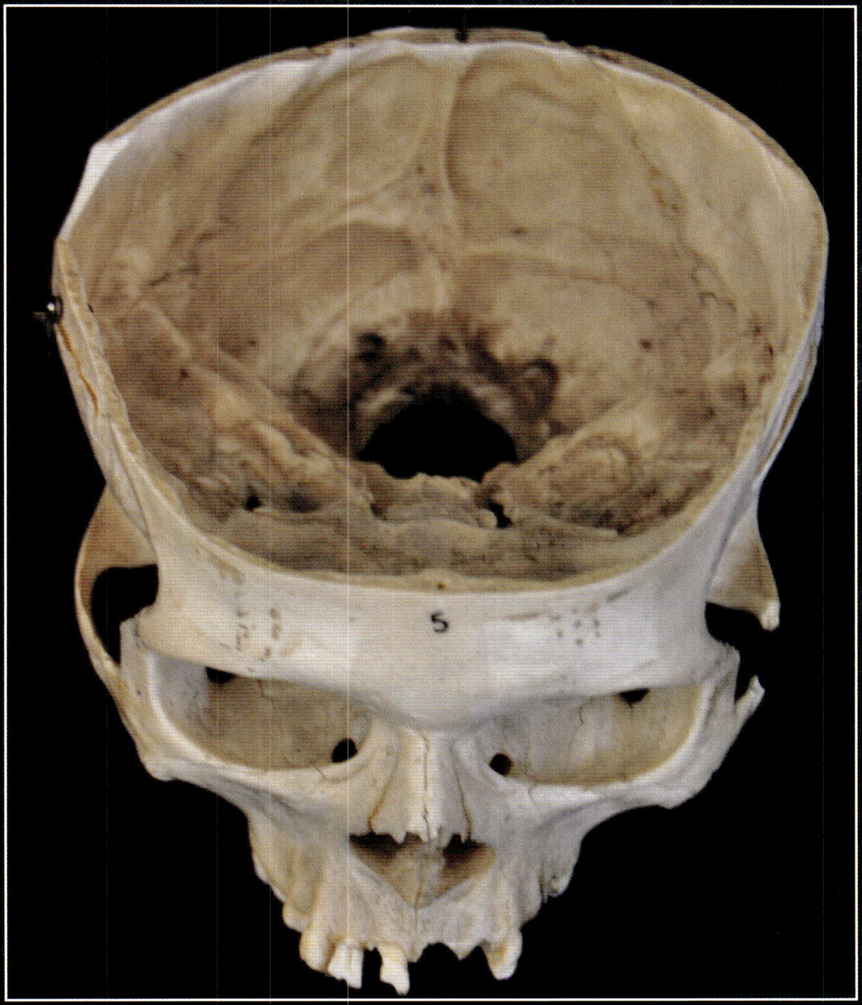

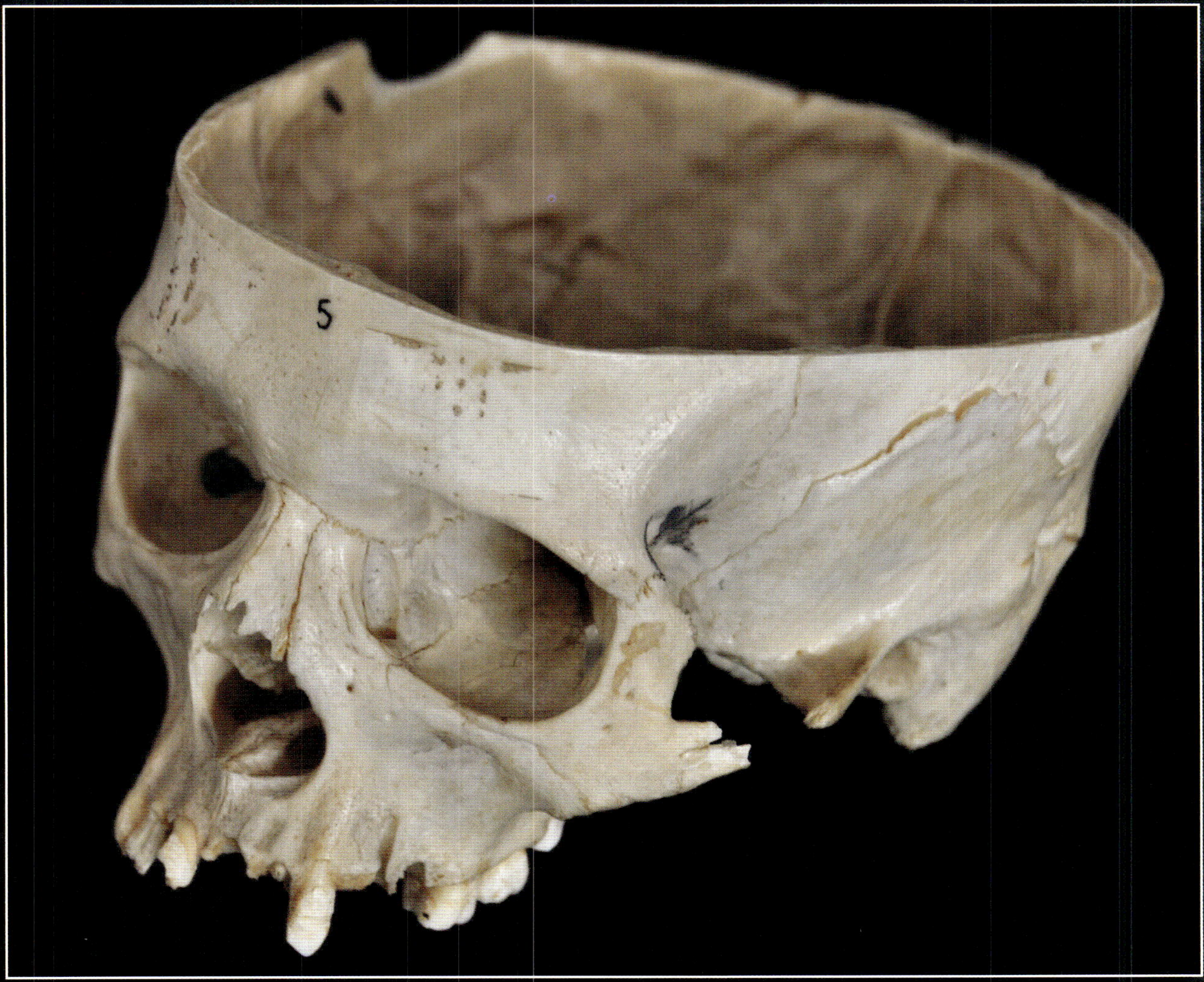

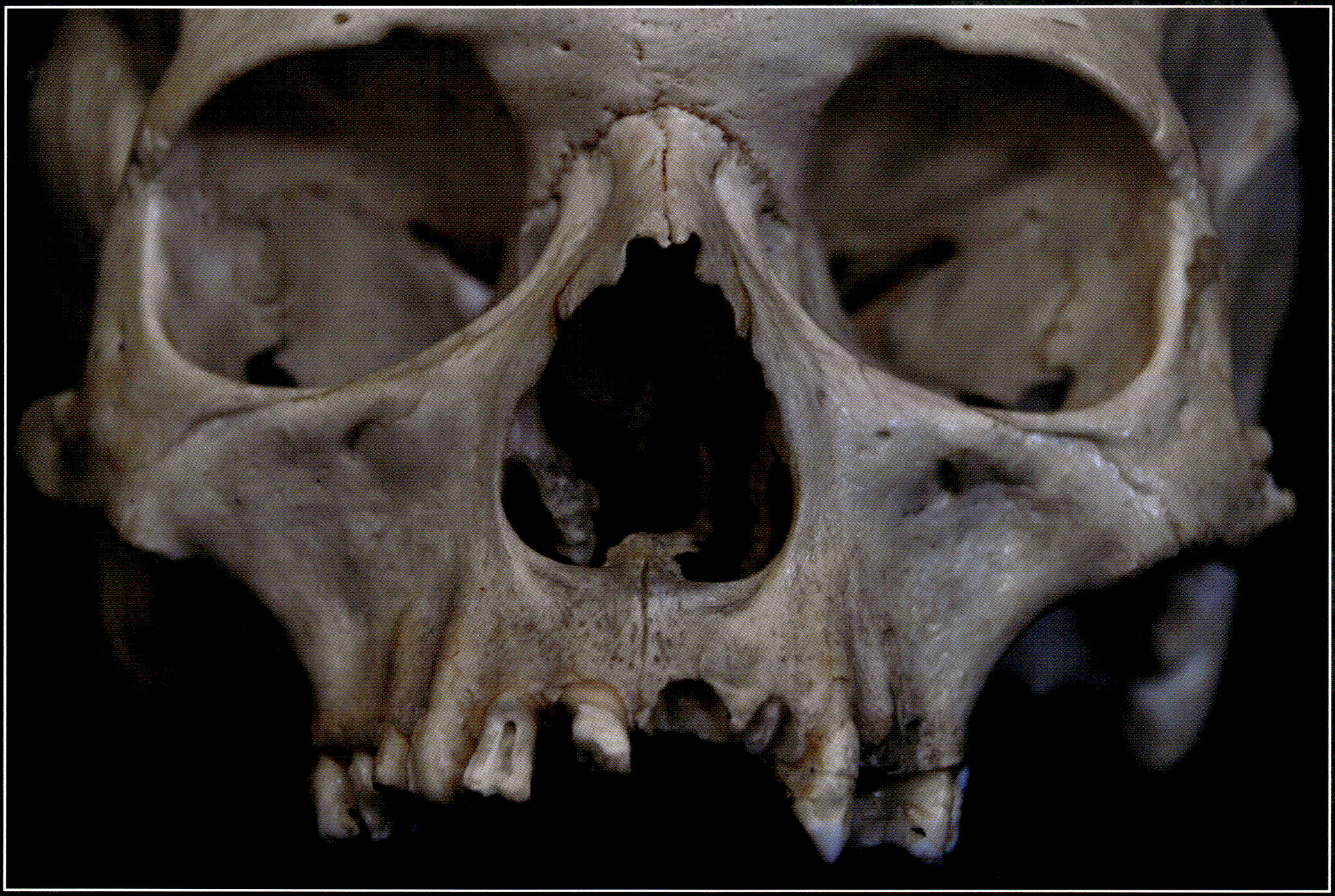

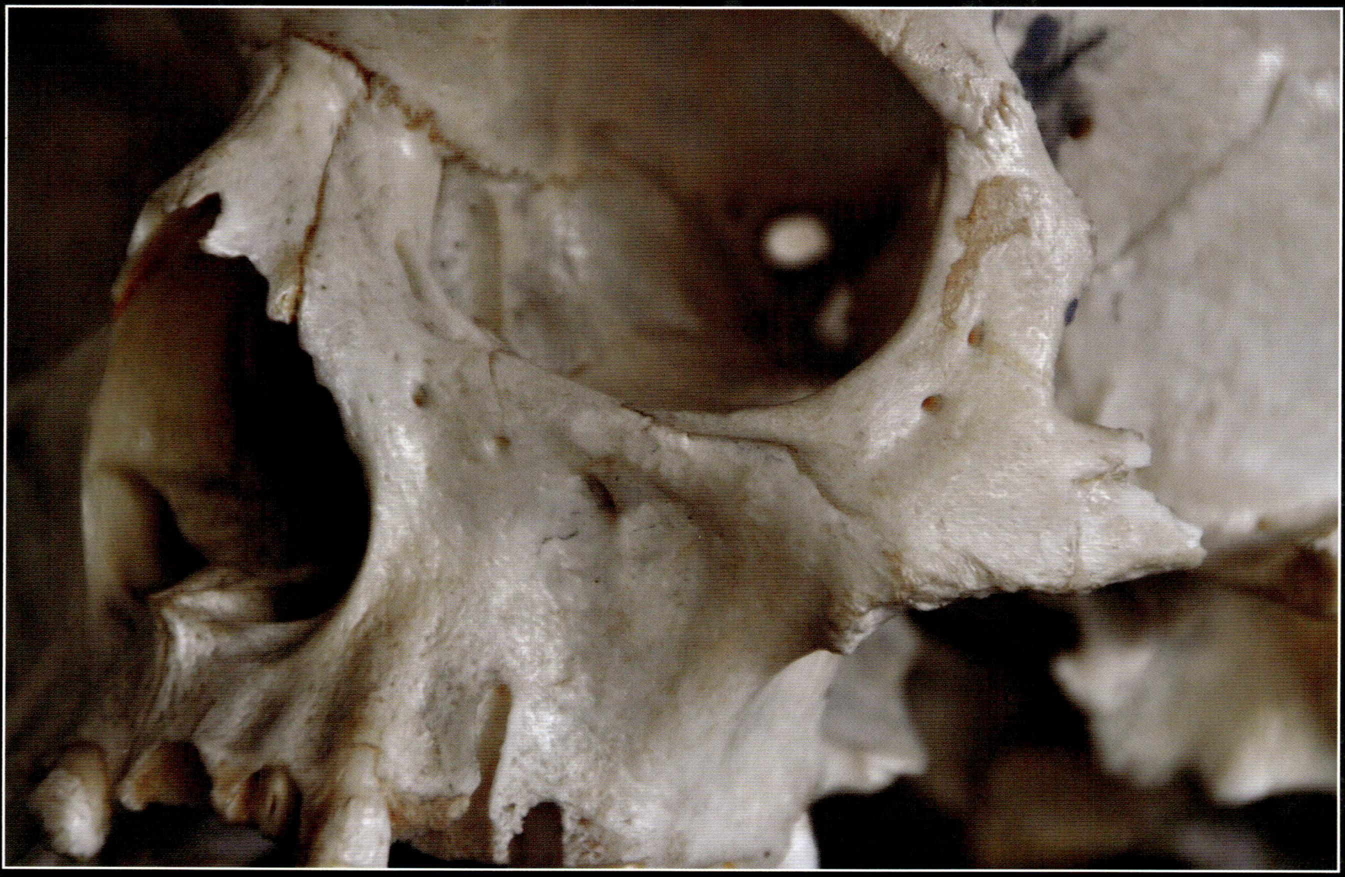

Mike DeVries

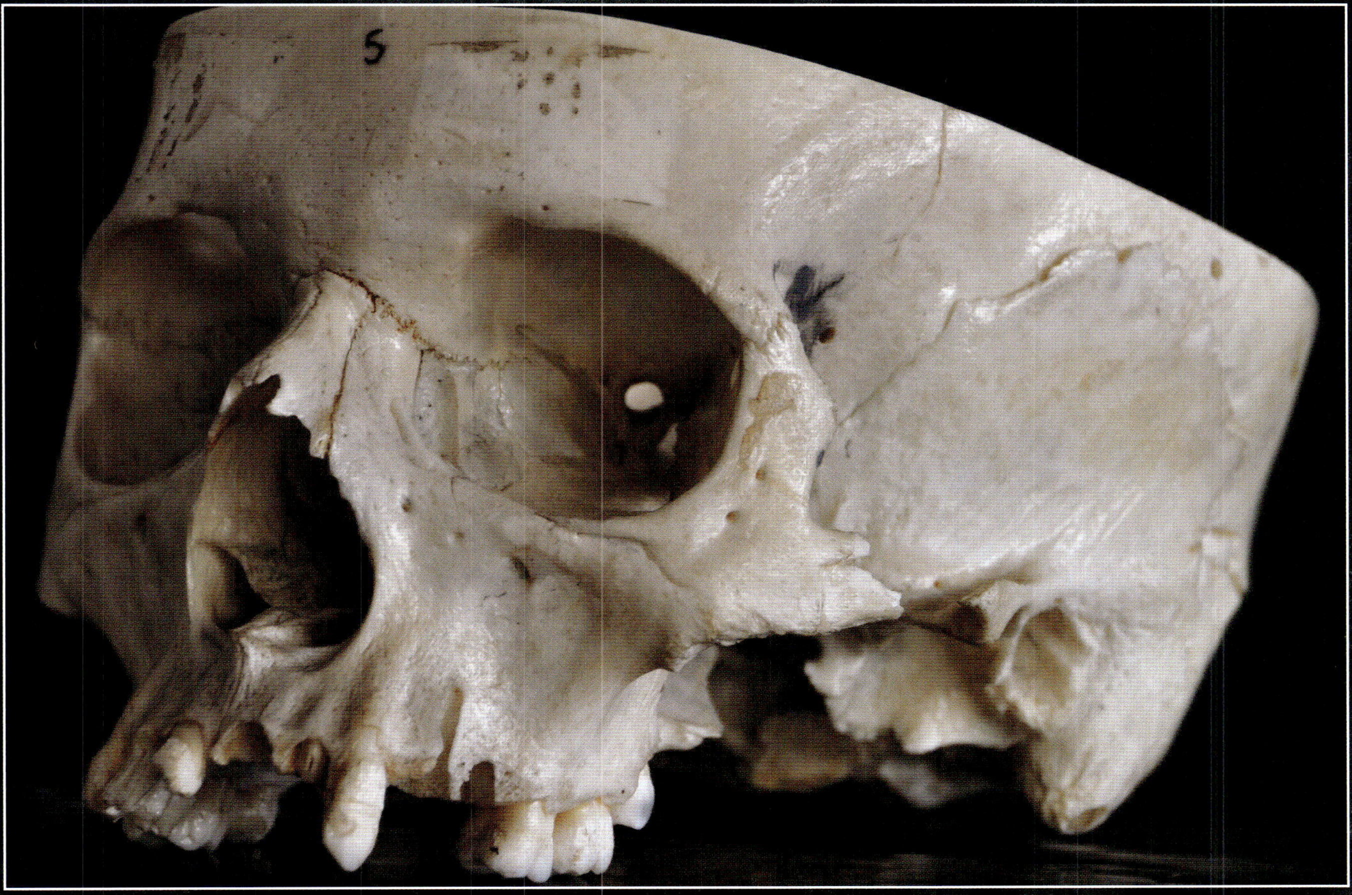

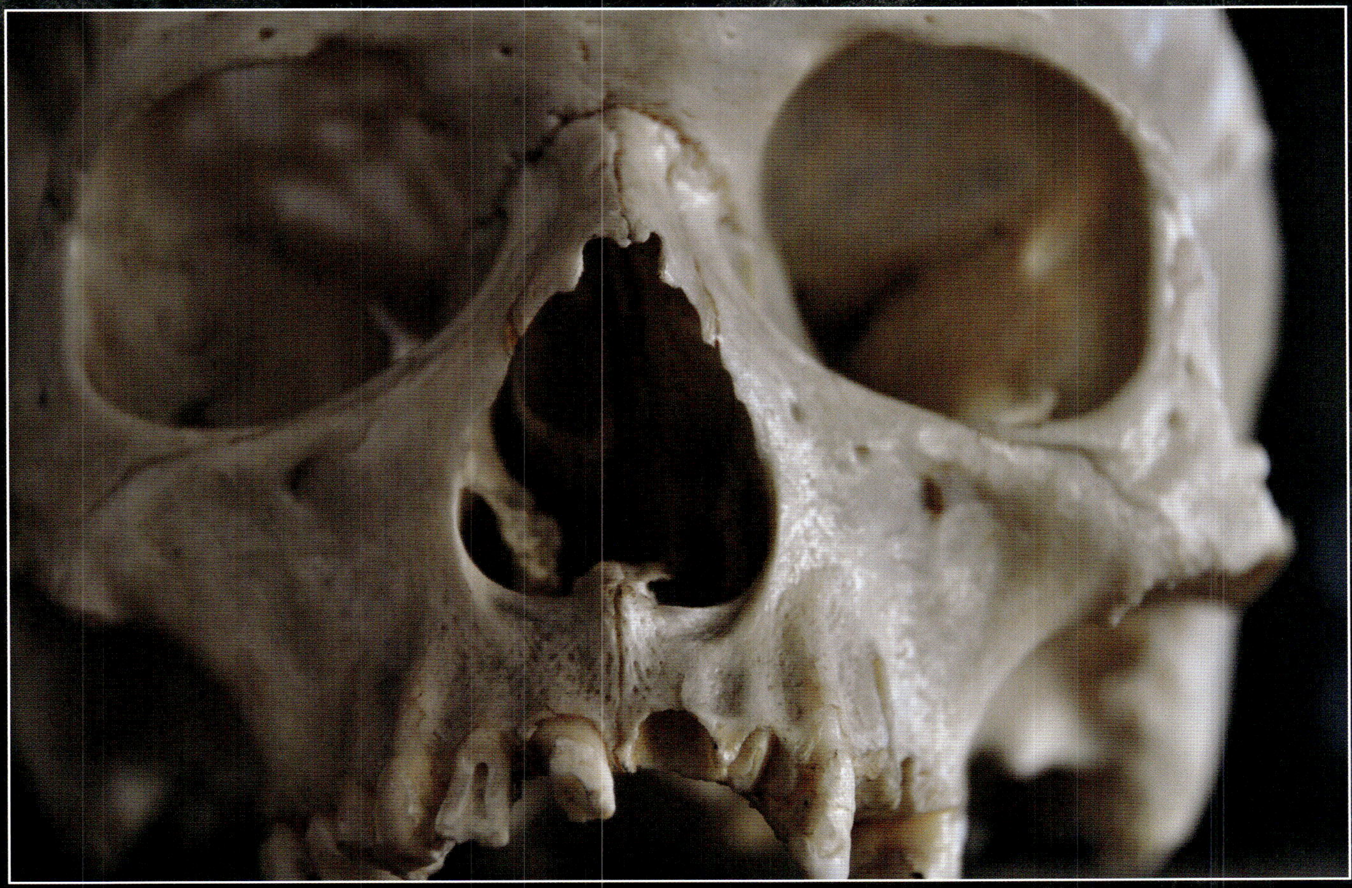

Mike DeVries

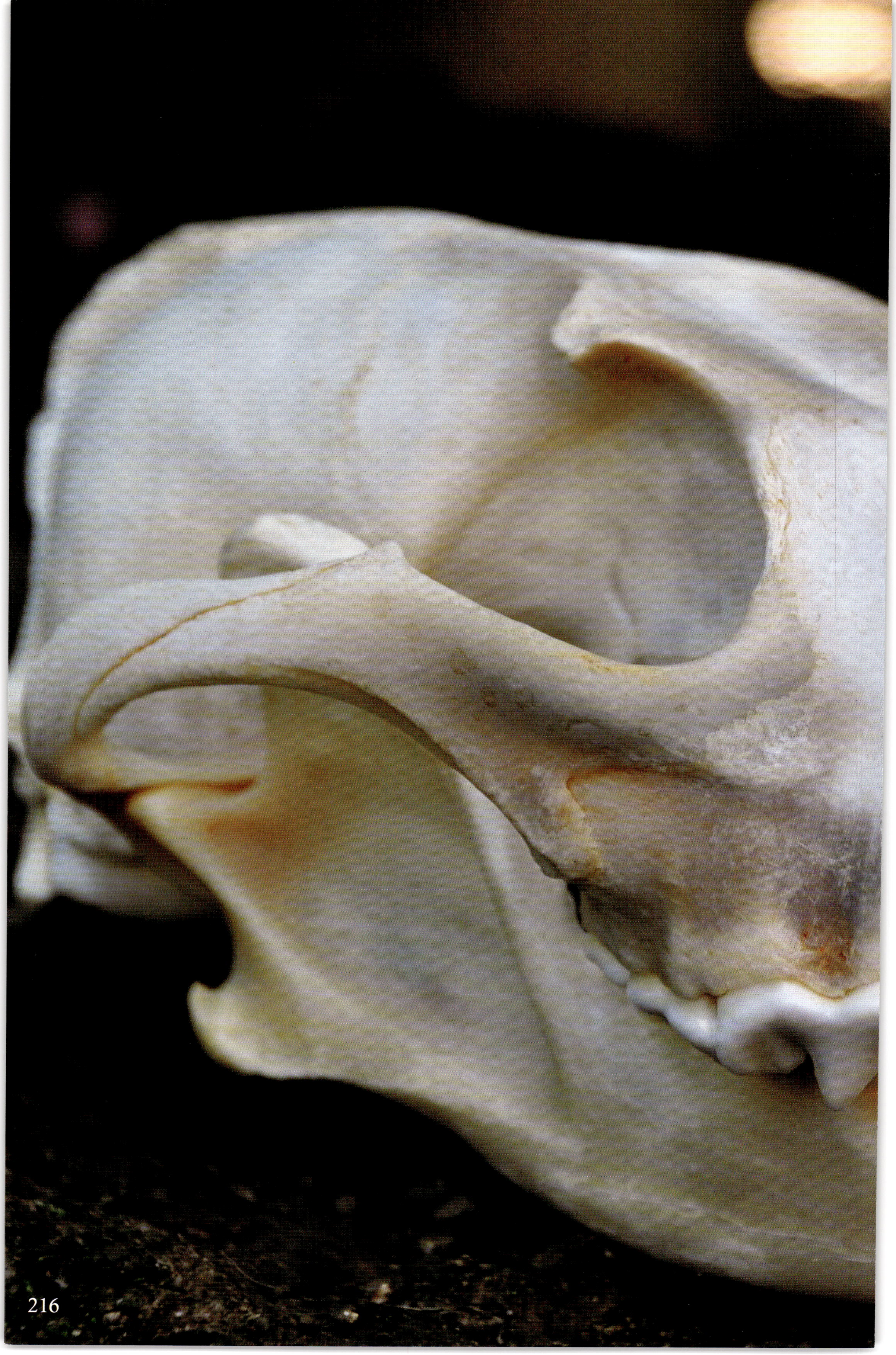

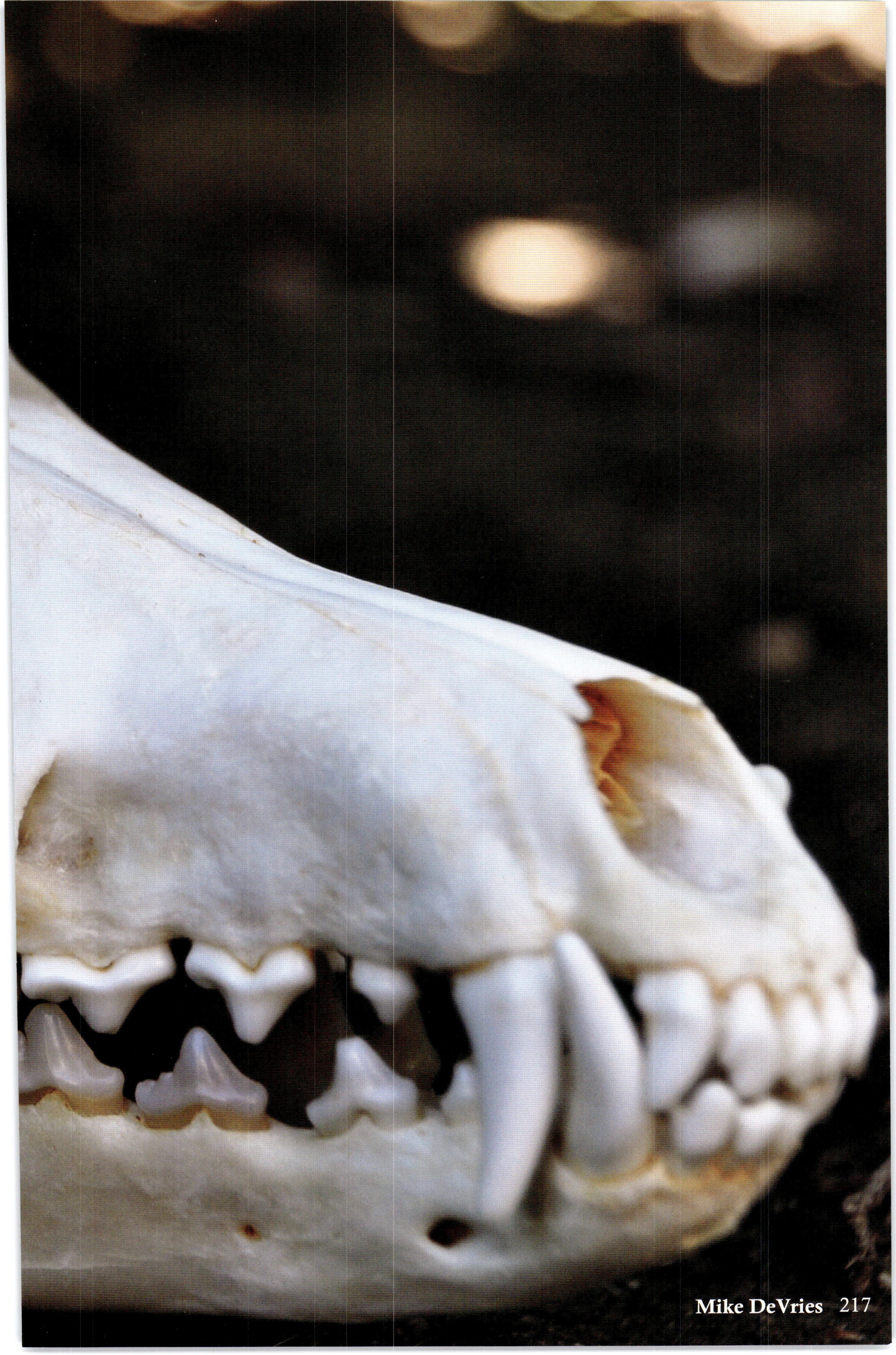

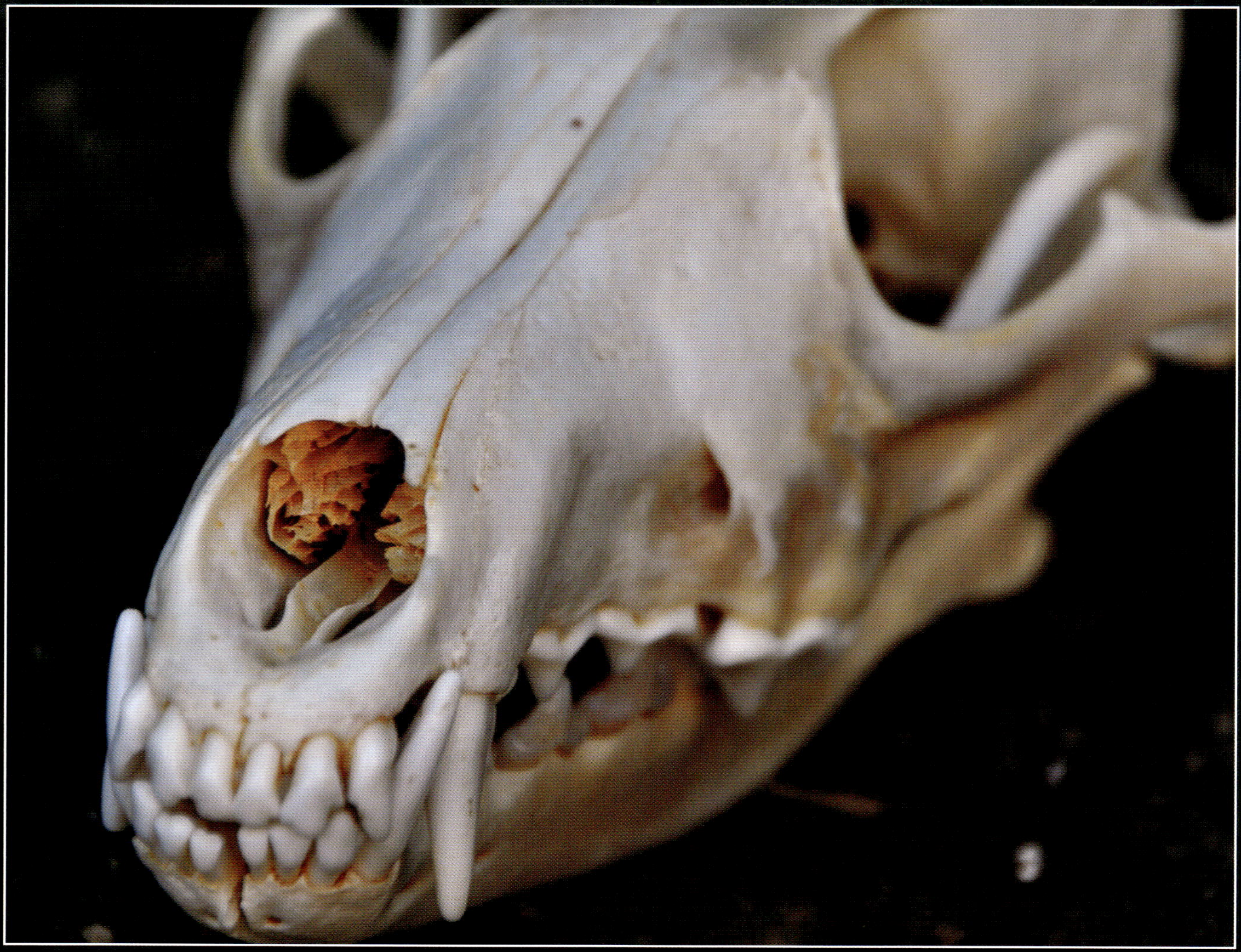

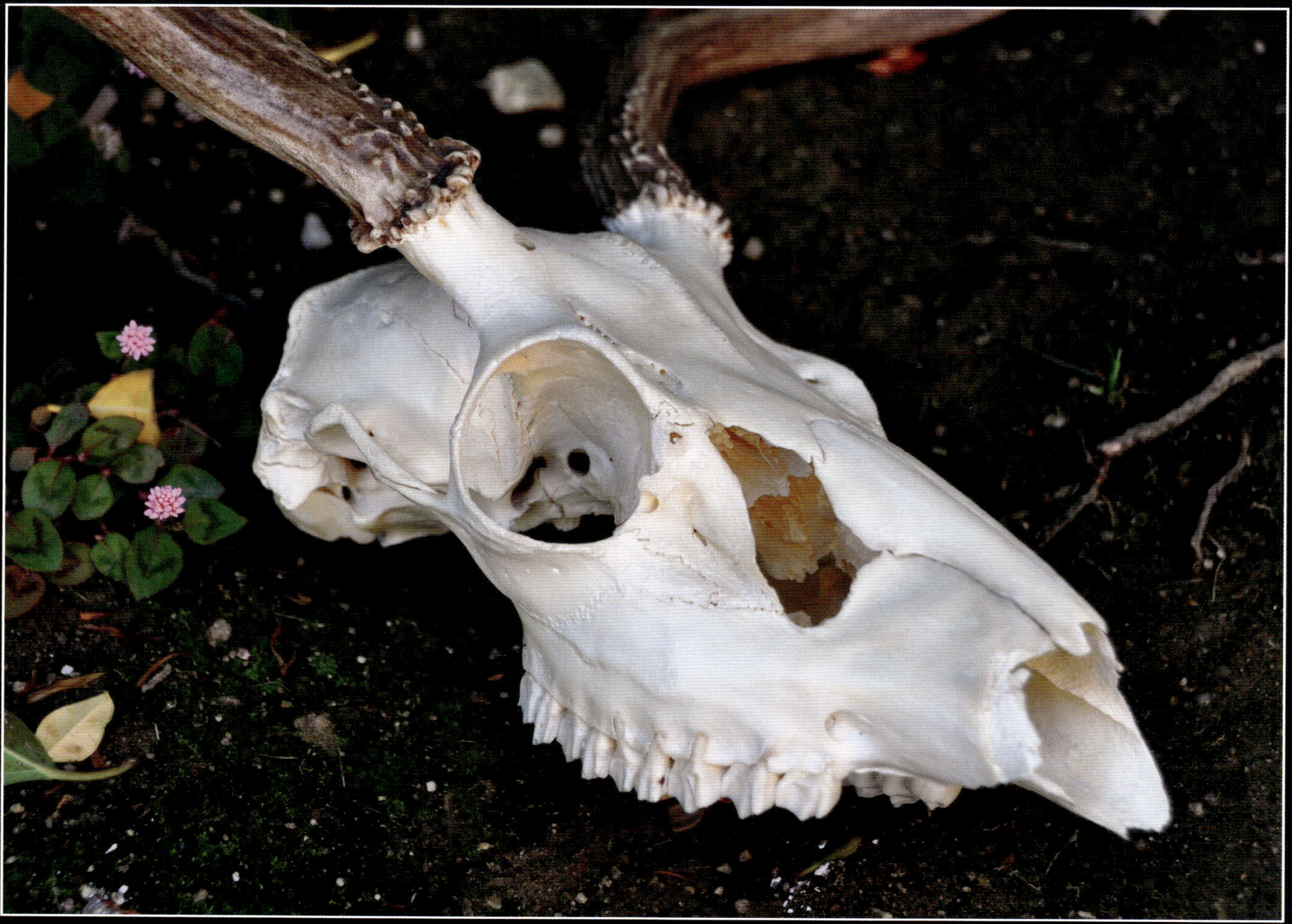

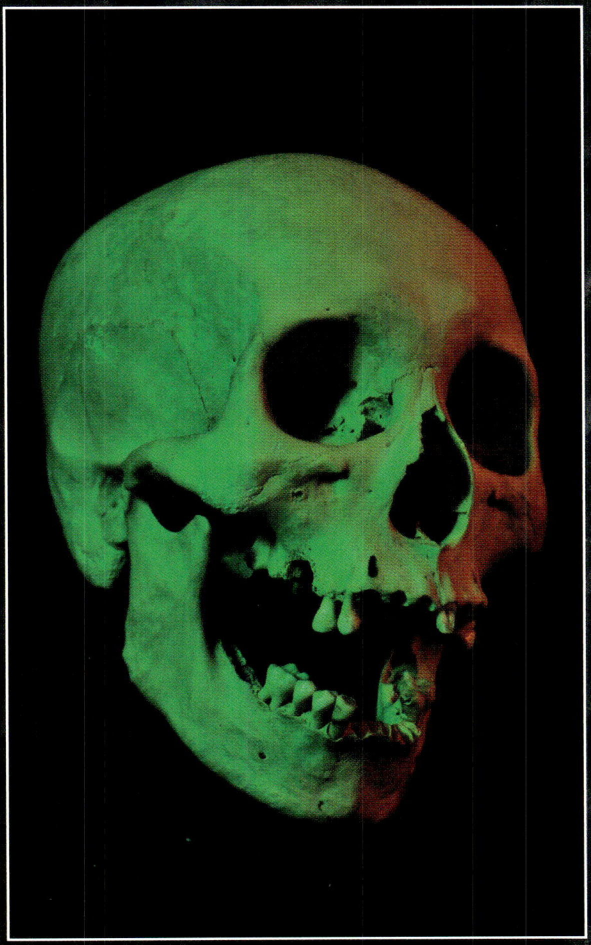

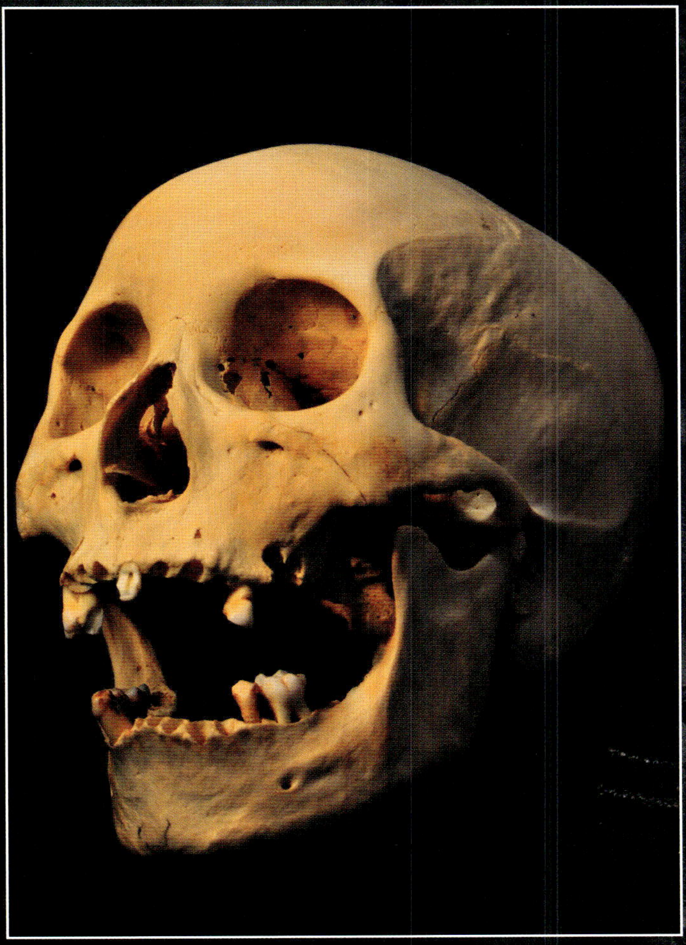

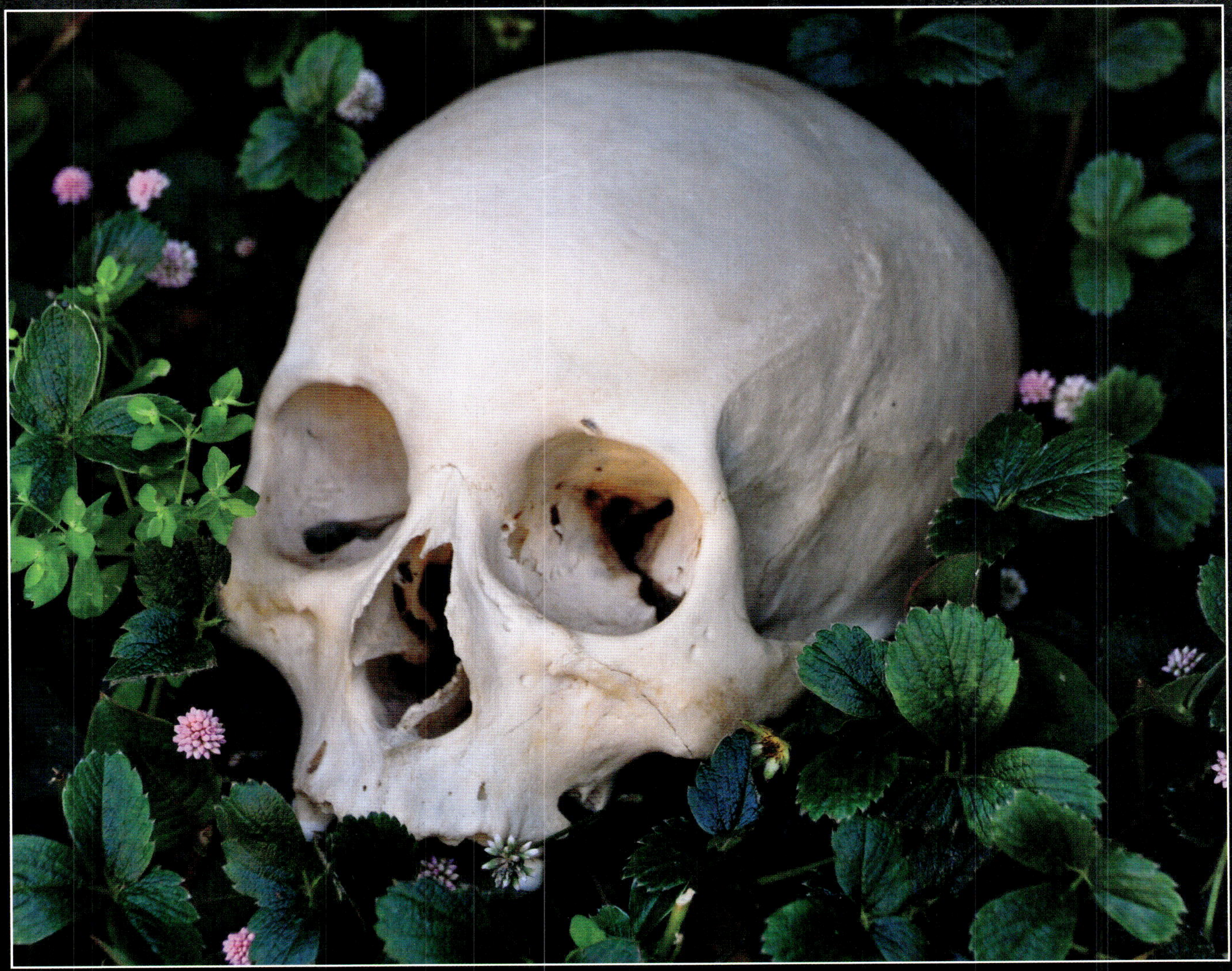

Mike DeVries

SKULLPTURES

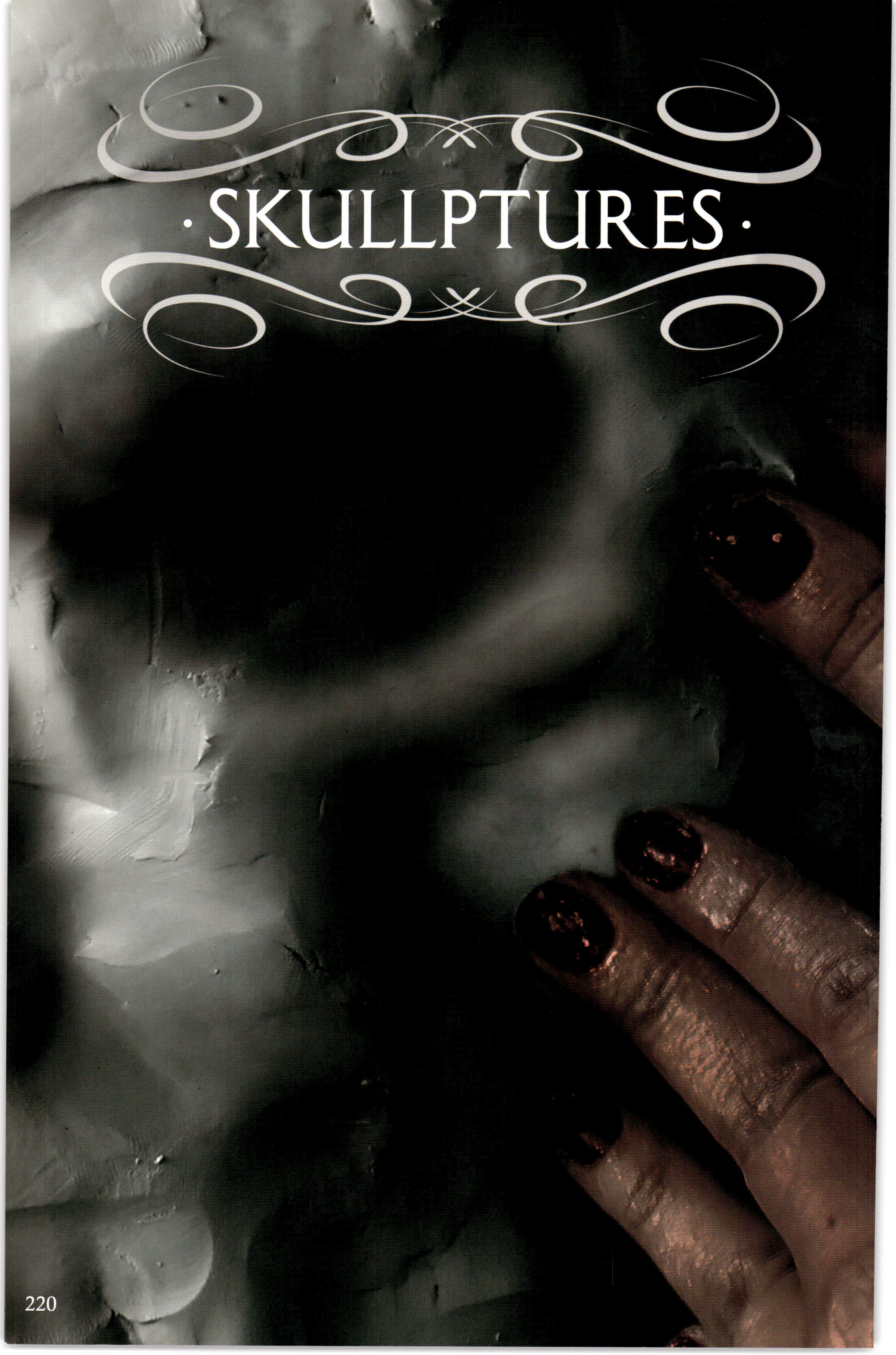

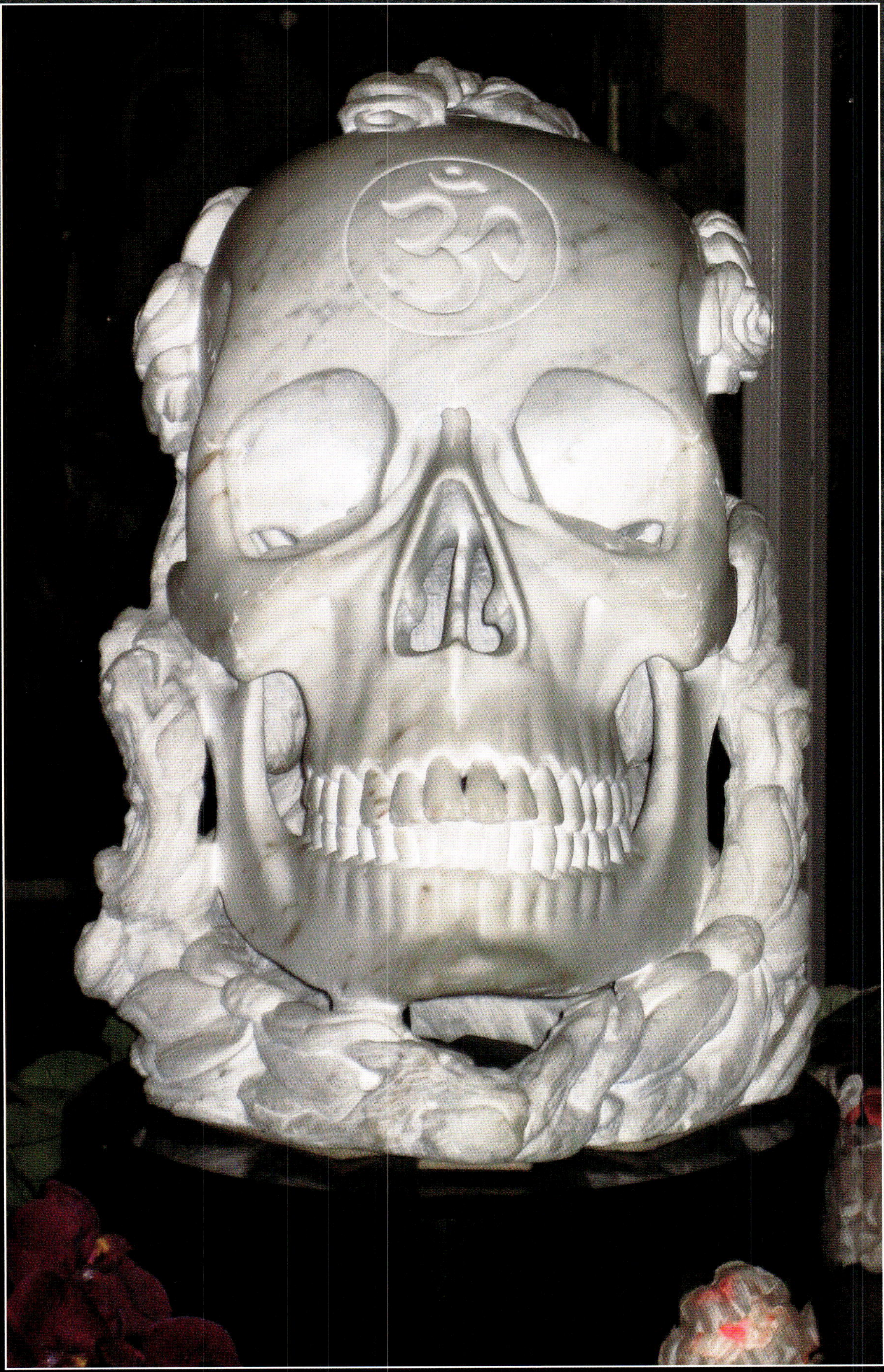

Paolo Acuna

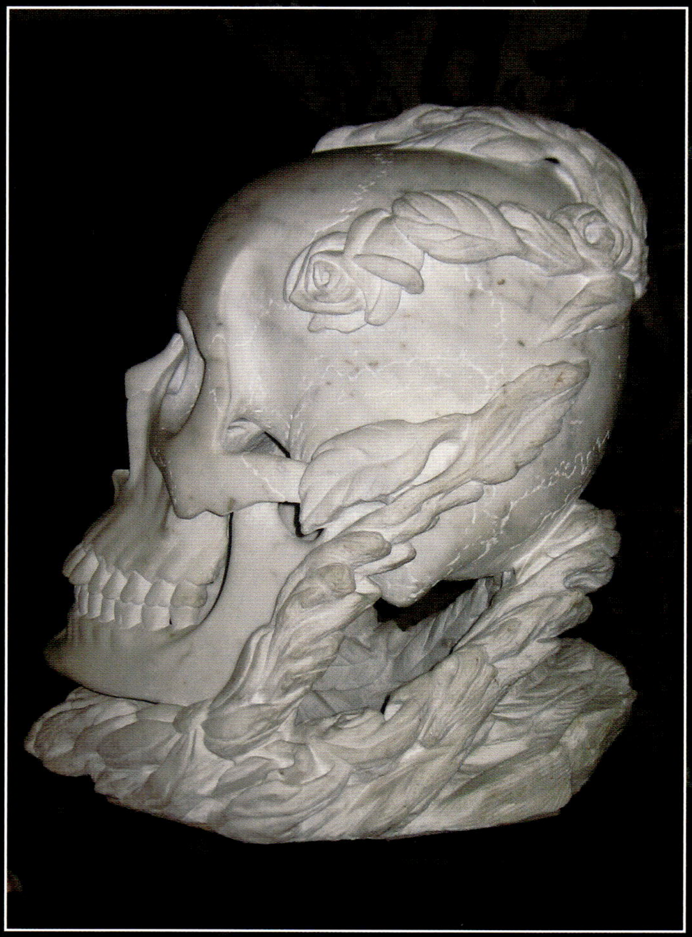

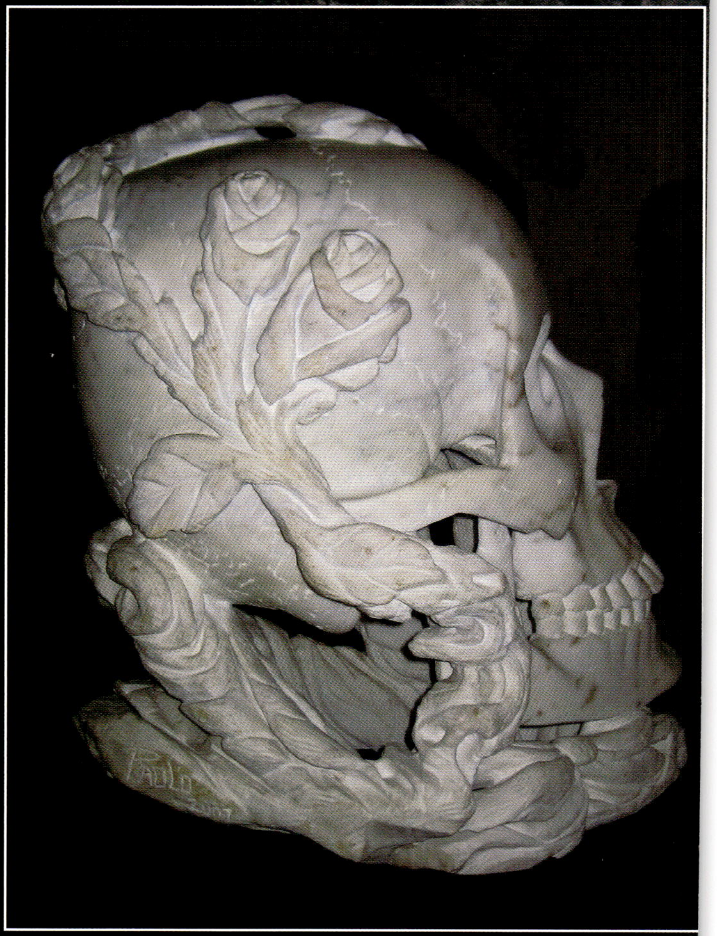

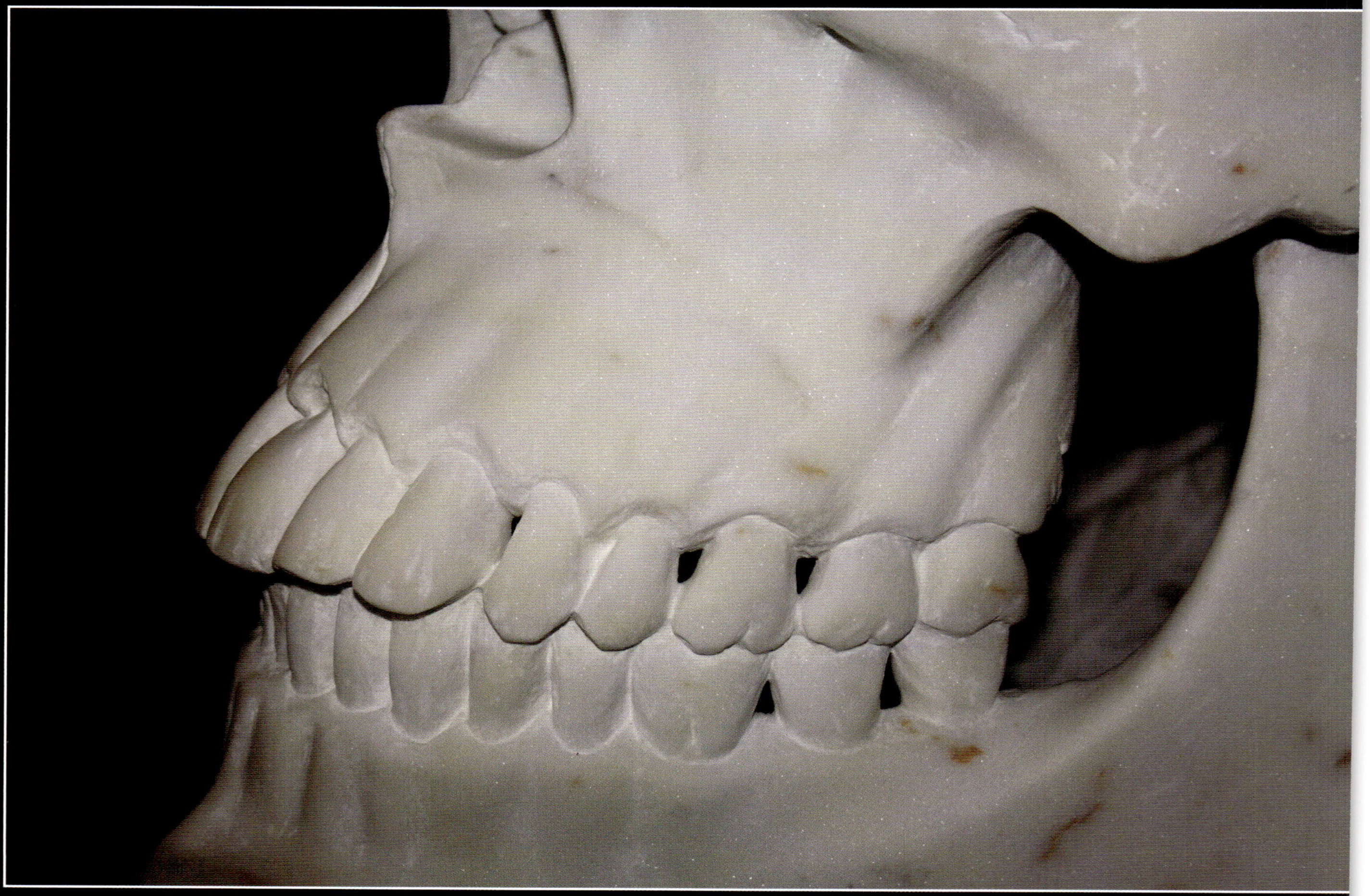

Paolo Acuna

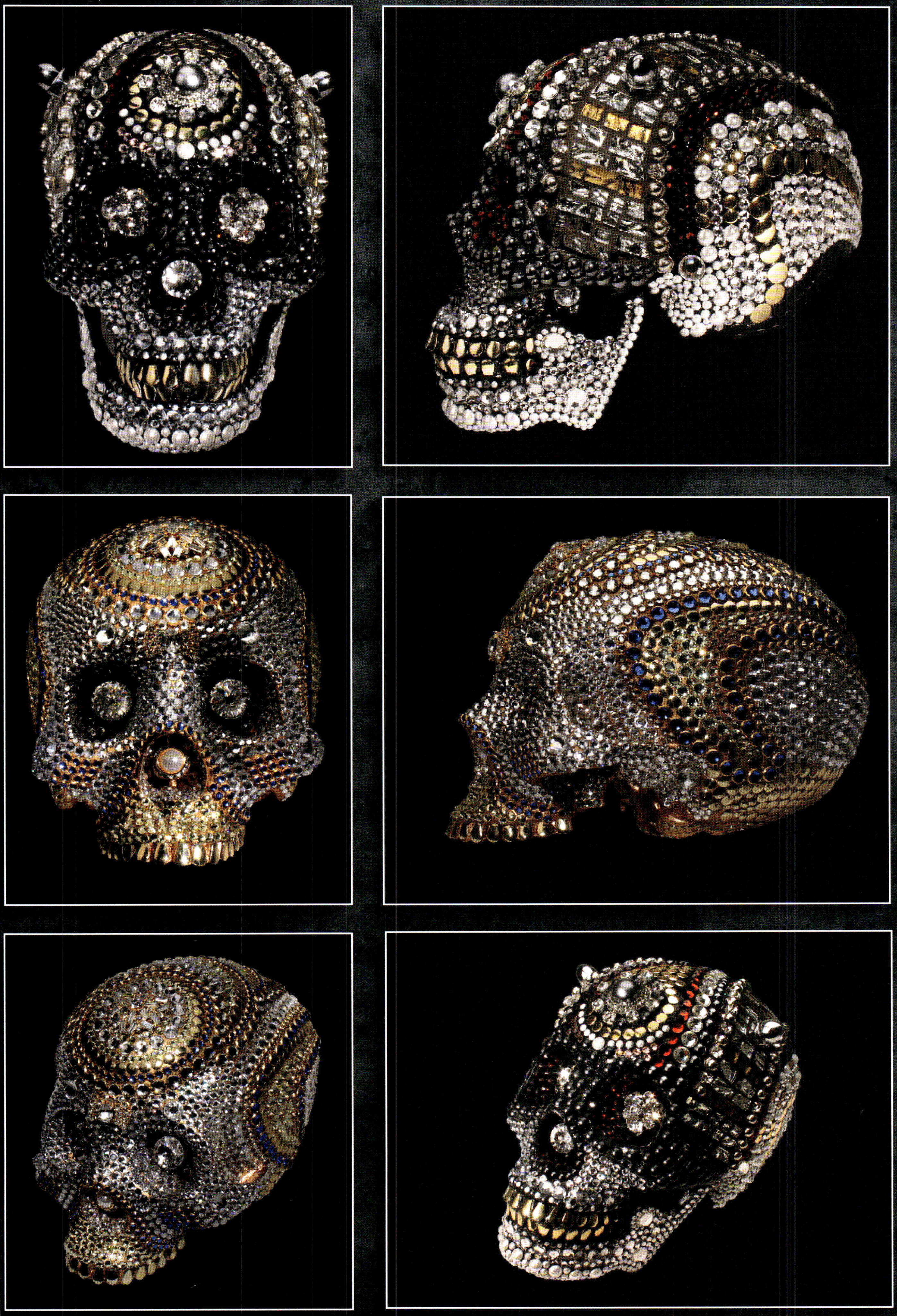

Stan Young

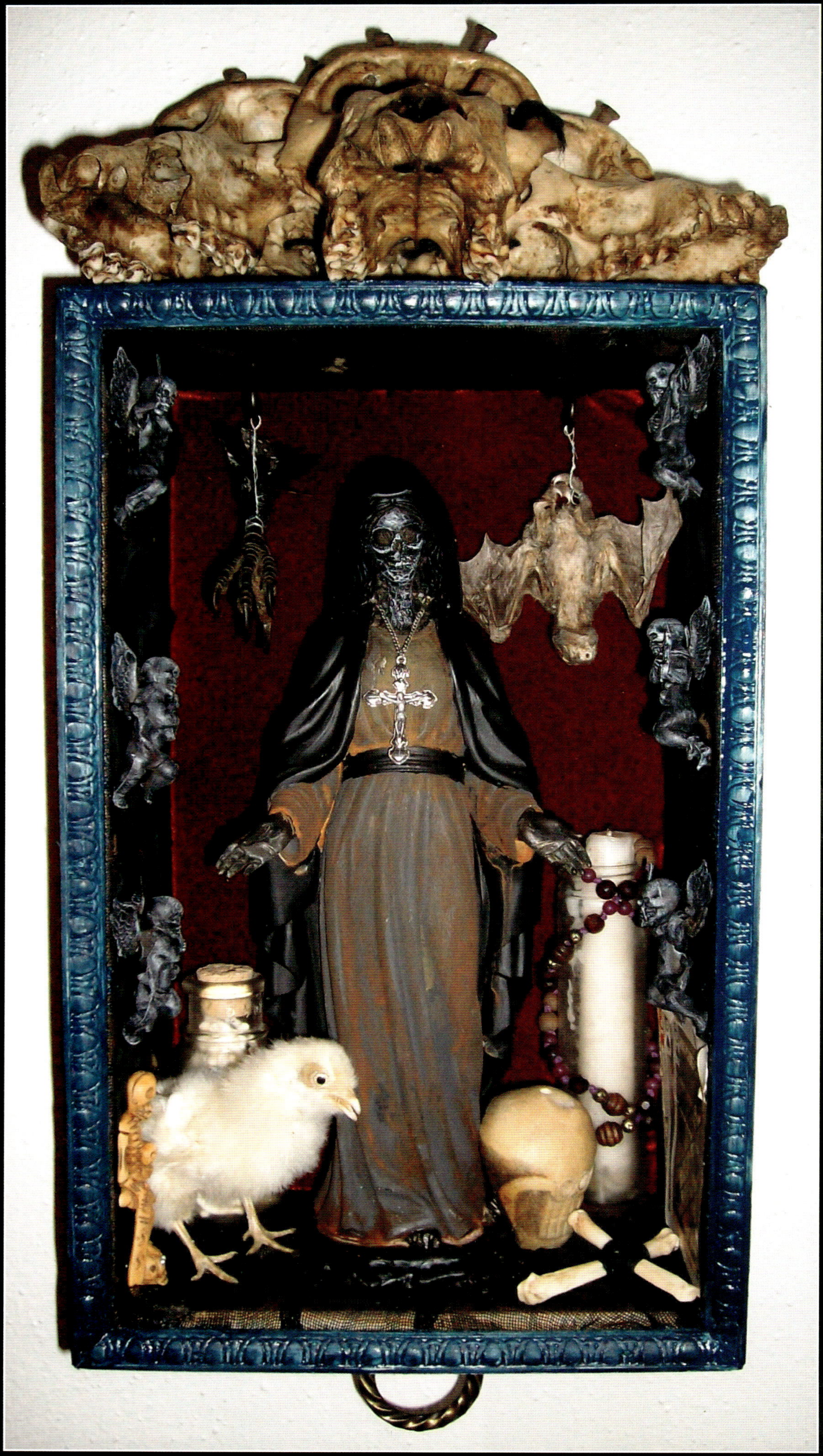

 Joey "The Skull Master" Williams

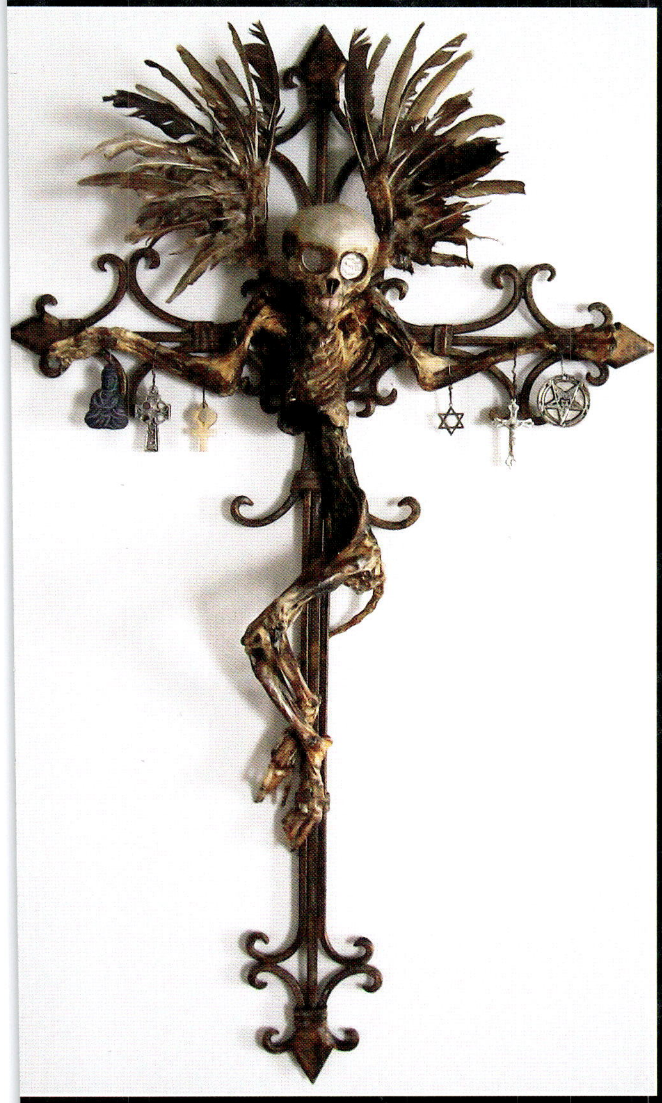

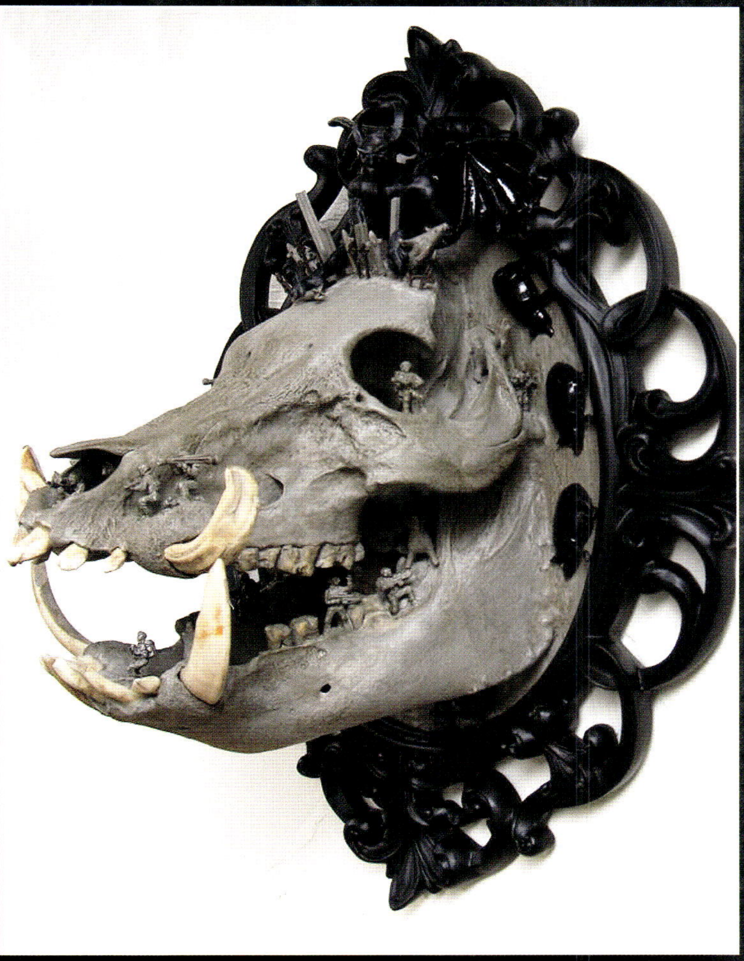

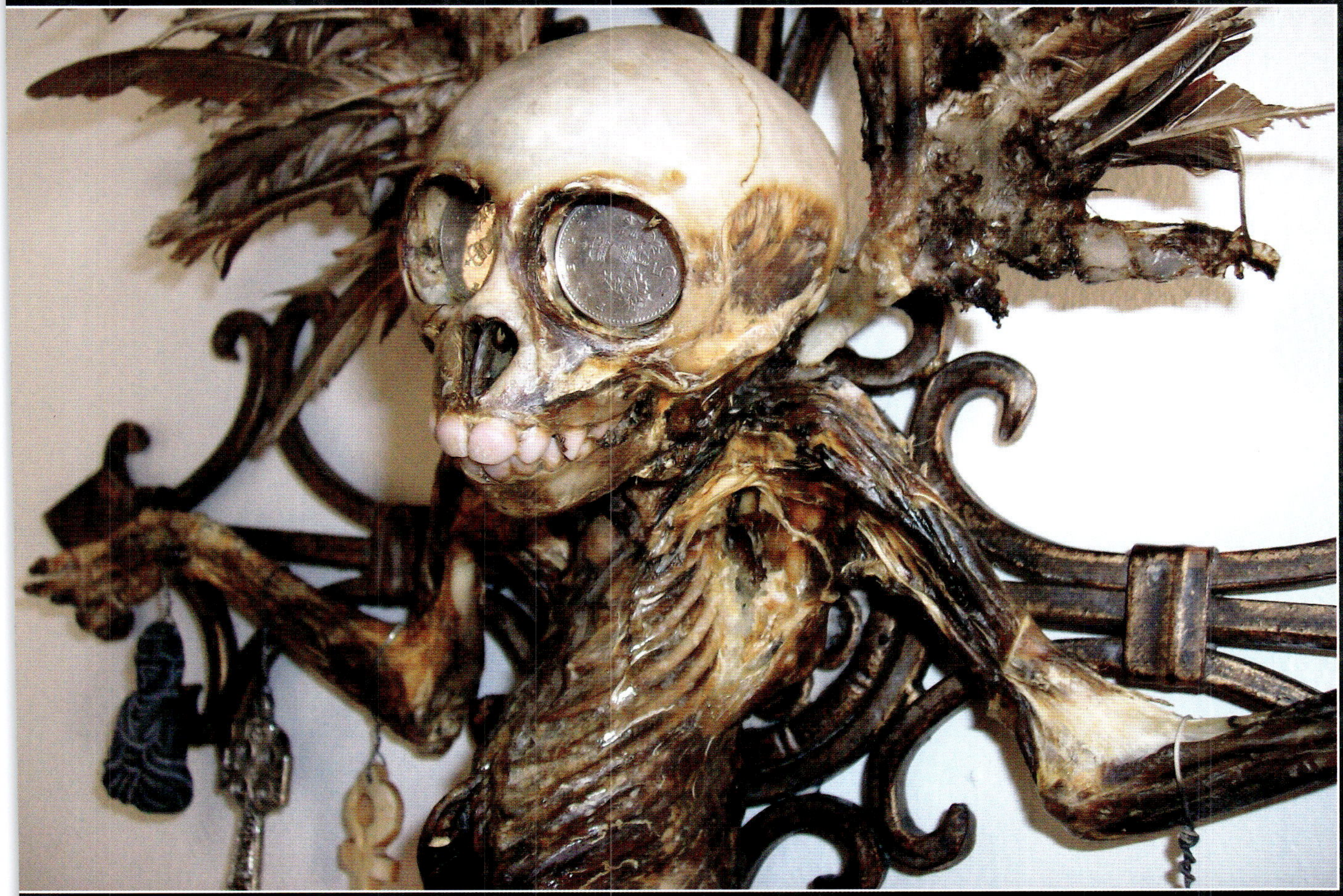

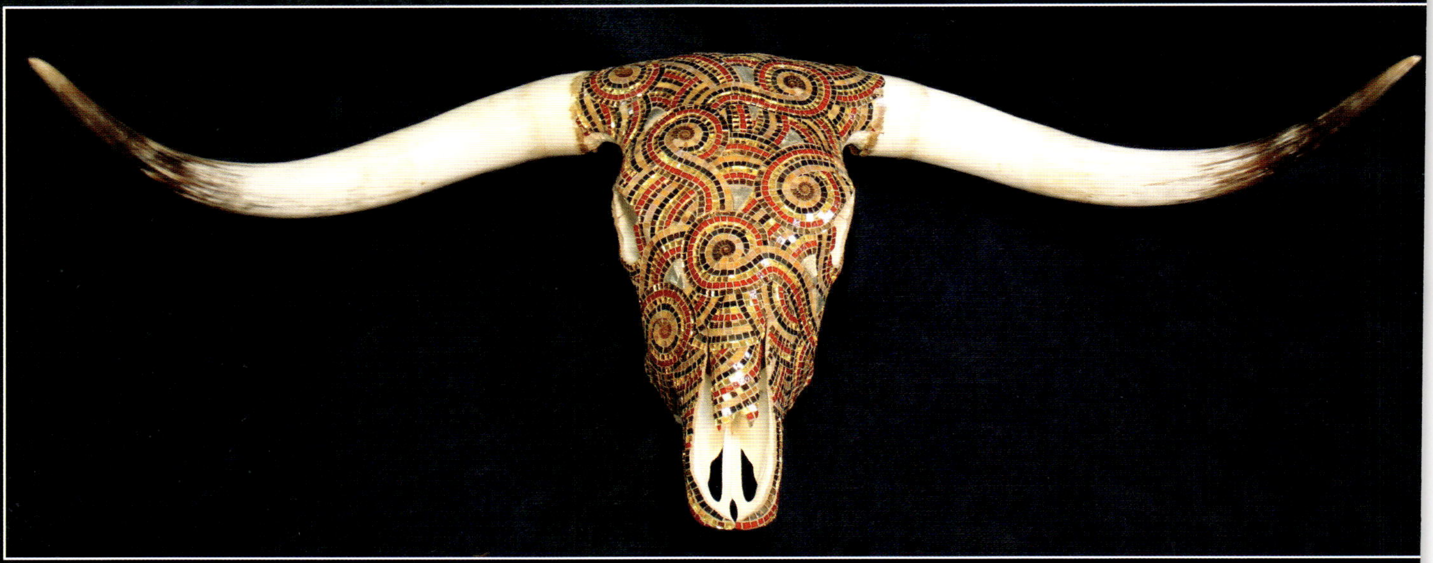

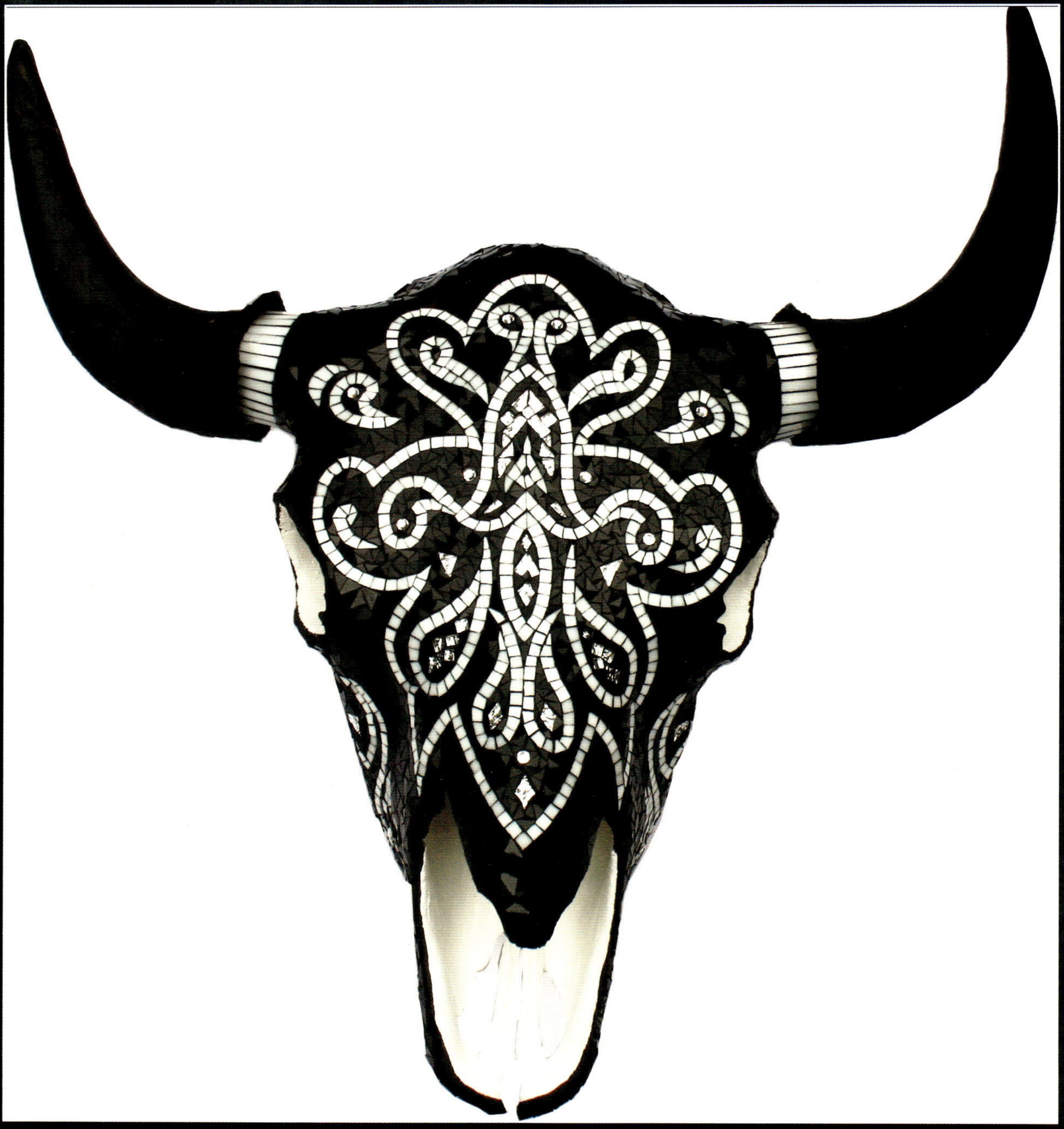

Crystal Thomas

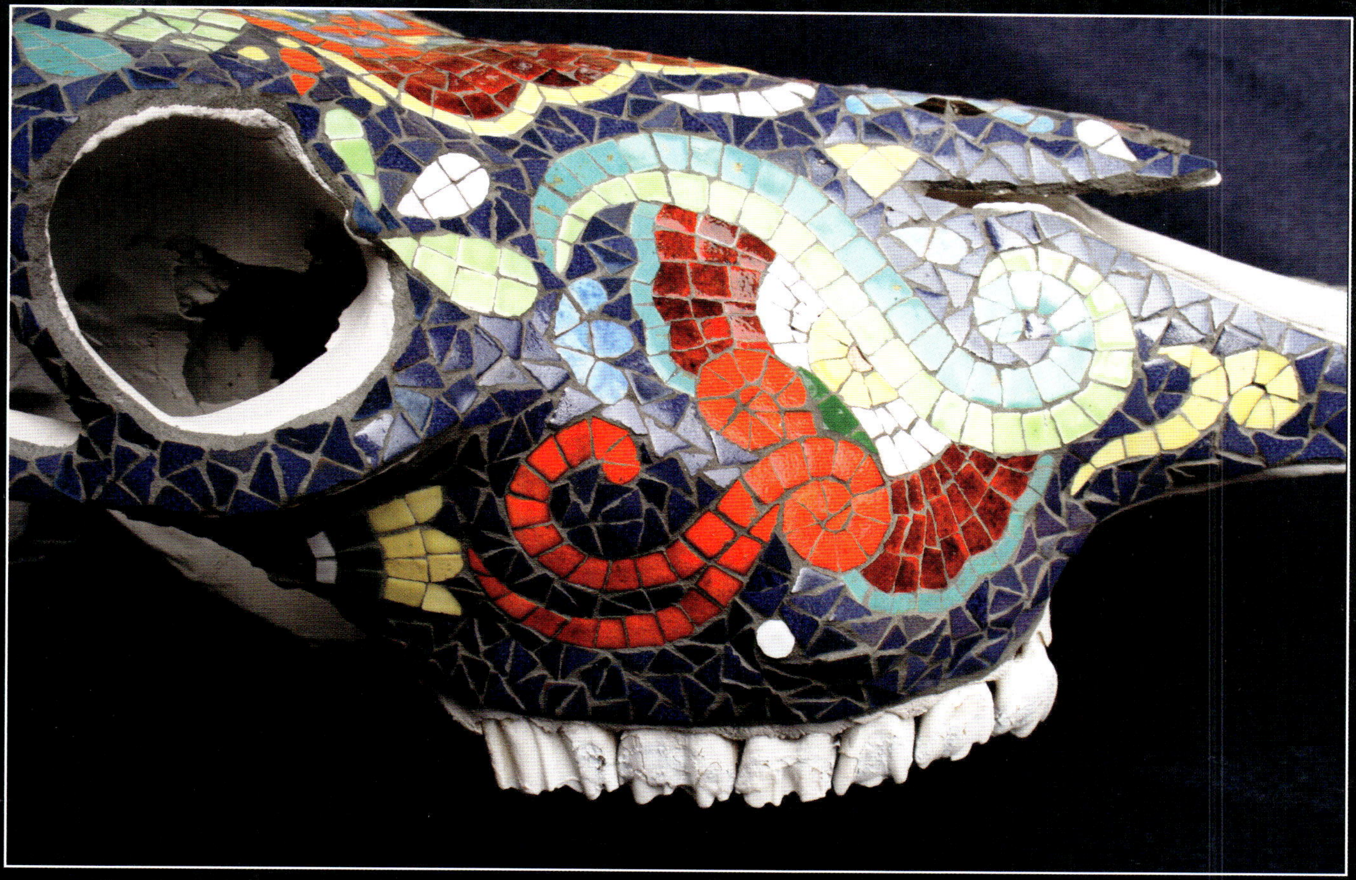

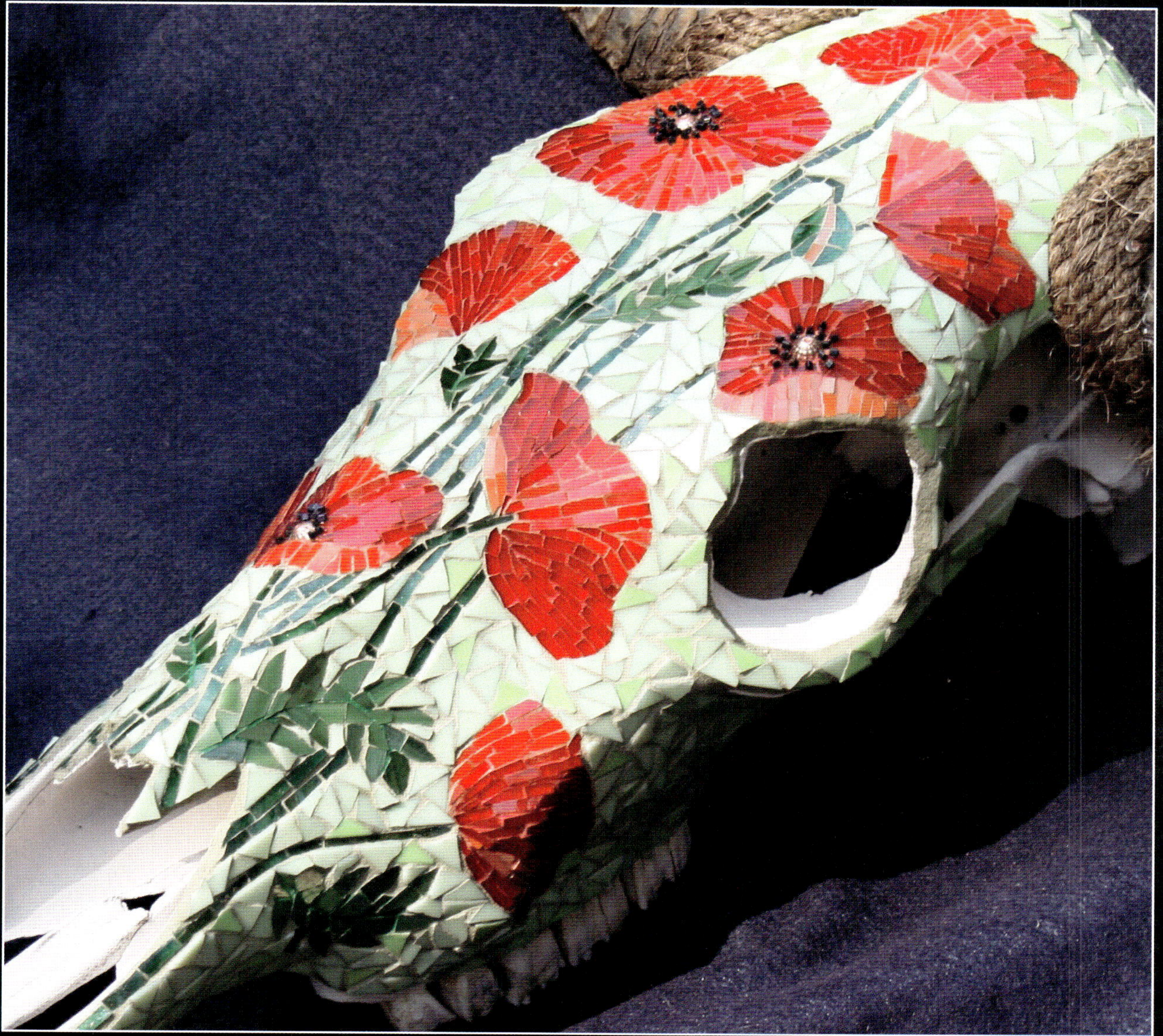

Crystal Thomas

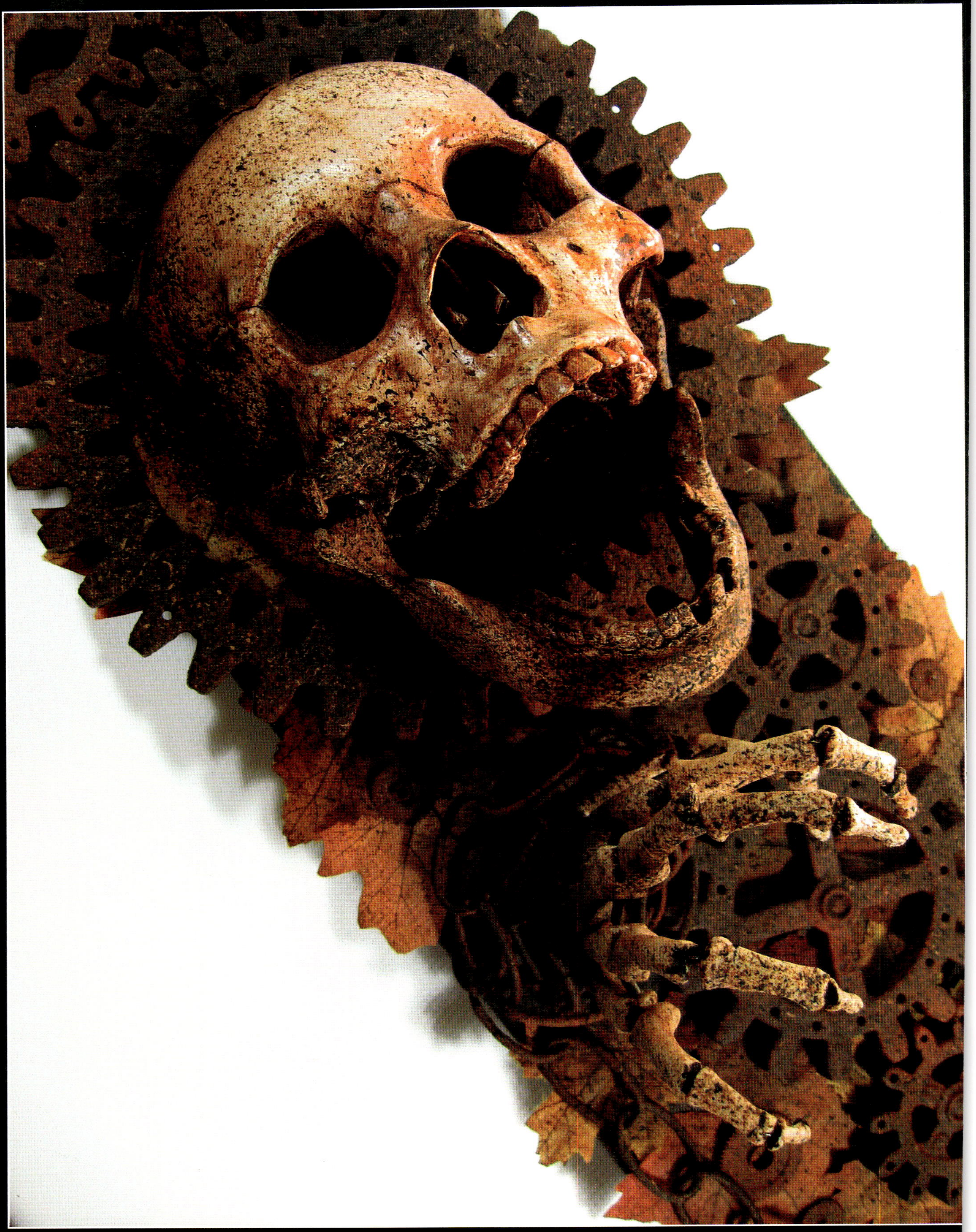

Brion Norwalk

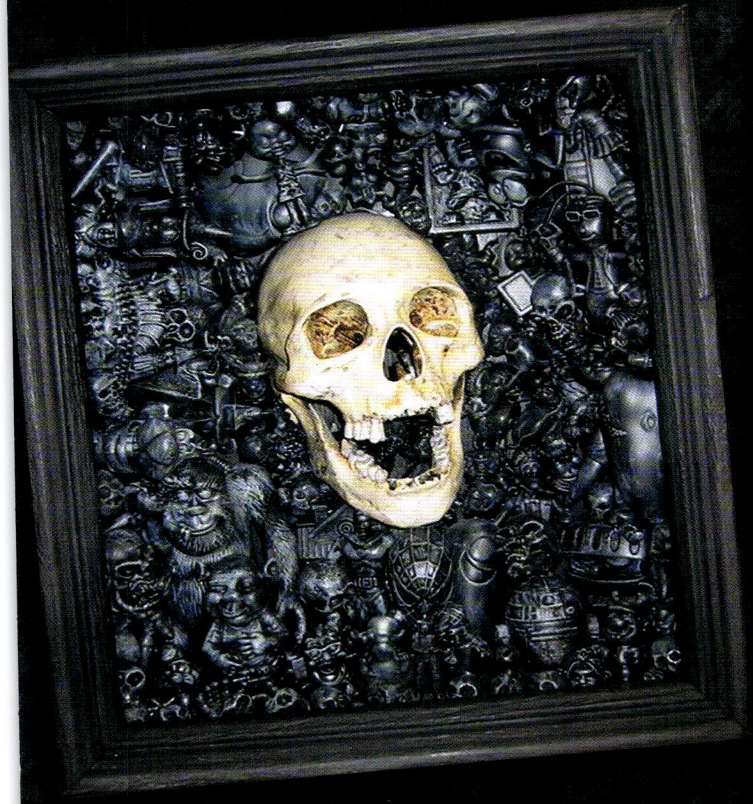

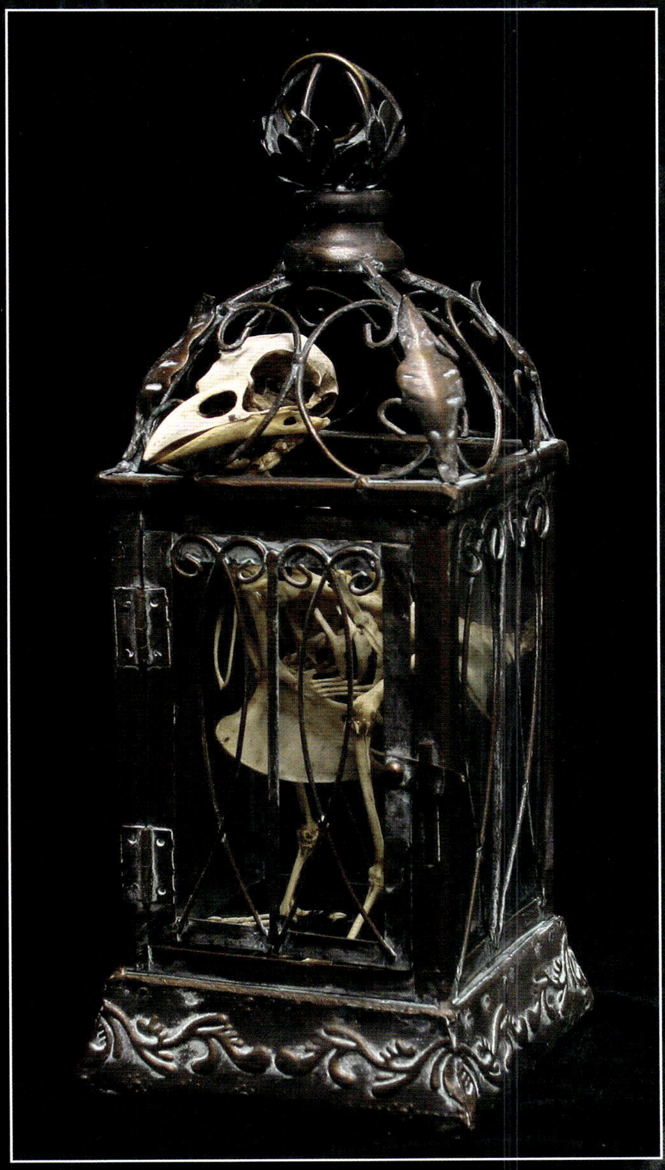

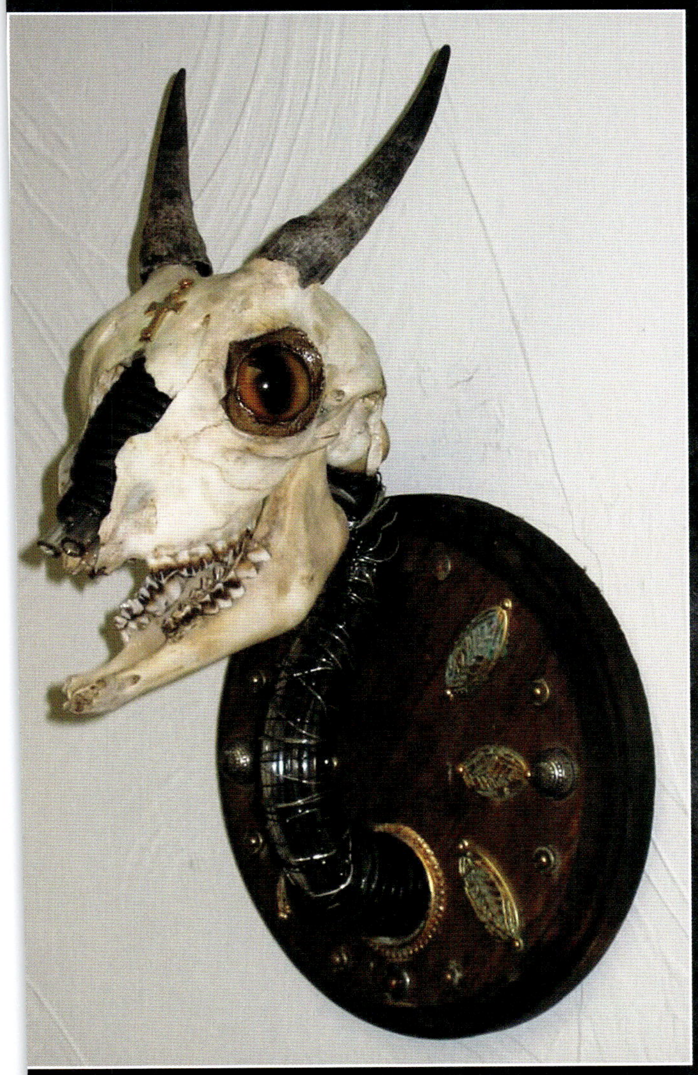

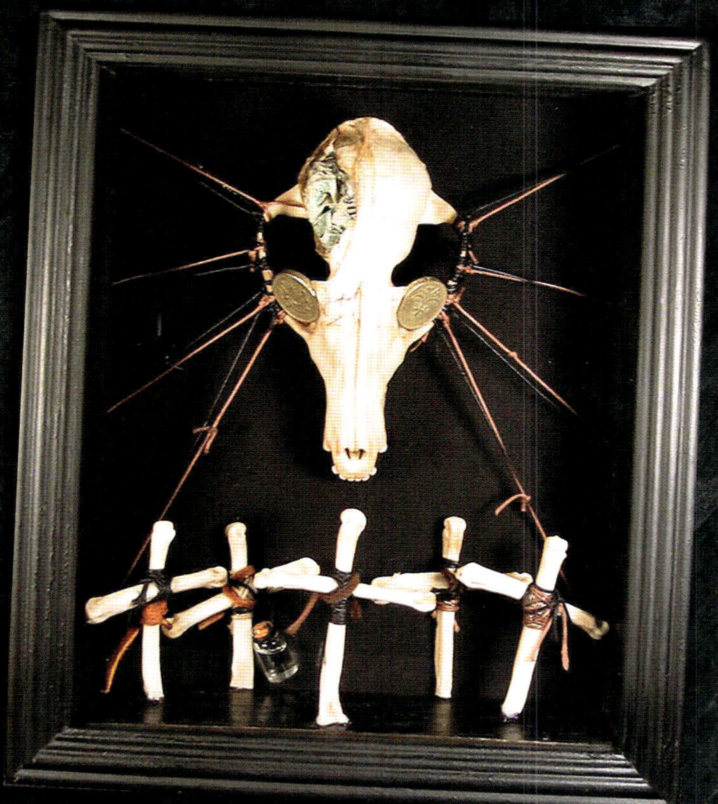

Joey "The Skull Master" Williams

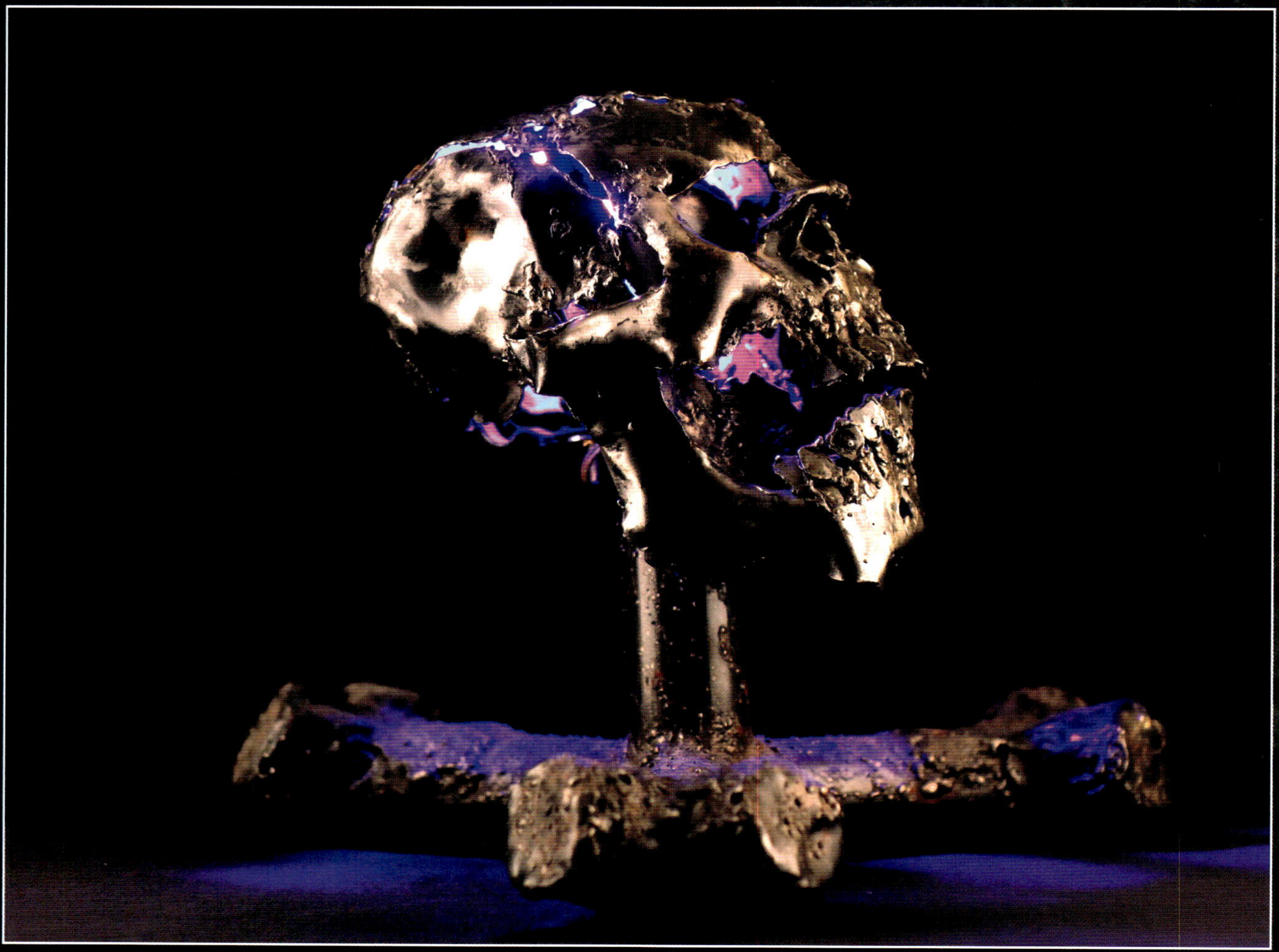

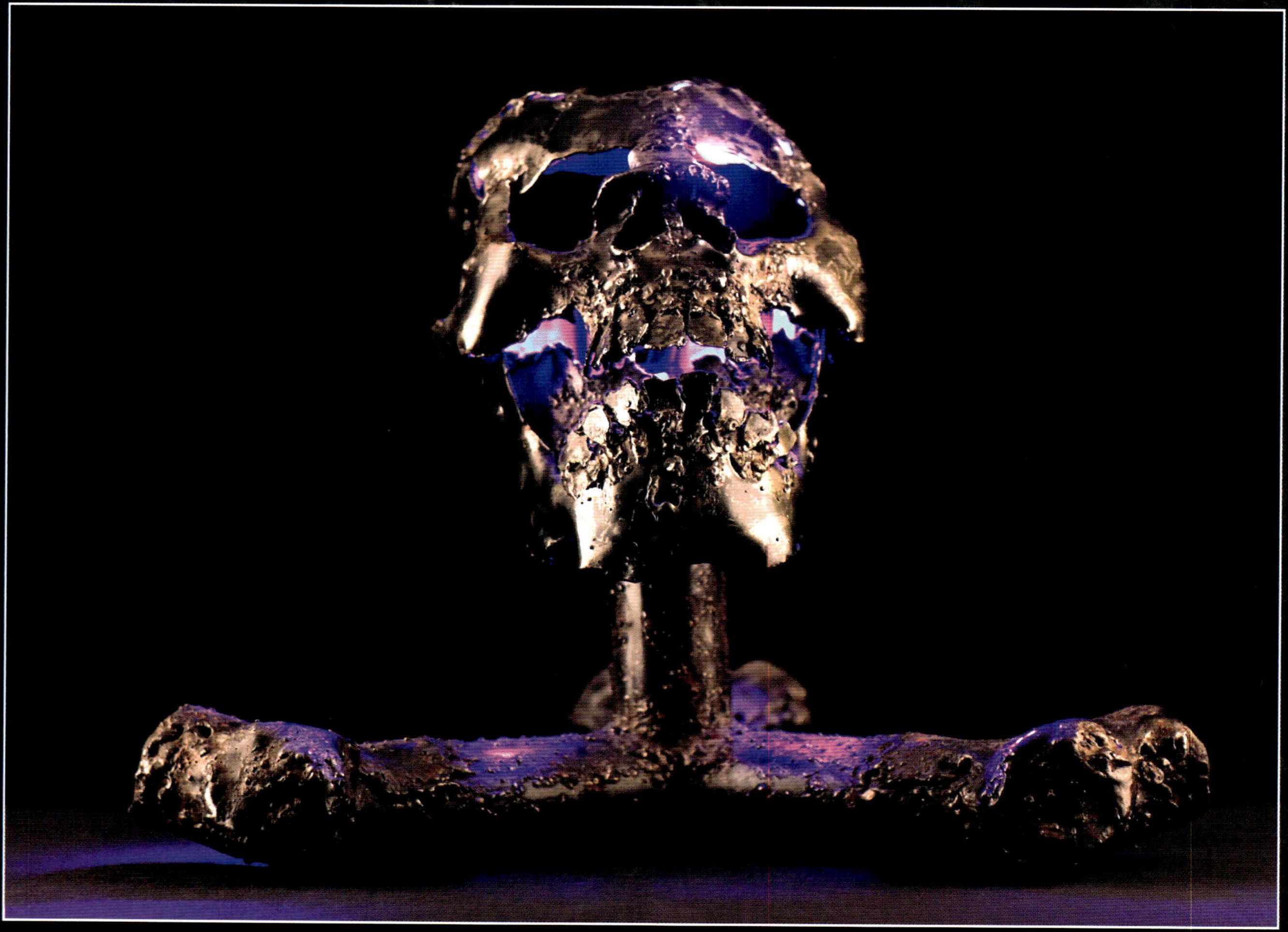

Durb Morrison

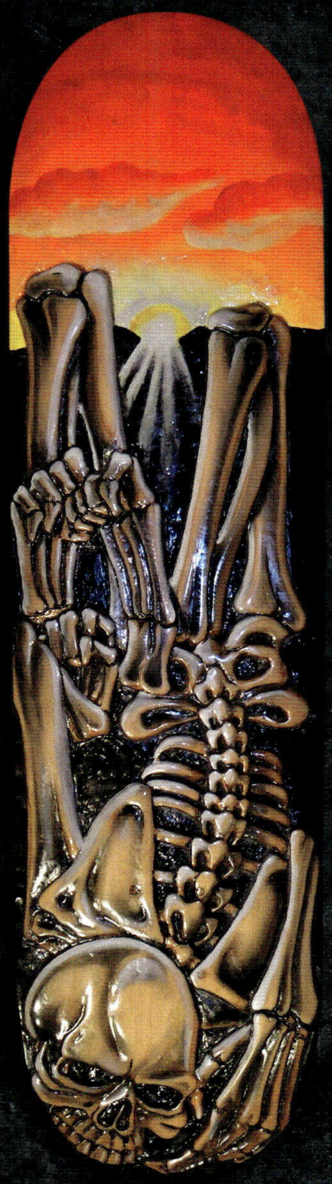

Brannon Grant

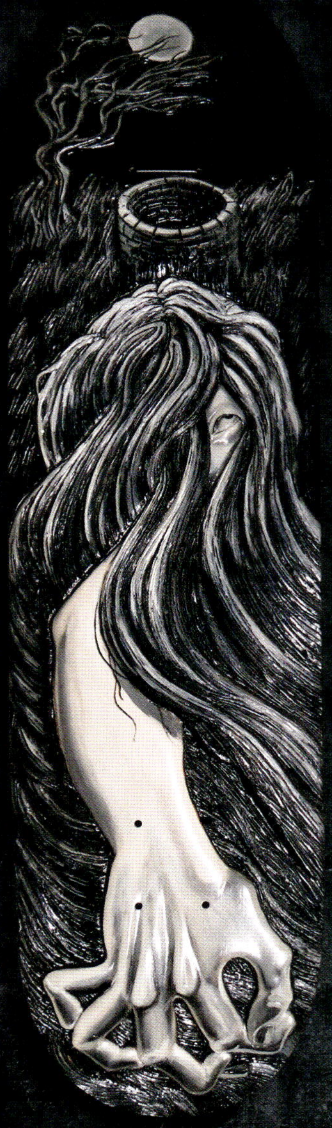

Brannon Grant

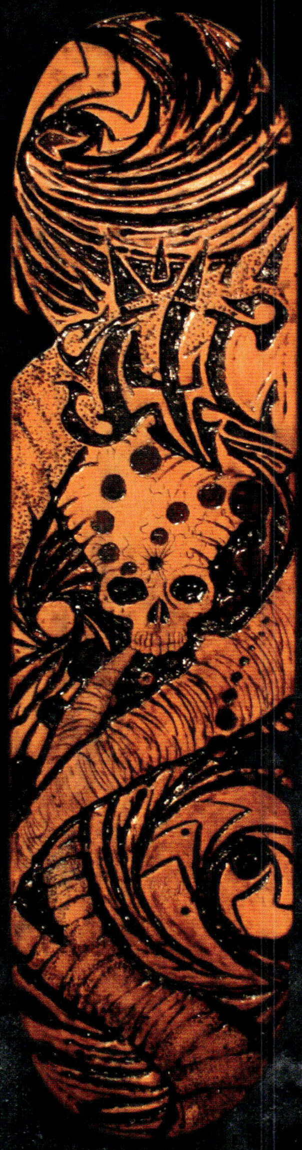

Alex Ansel

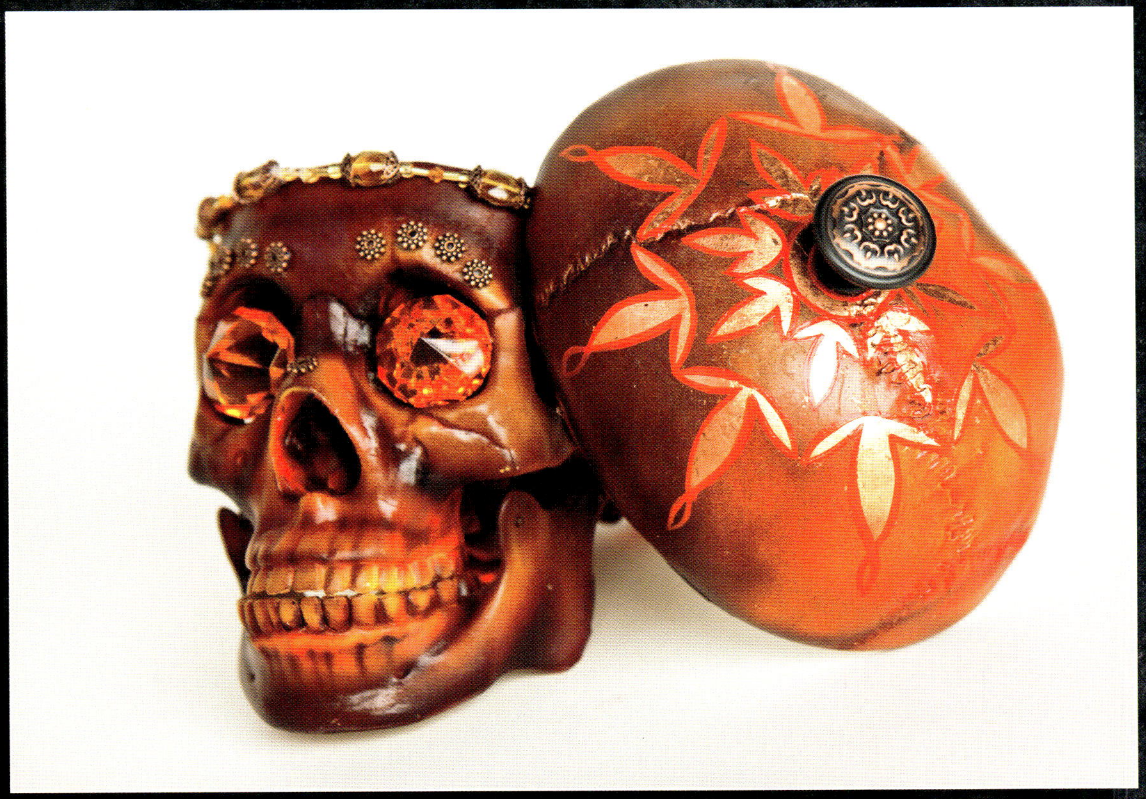

Steve Carian

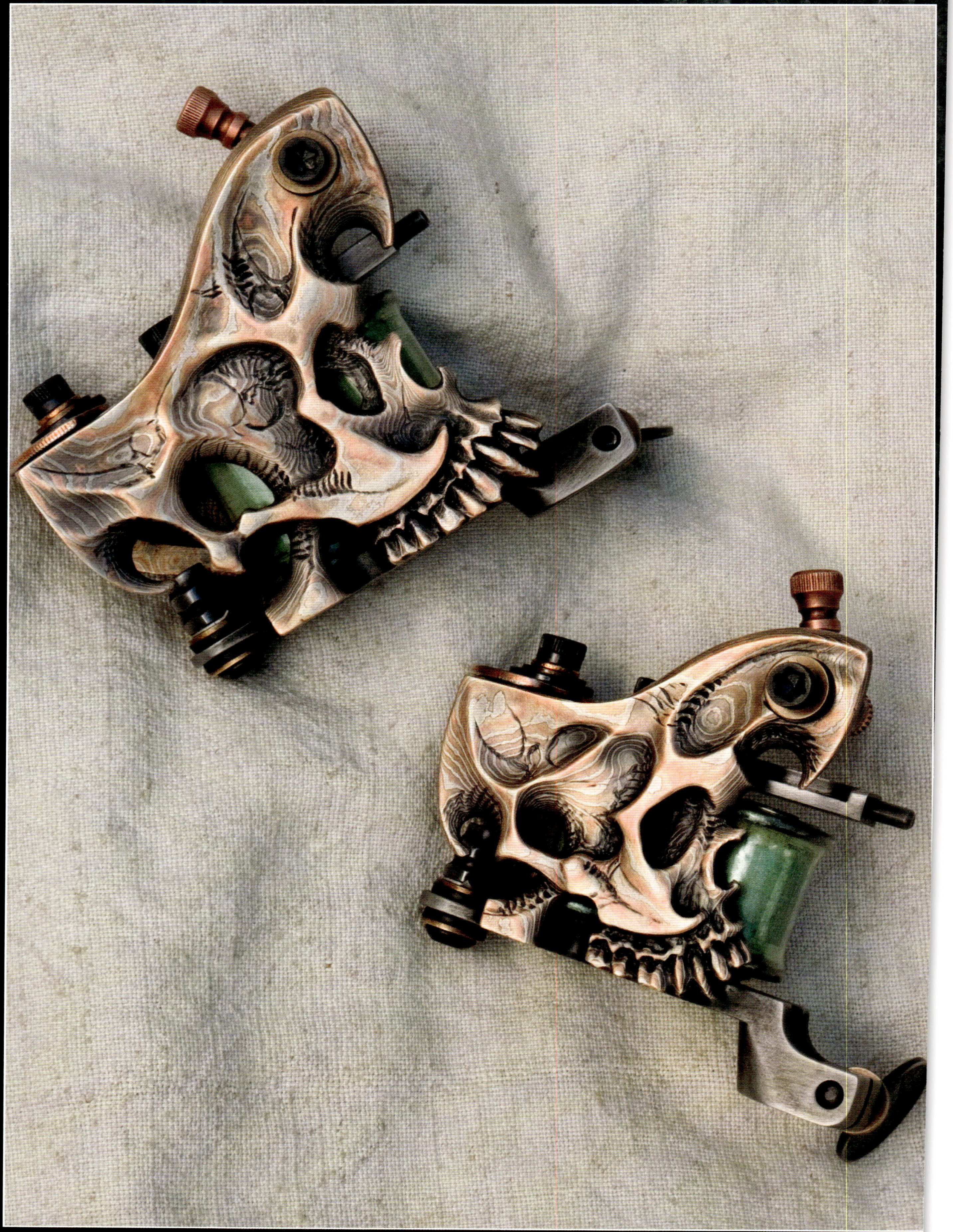

Aaron Cain

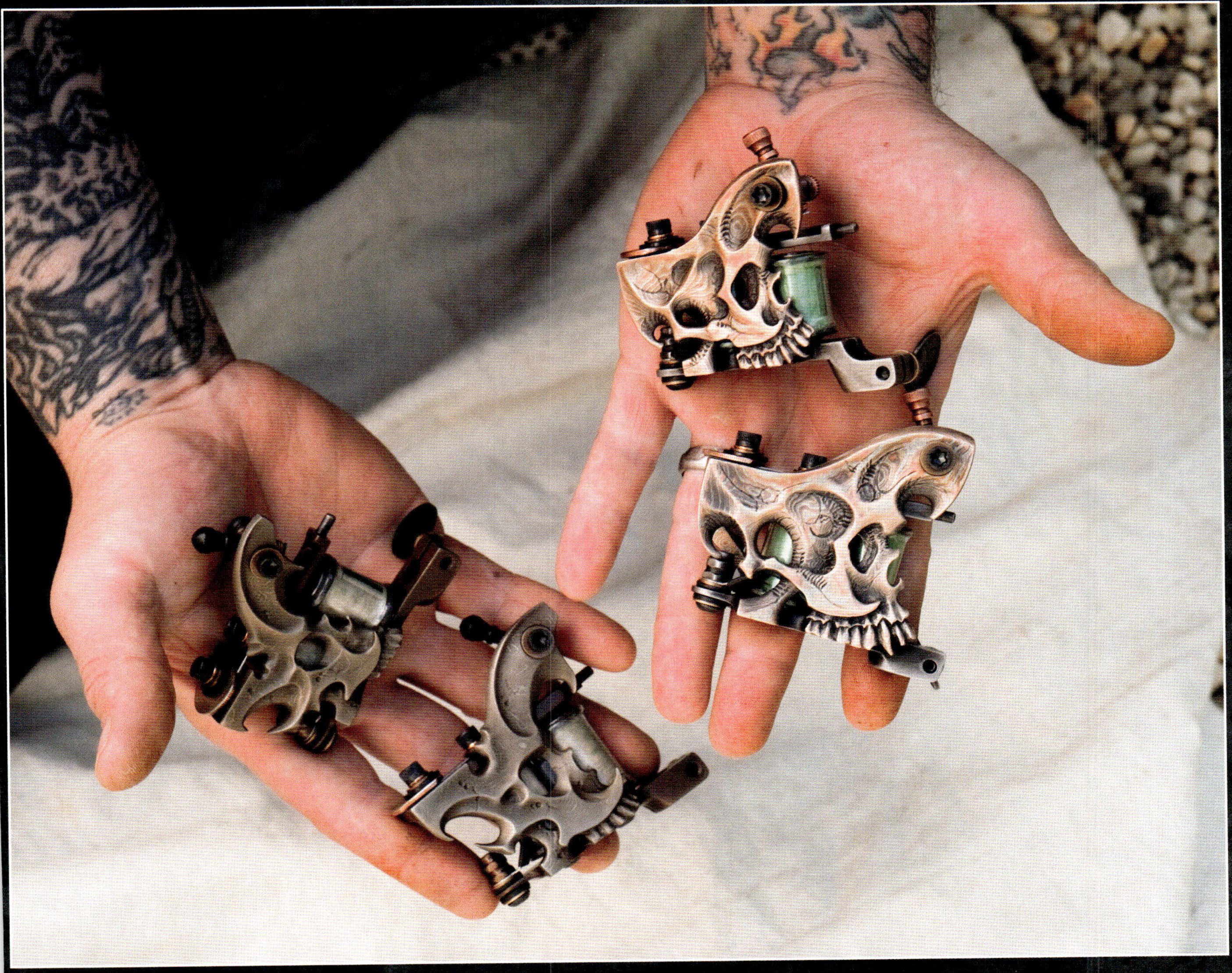

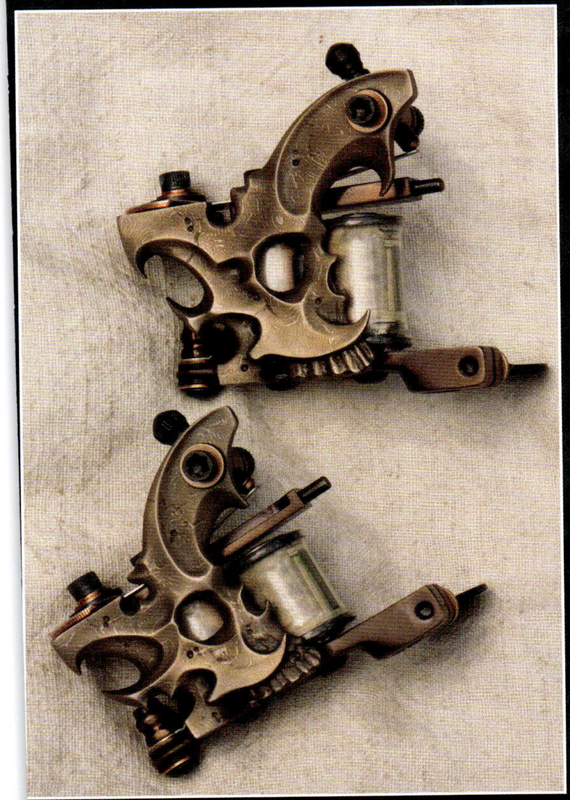

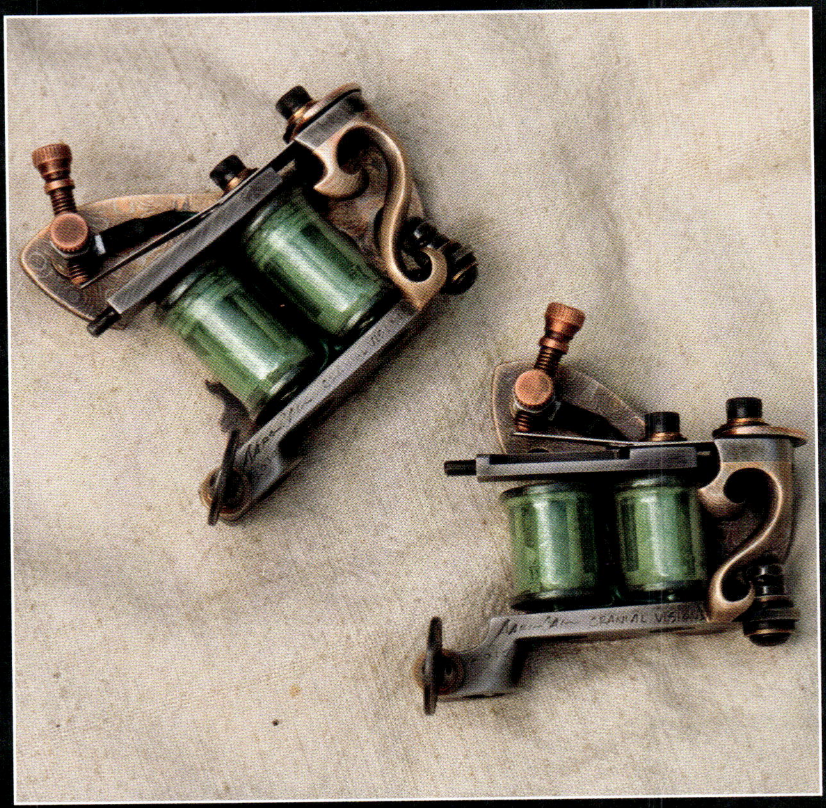

Aaron Cain

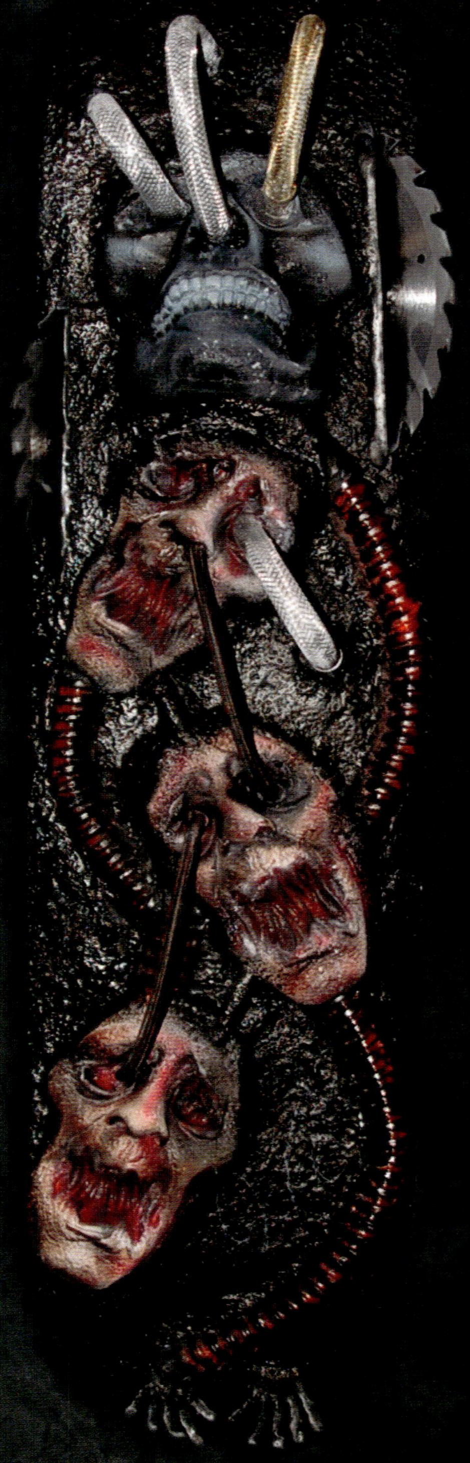

Brad Buehrle

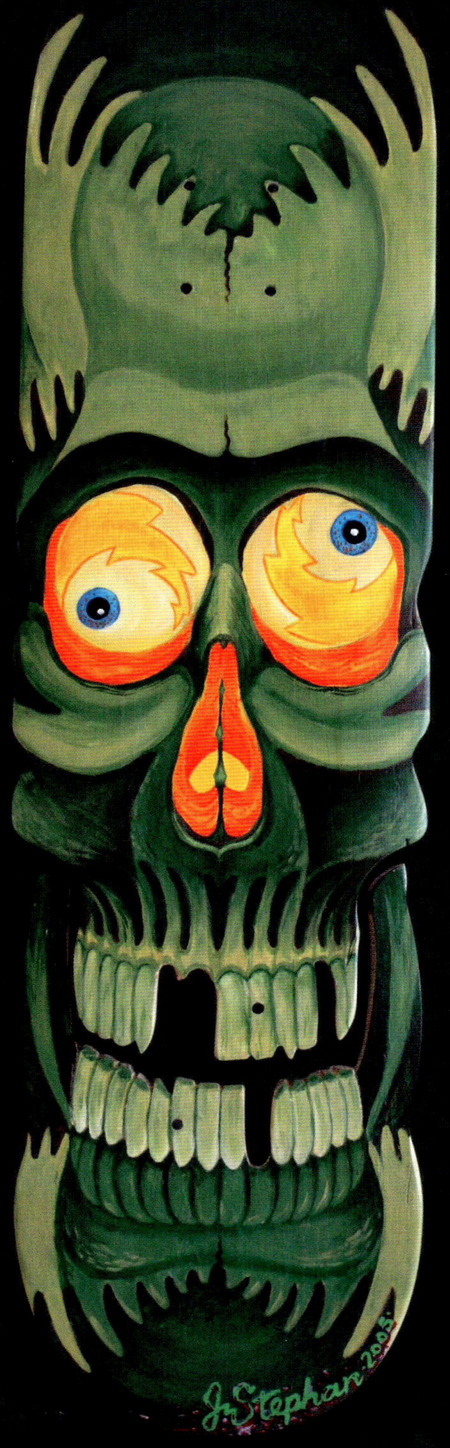

Jason Stephan

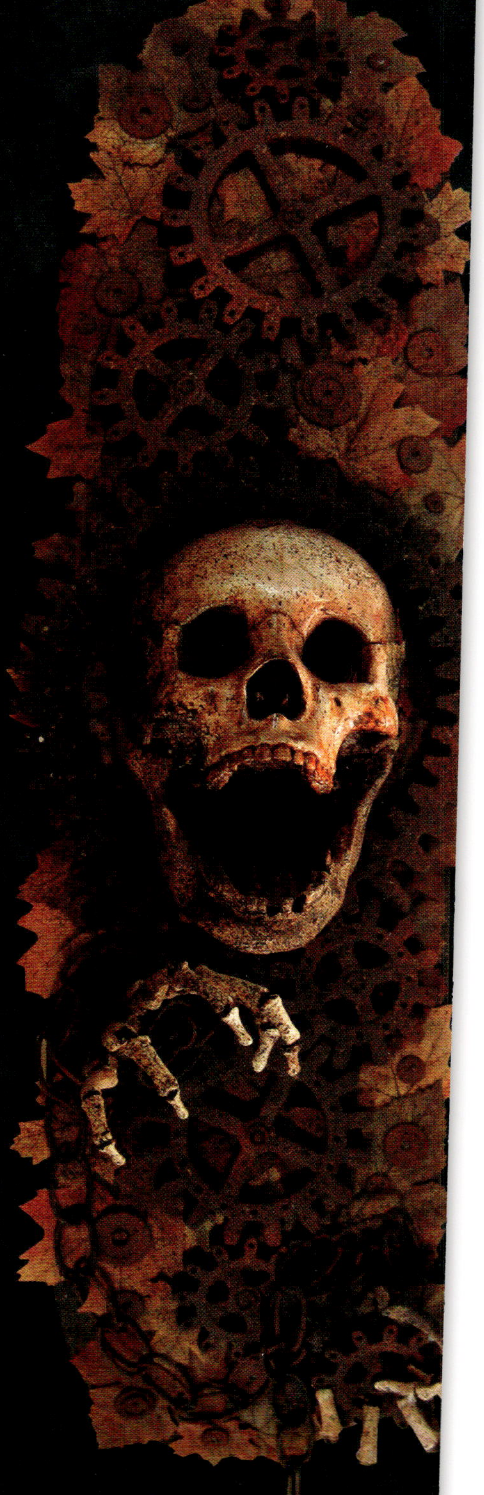

Brion Norwalk

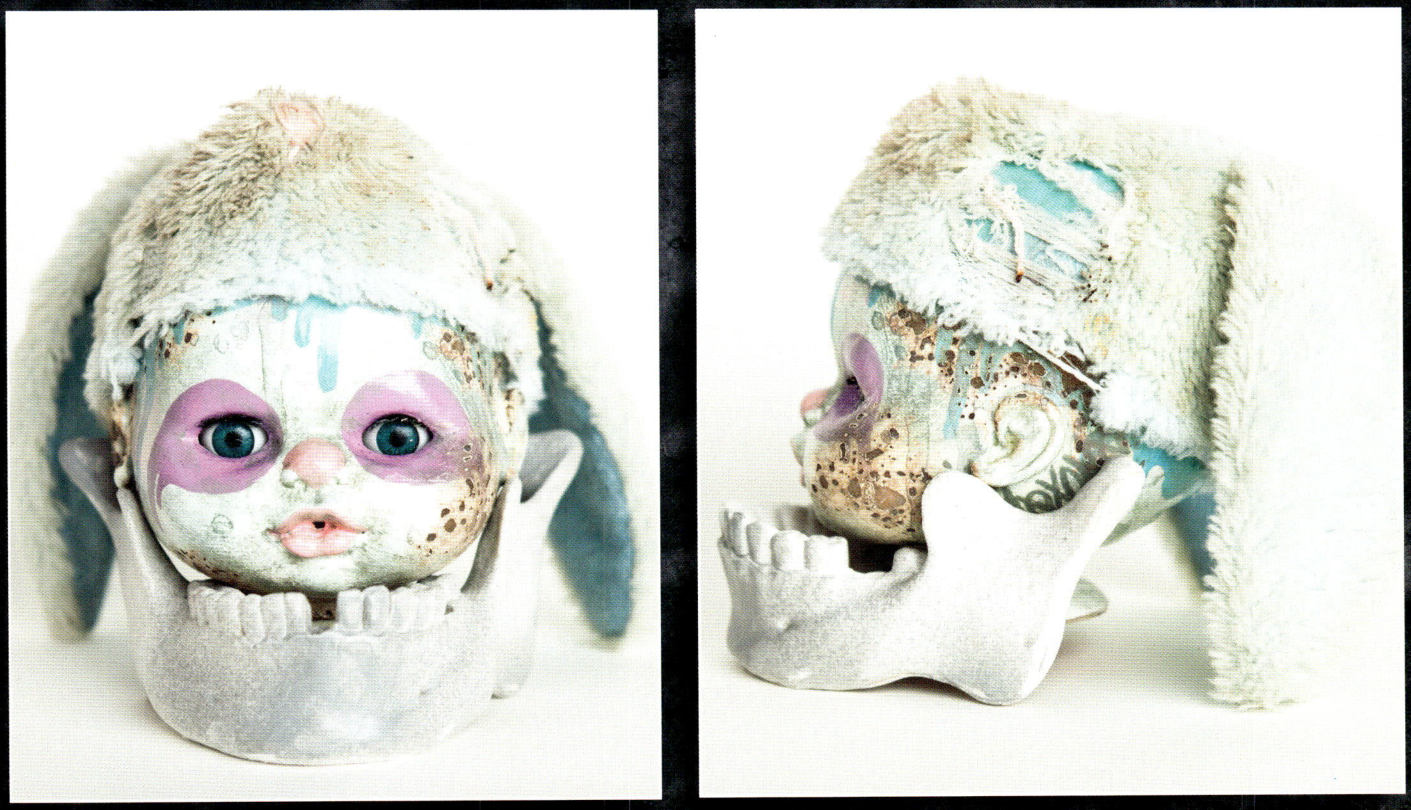

Paul Vetne

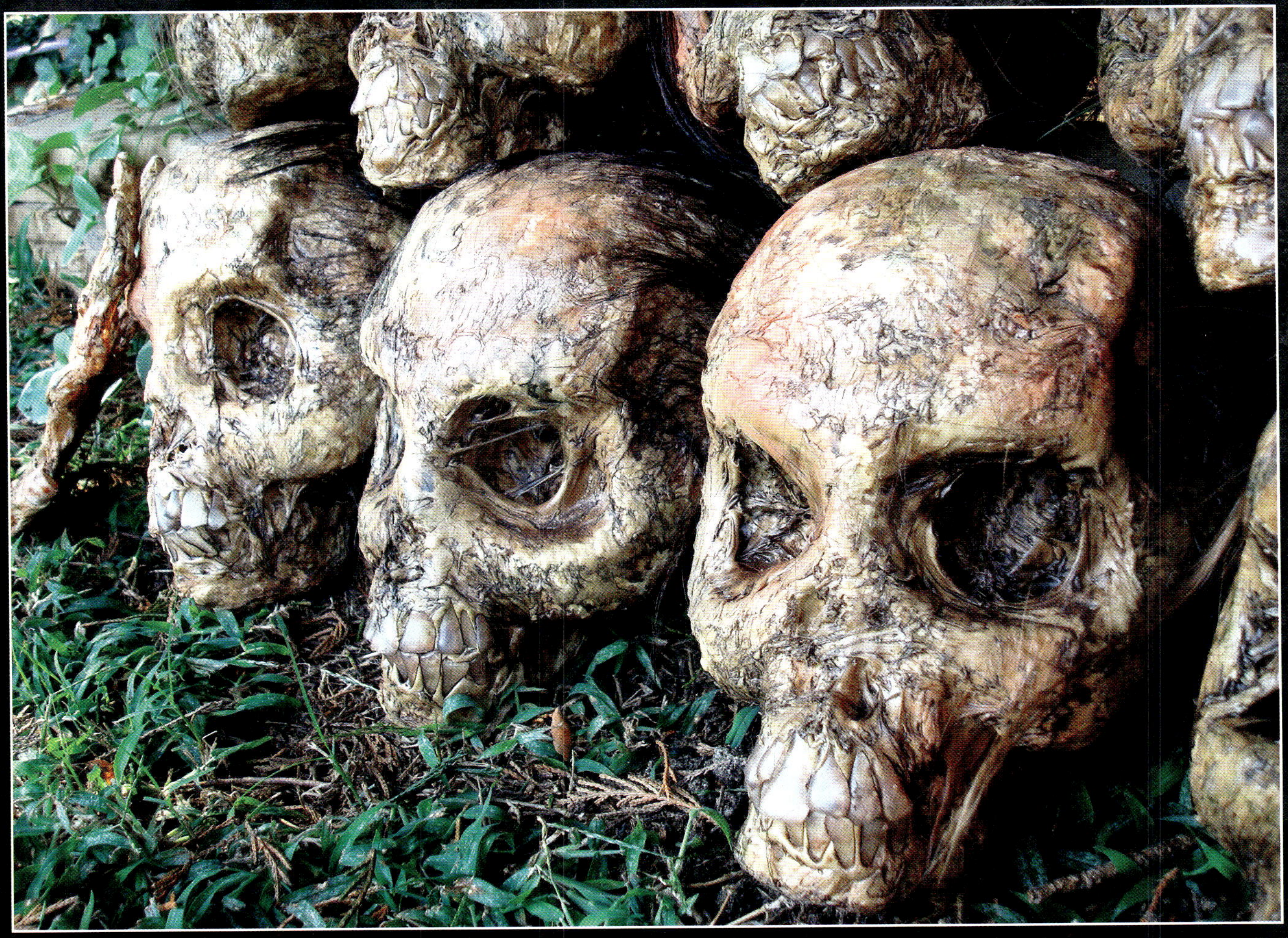

Nick DeRosa

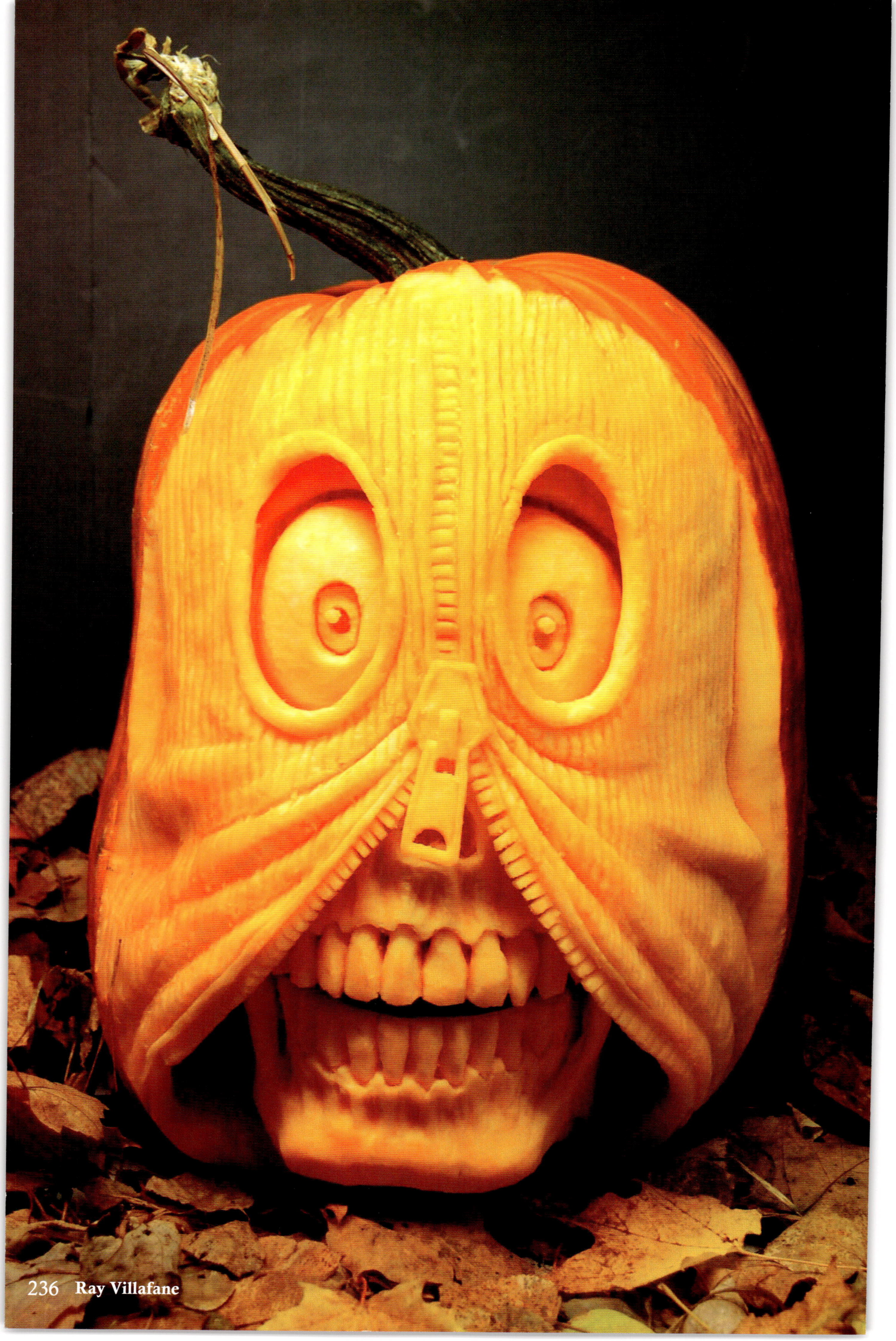

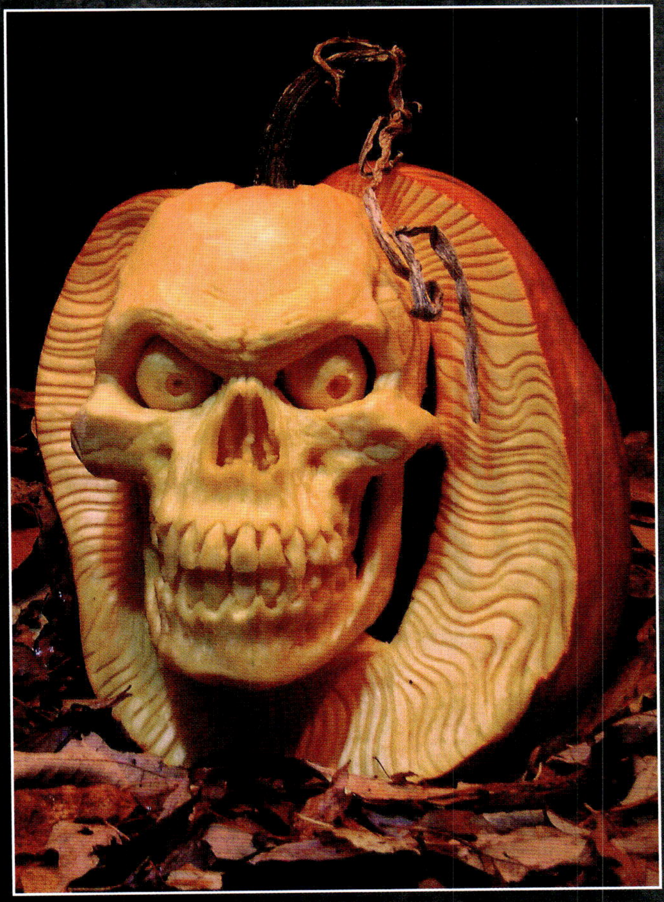

Ray Villafane

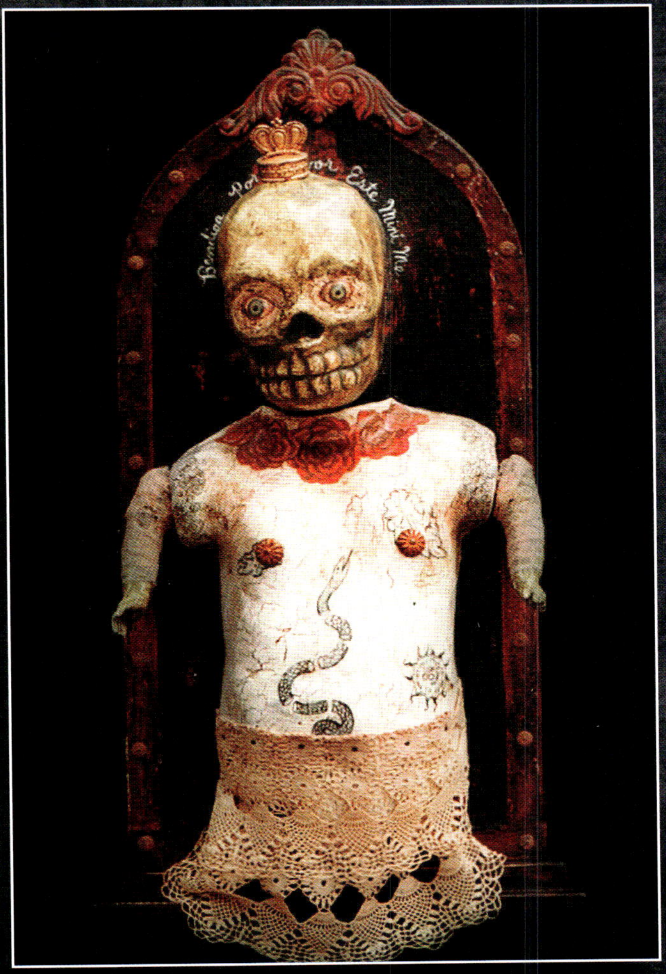

Craig Rotunda

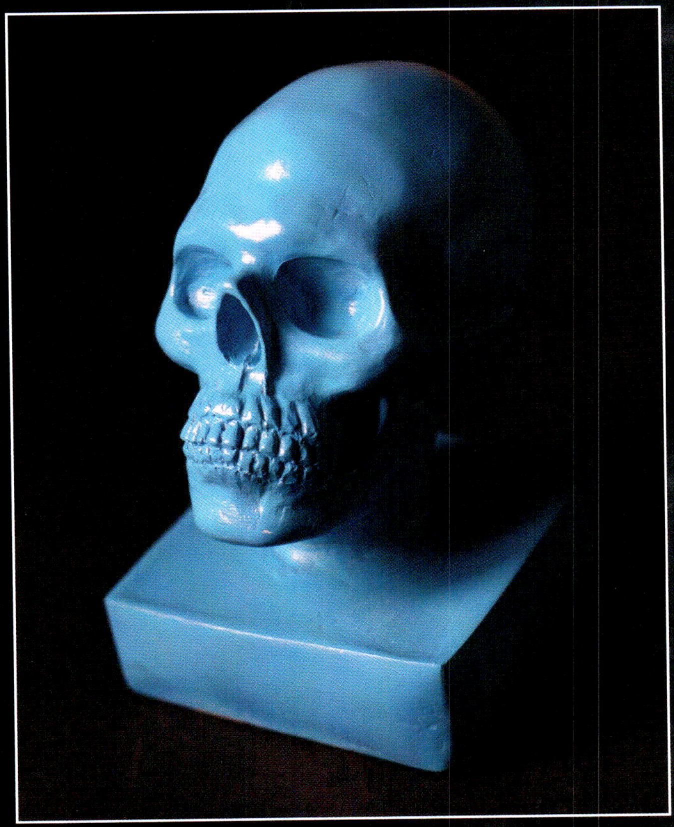

Cassie Stover

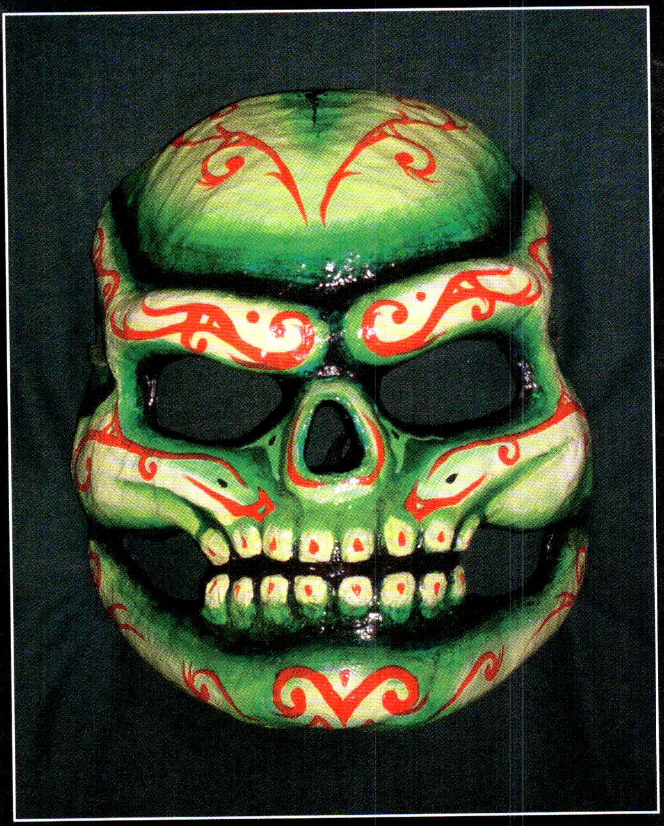

Gerardo Garcia

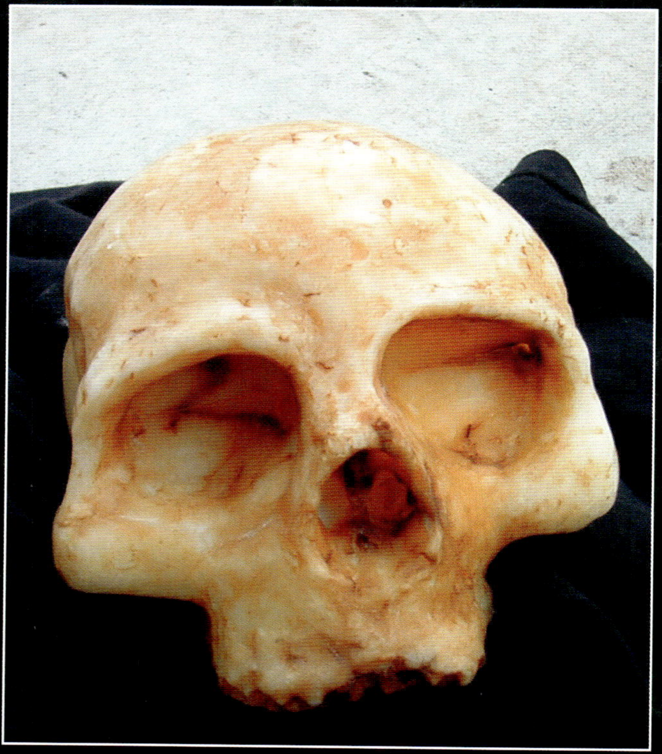
Jason Griffiths

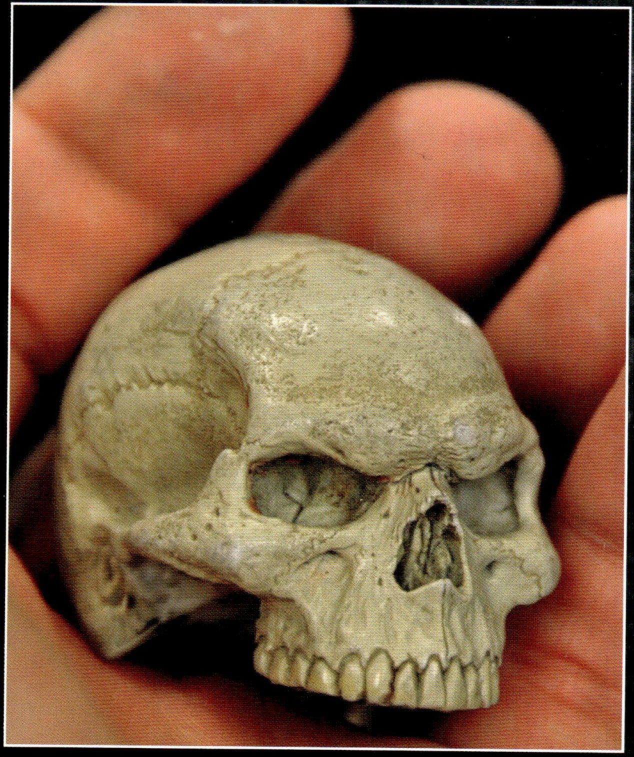
Ray Villafane

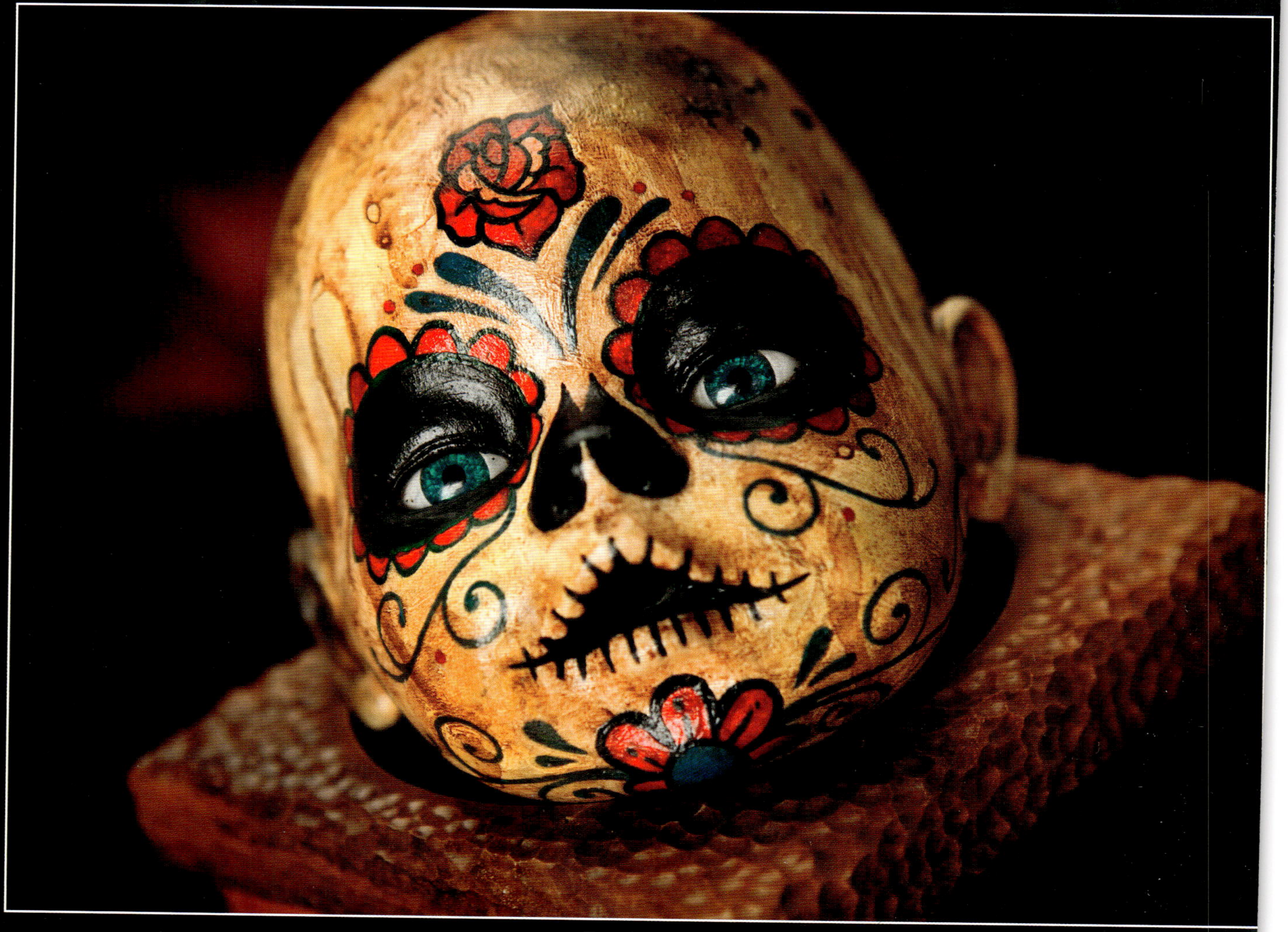
Paul Vetne

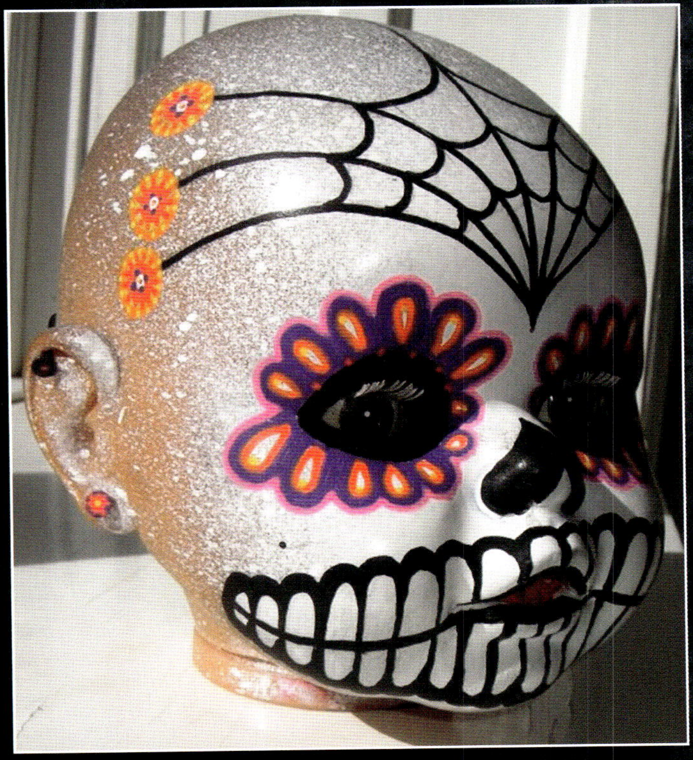

Jonas Bachmann

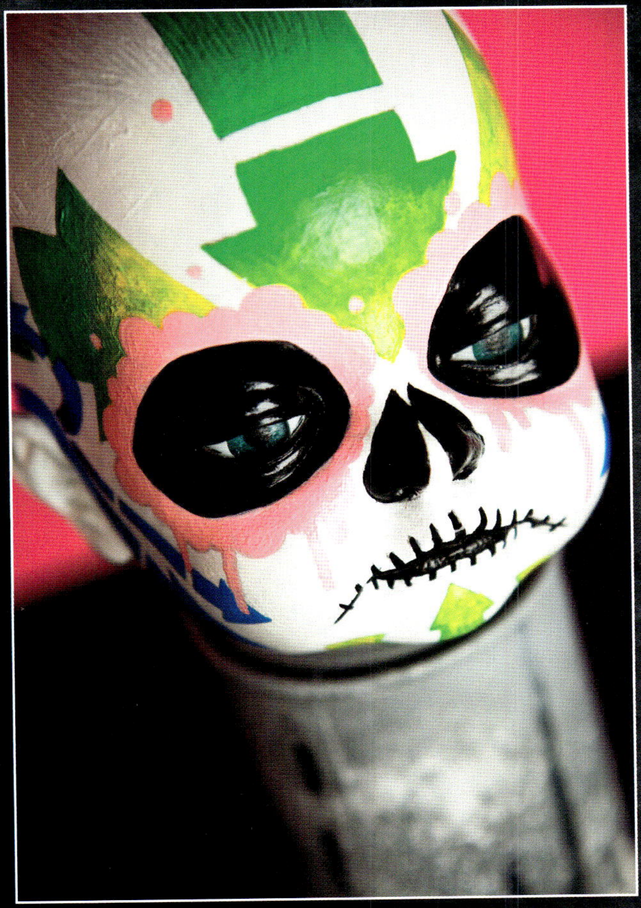

Paul Vetne

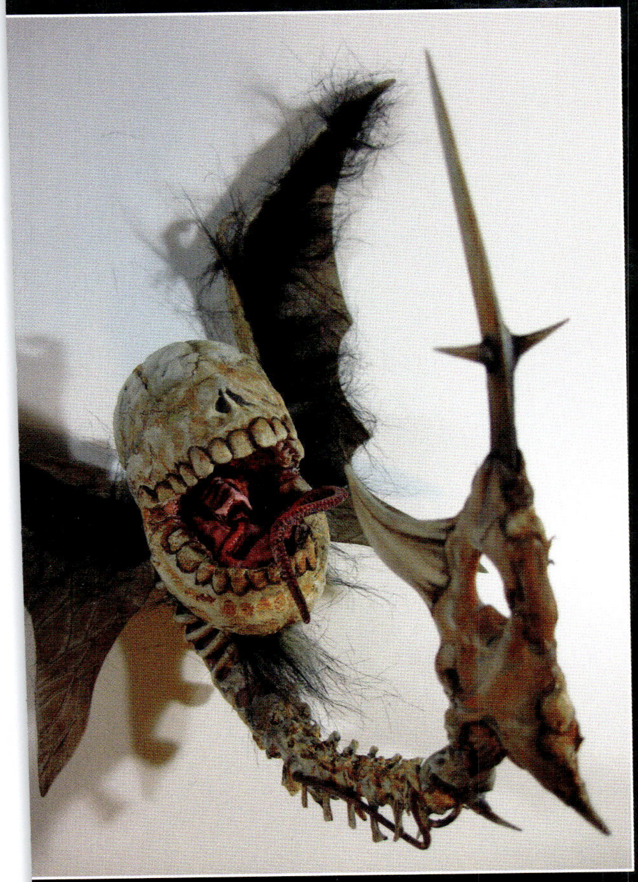

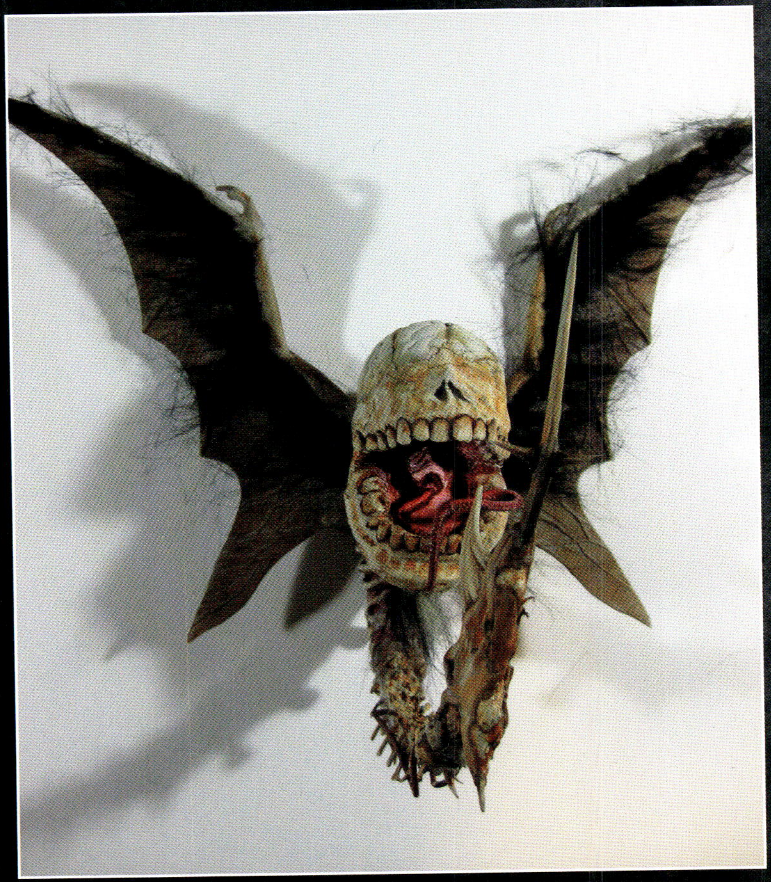

Brion Norwalk

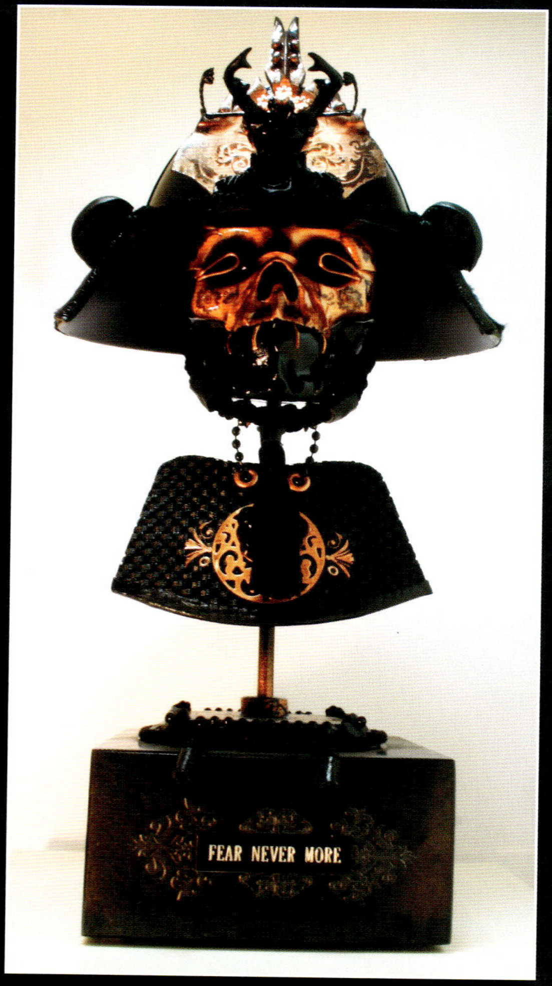
FEAR NEVER MORE

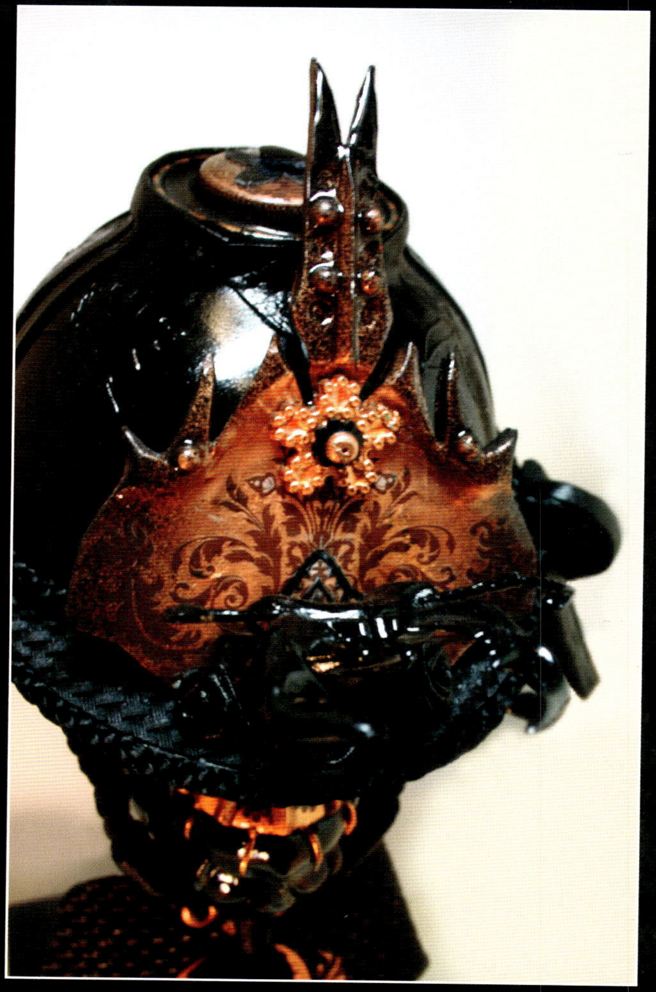

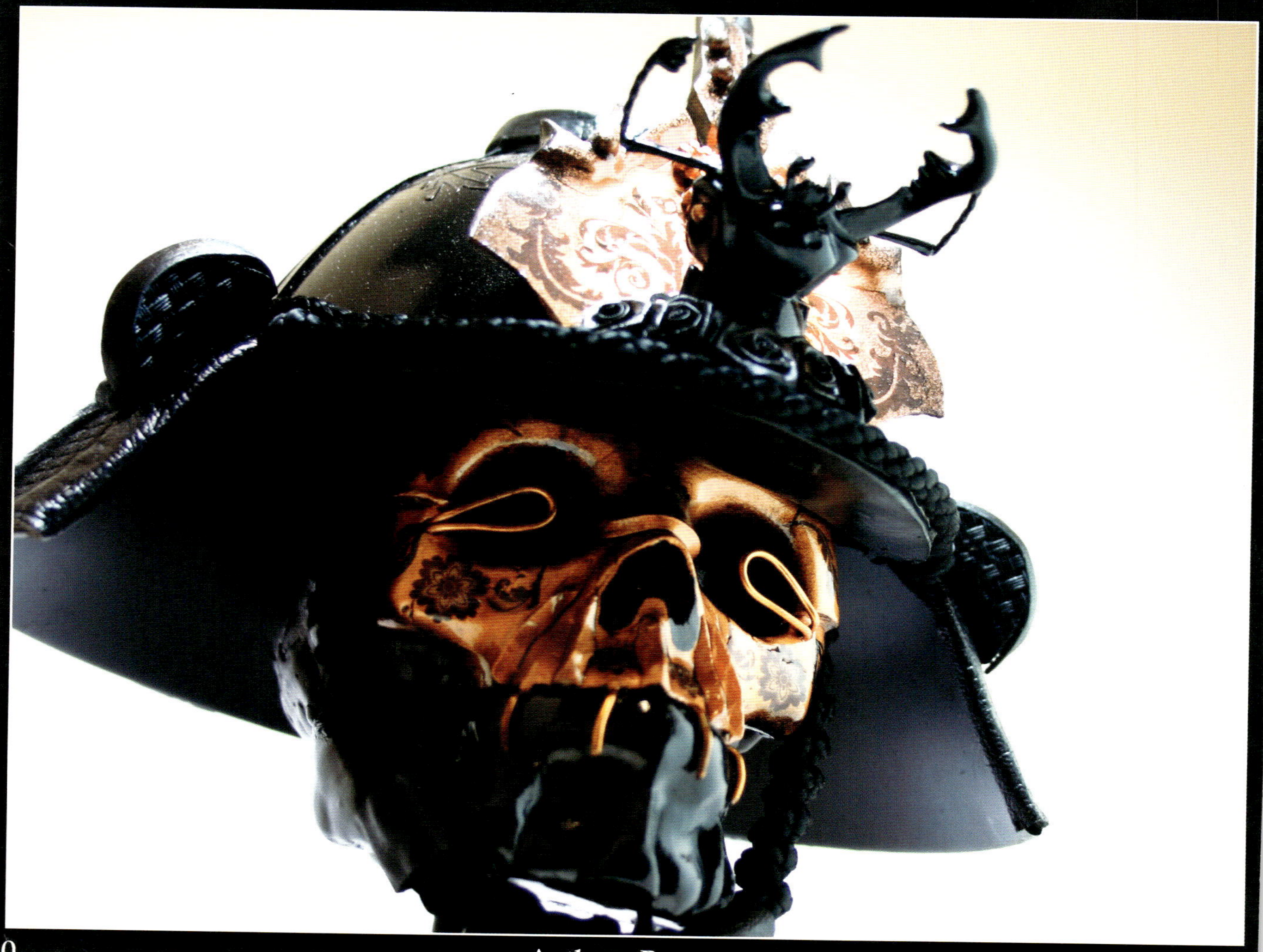

· CONCLUSION ·

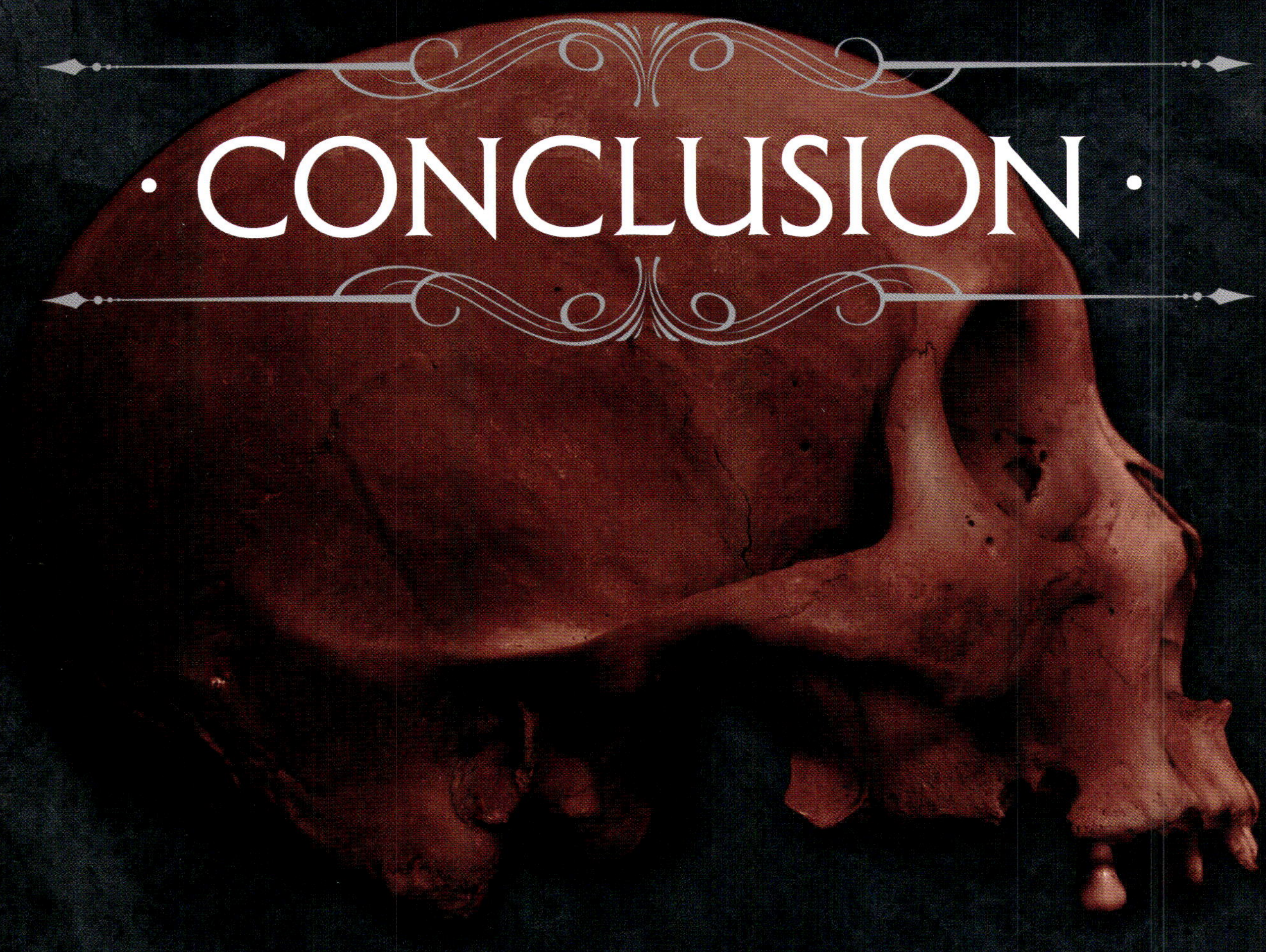

This cumulative journey through the creative interpretation of the skull has been a fascinating one. Discovering how different artists view this central theme and convey their vision to the world resulted in an incredible array of artwork, as diverse as it is inspiring. The submissions received validated our reasons for embarking on this book endeavor with each impressive and unique contribution.

Those participating submitted artwork in various forms: tattoos, paintings, sketches, drawings, digital art, photography, graffiti, sculptures and more. Because skulls have had such symbolic meanings throughout history, it was interesting to tie the assorted representations to the iconography of the art. All sorts of styles and approaches were used to capture the cranium in unique and bold ways.

We wish to express our gratitude to each artist who was a part of this project. The fusion of creativity and cooperation made this book possible and it was an honor to work with each of them. As with all efforts that are expressed in artistic form, we ask that you use the images within these pages as inspiration and not duplication. Learn from the contents, but please do not copy the artwork; out of respect for both artists and collectors.

With such a multitude of talent already on the scene, there is always room for additional influence and inspiration; and we hope that this book might be a device in the quest for ongoing cleverness and imagination. Whether it be the skull or any other influential idea, object, or theme; keep on creating and fill the world with ingenuity and brilliance for generations to come. As Henry Ward Beecher once said, “Every artist dips his brush in his own soul, and paints his own nature into his pictures.” Dip into your genius and paint your soul for all to see.